T0286234

Current Progress in Artificial Intelligence

Current Progress in Artificial Intelligence

Edited by Mick Benson

www.clanryeinternational.com

Clanrye International,
750 Third Avenue, 9th Floor,
New York, NY 10017, USA

ISBN: 978-1-63240-647-7

Cataloging-in-Publication Data

Current progress in artificial intelligence / edited by Mick Benson.
p. cm.
Includes bibliographical references and index.
ISBN 978-1-63240-647-7
1. Artificial intelligence. I. Benson, Mick.
TA347.A78 C87 2018
006.3--dc23

For information on all Clanrye International publications visit our website at www.clanryeinternational.com

Contents

Preface

Over the recent decade, advancements and applications have progressed exponentially. This has led to the increased interest in this field and projects are being conducted to enhance knowledge. The main objective of this book is to present some of the critical challenges and provide insights into possible solutions. This book will answer the varied questions that arise in the field and also provide an increased scope for furthering studies.

The intelligence displayed by machines is known as artificial intelligence. It aims to advance the reasoning, processing and learning abilities of machines through statistical models, computational intelligence, etc. Artificial intelligence has applications in multiple industries such as healthcare, automotive, finance, etc. This book unfolds the innovative aspects of artificial intelligence that will be crucial for the progress of this field in the future. The extensive content of this book provides the readers with a thorough understanding of the subject. As this field is emerging at a rapid pace, the contents of this book will help the readers understand the modern concepts and applications of the subject.

I hope that this book, with its visionary approach, will be a valuable addition and will promote interest among readers. Each of the authors has provided their extraordinary competence in their specific fields by providing different perspectives as they come from diverse nations and regions. I thank them for their contributions.

Editor

1

Numerical Deviation Based Optimization Method for Estimation of Total Column CO_2 Measured with Ground Based Fourier Transformation Spectrometer: FTS Data

Kohei Arai
Graduate School of Science and Engineering
Saga University
Saga City, Japan

Hiroshi Okumura
Graduate School of Science and Engineering
Saga University
Saga City, Japan

Takuya Fukamachi
Graduate School of Science and Engineering
Saga University
Saga City, Japan

Shuji Kawakami
JAXA, Japan
Tsukuba City, Japan

Hirofumi Ohyama
Nagoya University, Japan
Nagoya City, Japan

Abstract—**Numerical deviation based optimization method for estimation of total column CO_2 measured with ground based Fourier Transformation Spectormeter: FTS data is proposed. Through experiments with aircraft based sample return data and the ground based FTS data, it is found that the proposed method is superior to the conventional method of Levenberg Marquads based nonlinear least square method with analytic deviation of Jacobian and Hessean around the current solution. Moreover, the proposed method shows better accuracy and required computer resources in comparison to the internationally used method (TCCON method) for estimation of total column CO_2 with FTS data. It is also found that total column CO_2 depends on weather conditions, in particular, wind speed.**

Keywords—FTS; carbon dioxide; methane; sensitivity analysis; error analysis

I. INTRODUCTION

Greenhouse gases Observing SATellite: GOSAT carries TANSO CAI for clouds and aerosol particles observation of mission instrument and TANSO FTS[1]: Fourier Transformation Spectrometer[2] for carbon dioxide and methane retrieving mission instrument [1]. In order to verify the retrieving accuracy of two mission instruments, ground based laser radar and TANSO FTS are installed. The former is for TANSO CAI and the later is for FTS, respectively. One of the other purposes of the ground-based laser radar and the ground-based FTS is to check sensor specifications for the future mission of instruments to be onboard future satellite with extended mission. Although the estimation methods for carbon dioxide and methane are well discussed [2]-[6], estimation method which takes into account measurement noise is not analyzed yet. Therefore, error analysis for additive noise on estimation accuracy is conducted.

In order to clarify requirement of observation noises to be added on the ground-based FTS observation data, Sensitivity analysis of the ground-based FTS against observation noise on retrievals of carbon dioxide and methane is conducted. Experiments are carried out with additive noise on the real acquired data of the ground-based FTS. Through retrievals of total column of carbon dioxide and methane with the noise added the ground-based FTS signals, retrieval accuracy is evaluated. Then an allowable noise on the ground-based FTS which achieves the required retrieval accuracy (1%) is reduced [7].

In the paper, Numerical Deviation Based Optimization Method for Estimation of Total Column CO_2 Measured with Ground Based Fourier Transformation Spectormeter: FTS Data is proposed. Through experiments with aircraft based sample return data and the ground based FTS data, it is found that the proposed method is superior to the conventional method of Levenberg-Marquardt: LM [8] based nonlinear least square method with analytic deviation of Jacobian and Hessian around the current solution [9]. Moreover, the proposed method shows

[1] http://www.jaxa.jp/projects/sat/gosat/index_j.html

[2] http://ja.wikipedia.org/wiki/%E3%83%9E%E3%82%A4%E3%82%B1%E3%83%AB%E3%82%BD%E3%83%B3%E5%B9%B2%E6%B8%89%E8%A8%88

better accuracy and shorter required computer resources in comparison to the TCCON[3] data [10],[11]) for estimation of total column CO_2 with FTS data. It is also found that total column CO_2 depends on whether condition, in particular, wind speed with aircraft based sample return data [12] and ground based FTS data. The following section describes the proposed method for total column CO_2 estimation with FTS data followed by some experiments with ground based and aircraft based sample return data. Then concluding remarks with some discussions is described.

II. Proposed Method

A. Ground-based FTS

Figure 1 shows schematic configuration of the ground-based FTS which is originated from Michelson Interference Measurement Instrument. Light from the light source divided in to two directions, the left and the forward at the dichotic mirror of half mirror. The left light is reflected at the fixed hold mirror and reaches to the half mirror while the forward light is reflected at the moving mirror and reaches at the half mirror. Then interference occurs between the left and the forward lights. After that interference light is detected by detector. Outlook of the ground-based FTS is shown in Figure2.

Figure 3 (a) shows an example of the interferogram[4] (interference light detected by the detector of the ground-based FTS). By applying Fourier Transformation to the interferogram, observed Fourier spectrum is calculated as shown in Figure 3 (b). When the ground-based FTS observes the atmosphere, the observed Fourier spectrum includes absorptions due to atmospheric molecules and aerosol particles. By comparing to the spectrum which is derived from the radiative transfer code with atmospheric parameters, atmospheric molecules and aerosol particles are estimated.

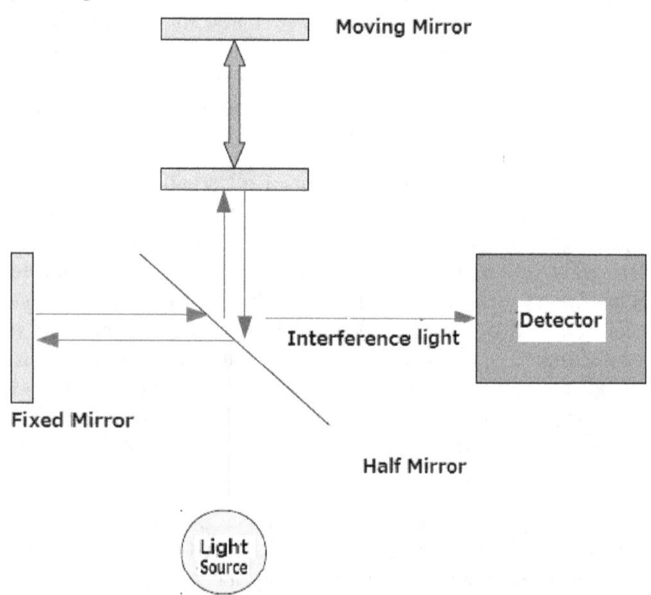

Fig. 1. Michelson Interference Measurement Instrument

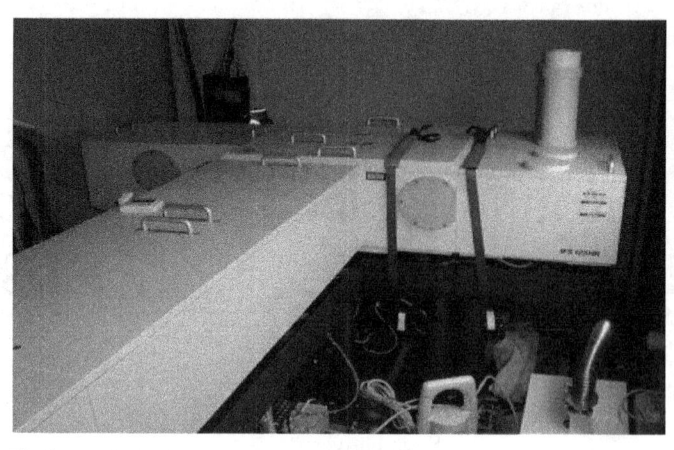

Fig. 2. Outlook of the FTS used

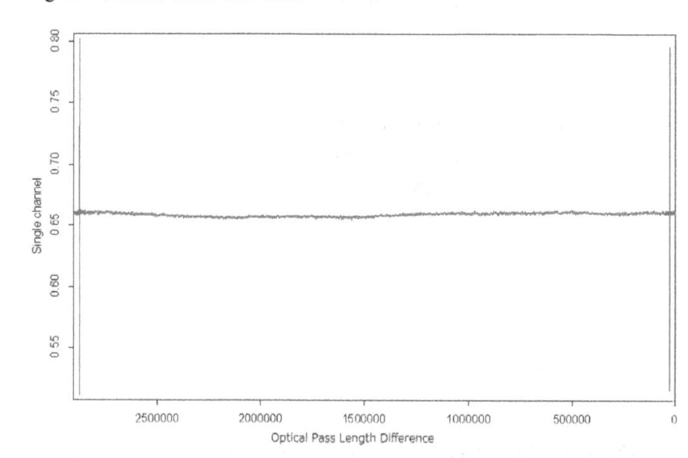

(a)Interferogram

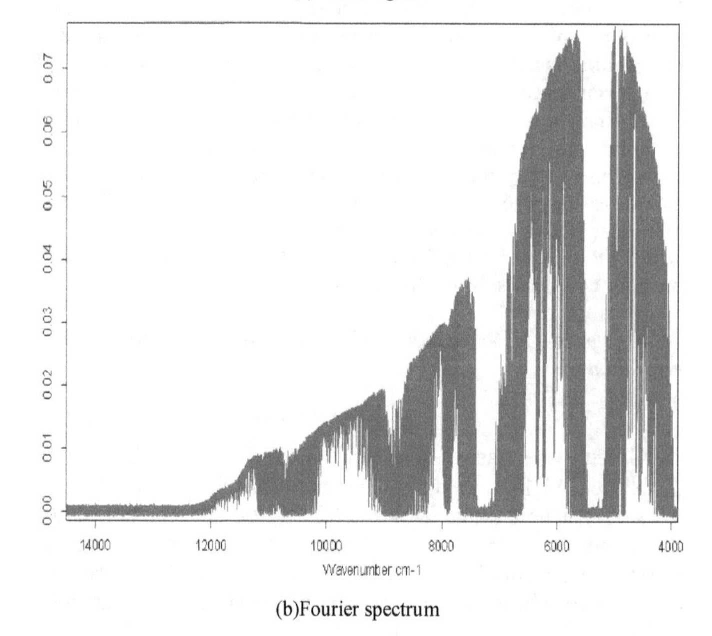

(b)Fourier spectrum

Fig. 3. Examples of interferogram and Fourier spectrum when FTS observes the atmosphere

[3] https://tccon-wiki.caltech.edu/Network_Policy/Data_Use_Policy#References_and_Contact_Information

[4] http://en.wikipedia.org/wiki/Interferometry

B. Principle for Carbon Dioxide and Methane Retrievals with TANSO FTS Data

Figure 4 shows a principle of the retrieval method for atmospheric continuants using GOSAT/TANSO data. Figure 4 (a) shows Top of the Atmosphere: TOA radiance in the wavelength ranges from 500 to 2500nm (visible to shortwave infrared wavelength regions). There are three major absorption bands due to oxygen (760-770nm), carbon dioxide and methane (1600-1700nm), and water vapor and carbon dioxide (1950-2050nm) as shown in Figure 4 (b), (c), and (d), respectively. These bands are GOSAT/TANSO spectral bands, Band 1 to 3, respectively. In addition to these, there is another wide spectrum of spectral band, Band 4 as shown in Figure 4(e) which covers from visible to thermal infrared regions..

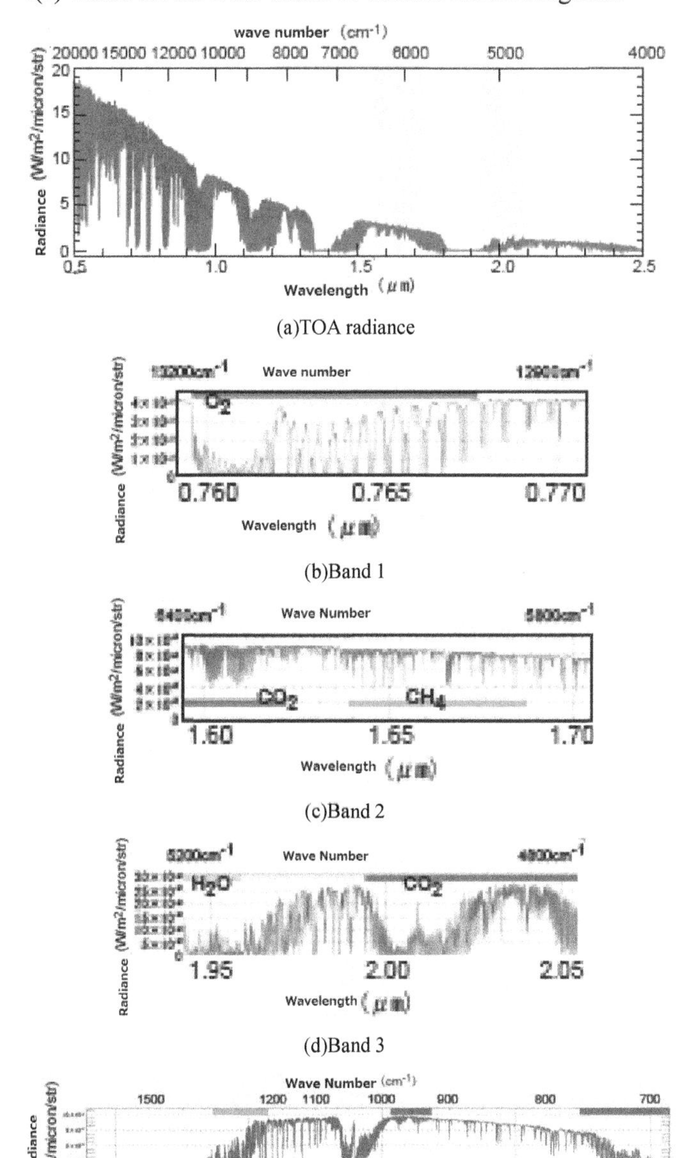

Fig. 4. Example of TOA radiance and absorption bands as well as spectral bands of GOSAT/TANSO instrument

C. Estimation Algorithm Description

The conventional method for estimation of CO_2 is as follows, (1) Estimated spectrum derived from atmospheric simulator with atmospheric parameters including CO_2 and the actual FTS derived spectrum is compared, (2) then the initial atmospheric condition are updated for minimizing the square of difference between spectra derived from simulator and the actual FTS data iteratively. In order to minimize the square of difference, LM method is used in the conventional method as follows,

$$min\ S = \sum_{i=1}^{N}(y_i - y(x_i; \mathrm{a}))^2 \tag{1}$$

where S denotes square difference, y_i and $y(x_i, a)$ denote actual spectrum and simulated spectrum with atmospheric condition of a,

$$\mathrm{a} = (a_1, a_2, \cdots, a_M) \tag{2}$$

A can be updated as follows,

$$\mathrm{a_{k+1}} = \mathrm{a_k} + \Delta \mathrm{a_k} \tag{3}$$

where $\Delta \mathrm{a_k}$ is determined as the following equation is satisfied.

$$(\mathrm{H}(\mathrm{a_k}) + \lambda \mathrm{H}(\mathrm{a_k}))\Delta \mathrm{a_k} = \mathrm{J}^T(\mathrm{a_k})\Delta \mathrm{y}(x; \mathrm{a_k}) \tag{4}$$

where H and J denote Hessian and Jocobian, respectively.

$$\Delta \mathrm{y}(x; \mathrm{a_k}) = (\ y_1 - y(x_1; \mathrm{a}), y_2 - y(x_2; \mathrm{a}), \cdots, y_N - y(x_N; \mathrm{a})\ ) \tag{5}$$

$$\mathrm{J}(\mathrm{a_k}) = \begin{bmatrix} \frac{\partial S}{\partial a_1} & \frac{\partial S}{\partial a_2} & \cdots & \frac{\partial S}{\partial a_M} \\ \frac{\partial S}{\partial a_1} & \frac{\partial S}{\partial a_2} & \cdots & \frac{\partial S}{\partial a_M} \\ \vdots & \vdots & \ddots & \vdots \\ \frac{\partial S}{\partial a_1} & \frac{\partial S}{\partial a_2} & \cdots & \frac{\partial S}{\partial a_M} \end{bmatrix} \tag{6}$$

$$\mathrm{H}(\mathrm{a_k}) = \begin{bmatrix} \frac{\partial^2 S}{\partial a_1 \partial a_1} & \frac{\partial^2 S}{\partial a_1 \partial a_2} & \cdots & \frac{\partial^2 S}{\partial a_1 \partial a_M} \\ \frac{\partial^2 S}{\partial a_2 \partial a_1} & \frac{\partial^2 S}{\partial a_2 \partial a_2} & \cdots & \frac{\partial^2 S}{\partial a_2 \partial a_M} \\ \vdots & \vdots & \ddots & \vdots \\ \frac{\partial^2 S}{\partial a_M \partial a_1} & \frac{\partial^2 S}{\partial a_M \partial a_2} & \cdots & \frac{\partial^2 S}{\partial a_M \partial a_M} \end{bmatrix} \tag{7}$$

where

$$\frac{\partial S}{\partial a_l} = -2\sum_{i=1}^{N}\{y_i - y(x_i; \mathrm{a})\}\frac{\partial y(x_i; \mathrm{a})}{\partial a_l} \tag{8}$$

$$\frac{\partial^2 S}{\partial a_l \partial a_m} = 2\sum_{i=1}^{N}\left[\frac{\partial y(x_i; \mathrm{a})}{\partial a_l}\frac{\partial y(x_i; \mathrm{a})}{\partial a_m} - \{y_i - y(x_i; \mathrm{a})\}\frac{\partial^2 y(x_i; \mathrm{a})}{\partial a_l \partial a_m}\right] \tag{9}$$

when the current solution is reached to one of minima, $y_i - y(x_i; \mathrm{a}) \approx 0$ so that the following approximation becomes appropriate,

$$\frac{\partial^2 S}{\partial a_l \partial a_m} = 2\sum_{i=1}^{N} \frac{\partial y(x_i; \mathrm{a})}{\partial a_l} \frac{\partial y(x_i; \mathrm{a})}{\partial a_m} \tag{10}$$

In the solution space, the updated solution can be determined with relatively small step size for all directions in isotropic manner in the LM method as shown in Figure 5.

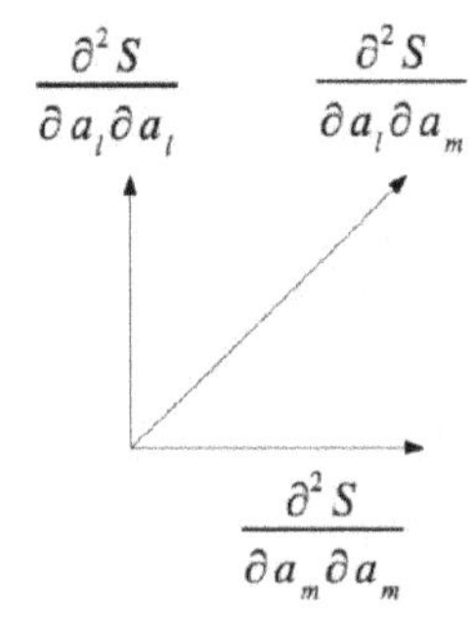

Fig. 5. Solution update directions for LM method

The method proposed here is the solution update direction can be determined arbitrary as shown in Figure 6.

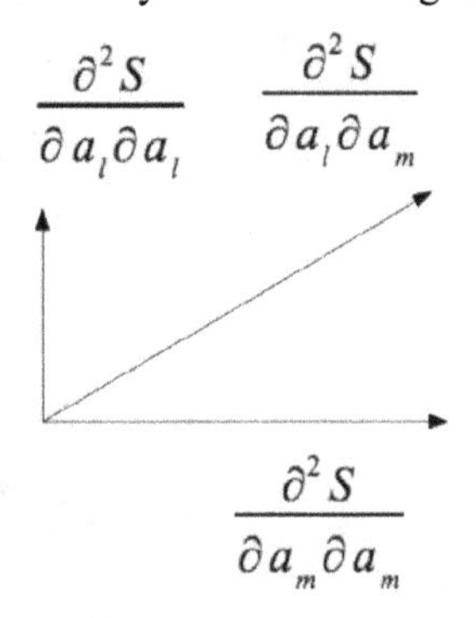

Fig. 6. Solution update directions for the proposed method

III. Experiments

A. Ground-based FTS Data Used

The ground-based FTS data used for experiments are acquired on November 14 and December 19 2011. Figure 7 shows the interferograms derived from the acquired the ground-based FTS data.

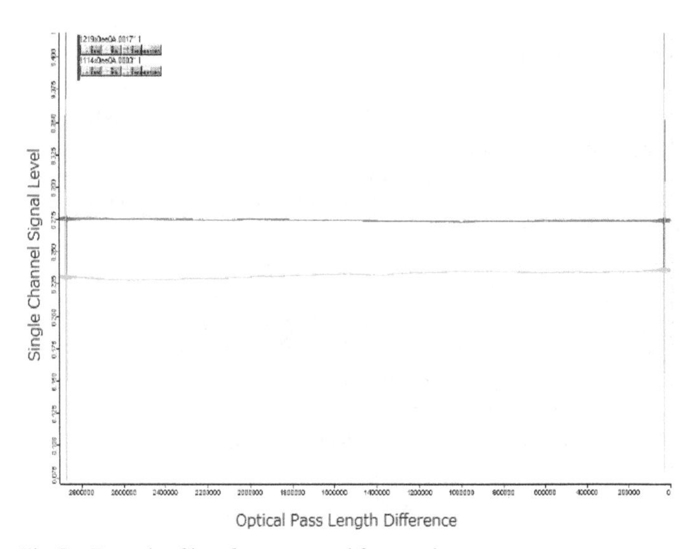

Fig. 7. Example of interferograms used for experiments

B. Experimental Method

Observation noise is included in the observed interferograms. In addition to the existing noise, several levels of additional noises which are generated by random number generator of Messene Twister with zero mean and several standard deviations is added on to the iterferograms as shown in Figure 8.

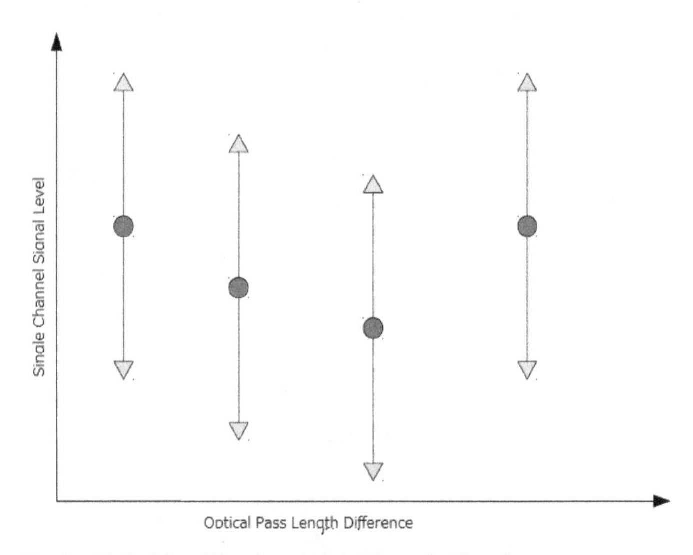

Fig. 8. Method for adding the noises to the acquired interferograms

C. Aircraft Based Sample Return Data Derived Total Column CO_2

Aircraft based sample return data is acquired with aircraft altitude of 500 m and 7km. Therefore, CO_2 for the atmosphere between two altitude can be retrieved. Using GlovalView-CO_2 model, CO_2 for the atmosphere above 7km is estimated. Also, it is assumed that CO_2 for the atmosphere below 500m can be the same CO_2 at 500m.

The conventional method utlizes the vertical profile model of GlobalView-CO_2. The profile can be estimated with the following equation,

$$c_s = \gamma c_a + \left(\frac{VC_{G,ak}{}^{airclaft} - \gamma VC_{G,ak}{}^{apriori}}{VC_{air}} \right) \quad (11)$$

where c_s denotes averaged column density of the dried atmosphere, $VC_{G,ak}$ denotes total column gas amount which is calculated with Rogers and Connor equation and vertical profile derived from the aforementioned model. Also, $VC_{G,ak}{}^{aircraft}$ denotes total column gas amount at the aircraft altitude.

D. Aircraft Based Sample Returnt Data Used

Aircraft based sample return data which are acquired on January 9 2012, January 13 2012 and January 15 2013 are used together with match-upped data of ground based FTS data. The number of data for each day is 116, 52, and 200 files, respectively.

E. Experimental Results

Total column CO_2 for three days of experiments is estimated with the proposed method and compared to the TCCON data, Ohyama (LM method based retrieval) as well as actual aircraft based sample return data derived total column CO_2. Figure 9 shows the results.

As shown in Table 1, it is found that the proposed method is superior to the other conventional methods in terms of estimation accuracy and the required computer resources. One of the reasons for this is that the proposed method allows update the next solution to the arbitrary directions with relatively large steps.

Table 2 shows weather conditions, atmospheric pressure, air temperature on the ground, Relative Humidity (RH), irradiant flax and the averaged wind speed on the ground. It was cloudy on January 9 and 13, 2012. In particular on January 13 2012, it was poor sun shine time period. Therefore, there are so many data missing.

On the other hand, it was fine on January 15 2013. Therefore, the number of data points is greater than the other two days.

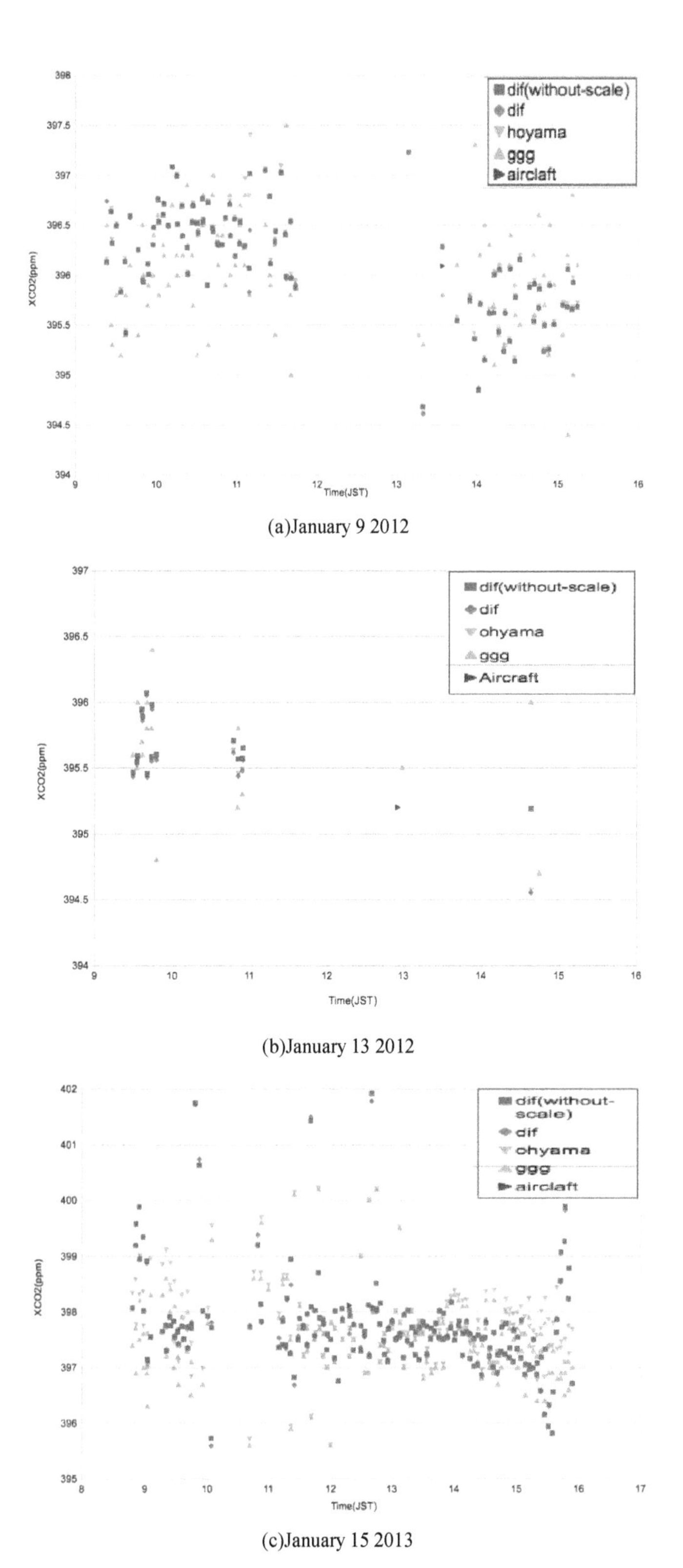

Fig. 9. Comparison among the estimated total column CO2 derived from the proposed method, the conventional methods, TCCON and LM method based retrievals

TABLE I. COMPARISON OF THE RESIDUAL RMS ERROR AGAINST TRUTH DATA DERIVED FROM AIRCRAFT WITH VERTICAL PROFILE MODEL AMONG THE PROPOSED, LM METHOD AND TCCON DATA

Date	Proposed	LM	TCCON
2012/1/9	0.429	0.438	0.467
2012/1/13			
2013/1/15	0.315	0.32	0.986

TABLE II. WEATHER CONDITIONS OF THE DATES FOR EXPERIMENTS

Date	Pressure (hPa)	Temp. (°C)	RH(%)	Irradiant Flax(MJ/m^2)	Wind(m/s)
2012/1/9	1022.8	7	66	9.23	2.7
2012/1/13	1020.1	4.8	65	4.99	1.8
2013/1/15	1019	6.7	55	12.39	4.2

IV. CONCLUSION

Numerical deviation based optimization method for estimation of total column CO_2 measured with ground based Fourier Transformation Spectormeter: FTS data is proposed. Through experiments with aircraft based sample return data and the ground based FTS data, it is found that the proposed method is superior to the conventional method of Levenberg Marquads based nonlinear least square method with analytic deviation of Jacobian and Hessean around the current solution. Moreover, the proposed method shows better accuracy and required computer resources in comparison to the TCCON data for estimation of total column CO_2 with FTS data. It is also found that total column CO_2 depends on weather conditions, in particular, wind speed.

Through the experiments with aircraft based sample return data (as the truth data) together with ground based FTS data, it is found that the proposed method is superior to the other conventional methods in terms of estimation accuracy and the required computer resources. One of the reasons for this is that the proposed method allows update the next solution to the arbitrary directions with relatively large steps.

ACKNOWLEDGMENT

The author would like to thank Dr. I. Morino, Dr. O. Uchino, G. Inoue, of National Institute of Environmental Studies and Dr. T.Tanaka of NASA/Ames Research Center for their effort to conduct the experiments and providing the aircraft data.

REFERENCES

[1] http://repository.tksc.jaxa.jp/dr/prc/japan/contents/AA0065136000/65136000.pdf?IS_STYLE=jpn(Accessed on September 14 2012)

[2] Clough, S. A., et al. [2006], IEEE Trans. Geosci. Remote Sens., 44, 1308–1323.

[3] Hase, F., et al. [2004], J. Quant. Spectrosc. Radiat. Transfer, 87(1), 25–52.

[4] Rodgers, C. D. [2000], Inverse Methods for Atmospheric Sounding: Theory and Practice.

[5] Tikhonov, A. [1963], Dokl. Acad. Nauk SSSR, 151, 501–504.J. Clerk Maxwell, A Treatise on Electricity and Magnetism, 3rd ed., vol. 2. Oxford: Clarendon, 1892, pp.68–73.

[6] Wunch, D, et al., [2001], The Total Carbon Cloumn Obesrving Network (TCCON), Phil., Tarns. R. Soc. A., 369, 2087-2112.

[7] K.Arai, T.Fukamachi, H.Okumura, S.Kawakami, H.Ohyama, Sensitivity analysis of Fourier Transformation Spectrometer: FTS against observation noise on retrievals of carbon dioxide and methane, International Journal of Advanced Computer Science and Applications, 3, 11, 58-64, 2012.

[8] William H. Press,et al.,:Numerical Recipes in C, Second Edition ,pp681-688, CAMBRIDGE UNIVERSITY PRESS, 1992

[9] Ohyama et al.,:Column-averaged volume mixing ratio of CO2 measured with ground-based Fourier transform spectrometer at Tsukuba, J. Geophys. Res., 114, D18303, doi:10.1029/2008JD011465, 2009

[10] D. Wunch, G.C. Toon, J.-F.L. Blavier, R.A. Washenfelder, J. Notholt, B.J. Connor, D.W.T. Griffith, V. Sherlock, P.O. Wennberg. The Total Carbon Column Observing Network. Phil. Trans. R. Soc. A (2011) 369, doi:10.1098/rsta.2010.0240 2011

[11] Wunch et al.,:Calibration of the Total Carbon Column Observing Network using aircraft pro_le data, Atmos. Meas. Tech., 3, pp351-1362, 2010

[12] Tanaka et al.,:Aircraft measurements of carbon dioxide and methane for the calibration of ground-based high-resolution Fourier Transform Spectrometers and acomparison to GOSAT data measured over Tsukuba and Moshiri,Atmos. Meas.Tech., 5, pp2003-2012, 2012

2

Improved Fuzzy C-Mean Algorithm for Image Segmentation

Hind Rustum Mohammed
CS dept. Faculty of Computer Science and Mathematics
University of Kufa
Najaf, Iraq

Husein Hadi Alnoamani
MS dept. Faculty of Computer Science and Mathematics
University of Kufa
Najaf, Iraq

Ali AbdulZahraa Jalil / M.Sc. Student
CS dept. Faculty of Computer Science and Mathematics
University of Kufa, Najaf, Iraq

***Abstract*—The segmentation of image is considered as a significant level in image processing system, in order to increase image processing system speed, so each stage in it must be speed reasonably. Fuzzy c-mean clustering is an iterative algorithm to find final groups of large data set such as image so that is will take more time to implementation. This paper produces an improved fuzzy c-mean algorithm that takes less time in finding cluster and used in image segmentation.**

Keywords—pattern recognition; image segmentation; fuzzy c-mean; improved fuzzy c-mean; algorithms

I. INTRODUCTION

To recognize pattern and analysis an image the main process is segmentation of image[1-3]. Is an operation of dividing an image into parts that have same features and the collection of these parts form the original image[4]. Fig.1. illustrate variant levels of processing of image and technique of analyzing [5], and it shows clearly segmentation stage.

There are many types of image's pattern recognition and segmentation, but there are two mainly types of classification which are used: Supervised classification and unsupervised classification, in the first one the classes are defined in advance and in the second they are not defined in advance which known as clustering. There are two types of clustering: hard clustering and fuzzy clustering, in hard clustering, the data item is belong exactly to one cluster but in fuzzy clustering, the data item belong by the degree of membership to each cluster of clusters, and the summation of all memberships values to one of data items is equal to one.

Fuzzy c-mean clustering is one of unsupervised clustering algorithms that is widely used in image processing and computer vision because it easy to implement and clustering performance[6], [7]. It's used to segment an image by grouping pixels that have similar or nearly similar values into a cluster, where each group of pixel's values that belong to one cluster are similar to each other and different from pixel's values that belong to other clusters, and then these clusters represent the segments of the segmented image. The traditional fuzzy c-mean suffers from some limitations, it's not accurate in the segmentation of noisy image and time consuming because it's iterative nature. Our proposed algorithm which named Improved fuzzy c-mean algorithm offers an overcoming of one limitation of traditional fuzzy c-mean which is time-consuming.

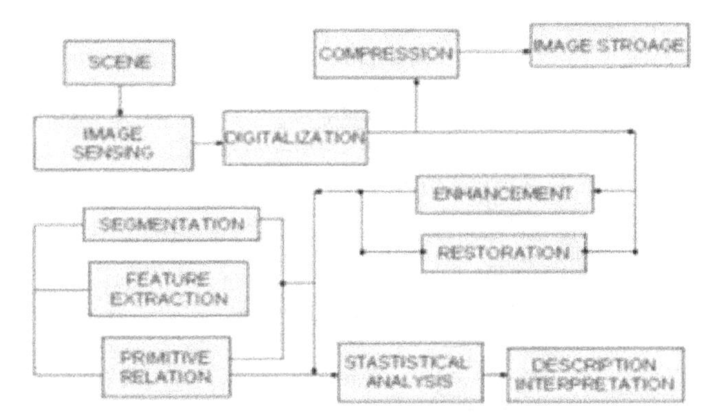

Fig. 1. Schema of variant levels of processing of image and technique of analyzing

In our proposed algorithm we use frequency of each data item of image and processing these frequencies instead of processing whole data items of the image. That is reduce processing time in the great form. This paper contains five parts and arranged as follow: Section 2 talking about time complexity, section 3 the traditional fuzzy c-mean, section 4 proposed an algorithm, section 5 Experimental Results, and in section 6 the conclusion.

II. TIME COMPLEXITY

It's time that required to run or execute an algorithm[11].the notation big O is used to express time complexity. it's proportional to the size of input data. If the input size is n, then the time complexity is the time required by the algorithm to process these input. Each algorithm has a primitive operation(s), so the time of the algorithm is determined by computing the summation of times that required to run each of these operations. It's always expressed by the prevalent term, which is the term have exponent with the highest value. It also ignores constant of multiplication and constant of the division. if the time required to accomplish an algorithm of n input size is $10n^3 + 6n$, then the expression of its time complexity is $O(n^3)$ also if the time required to accomplish an algorithm is $c*n^2$ or n^2/c where c is constant then the time complexity is $O(n^2)$. If the algorithm processed all inputted data to get the desired solution, then the time complexity called the worst-case of time complexity, which is show, the algorithm take maximum time to achieve the required process. There are many types of time complexities which depend on algorithm's function nature. Some common types of time complexities are constant time

O(1), linear time O(n), quadratic time $O(n^2)$, exponential time $O(c^n)$ where $c \geq n > 1$. In our proposed algorithm we suggest an algorithm that consumes so little time amount compared with the traditional fuzzy C-mean algorithm.

III. TRADITIONAL FUZZY C-MEAN

The fuzzy c-mean algorithm is one of the common algorithms that used to image segmentation by dividing the space of image into various cluster regions with similar image's pixels values. For medical images segmentation, the suitable clustering type is fuzzy clustering. The Fuzzy c-means (FCM) can be seen as the fuzzified version of the k-means algorithm. It is a clustering algorithm which enables data item to have a degree of belonging to each cluster by degree of membership. It's developed by Dunn [9] and changed by Bezdek [10]. The algorithm is an iterative clustering method that produces an optimal c partition by minimizing the weighted within group sum of squared error objective function [10]. Is widely used in image segmentation and pattern recognition. Following are steps of traditional fuzzy c-mean:

Step1:Choose random centroid at least 2 and put values to them randomly.

Step2:Compute membership matrix:

$$U_{ij} = \frac{1}{\sum_{k=1}^{c}\left[\frac{|x_i - c_j|}{x_i - c_k}\right]^{\frac{2}{m-1}}} \text{, where m > 1, c cluster's No.} \quad (1)$$

Step3: calculate the clusters centers:

$$C = \frac{\sum_{i=1}^{n} U^m{}_{ij} * x_i}{\sum_{i=1}^{n} U^m{}_{ij}} \quad (2)$$

Step4: if $C^{(k-1)} - C^k < \varepsilon$ then Stop else go to Step2.

This traditional algorithm is an iterative algorithm that suffers from time and memory consuming because it computes membership value for each item in the data.

IV. PROPOSED ALGORITHM

In the following section we provide the improved fuzzy c-mean algorithm:

Step1: Let H represent the frequency of each item in Data.

Step2:create vector $I = \min(\text{Data}) : \max(\text{Data})$

Step3:Choose random centroid at least 2.

Step4:Compute membership matrix:

$$U_{ij} = \frac{1}{\sum_{k=1}^{c}\left[\frac{|I_i - c_j|}{|I_i - c_k|}\right]^{\frac{2}{m-1}}} \quad (3)$$

Step5: calculate the cluster center:

$$C = \frac{\sum_{i=1}^{n} U^m * H * I}{\sum_{i=1}^{n} U^m * H} \quad (4)$$

Step6: if $C^{(k-1)} - C^k < \varepsilon$ then Stop else go to Step4.

The improved fuzzy c-mean use values that represent the frequency of items instead of actual values, in gray images the number of values of it may be reached to 256*256=65,536 and that is will take more time in processing, but in improved algorithm will take, at worst case, 256 item to process it. The proposed algorithm does not depend on whole data of image, it actually depends on data that represent the frequency of each data item in original image's data. A number of frequencies at most is 256.

V. EXPERIMENTS RESULTS

We tested Improved fuzzy c-mean by implemented by using MATLAB and compared it with implementation of fuzzy c-mean algorithm that used by MATLAB by calling command fcm, we try algorithm in database of images contains 100 images, in the following we provide a sample from tested images, in this testing sample we use C=3:

Fig. 2. Orginal image, "football"

Fig. 3. Orginal image, "office"

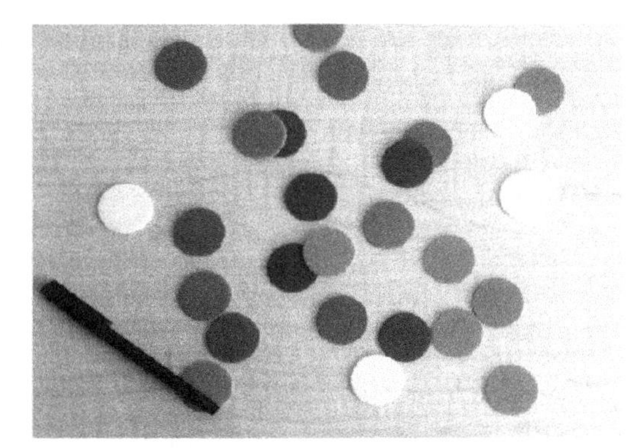

Fig. 4. Orginal image, “coloredChips”

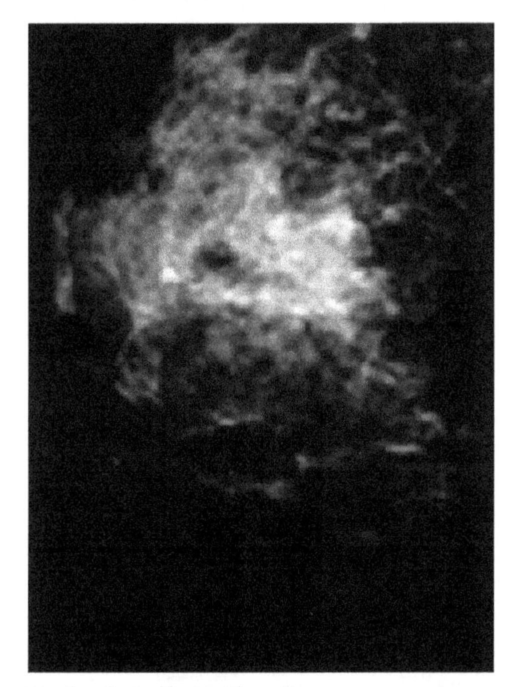

Fig. 5. Orginal image, “breast”

Fig. 6. Orginal image, “house”

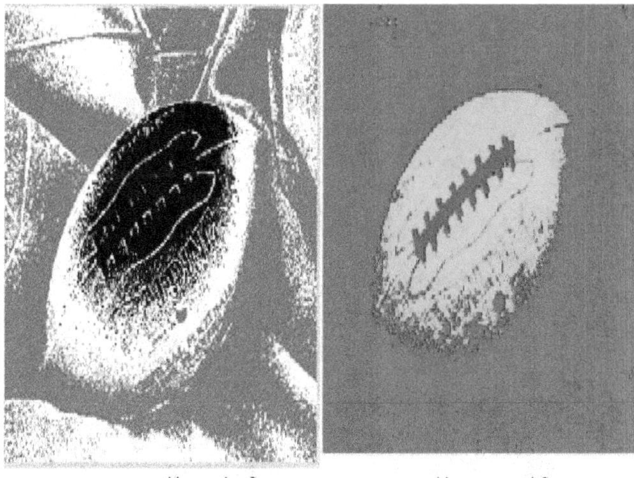

Fig. 7. Comparison between fcm and proposed fcm on “football” image

Fig. 8. Comparison between fcm and proposed fcm on “office” image

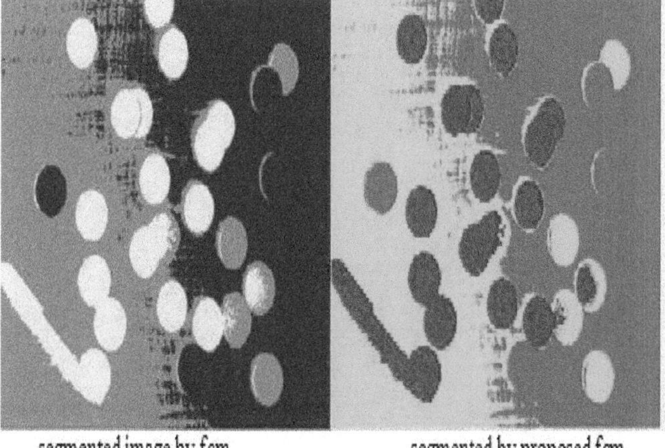

Fig. 9. Comparison between fcm and proposed fcm on “coloredChips” image

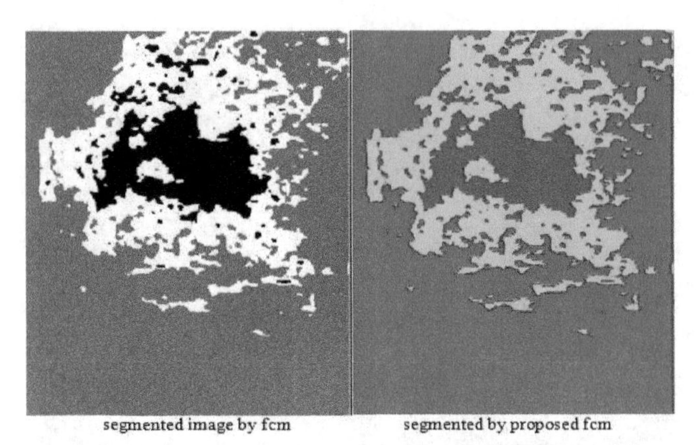

Fig. 10. Comparison between fcm and proposed fcm on "breast" image

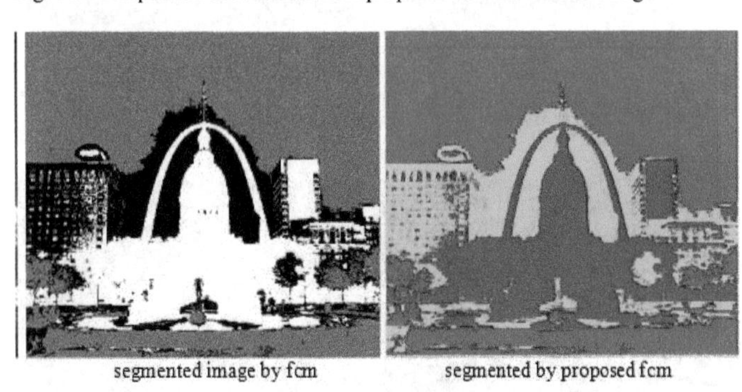

Fig. 11. Comparison between fcm and proposed fcm on "house" image

TABLE I. TIME COMPARISON BETWEEN TRADITIONAL FCM AND PROPOSED FCM

Image name	Segmentation time by fcm (sec)	Segmentation time by proposed fcm (sec)
football	3.828125	0.015625
office	22.796875	0.140625
coloredChips	8.296875	0.031250
breast	14.00000	0.062500
house	24.828125	0.078125

VI. CONCLUSION

From above results in accuracy and speed of proposed fuzzy c-mean algorithm compared with the traditional fuzzy c-mean algorithm, we conclude this algorithm is a great enhancement in implementation and performance of traditional fuzzy c-mean.

REFERENCES

[1] K.S. Deshmukh, G.N. Shinde, An adaptive color image segmentation, Electron.Lett. Comput. Vis. Image Anal. 5 (4) (2005) 12–23.

[2] Y. Zhang, A survey on evaluation methods for image segmentation, Pattern Recognition 29 (8) (1996) 1335–1346.

[3] V. Boskovitz, H. Guterman, An adaptive neuro-fuzzy system for automaticy image segmentation and edge detection, IEEE Trans. Fuzzy Syst. 10 (2) (2002) 247–262.

[4] C.Harris and M.Stephens, "A Combined Corner and Edge Detection," Proc.Fourth Alvey Vision Conf., pp.147-151, 1988.

[5] Cahoon, T.C .Sutton, M .A. Bezdek "Brest cancer detection using image processing techniques", J.C.Dept.of Comp.Sci.Univ.of West Florida, Pensacola, FL Fuzzy IEEE 2000.The ninth IEEE conference.

[6] D. L. Pham and J. L. Prince, "An adaptive fuzzy c-means algorithm for image segmentation in the presence of intensity in homogeneities," Pattern Recognition. Lett. vol. 20, pp 57-68, 1999.

[7] W. J. Chen, M. L. Giger, and U. Bick, "A fuzzy c-means (FCM)-based approach for computerized segmentation of breast lesions in dynamic contrast enhanced MRI images," Acad. Radiol, vol. 13, pp. 63-72, 2006.

[8] J. M. Gorriz, J. Ramirez, E. W. Lang, and C. G. Puntonet, "Hard c-means clustering for voice activity detection," Speech Commun, vol. 48, pp. 1638-1649, 2006

[9] J. C. Dunn, "A fuzzy relative of the ISODATA process and its use in detecting compact well-separated clusters," J. Cybernetics, vol. 3, no. 3,pp. 32-57, 1973.

[10] J. C. Bezdek, "Pattern recognition with fuzzy objective function algorithms," New York, Plenum, 1981.

[11] Sipser, Michael, "Introduction to the Theory of Computation". Course Technology Inc. ISBN 0-619-21764-2, 2006.

3

Military Robotics: Latest Trends and Spatial Grasp Solutions

Peter Simon Sapaty
Institute of Mathematical Machines and Systems
National Academy of Sciences
Kiev, Ukraine

***Abstract*—A review of some latest achievements in the area of military robotics is given, with main demands to management of advanced unmanned systems formulated. The developed Spatial Grasp Technology, SGT, capable of satisfying these demands will be briefed. Directly operating with physical, virtual, and executive spaces, as well as their combinations, SGT uses high-level holistic mission scenarios that self-navigate and cover the whole systems in a super-virus mode. This brings top operations, data, decision logic, and overall command and control to the distributed resources at run time, providing flexibility, ubiquity, and capability of self-recovery in solving complex problems, especially those requiring quick reaction on unpredictable situations. Exemplary scenarios of tasking and managing robotic collectives at different conceptual levels in a special language will be presented. SGT can effectively support gradual transition to automated up to fully robotic systems under the unified command and control.**

Keywords—military robots; unmanned systems; Spatial Grasp Technology; holistic scenarios; self-navigation; collective behavior; self-recovery

I. INTRODUCTION

Today, many military organizations take the help of military robots for risky jobs. The robots used in military are usually employed within integrated systems that include video screens, sensors, grippers, and cameras. Military robots also have different shapes and sizes according to their purposes, and they may be autonomous machines or remote-controlled devices. There is a belief that the future of modern warfare will be fought by automated weapons systems.

The U.S. Military is investing heavily in research and development towards testing and deploying increasingly automated systems. For example, the U.S. Army is looking to slim down its personnel numbers and adopt more robots over the coming years [1, 2]. The Army is expected to shrink from 540,000 people down to 420,000 by 2019. To keep things just as effective while reducing manpower, the Army will bring in more unmanned power, in the form of robots. The fact is that people are the major cost, and first of all their life. Also, training, feeding, and supplying them while at war is pricey, and after the soldiers leave the service, there's a lifetime of medical care to cover.

Military robots are usually associated with the following categories: *ground*, *aerial*, and *maritime*, with some of the latest works in all three discussed in the paper, including those oriented on collective use of robots.

Most military robots are still pretty dumb, and almost all current unmanned systems involve humans in practically every aspect of their operations. The Spatial Grasp ideology and technology described in the rest of this paper can enhance individual and collective intelligence of robotic systems, especially distributed ones. It can also pave the real way to massive use of advanced mobile robotics in human societies, military systems including and particularly.

II. SOME LATEST DEVELOPMENTS AND DEMANDS TO MILITARY ROBOTICS

A. Ground Robots

The ability of robots to save lives has secured future path for ground robotics alongside the warfighter. Ground robotics can be engaged in different missions including Explosive Ordnance Disposal (EOD), Combat Engineering, Reconnaissance, and many others. The US Army plans to refurbish 1,477 of its ground robots, which is about 60 percent of the total fleet [3]. The following may be named among the latest developments in ground robotics.

Boston Dynamics designed the LS3 *"robot mules"* to help soldiers carry heavy loads [4], see Fig. 1*a-c*. LS3 is a rough-terrain robot designed to go anywhere Marines and Soldiers go on foot, helping carry their load. Each LS3 carries up to 400 lbs of gear and enough fuel for a 20-mile mission lasting 24 hours. LS3 automatically follows its leader using computer vision, so it does not need a dedicated driver. It also travels to designated locations using terrain sensing and GPS.

a) b) c)

Fig. 1. Boston Dynamics robot mules: a) Carrying heavy loads; b) Following soldiers; c) Moving through complex terrains

The Boston Dynamics' *Cheetah robot* (Fig. 2*a-b*) is the fastest legged robot in the World, surpassing 29 mph, a new land speed record for legged robots [5]. The Cheetah robot has an articulated back that flexes back and forth on each step, increasing its stride and running speed, much like the animal does. The current version of the Cheetah robot runs on a high-speed treadmill in the laboratory where it is powered by an off-board hydraulic pump and uses a boom-like device to keep it running in the center of the treadmill.

a) b) c)

Fig. 2. Boston Dynamics robots: a) The Cheetah concept; b) Cheetah on a high-speed treadmill; c) Cheetah becoming Wild Cat running untethered

The next generation Cheetah robot, *WildCat*, Fig. 2*c*, is designed to operate untethered. WildCat is an early model for field testing. It sports a noisy combustion onboard engine. Named the WildCat, the outdoor runner is funded by the Defense Advanced Research Projects Agency (DARPA), and is being developed for military use. With a large motor attached, WildCat isn't as fast as its 28mph-plus cousin, being currently limited to around 16mph on flat terrain.

New military technology 2014 *supersoldier* robot has been developed [6]: all-terrain, highly mobile, and with high precision shooting (Fig. 3*a-c*). It is logical to assume that killer robots are already here, and the new science discoveries of 2014 may be used to create real terminators.

a) Ammunition b) All terrain chassis c) Field trials

Fig. 3. Supersoldier robot

B. Aerial Robotics

The US Army, Air Force, and Navy have developed a variety of robotic aircraft known as unmanned flying vehicles (UAVs). Like the ground vehicles, these robots have dual applications: they can be used for reconnaissance without endangering human pilots, and they can carry missiles and other weapons [7].

The best known armed UAVs are the semi-autonomous Predator Unmanned Combat Air Vehicles (UCAV) built by General Atomics which can be equipped with Hellfire missiles. The military services are also developing very small aircraft, sometimes called Micro Air Vehicles (MAV) capable of carrying a camera and sending images back to their base. Some newest UCAV developments are mentioned below.

The Northrop Grumman X-47B is a demonstration unmanned combat air vehicle (UCAV) designed for carrier-based operations [8], see Fig. 4*a-c*. Developed by the American defense technology company Northrop Grumman, the X-47 project began as part of DARPA's J-UCAS program, and is now part of the United States Navy's Unmanned Combat Air System Demonstration (UCAS-D) program.

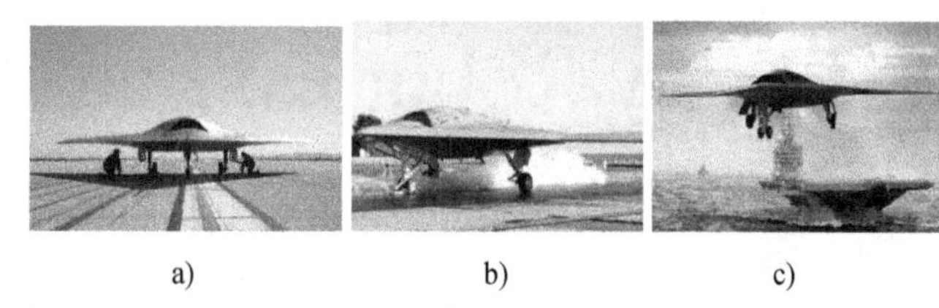

a) b) c)

Fig. 4. Northrop Grumman X-47B: a) Front view; b) Land-launched; c) Carrier-launched

The X-47B first flew in 2011, and as of 2014, it is undergoing flight and operational integration testing, having successfully performed a series of land- and carrier-based demonstrations. In August 2014, the US Navy announced that it had integrated the X-47B into carrier operations alongside manned aircraft. Northrop Grumman intends to develop the prototype X-47B into a battlefield-ready aircraft, the Unmanned Carrier-Launched Surveillance and Strike (UCLASS) system, which will enter service around 2019. X-47B can stay in the air for 50 hrs, carry 2 tons of weaponry, and be refuelled in the air.

Doubling the Threat: Drones + Lasers. The research and development arm of the US Department of Defense plans to establish drone-mounted laser weapons, a scheme referred to as 'Project Endurance' in the agency's 2014 budget request [9], see Fig. 5*a-c*. The Pentagon edged closer to mounting missile-destroying lasers on unmanned and manned aircraft, awarding $26 million to defense contractors to develop the technology.

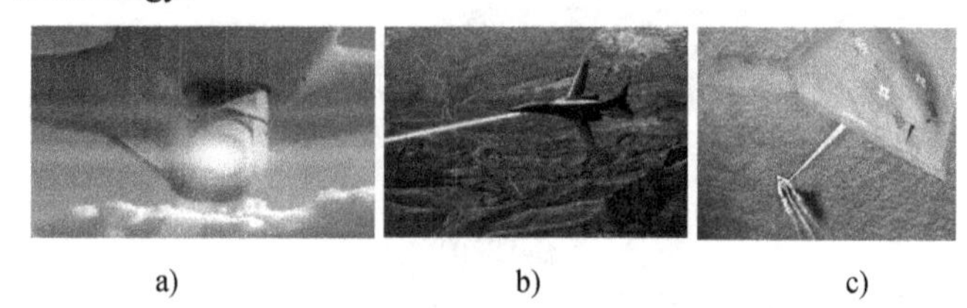

a) b) c)

Fig. 5. Drones with lasers: a) HELLADS mounted on a drone, b-c) Drone laser in operation

General Atomics is getting increasingly excited by the HELLADS— the High-Energy Liquid Laser Defense System. It is designed to shrink a flying laser into a package small enough to cram into an aircraft. This will give a potentially unlimited shooting magazine to the drone.

Hypersonic aircraft. The SR-72 [10] could fly as fast as Mach 6, will have the ability to gather intelligence, conduct surveillance and reconnaissance, and launch combat strikes at an unprecedented speed, see Fig. 6*a*. SR-72 could be operational by 2030. At this speed the aircraft would be so fast that adversary would have no time to react or hide.

a) b)

Fig. 6. Hypersonic vehicles: a) SR-72 with Mach 6; b) DARPA HTV-2 with Mach 20

DARPA rocket-launched HTV-2, 13,000 mph *Hypersonic Glider* [11] (see Fig. 6*b*), was designed to collect data on three technical challenges of hypersonic flight: aerodynamics, aerothermal effects, and guidance, navigation and control. A technology demonstration and data-gathering platform, the HTV-2's second test flight was conducted to validate current models and increase technical understanding of the hypersonic regime. The flight successfully demonstrated stable aerodynamically-controlled flight at speeds up to Mach 20.

C. Maritime Robotics

Sea-based robots—unmanned maritime systems, or UMSs, can be either free-swimming or tethered to a surface vessel, a submarine, or a larger robot [12], see examples in Fig. 7. Tethers simplify providing power, control, and data transmission, but limit maneuverability and range. Recently developers have built highly autonomous systems that can navigate, maneuver, and carry out surprisingly complex tasks. UMSs can operate on the ocean's surface, at or just below the surface, or entirely underwater. Operating above or near the surface simplifies the power and control, but compromises stealth. The U.S. Navy has devoted particular attention to unmanned underwater vehicles (UUSs) during the past 10-15 years. Its unmanned surface vehicles (USVs) are much less far along (Fig. 7*a*); the Navy has put a higher priority on using automation to reduce crew size in U.S. warships. Some latest works on UUSs follow.

Large Displacement Unmanned Undersea Vehicle (LDUUV) [13], see Fig. 7*b*, is to conduct missions longer than 70 days in open ocean and littoral seas, being fully autonomous, long-endurance, land-launched, with advanced sensing for littoral environments. The vehicle's manufacturing and development phase will begin in 2015 with testing planned for 2018. According to the Navy's ISR Capabilities Division, LDUUV will reach initial operating capability as a squadron by 2020 and full rate production by 2025.

Fig. 7. a) Unmanned surface vehicle; b) Large Displacement Unmanned Undersea Vehicle, LDUUV; c) Underwater glider

Underwater gliders [14], see Fig. 7*c*, will not require fuel but will instead use a process called "hydraulic buoyancy," which allows the drone to move up and down and in and out of underwater currents that will help it move at a speed of about one mile per hour. Carrying a wide variety of sensors, they can be programmed to patrol for weeks at a time, surfacing to transmit their data to shore while downloading new instructions at regular intervals.

D. Collectively Behaving Robots

To be of real help in complex military applications, robots should be integral part of manned systems, they should also be capable of being used massively, in robotic collectives. The tests on Virginia's James River represented the first large-scale military demonstration of a *swarm of autonomous boats* designed to overwhelm enemies [15], see Fig. 8*a*. The boats operated without any direct human control: they acted as a robot boat swarm. This capability points to a future where the U.S. Navy and other militaries may deploy multiple underwater, surface, and flying robotic vehicles to defend themselves or attack a hostile force.

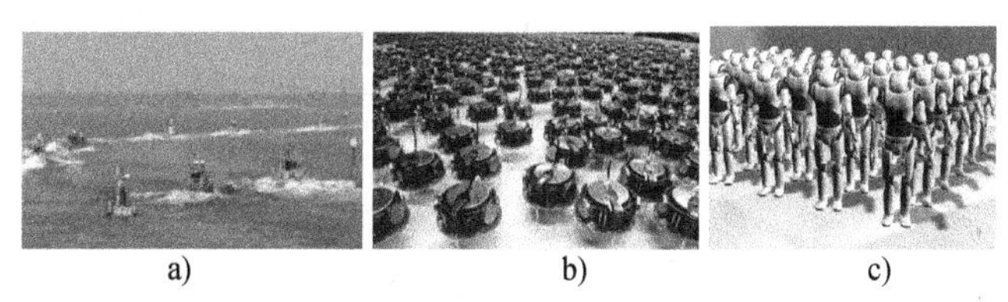

Fig. 8. a) Swarm of autonomous boats; b) Harvard University multiple robots operating without central intelligence; c) Sci-fi image of future robotic armies

Harvard University scientists have devised a swarm of 1,024 tiny robots that can work together without any guiding central intelligence [16], see Fig. 8*b*. Like a mechanical flash mob, these robots can assemble themselves into five-pointed stars, letters of the alphabet and other complex designs. Swarm scientists are inspired by nature's team players—social insects like bees, ants and termites; schools of fish; and flocks of birds. These creatures collaborate in vast numbers to perform complicated tasks, even though no single individual is actually in charge. These results are believed to be useful for the development of advanced robotic teams even armies, (with futuristic image in Fig. 8c).

E. General Demands to Military Robotic Systems

A thorough analysis of aims and results of the development and implementation of military robots, including the ones briefed above, helps us formulate general demands with regard to their overall management and control, which may be as follows.

- Despite the diversity of sizes, shapes, and orientations, they should all be capable of operating in distributed, often large, physical spaces, thus falling into the category of distributed systems.
- Their activity is to include navigation, movement, observation, gathering data, carrying loads which may include ammunitions or weapons, and making impact on other manned on unmanned units and the environment.
- They should have certain, often high, degree of autonomy and capability of automatic decision making to be really useful in situations where human access and activity are restricted.
- They should effectively interact with manned components of the systems and operate within existing command and control infrastructures, to be integral parts of the system.
- They should be capable of effective swarming for massive use, and this swarming should be strongly controlled from outside -- from manned parts of the system or from other, higher-level, unmanned units.
- Their tasking and retasking (including that of swarms) should be flexible and convenient to humans to

guarantee runtime reaction on changing goals and environments, especially on battlefields.

- The use of unmanned units should be safe enough to humans and systems they are engaged in.
- Their behaviour should satisfy ethical and international norms, especially in life-death situations.

III. SPATIAL GRASP TECHNOLOGY FOR MANAGEMENT OF ROBOTIC SYSTEMS

The developed high-level Spatial Grasp ideology and Technology, SGT, for coordination and management of large distributed systems [17] allows us to investigate, develop, simulate, and implement manned-unmanned systems in their integrity and entirety. Also gradually move to fully unmanned systems with dynamic tasking and managing individual robots and their groups, regardless of the group's size. SGT can believably satisfy most of the demands to military robotic systems formulated above.

A. SGT General Issues

SGT is based on coordinated integral, seamless, vision & navigation & coverage & surveillance & conquest of physical, virtual, or execution spaces, as shown in Fig. 9*a-b*.

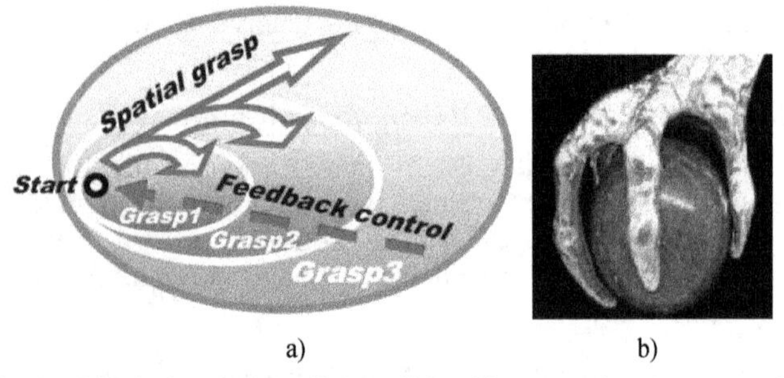

a) b)

Fig. 9. SGT basics: a) Controlled parallel and incremental space grasp; b) Symbolic physical analogy

It has a strong psychological and philosophical background reflecting how humans, especially top commanders, mentally plan, comprehend and control operations in complex and distributed environments. SGT pursues *holistic*, *gestalt* [18], or *over-operability* [19] ideas rather than traditional multi-agent philosophy [20], with *multiple agents and their interactions appearing and disappearing dynamically,* on the implementation level, and only if and when needed in particular places and moments of time.

SGT can be practically implemented in distributed systems by a network of universal control modules embedded into key system points (humans, robots, sensors, mobile phones, any electronic devices, etc.), which altogether, collectively, understand and interpret mission scenarios written in a special high-level Spatial Grasp Language, SGL [17], see Fig. 10.

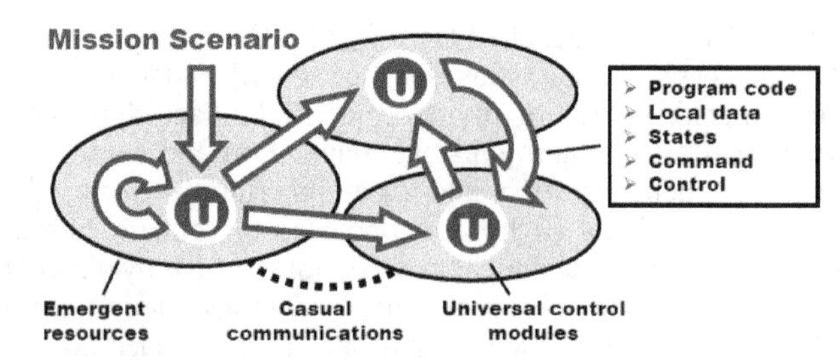

Fig. 10. Collective spatial interpretation of SGL scenarios

Capable of representing any parallel and distributed algorithms, these scenarios can start from an arbitrary node, covering at runtime the whole system or its parts needed with operations, data, and control, as shown in Fig. 11. Different scenarios can intersect in the networked space while cooperating or competing (Fig. 11).

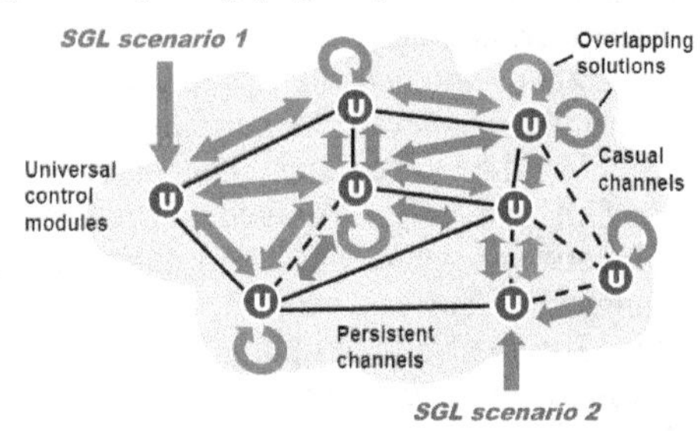

Fig. 11. Spreading scenarios intersection & cooperation,

They can establish distributed runtime information and control infrastructures that can support distributed databases, command and control, situation awareness, autonomous decisions, also any other existing or hypothetical computational and/or control models (Fig. 12).

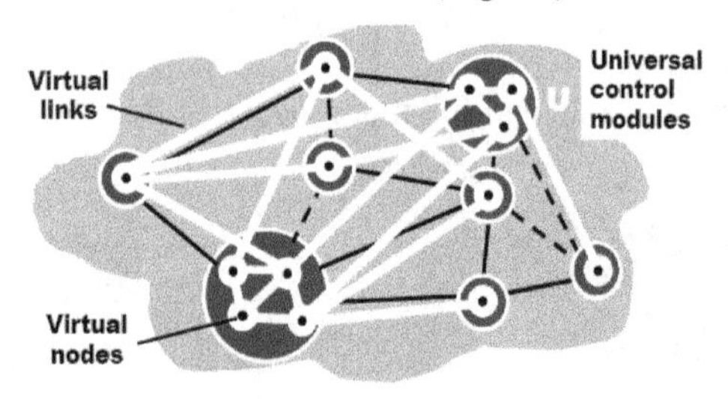

Fig. 12. Creating spatial infrastructures

B. Spatial Grasp Language, SGL

SGL allows us to directly move through, observe, and make any actions and decisions in fully distributed environments. SGL scenario develops as parallel transition between sets of progress points (or *props*) reflecting progressive spatial-temporal-logical stages of the scenario development, which may be associated with different physical, virtual or execution locations in distributed worlds. *Any sequential or parallel, centralized or distributed, stationary or mobile algorithm operating with information and/or physical matter can be written in SGL at any levels.*

SGL directly operates with the following worlds:

- *Physical World* (PW), infinite and continuous, where each point can be identified and accessed by physical coordinates, with certain precision.
- *Virtual World* (VW), which is finite and discrete, consisting of nodes and semantic links between them.
- *Executive world* (EW) consisting of active doers which may be humans, robots, sensors or any intelligent machines capable of operations on matter, information, or both, i.e. on the previous two worlds.

Directly working with different worlds, SGL can provide high flexibility, convenience, and compactness in expressing complex scenarios within the same formalism. From one side, it can support high level, semantic descriptions abstracting from physical resources which can vary and be assigned at runtime, and from the other side, detailing some or all such resources, and to the full depth, if necessary.

For example, working directly with PW, like moving through and impacting it, can be free from naming physical devices which can do this (e.g. humans, robots), the latter engaged and disengaged automatically upon necessity, availability, or uselessness. Directly working with VW, like creating knowledge, operational, or C2 infrastructures, can also abstract away from physical resources (humans or computing facilities) which can be assigned or reassigned at runtime. Working directly with EW, can bring any necessary details for execution of missions, like particular human, robotic or sensor units and their interactions and subordination. Any combination and integration of these three worlds can be possible, with direct management of the mixture in SGL too. Integration between PW and VW can be named as PVW, with other cases presented as PVW, PEW, VEW, and all three together as PVEW.

SGL has universal recursive syntactic structure shown in Fig. 13 capable of representing any parallel, distributed and spatial algorithm working with arbitrary complex data. This structure, following the spatial grasp ideology of SGT mentioned above, also allows any language obeying it to be arbitrarily extended with new operations, data and control.

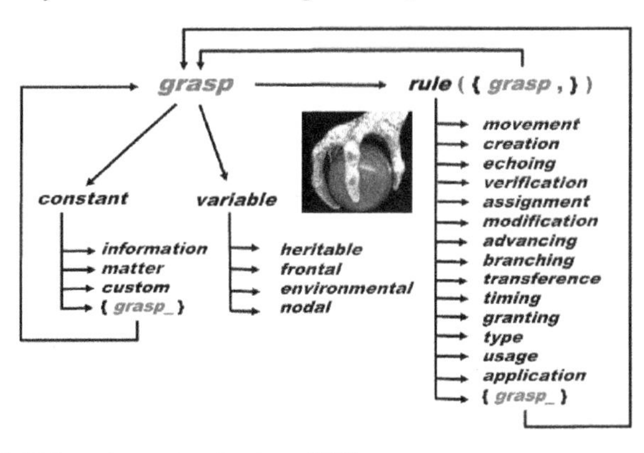

Fig. 13. Universal recursive structure of SGL

Mentioning some SGL details may be helpful for understanding the rest of this paper, as follows. The basic language construct, *rule*, can represent, for example, the following categories (this list being far from complete):

- Elementary arithmetic, string or logic operation.
- Hop or move in a physical, virtual, or combined space.
- Hierarchical fusion and return of local or remote data.
- Distributed control, both sequential and parallel.
- A variety of special contexts for navigation in space (influencing embraced operations and decisions).
- Type or sense of a value or its chosen usage, assisting automatic interpretation.
- Creation or removal of nodes and links in distributed knowledge infrastructures.
- Composition of other rules.

Working in fully distributed physical, virtual, executive or combined environments, SGL has different types of variables, called *spatial*, effectively serving multiple cooperative processes. They belong to the following four categories:

- *Heritable variables* – starting in a prop and serving all subsequent props which can share them in read & write operations.
- *Frontal variables* – individual and exclusive prop's property (not shared with other props), being transferred between consecutive props and replicated if from a single prop a number of other props emerge – thus propagating together with the evolving spatial control.
- *Environmental variables* – accessing different elements of the physical and virtual words when navigating them, also basic parameters of the internal world of SGL interpreter.
- *Nodal variables* – adding individual temporary property to VW, PW, EW or combined nodes; they can be accessed and shared by all activities currently associated with these nodes.

For simplifying and shortening complex scenarios (say, reducing nested parentheses in them), SGL programs can additionally use syntactic constructs common for traditional languages, as will be seen from the forthcoming examples of this paper, always remaining, however, within the general structure depicted in Fig. 13.

C. Elementary Examples in SGL

Let us consider some elementary scenarios from the mentioned three worlds (PW, VW, and EW), as shown in Fig. 14*a-f*.

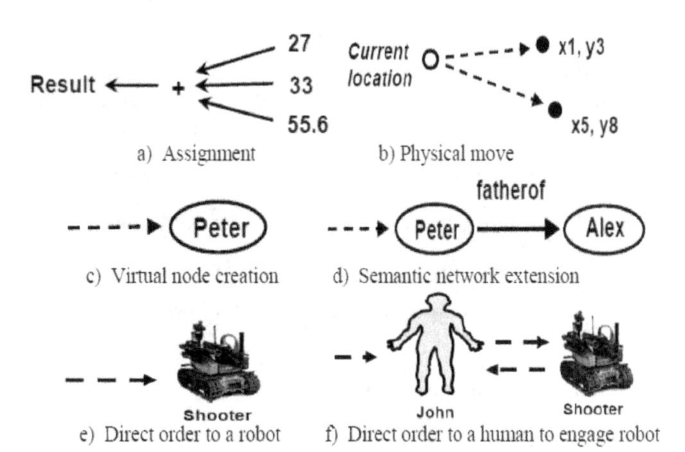

Fig. 14. Some elementary scenarios for programming in SGL

They all can be expressed within the same spatial grasp ideology and unified SGL syntax, as follows.

- Assignment (Fig.14*a*):

  `assign(Result, add(27, 33, 55.6))` or

  `Result = 27+33+55.6`

- Moves in physical space to coordinates (x1, y3), and (x5, y8) independently or in parallel (Fig.14*b*):

  `move(location(x1,y3), location(x5,y8))`

- Creation of a virtual node (Fig.14*c*):

  `create('Peter')`

- Extending virtual network with a new link-node pair (Fig.14*d*):

  `advance(hop('Peter'), create(+'fatherof','Alex'))` or

  `hop('Peter'); create(+'fatherof','Alex')`

- Giving direct command to robot Shooter to fire at coordinates (x, y) (Fig. 14*e*):

  `hop(robot(Shooter)); fire(location(x,y)`

- Order soldier John to fire at coordinates (x, y) by using robot Shooter and confirm robot's action in case of its success (Fig. 14*f*):

```
hop(soldier:John);
if((hop(robot:Shooter);
    fire(location:x,y)),
   report:done)
```

D. SGL Interpreter Architecture

SGL interpreter consists of specialized modules handling & sharing specific data structures, as shown in Fig. 15.

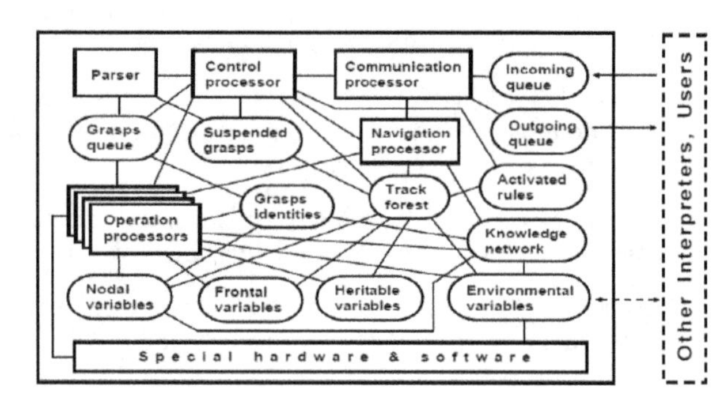

Fig. 15. SGL interpreter architecture

The network of the interpreters can be mobile and open, runtime changing the number of nodes and communication structure between them. The SGL interpreters can be concealed if to operate in hostile environments.

The dynamically networked SGL interpreters are effectively forming a sort of a *universal parallel spatial machine* capable of solving any problems in a fully distributed mode, without any special central resources. "Machine" rather than a computer or "brain" because it can operate with physical matter too, and can move partially or as a whole in physical environment, possibly, changing its distributed shape and the space coverage. This machine can operate simultaneously on many mission scenarios which can be injected at any time from its arbitrary nodes/interpreters.

Tracks-Based Automatic Command & Control. The backbone and "nerve system" of the distributed interpreter is its spatial track system covering the spaces navigated and providing overall awareness, ad hoc automatic command and control of multiple distributed processes, access to and life of different types of spatial variables, as well as self-optimization and self-recovery from damages. Different stages of its operation during parallel space navigation are shown in Fig. 16*a-d*.

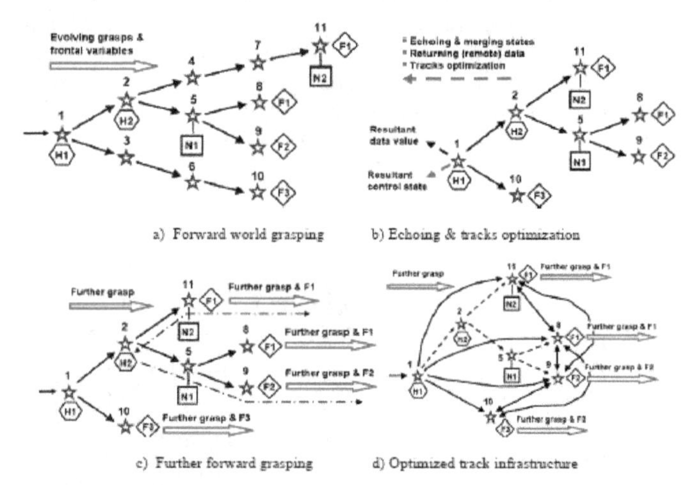

Fig. 16. The evolving track-based automatic command and control infrastructure

The symbols in Fig. 16 have the following meanings: □ — nodal variables, ◇ — frontal variables, ○ — heritable variables, ☆ — track nodes, and ⟶ — track links.

E. Integration with Robotic Functionalities

By embedding SGL interpreters into robotic vehicles, as in Fig. 17, we can provide any needed behavior of them, on any levels, from top semantic to detailed implementation. The technology can be used to task and control single robots as well as their arbitrary groups, with potentially unlimited number and diversity of individual robots (some hypothetic group scenarios shown in Fig. 17). For the robotic teams (or even possible future armies) it can describe and organize any collective behavior needed — from semantic task definition of just what to do in a distributed environment — to loose swarming — to a strongly controlled integral unit strictly obeying external orders. Any mixture of different behaviors within the same scenario can be guaranteed too.

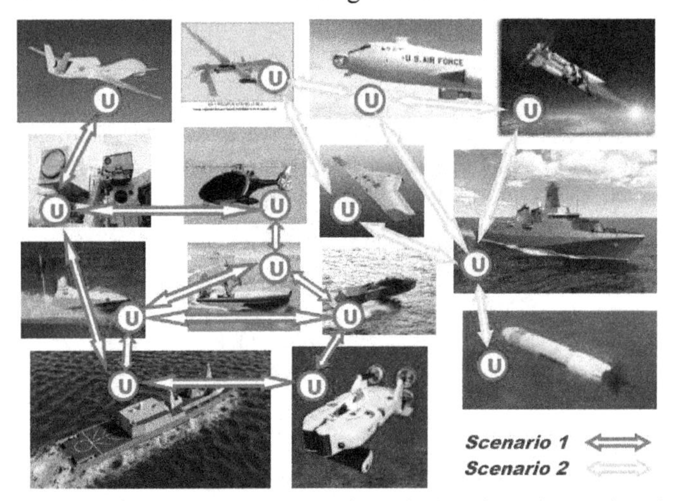

Fig. 17. Embedding SGL interpreters into robotic units and examples of collective scenarios

IV. Application of SGT to Robotics

A. Collective Spatial Task Execution, Purely Semantic Level

At the semantic level we can describe in SGL only what to do in a distributed space and the top decisions needed, regardless of a possible hardware or even system organization to accomplish this — these can be effectively shifted to intelligent automatic networked interpretation of the language. Let us consider the following task:

Go to physical locations of the disaster zone with coordinates:

(50.433, 30.633), (50.417, 30.490), and (50.467, 30.517).

Evaluate damage in each location and return the maximum damage value on all locations.

The corresponding SGL program will be as follows:

```
maximum(
   move((50.433, 30.633),
         (50.417, 30.490),
         (50.467, 30.517));
   evaluate(damage))
```

This task can be executed by different number of available mobile robots (actually from one to four, using more robots will have no much sense), and let three robots be available in the area of interest for our case, as in Fig. 18. The semantic level scenario can be initially injected into any robot (like R1), Fig. 18*a*, and then the distributed networked SGL interpreter installed in all robots automatically takes full care of the distributed task solution, with different stages depicted in Fig. 18*b-d*.

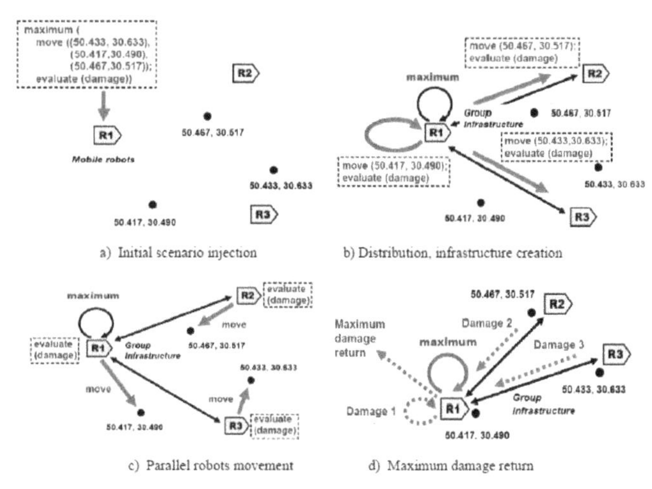

Fig. 18. Solving the task with three robots

The robots, with installed SGL interpreters and communicating with each other, are effectively forming integral distributed spatial machine that solves the problem defined purely semantically, with runtime partitioning, modifying, distributing, replicating and interlinking the emerging scenario parts automatically.

B. Explicit Collective Behavior Set Up

In contrast to the previous task defined on the level "what to do" only, different kinds of explicit behaviours can be expressed in SGL too, which, when integrated with each other, can provide very flexible, powerful, and intelligent global behaviour. Imagine that a distributed area needs to be investigated by multiple unmanned aerial vehicles that should search the space in a randomized way (preserving, however, some general direction of movement), create and update ad hoc operational infrastructure of the group (for it to follow global goal and be controlled from outside if needed), collect information on the discovered objects throughout the region covered, classifying them as targets, and organize collective reaction on the targets, as in Figure 19*a-d*.

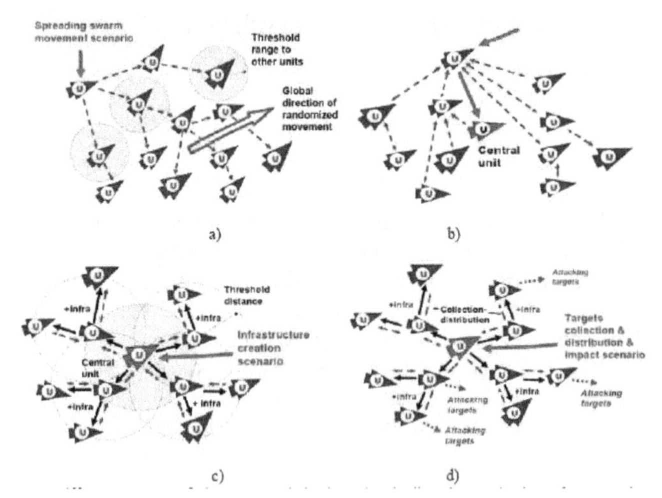

Fig. 19. Different aspects of the group's behavior: a) Distributed organization of cooperative swarm movement; b) Updating & hopping to the topological center; c) Creating & updating of spatial runtime infrastructure starting from the updated center; d) Collecting, distributing, selecting & attacking targets

The different stages depicted in Fig. 19 can be easily expressed in SGL and altogether integrated into the resultant holistic group scenario, as follows.

- Randomized swarm movement (starting in any node, with minimum, threshold, distance between moving nodes allowed), naming it ***swarm***:

```
hop(all_nodes);
frontal(Limits = (dx(0, 8), dy(-2, 5)),
Threshold = 50);
repeat(
  nodal(Shift) = random(Limits);
  if(empty(hop(WHERE + Shift, Range, all)),
     shift(Shift)))
```

- Regular updating and subsequent hopping to topological center (as the latter may change in time due to randomized movement resulting in varying distances between nodes and also possible spatial shape of the group); starting from any node, including the current center), naming it ***center***:

```
frontal(Average) =
   average(hop(all_nodes); WHERE);
min_destination(
   hop(all_nodes);
   distance(Average, WHERE))
```

- Regular creating & updating of spatial runtime infrastructure (starting from the updated central node, using semantic links "infra" and maximum allowed physical distance, or range, between nodes to form direct links), naming the program as ***infra***:

```
stay(
  frontal(Range) = 100;
  repeat(
    remove(previous_links);
    linkup(+infra, first_come, Range)))
```

- Collecting & selecting & attacking targets on the whole territory controlled (starting from the updated central node and using the updated spatial infrastructure leading to all nodes, to be used repetitively until the infrastructure is updated again), let it be called ***targets***:

```
nonempty(frontal(Seen) =
  repeat(
    free(detect(targets)), hop(+infra)));
repeat(
  free(select_move_shoot(Seen)),
  hop(+infra))
```

- Using these SGL scenarios for different behavioral stages, we can easily integrate them within the global one, as follows.

```
independent(
  swarm,
  repeat(center; infra;
         or_parallel(
           loop(targets),
           wait(time_delay))))
```

The obtained resultant scenario, which can start from any mobile unit, combines loose swarm movement in a distributed space with regular updating of topologically central unit and runtime hierarchical infrastructure between the units. The latter regularly controls observation of the distributed territory, collects data on targets and distributes them back to all units for individual selections and impact operations. The resultant scenario is setting certain time interval (*time_delay*) for preserving status of the current central node and emanating from it infrastructure before updating them due to possible change of distances between freely moving nodes.

V. Other Applications: Formalizing & Automating Command and Control

Formalization of Command Intent (CI) and Command and Control (C2) are among the most challenging problems on the way to creation of effective multinational forces, integration of simulations with live control, and transition to robotized armies. The existing specialized languages for unambiguous expression of CI and C2 (BML, C-BML, JBML, geoBML, etc.) [21] are not programming languages themselves, requiring integration with other linguistic facilities and organizational levels. Working directly with both physical and virtual worlds, SGL, being a universal programming language, allows for effective expression of any military scenarios and orders, drastically simplifying their straightforward implementation in robotized systems. SGL scenarios are much shorter and simpler than in BML or related languages, and can be created at runtime, on the fly. Typical battlefield scenario example, borrowed from [21], is shown in Fig. 20

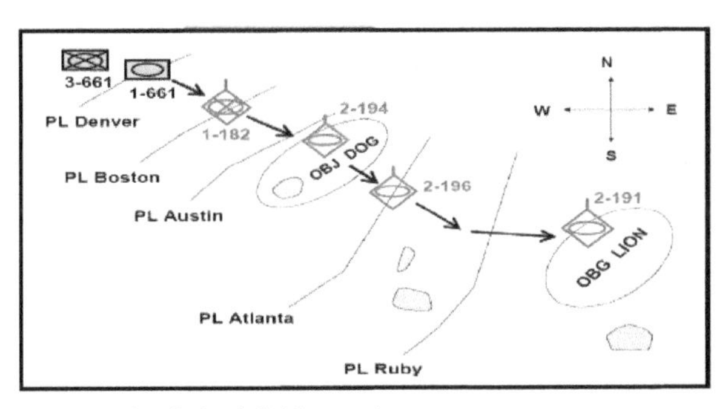

Fig. 20. Example of a battlefield scenario

The task is to be performed by two armoured squadrons BN-661 Coy1, and BN-661 Coy3, which are ordered to cooperate in coordination. The operation is divided into four time phases: from TP0 to TP1, from TP1 to TP2, from TP2 to TP3, and from TP3 to TP4, to finally secure objective LION, and on the way to it, objective DOG. Their coordinated advancement should be achieved by passing Denver, Boston, Austin, Atlanta, and Ruby lines, while fixing and destroying enemy units Red-1-182, Red-2-194, Red-2-196, and Red-2-191.

This scenario can be presented in SGL as follows.

```
FIXER:BN_661_Coy1;
SUPPORTER_DESTROYER:BN_661_Coy3;
deploy(Denver, T:TP0);
advance_destroy(
 (PL:Boston, TARGET:Red_1_182, T:TP1),
 (PL:Austin, OBJ:DOG, TARGET:Red_2_194, T:TP2),
 (PL:Atlanta, TARGET:Red_2_196, T:TP3),
 (PL:Ruby, OBJ:LION, TARGET:Red_2_191, T:TP4));
seize(LION, T:TP4)
```

This description is much clearer, and more compact (about 10 times) than if written in BML on the level of interacting individual units, as in [21]. This simplicity may allow us redefine the whole scenario or its parts at runtime, on the fly, when the goals and environment change rapidly, also naturally engage robotic units instead of manned components. Similar to possibility of expressing different levels of organization of robotic swarms in the previous section, we may further represent this current battlefield scenario at different levels too, for example, moving upwards with its generalization, as follows:

- Not mentioning own forces, which may become clear at runtime only:

```
deploy(Denver, T:TP0);
advance_destroy(
 (PL:Boston, TARGET:Red_1_182, T:TP1),
 (PL:Austin, OBJ:DOG, TARGET:Red_2_194, T:TP2),
 (PL:Atlanta, TARGET:Red_2_196, T:TP3),
 (PL:Ruby, OBJ:LION, TARGET:Red_2_191, T:TP4));
seize(LION, T:TP4)
```

- Further up, not mentioning adversary's forces, which may not be known in advance but should be destroyed if discovered, to move ahead:

```
deploy(Denver, T:TP0);
advance(
 (PL:Boston, T:TP1),
 (PL:Austin, OBJ:DOG, T:TP2),
 (PL:Atlanta, T:TP3),
 (PL:Ruby, OBJ:LION, T:TP4));
seize(LION, T:TP4)
```

- Further up, setting main stages only, with starting and final time only known:

```
deploy(Denver, T:TP0);
advance(PL:(Boston, Austin, Atlanta, Ruby));
seize(LION, T:TP4)
```

- And final goal only:

```
seize(LION, T:TP4)
```

Having the same formal language for any system levels and their any mixtures, provides us with high flexibility for organization of advanced missions, especially with limited or undefined resources and unknown environments; also possibility of potentially unlimited engagement of robotic components under the unified command and control philosophy.

VI. Conclusions

Robots can assist humans in many areas, especially in dangerous and hazardous situations and environments. But the fate of robotics, military especially, will depend on *how it conceptually and organizationally integrates with manned systems within overall management and command and control.*

The developed high-level distributed control technology, SGT, based on holistic and gestalt principles can effectively support a unified transition to automated up to fully unmanned systems with massive use of advanced robotics. The practical benefits may be diverse and numerous. One of them, for example, may be effective management of advanced robotic collectives, regardless of their size and spatial distribution, by a single human operator only, due to high level of their internal self-organization and integral responsiveness provided by SGT. More on the SGT philosophy and history, details of SGL with its networked implementation, and the researched applications, some of which have been mentioned throughout this paper, can be found elsewhere [22-28].

References

[1] "U.S. Army Considers Replacing Thousands of Soldiers with Robots", U.S.S. Enterprise, IEEE Starship U.S.S Enterprise Section, 2015, http://sites.ieee.org/uss-enterprise/u-s-army-considers-replacing-thousands-of-soldiers-with-robots/.

[2] E. Ackerman, "U.S. Army Considers Replacing Thousands of Soldiers with Robots", IEEE Spectrum, 22 Jan 2014, http://spectrum.ieee.org/automaton/robotics/military-robots/army-considers-replacing-thousands-of-soldiers-with-robots.

[3] "US Army Works Toward Single Ground Robot", Defense News, Nov. 15, 2014, http://archive.defensenews.com/article/20141115/DEFREG02/311150033/US-Army-Works-Toward-Single-Ground-Robot.

[4] "LS3 - Legged Squad Support Systems", Boston Dynamics, http://www.bostondynamics.com/robot_ls3.html.

[5] "CHEETAH - Fastest Legged Robot", Boston Dynamics, 2013, http://www.bostondynamics.com/robot_cheetah.html.

[6] W. Rodriguez, "New Military Technology 2014 Supersoldier Robot Developed", Latest New Technology Gadgets, Sept. 28, 2014, http://latestnewtechnologygadgets.com/wp/new-military-technology-2014-supersoldier-robot-developed/.

[7] P. Lin, G. Bekey, K. Abney, "Autonomous Military Robotics: Risk, Ethics, and Design". US Department of Navy, Office of Naval Research

December 20, 2008.http://www.unog.ch/80256EDD006B8954/(httpAssets)/A70E329DE7B5C6BCC1257CC20041E226/$file/Autonomous+Military+Robotics+Risk,+Ethics,+and+Design_lin+bekey+abney.pdf.

[8] "X-47BUCAS, Capabilities", Northrop Grumman, 2015, http://www.northropgrumman.com/Capabilities/X47BUCAS/Pages/default.aspx.

[9] A. McDuffee, "DARPA Plans to Arm Drones With Missile-Blasting Lasers", WIRED, 11.01.13, http://www.wired.com/2013/11/drone-lasers/.

[10] "Meet the SR-72", Lockheed Martin, 2013, http://www.lockheedmartin.com/us/news/features/2013/sr-72.html.

[11] Engineering Review Board Concludes Review of HTV-2 Second Test Flight, DARPA, April 20, 2012, http://www.darpa.mil/newsevents/releases/2012/04/20.aspx.

[12] B. Berkowitz, "Sea Power in the Robotic Age", ISSUES in Science and Technology, 2015, http://issues.org/30-2/bruce-2/.

[13] "Large Displacement Unmanned Underwater Vehicle Innovative Naval Prototype (LDUUV INP)", in Naval Drones, http://www.navaldrones.com/LDUUV-INP.html.

[14] Wood, Stephen, "Autonomous Underwater Gliders", Florida Institute of Technology, http://my.fit.edu/~swood/26_Wood_first.pdf.

[15] J. Hsu, "U.S. Navy Tests Robot Boat Swarm to Overwhelm Enemies", IEEE Spectrum, 5 Oct 2014. http://spectrum.ieee.org/automaton/robotics/military-robots/us-navy-robot-boat-swarm.

[16] R. L. Hotz, "Harvard Scientists Devise Robot Swarm That Can Work Together". The Wall Street Journal. Aug. 15, 2014. http://www.wsj.com/articles/harvard-scientists-devise-robot-swarm-that-can-work-together-1408039261.

[17] P. Sapaty, "The World as an Integral Distributed Brain under Spatial Grasp Paradigm", book chapter in Intelligent Systems for Science and Information, Springer, Feb 4, 2014. http://link.springer.com/chapter/10.1007/978-3-319-04702-7_4.

[18] M. Wertheimer, Gestalt theory. Erlangen, Berlin. 1924.

[19] P.S. Sapaty, "Over-Operability in Distributed Simulation and Control", The MSIAC's M&S Journal Online, Winter 2002 Issue, Volume 4, No. 2, Alexandria, VA, USA.

[20] M. Minsky, The Society of Mind, Simon & Schuster, 1988.

[21] U. Schade, M. R. Hieb, M. Frey, K. Rein, "Command and Control Lexical Grammar (C2LG) Specification", FKIE Technical Report ITF/2010/02, July 2010.

[22] P.S. Sapaty, "Unified Transition to Cooperative Unmanned Systems under Spatial Grasp Paradigm", International journal Transactions on Networks and Communications (TNC), Vol.2, Issue 2, Apr 2014. http://scholarpublishing.org/index.php/TNC.

[23] P.S. Sapaty, "Distributed Human Terrain Operations for Solving National and International Problems", International Relations and Diplomacy, Vol. 2, No. 9, September 2014. http://www.davidpublishing.com/journals_info.asp?jId=2094.

[24] P.S. Sapaty, "From Manned to Smart Unmanned Systems: A Unified Transition", SMi's Military Robotics, Holiday Inn Regents Park London, 21-22 May 2014. http://www.smi-online.co.uk/defence/archive/5-2014/conference/military-robotics.

[25] P.S. Sapaty, "Integration of ISR with Advanced Command and Control for Critical Mission Applications", SMi's ISR conference, Holiday Inn Regents Park, London, 7-8 April 2014. http://www.smi-online.co.uk/defence/archive/4-2014/conference/isr.

[26] P.S. Sapaty, Ruling distributed dynamic worlds. John Wiley & Sons, New York, 2005.

[27] P.S. Sapaty, Mobile processing in distributed and open environments. John Wiley & Sons, New York, 1999.

[28] P.S. Sapaty, A distributed processing system. European Patent No. 0389655, Publ. 10.11.93, European Patent Office, 1993.

Software Requirements Management

Ali Altalbe
Deanship of e-Learning and Distance Education
King Abdulaziz University
Jeddah, Saudi Arabia

***Abstract*—Requirements are defined as the desired set of characteristics of a product or a service. In the world of software development, it is estimated that more than half of the failures are attributed towards poor requirements management. This means that although the software functions correctly, it is not what the client requested. Modern software requirements management methodologies are available to reduce the occurrence of such incidents. This paper performs a review on the available literature in the area while tabulating possible methods of managing requirements. It also highlights the benefits of following a proper guideline for the requirements management task. With the introduction of specific software tools for the requirements management task, better software products are now been developed with lesser resources.**

Keywords—Software; Software Requirements; Software Development; Software Management

I. Introduction

The term 'Requirement' is used to describe the set of desired characteristics and attributes possessed by a particular product or a service. These requirements may take the form of a formal statement of a function that needs to be performed, or it may take the form of an attribute that needs to be an integral part of the said entity. It is obvious that understanding the customer requirement is the number one priority in the designing phase of a system or an application. It is estimated that about 50% - 60% of software malfunctions are caused as a consequence of bad software management. This means that the problem is not associated with programming and the development team is by no means at blame, rather, the problem is purely due to the final product not meeting the customer's requirement. Inadvertently, it is possible for prerequisites to bear blunders due to vague or inadequate descriptions, however, by following modern software requirements management techniques, it is possible to successfully face these pitfalls [1].

Analyzing requirements is inherently a time consuming task which involves a considerable amount of observation. it involves the determination of requirements for an innovative or a changed organization while being mindful about probable incompatibilities. Basically, this process can be segmented as the congregation, incarceration and the specification of the particular project requirements specification.

The rest of this paper is divided into several segments. The literature review presents an overview of the main principle based on current literature available. The next chapter discusses the practices in the field of requirements management. The fourth chapter introduces the requirements traceability matrix, which is essentially one of the best methods available for the task at hand. It also discusses the advantages posed by the use of various tools for the task. The final chapter concludes the paper while briefly noting the potential work for the future.

II. Literature Review

A. Overview of Software Requirements Management Principles and Practices

As mentioned before, requirements management is an integral part of the standard project management life cycle. The main purpose of requirement management is to maintain a good relationship between the client and the developer of the project. The main objective of requirement management is therefore to guarantee that the organization, documentation and the verification of a project meets the requirements of the client. Furthermore, the involvement of this process within the life cycle will further enhance the software development by incorporating various management principles that aids to capture, evaluate, articulate, persuade, correspond, supervise and handle the needs for the required outfitted competence. These requirements can be broadly categorized as technical and non-technical [2], [3].

B. Key Principles

1) Early engagement of stakeholders in process: Management techniques must be capable of identifying stakeholders while guaranteeing a wide focus during the requirement analysis

2) Requirements analysis in the working context: Management techniques must foresee and analyze the initial requirements into product and process requirements or functional and non-functional requirements and carefully plan the situations at which each service must be delivered

3) Discriminating evidently between prepared solution and requirement: In order to facilitate any disparity that needs to be determined and supervised, the operational requirement of the particular need has to be well defined. Both descriptions necessitate their own varying classifications, drivers and metrics that develop at different speeds

4) Distinguishing between the roles of customer, supplier and user: These different roles necessitate different mind-sets, ethics, proficiencies, responsibilities, intentions and working measures. The distinguishment between these roles therefore enables expertise to be enhanced and the provision of a strong review of all related activities

5) Early identification and addressing of the interoperability : Few services can be installed secludedly. Due to the this nature, most of them are required to amalgamate, interface and operate with one another. Management techniques need to recognize and tackle with such interoperability issues from the beginning. This ensures the seamless integration of products that will eventually provide an elucidated system of systems

6) Review achievability: Requirements management should sustain the constant estimation of the planned solution in order to make sure that it is attainable and the predefined criteria are met with

7) Trade-off between presentation, capital investment and time: To reduce the acquisition risk, either presentation, capital or time has to be negotiated, which in turn might have an impact on the return on investment

8) Aid in managing jeopardy and ambiguity: By implementing a cyclic approach to the development process, it aids in providing lucidity and can tremendously reduce ambiguities which may prevail in final products. Through the use of requirements engineering, if any ambiguities do arise, necessary provisions can be made to focus the attention of stakeholders on such matters immediately. These ambiguities are often the result of impractical user requirements and the handling of such is known as risk management

9) Familiarizing or scale to suit: Requirements management should be pliable to all attainments of all final goods and services regardless of the intricacy

10)Help in identifying, analyzing and final s election of choices: In practical scenarios, it is often the case that the primarily identified 'best' approach later turns out to be meagre in comparison with others. By doing a thorough requirements analysis, it is possible to choose an approach which provides the finest returns for capital, time and investment

III. Practices in Requirements Management

The administration of software requirements is tricky even under the finest of conditions. Therefore, it is often necessary to reduce the complexity and improve the possibilities of achievement prior to the application of such methods. Top practices followed in the industry can be categorized as follows [4].

- Project Planning
- Work Estimation
- Progress Tracking
- Learning for the future

A. Defining Requirements for Requirements Management Practice

Three very important considerations have to be kept in mind to successfully implement such a practice. These three factors are people, process and technology. The goal is to effectively achieve efficiency by determining which tool or technique should be used for each occasion [5].

1) From a Process perspective: Resource Management strategies assign tasks for both the resource management process itself as well as for the closely related change management process

2) From a People perspective: The identification and definition of system development roles as a source of requirement falls into this category. These categories might deliver roles such as designing, coding or testing and the allocated person is responsible for eliciting and analyzing the requirement.

3) From a Technology perspective: Identification of the tool of choice is required through this category. The tool can be diversified, ranging from simple pen and paper to software packages such as Microsoft Word or Excel to even sophisticated management repository packages.

B. Managing Change Requests

Change management is the task during which the changes in the requirements are executed in a well managed mode presented through a pre-described structure with sensible changes. Software Configuration Management (SCM) or the change control process is considered as the perfect solution for tackling changes in the software development life cycle. In its essence, SCM is a job involving the tracing and the management of modifications of the software itself. Due to the vast benefits this process can deliver, it is considered to be vital in the industry. The change management process defines the requirement for change tracing and teh capability to validate that the final deliverable software has all of the desired augmentations that are required to be contained within the next release. The methods listed below are described for each and every software project in order to ensure that a proper change management procedure is employed.

- Identification of change in requirement is the procedure of recognizing the functionality that describes everything of a configation item. A configuration element is an item for consumption that has a consuming user function
- Change status explanation is the facility to record and account for on the configuration items at any given time instance
- Change control is a collection of processes and agreed upon stages that are necessary to change a configuration item's features and reestablishing the baseline
- Configuration assessments are busted into physical and functional configuration assessments. They arise either at the time of effecting a modification to the project or at the time of delivery

C. Potential Issues of Managing Change

The changes made in software usually results in affecting many users as well the stored data. If these changes are not well managed, the end result will be disastrous [6], [7].

1) Data Storage: Database changes are introduced that will require application modifications. Lack of a defined change in the management process often makes it difficult for the development team to troubleshoot or resolve an issue clearly

2) Data Movement: If not managed properly, this may lead to a issues of compromised information continuity. The requested change may often require changing a data feed into a data warehouse. If saving is not properly carried out, data recovery is extremely difficult

3) Security: The new changes made to the code might induce security breaches. If the changes are not properly managed and untested, it can lead to potential data catastrophes

4) Metadata Problem: A new software addition without updating the metadata repository will lead to incomplete analyses of future projects as well. If new processes are added without updating the metadata, similarly, it might lead to providing an incomplete picture of the operation of the system

5) Data Quality Problem: If the changes made are not compatible or comparable with the existing quality, it will lead to failures of the process in the future

6) Documentation: Changes made must also be updated in the documents with utmost care and if failed, will often result in leaving the production team in misery

7) Configuration Management: Changes may be introduced without updating the configuration management database, thus creating problems with the production and production support

D. Change Control Approach

Changes made to the baseline of the system are subject to the approval of the configuration control board. Following are the steps that are followed to handle a request to change the baseline [7].

1) Submit the change request along with the information about financial cost and resources required, etc. These changes are then submitted to the change control board
2) Access the change request after the change of evaluation
3) Depending on the outcome of step 2, outpour request is either accepted or rejected. If the submitted information is incomplete, it can be deferred
4) If the change request is accepted, necessary planning is done to implement the change by assigning work to the developer team
5) Once the changes are validated and examined by the quality control team, all configuration items are then updated throughout the entire library and the baseline of the system

IV. Requirements Traceability Matrix

The creation and implementation of requirements traceability techniques are done for the completeness, consistency, and traceability of the system requirements. It can be defined as the capability to define and trace the life cycle of a desired requirement, in both the backward and forward phases. Two methods are used for traceability cross referencing. One method can have cross referencing phrases such as 'see section A'. This method implements techniques such as numbering or tagging of requirements and changes in specialized tables or matrices. Each new addition is cross referenced at the older location in order to point towards these changes. The other method involves the restructuring of the documentation in terms of an underlying network of a graph that is used to keep track of the changes in the requirements. A traceability matrix is obtained by correlating requirements with the products in the software life cycle that satisfies them. Tests are correlated with the requirements of the designed product and the desired. These traceability matrices can be generated using multiple tools incorporating management software packages, spreadsheets, database packages or with hyper-linked tables on a processor. Configuration management plans are used on products since traceability is a key component of managing changes. Traceability ensures completeness such that all higher level requirements are assigned to lower levels and and all the lower level requirements are derived from higher level requirements [8]. The traceability matrix is a table which is 2×2 in size for most scenarios. These tables show a relationship between any of the two baseline files about the changes in requirements. These relationships maybe be one to many or bijection and displays the complete relationship. This method is used continuously in high-level projects that require specific requirements according to the high-level plan, test plan and test cases. The method commonly used is, during the first step, an identifier for each requirement of a file is taken and placed on the leftmost top column of the 2×2 matrix. The identifiers of the most related files are placed across the topmost row horizontally. When a requirement in the left column is related to a requirement in the topmost horizontal, a cross marx (×) is placed in the respective position where the intersection takes place. The number of relationships is summed up for each row and each column and written on the 2nd row and the 2nd column that indicated the value of the two mapping items. A sample traceability matrix is shown in table I [8].

TABLE I: Sample Traceability Matrix

Identifier	Reqs. Tests	A 0.3	A 0.4	B 1.1	B 1.2	C 2.1	C 2.2	D 1.0
Test Cases	3	2	1	1	2	2	2	3
Implicit Tests	7							
1.0.1	2			×		×		
1.0.2	2	×				×		
1.1.1	1					×		
1.1.2	1			×				
1.2.1	1	×						
1.2.2	2		×			×		
1.3.1	3			×	×		×	

The purpose of this matrix is to assist in ensuring that the requirement objectives are met by correlating each requirement with the object through the traceability matrix made for the requirement. In the forward trace, the matrix is used to validate that the affirmed requirements are distributed to system components and other deliverable products. It can also be used to conclude the source of the requirement in the backward trace. Requirement traceability matrix embraces outlining to things that convince the requirements such as design materials, capabilities, manual processes and analysis. This matrix is also utilized to make sure that all expected requirements are gathered. It is also used to establish the influence between system components when a modification is required. The capability to directly locate influenced components in the project allows the designers and developers to modify the project while ensuring maximum benefits while reducing costs and providing proper to-do estimates [8].

A. Tools for Requirements Management

Requirements management tools are used to facilitate the process of requirements management. A toolkit can consist of one or more tools, each designed to keep track of processes that the human mind is simply incapable of doing. Following are a list of criteria that needs to be considered when designing a requirements management tool [1].

- Identification of 'individual' requirements
- Assignment to a destination and sorting of requirements
- Identification of requirement groups (collection), revision and base lining
- Provision of a basic data interface

B. Benefits of using Tools

Following is a list of benefits that can be achieved through the use of specific requirements management tools instead of general purpose tools [9].

1) Structured Requirements: Only specific tools allow the gathering of 'structured' requirements. This means that it is possible to define attributes that would be helpful to track individual requirements and make sure that each requirement has its own set of attributes

2) Save Time: Good requirements management tools can save a lot of time by properly managing the software requirements. Since these tools are automated for most of the management tasks such as creating automated documentation, the time saving will be considerable

3) Less Stress: Most gathering and tracking of requirements are very chaotic in nature. A good requirement management tool can eliminate a lot of the unnecessary stress associated with the process

4) Work flow and Best Practices: Built-in methods in these specialized tools enable an efficient work flow while adhering to best practices in the field automatically

5) Easy to Collaborate: A good requirements management tool enables collaboration among external and internal stakeholders effectively. This is one area where general purpose tools lack significantly

6) Increase in Precision of Requirement: Good tools increase the exactness between customer requirement and the project output. If offers trouble free implementation in such a way that the recording and the supervision of the customer requirement is easily understandable

7) Cost Reduction: Tools have the ability to reduce maintenance costs, training costs, deployment costs and well as reduce the deployment risks and the time required

8) Added Benefits: These tools can increase the collaboration between the team and support services and thereby, deliver better solutions in lesser times

V. Conclusion

Depending on the literature available, it is clear that the use of requirements management techniques is highly beneficial to the software industry. Usually, the process begins with the analysis and elicitation of the objectives and the constrains of the project and the organization. Once this portion is complete, the next step is using change control management. Change control management is an important process that can deliver immense benefits. By using the traceability matrix, it is easy to identify the relationship between different requirements of the project. Once these objectives are clear, it is possible to address these requirements issues both in the forward and backward traces. The process of requirements management can further be enhanced through the use of specific software tools that can be utilized to save both time and money while increasing the quality of the output.

A. Recommendations

While it is difficult to present a global recommendation for all projects, it is obvious that the use of these techniques have a direct positive impact on the productivity. With this in mind, each software company must try their best to enhance their capabilities using these methods. However, it is doubtful whether sophisticated software tools will be beneficial to all software projects. One major issue regarding these tools is the cost factor. It is possible for the tool cost to be as large as the budget in some cases. Due to this reason, it is important to properly understand the scope of the project. If the project involves the designing and implementation of a large system which requires regular updates, or if the same system is to be sold to multiple markets with minor customisations, it is important to invest on these specific tools. However, regardless of the project size, it is always important to invest time on the requirements management process. If the budget is small, it is best to use an all-purpose software such as Microsoft Excel, but nonetheless, the task must still be done. As it was mentioned, requirements management has a direct impact on the documentation process, and even if the developers are developing something as small as a mobile application, it is important to understand the significance of the documentation. Based on these points, it is highly recommended for all software developers to adopt principles of software requirements management to their development life cycle, regardless of the size of the project.

B. Future Work

This paper presented a review of literature in order to highlight the pros and cons of software requirements management. Through the review, it is clear that software requirements management has become an integral part of the software development life cycle. Future work in this regard should include a review of the actual software packages that are being used in the industry at the moment for the task of software requirements management. Such an analysis should quantify the possible parameters of these applications. Given the highly dynamic nature of the field, it will obviously be futile to attempt naming the best software available, however, it is possible to obtain some qualitative data from the end users from various fields and present a recommendation as to which software package is the best for a given task.

REFERENCES

[1] D. Leffingwell and D. Widrig, *Managing software requirements: a use case approach.* Pearson Education, 2003.

[2] K. L. Evans, R. P. Reese, and L. Weldon, "Unit information management practices at the joint readiness training center," DTIC Document, Tech. Rep., 2007.

[3] K. Pohl, *Requirements engineering: fundamentals, principles, and techniques.* Springer Publishing Company, Incorporated, 2010.

[4] B. W. Boehm, "Software risk management: principles and practices," *Software, IEEE*, vol. 8, no. 1, pp. 32–41, 1991.

[5] M. Dumas, W. M. Van der Aalst, and A. H. Ter Hofstede, *Process-aware information systems: bridging people and software through process technology.* John Wiley & Sons, 2005.

[6] S. G. Eick, T. L. Graves, A. F. Karr, J. S. Marron, and A. Mockus, "Does code decay? assessing the evidence from change management data," *Software Engineering, IEEE Transactions on*, vol. 27, no. 1, pp. 1–12, 2001.

[7] R. S. Pressman, *Software engineering: a practitioner's approach.* Palgrave Macmillan, 2005.

[8] B. Ramesh and M. Jarke, "Toward reference models for requirements traceability," *Software Engineering, IEEE Transactions on*, vol. 27, no. 1, pp. 58–93, 2001.

[9] S. J. Andriole, *Managing systems requirements: methods, tools, and cases.* McGraw-Hill Companies, 1996.

5

Case-based Reasoning with Input Text Processing to Diagnose Mood [Affective] Disorders

Sri Mulyana, Sri Hartati, Retantyo Wardoyo, Edi Winarko
Department of Computer Sciences and Electronics
Gadjah Mada University
Yogyakarta, Indonesia

***Abstract*—Case-Based Reasoning is one of the methods used in expert systems. Calculation of similarity degree among the cases has always been an important aspect in CBR as the system will attempt to identify cases with the highest of similarity degree in a case-base to provide solutions for new problems. In this research, a CBR model with input text processing for diagnosing mood [affective] disorder is developed. It correlates with the increased tendency of mood disorder in accordance with the dynamics of the economic and political situation.**

Calculation of similarity degree among the cases is one of the main focuses in this research. This study proposed a new method to calculate similarity degree between cases, Modified-Tversky. The analysis performed to assess the method used in measuring case similarity reveals that the Modified-Tversky Method surpasses the other methods. In the all tests conducted, the results of case similarity measures using the Modified-Tversky method is greater than or equal to the calculations performed using the Jaccard dan Tversky methods. The test results also provide an average level of performance in processing text input is 89.3 %.

Keywords*—*Case-Based Reasoning; mood disorder; case similarity; Jaccard Method; Tversky Method; Modified-Tversky Method

I. INTRODUCTION

Case-Based Reasoning (CBR) is a method adopted from knowledge-based system in various domains. The method uses experiences from previous similar cases to solve new problems. The main idea is the assumption that similar cases have similar solutions [1].

The development of CBR possesses vast research opportunities including those related to case similarity calculation algorithms, indexing techniques to increase the efficiency of retrieval processes, case representation techniques, and methods to append new cases to the case base [2].

Researches could also be conducted by combining existing techniques through improvement or addition of new algorithms to achieve a need-fulfilling case-based computerized reasoning system. To test the validity of the techniques developed, implementation in a certain field is required.

In this research, a CBR model to diagnose mood [affective] disorder is developed. Mood [affective] disorder is caused by the inability of an individual to adapt to social evolution. There is an increased tendency of mood disorder parallel to economic and political change.

That choice of field of implementation is based on the increased tendency of mood disorder due to the occurance of economic and political change. The cause is the inability of an individual to adapt to social evolution.

In life, every single human being exhibits different kinds of emotions. Those expressions and feelings vary and is usually temporary, but could pose a problem if not immediately treated, resulting in emotional disorder including heavy depression rendering the subject devoid of motivation, happiness, and empathy.

Bitter experiences could cause an individual to feel insecure, void, depressed, and even hateful. That in turn results in uncontrollable anger planted deeply in one's inner conscience. What is felt as an unpleasant experience leads to extreme grief and all other forms of depression [3]. This condition could formulate as a chronic and repetitive problem interfering with an individual's ability to live out his every-day responsibilities.

II. CBR RESEARCH IN HEALTH

The application of CBR has been vastly developed in various fields, particularly in health. Those is deployed to diagnose infection related diseases [4]. Explained in those researches is that such cases possess certain attributes such as temperature, headaches including its intensity and area of effect, coughing, bowel movement frequency, nausea, and urination. All those attributes have a numerical value in a scale of 1 to 4 to represent its intensity ranging from none or never, low, average, and high. Euclidean distance is employed in this research to measure similarity degrees between cases.

Current ongoing researches show that the implementation of CBR in the medical field has seen great development. Chakraborty at al. introduced a CBR system to detect swine flu dubbed SFDA (*Swine Flu Diagnostic Assistant*) [5], and Tomar at al. developed a CBR system to diagnose heart attacks [6]. Pant and Joshi have also done the same in the field of neurology in the form of NDS (*Neurology Diagnosis System*) [7]. The method used to measure similarity was the Nearest-neighbor method. Other CBR-based system was developed to advertise the dangers of smoking. It contained early warnings about the health effects of smoking broadcasted through mobile phones [8].

In the future, CBR systems could help provide better services in the medical field and perhaps have tighter integration with the clinical environment.

III. Mood [Affective] Disorders

An individual's mood is the internal emotion that he or she dominantly feels, affects one's behavior and perception, and tends to be stable for a period of time. On the other hand, an affect is an external expression of a mood. Moods can be in a normal state, increasing, and deppressive. A healthy mind has a wide range of moods, a good balance of affect expression, and is able to control mood and affects [9]. Mood disorders are triggered by extreme and uncontrollable mood fluctuation, resulting in a lack of adaptive ability, and could cause difficulty or discomfort.

Mood disorders are emotional illnesses accompanied by symptoms of mania and depression [10]. It is the result of the presence of complete negativity in an individual's mind, where said individual views himself, life, and the future as a chaotic mess of failure. Someone experiencing such a mood [affective] disorder would feel a lack of connection with other people or devoid of an influential role in their lives [3].

Mania is a certain mood disorder indicated by an abnormal increase of a certain mood. This condition is also accompanied by behavioral symptoms in the form of over-activity, talkativeness, great enthusiasm, euphoria, and even sexual deviation [10]. There exists two groups of maniacal symptoms, one being euphoric (elation, increased enthusiasm, heightened optimism), and the other being irritability (anger, aggressiveness, and violence). Hypomania is the condition of a slightly lower increase of mood compared to mania, and that usually does not reach the level where day to day functionality is threatened. Related to the time needed for diagnosis, hypomania can be identified as quickly as four days since the symptoms arrive [11].

Depression is a common emotional disorder indicated by constant sadness, loss of desire, a feeling of guilt, insecurity, sleeping problems or a decrease of appetite, low enthusiasm and energy, and lack of concentration [10]. It can also be associated with continuous grief in a dangerous amount, and is the most common emotional disorder. One in seven all around the globe have experienced an episode of depression at least once in their lives. It is the most common cause of death in developed nations and is the fourth most common cause in developing countries. Bipolar disorder is a chronic type of mood disorder with recurring episodes of mania or hypomania that strikes in turn with depression [11]. It also includes depression symptoms major or minor, usually identifiable by regression or even loss of ability to feel happy and positive during fun activities. This is constantly felt every day for at least two weeks [11]. People suffering from bipolar disorder can appear as overly optimistic individuals, frequently boasting about themselves, and even to the extent of making dangerous or risky decisions without calculating drawbacks and consequences. Sometimes those individuals can turn up brilliant ideas with the need for it to be immediately fulfilled. Unfortunately, those ideas tend to end in failure because of lack of thought, preparation, and some are just too unrealistic.

As a result of their easily irritable mood and the tendency to be easily offended in a state of either mania or depression, those with bipolar disorder often experience interpersonal communication problems with family and people around them. Sometimes they suspect that others are envious of them, and endeavor to prevent those bipolar individuals from achieving great things. During episodes of depression, they tend to blame family members and close friends as the root of their failure and suffering because in their minds they think that they are not provided with enough love, care, and support.

IV. Text Processing

Text processing is often required to simplify search process. The main objective is to convert various word forms into much more consistent index terms. An *Index term* is the representation of the essence of a document needed for search purposes. Text processing includes:

a) Information Extraction: extracting important information within a document.

b) Text Summarization: to automatically produce a summary of a document.

c) Data Mining: A validation identification process to understand data patterns recorded in a structured database.

d) Text Mining: also known as text data mining or information searching.

e) Information retrieval: a document search.

f) Document Clustering: similar to document classification, but document class is not determined beforehand.

The steps to process a document text is usually specific according to the process that needs to be carried out. One of the steps of processing a text document is indexing. The aim is to identify the best term to represent a certain document, so during document processing it could be obtained accurately. Automatic indexing consists of:

1) Tokenization: a stage of processing where the input text is divided into units or "tokens" in the forms of words, numbers or punctuation marks.

2) Deletion of stop words: erasing less meaningful words. The choice of stop words are assessed using specific dictionaries, or lists of stops words.

3) Synonym identification: locating synonyms from words obtained from a document. This can be conducted with the use of a thesaurus or other similar methods. Similar words can be replaced with general terms.

4) Stemming: The process of text normalization within a document or changing an expression to its simplest form that in a literal view may not have a concrete meaning [12].

5) Frequency measures of stemming results: One of the methods that can be applied is the n-gram or a series of linguistic n item orders. Those items can be in the form of letters, vocabulary, or words. It is one of the implementations basic word features used in machine learning [13].

6) Selection or comparison of word frequency measure results.

V. CALCULATION OF CASE SIMILARITY DEGREE

Let $\text{Sim}(X, Y)$ denotes the similarity between two cases X dan Y where each have a finite number of features. The following are techniques to calculate the similarity degree:

1) Jaccard Method [14]

Case similarity using the Jaccard Method can be formulated as:

$$\text{Sim}(X, Y) = \frac{|X \cap Y|}{|X \cup Y|} \quad (1)$$

which also means :

$$\text{Sim}(X, Y) = \frac{\#\text{SAME}(X, Y)}{\#\text{SAME}(X, Y) + \#\text{DIFFER}(X, Y)} \quad (2)$$

Where:

#SAME(X,Y): The number of same features between X and Y

#DIFFER(X,Y) : The number of different features between X and Y

This decision usually requires a specific threshold to state that the features of both cases are same or not.

2) Tversky Method[15]

Case similarity with the original Tversky method can be formulated as:

$$\text{Sim}(X, Y) = \frac{\alpha|X \cap Y|}{\alpha|X \cap Y| + \beta|X \oplus Y|} \quad (3)$$

Where:

$$|X \oplus Y| = |X \cup Y| - |X \cap Y| \quad (4)$$

which also means :

$$\text{Sim}(X, Y) = \frac{\alpha(\#\text{SAME}(X, Y))}{\alpha(\#\text{SAME}(X, Y)) + \beta(\#\text{DIFFER}(X, Y))} \quad (5)$$

The value of α dan β is parallel with their respective significance of value, usually formulated by experts or produced using machine based learning techniques.

3) Modified-Tversky Method

The Modified-Tversky Method is a modification of the Tversky Method formulated as:

$$\text{Sim}(X, Y) = \frac{2|X \cap Y|}{|X| + |Y|} \quad (6)$$

By adding weight of Tversky-Method, the equation can be writen as :

$$\text{Sim}(X, Y) = \frac{2\alpha(\#\text{SAME}(X, Y))}{2\alpha(\#\text{SAME}(X, Y)) + \beta(\#\text{DIFFER}(X, Y))} \quad (7)$$

VI. METHODOLOGY

The implementation of Case-based Reasoning method to diagnose mood [affective] disorder consists of three parts: contruction of case base, input text processing and case retrieval process, as shown in Figure-1:

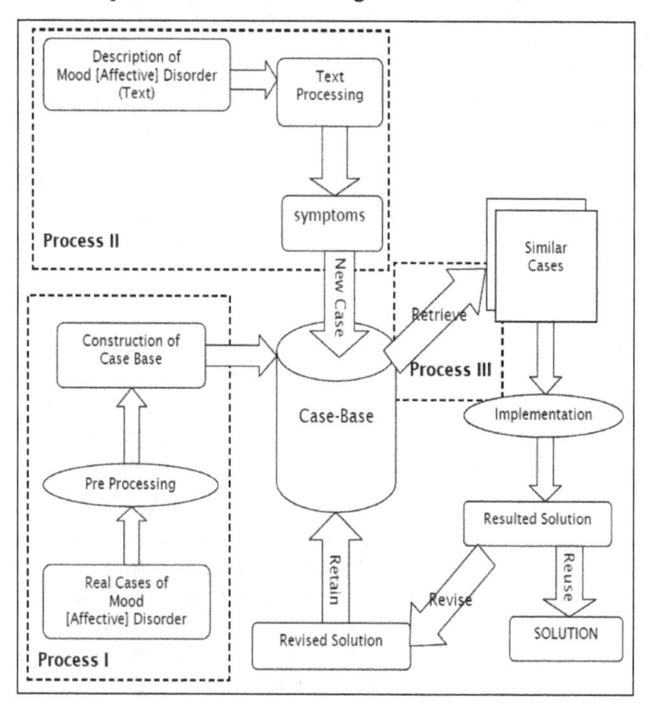

Fig. 1. The proposed CBR implementation method

Contruction of case base is done by real life cases from various sources including the internet. The case base is then acquired as shown in Table-1:

TABLE I. CASE BASIS

No	Codes	Name of Mood Disorder	Symptoms Codes
1	F30.0	Hipomania	G13, G14, G15, G16, G25, G26, G27, G28
2	F30.0	Hipomania	G13, G14, G17, G19, G25, G26, G27, G28, G46
3	F32.1	Depression II	G36, G37, G38, G39, G40, G41, G42, G43, G44, G45, G46
4	F32.0	Depression I	G36, G37, G39, G44, G53
5	F32.0	Depression I	G36, G38, G41, G44, G53
6	F32.0	Depression I	G37, G39, G41, G42, G44, G45, G46
7	F32.0	Depression I	G35, G36, G37, G38, G39, G41, G42, G43, G44
8	F32.1	Depression II	G36, G37, G38, G40, G41, G42, G44, G45, G46, G47, G48
9	F32.1	Depression II	G35, G37, G38, G39, G40, G41, G43, G44, G46, G47, G50, G53
10	F31.1	Bipolar II	G13, G14, G15, G17, G21, G24, G25, G26, G27, G28, G36, G46, G49, G52
11	F31.1	Bipolar II	G13, G14, G15, G17, G21, G24, G25, G26, G27, G28, G36, G46, G48, G52
12	F31.0	Bipolar I	G13, G17, G25, G27, G37, G38, G39, G40, G41, G42, G44, G48, G52
13	F31.0	Bipolar I	G13, G14, G17, G26, G30, G36, G38, G39, G40, G44, G46, G52
14	F30.0	Hipomania	G13, G14, G15, G17, G18, G19, G21

The explanation meaning of the symptoms codes are shown in Table -2.

TABLE II. THE NAMES OF SYMPTOMS CODES

Id	Name of symptom
G13	increased mood
G14	increased energy and activity
G15	increased sociability, talkativeness and over familiarity
G16	increased sexual energy
G17	decreased need for sleep
G18	concentration and attention maybe impaired
G19	diminished the ability to settle down to work or to relaxation and leisure
G21	joviality
G23	easily distracted attention
G24	increased motoric activity
G25	talk too much
G26	self-esteem is inflated
G27	grandiose ideas
G28	over-optimistic ideas
G36	deminished energy and activity
G37	depressed mood
G38	loss of interest and enjoyment
G39	reduced concentration and attention
G40	reduced self-esteem and self-confidence
G41	idea of guilt and useless
G42	pessimistic
G43	idea of acts of self-harm
G44	disturbed sleep
G45	diminished appetite
G46	shows considerable distress
G47	loss of self-esteem
G48	attempted suicide
G52	decreased mood
G53	tiredness

Input text processing follow the steps above: tokenization, deletion of stop words, synonym search and replacement, N-gram configuration and symptoms diagnosis. Case retrieval is conducted by selecting a case from case base with the highest of case similarity degree. We implemented 3 methods to calculate the case similarity degree: Our Modified-Tversky, the original Tversky and Jaccard methods.

VII. RESULTS AND DISCUSSION

During text processing, testing is conducted through comparison of the number of symptoms used as the foundation of the input text configuration with those produced by text processing. The performance value is then formulated as:

$$P = \frac{NS}{NB} x100\% \tag{8}$$

Where:

P = Performance

NS = The number of same symptoms obtained from text processing

NB = The number of symptoms in the text (ground truth)

In the test conducted, text processing performance is shown in Table-3.

TABLE III. TEXT PROCESSING PERFORMANCE

No	Symptoms in the text (ground truth)	NB	Symptoms obtained from text processing	NS	P(%)
1	G36, G38, G39, G44, G53	5	G36, G38, G39, G44, G53	5	100
2	G36, G37, G39, G42, G53	5	G36, G37, G39, G42, G53	5	100
3	G13, G14, G16, G17, G18, G20	6	G13, G14, G16, G17, G18, G20	6	100
4	G36, G37, G38, G44	4	G36, G37, G38, G44	4	100
5	G13, G14, G15, G17, G18, G19, G20, G25	8	G13, G14, G15, G17, G18, G19, G20	7	87,5
6	G13, G14, G26, G27, G28, G36, G37, G38, G42, G43, G52	11	G13, G14, G27, G36, G37, G38, G42, G43, G48, G52	9	81,8
7	G13, G14, G16, G17, G18, G36, G46, G47, G48, G52, G53	11	G13, G14, G16, G17, G36, G43, G46, G48, G52, G53	9	81,8
8	G36, G37, G38, G39, G40, G41, G42, G46, G47, G48, G53	11	G37, G38, G39, G41, G43 G46, G47, G48, G53	8	72,7
9	G37, G38, G39, G42, G44, G53	6	G37, G38, G39, G44, G53	5	83,3
10	G36, G37, G38, G39, G43, G44, G53	7	G36, G37, G38, G39, G44, G53	6	85,7
Average Performance of Text Processing					**89,3**

Based on Table-2, the average performance of text processing of the test data is 89.3%. In several tests show that the performance of text processing does not reach 100 %, it indicates that the text processing algorithms in this study cannot always produce the correct result due to different interpretation of words, therefore the algorithm can be developed in order to improve performance.

During case retrieval, experiments are conducted by evaluating the correspondence of the diagnosis given by experts with the one produced by the CBR system and the case similarities degree generated. The entire results of the case retrieval experiment of 10 new problems are shown in Table-4:

TABLE IV. CASE RETRIEVAL TEST RESULTS

No.	Expert Diagnose	CBR Diagnose	Case Similarity degree (%)		
			Modified-Tversky	Jaccard	Tversky
1	F32.0	F32.0	88,89	66,67	80,00
2	F32.0	F32.0	88,89	66,67	80,00
3	F30.0	F30.0	44,44	44,44	28,57
4	F32.0	F32.0	36,44	36,44	22,22
5	F30.0	F30.0	83,33	55,56	71,43
6	F31.0	F31.0	43.75	43,75	28,75
7	F31.1	F31.1	41,18	41,18	25,93
8	F32.1	F32.1	86,49	61,54	76,19
9	F32.0	F32.0	88,89	66,67	80,00
10	F32.0	F32.0	95,24	83,33	90,91

The table shows that out of the three methods used in calculating of case similarity degree, the Modified-Tversky Method produces the best result. The statement is based on the all results of the test, as the suggested Modified-Tversky Method consistently generates better or at least equal outcome compared to other methods. This could be used as a refference in determining the method to calculate case similarity degree in future CBR researches. Based on Tebel-4 is also known that the diagnosis is generated by the CBR system in accordance with the diagnosis given by experts.

VIII. Conclusion

In compliance with the test results and analysis carried out in this research, conclusions can be drawn as such: The modified-Tversky method outperforms the Jaccars and Tversky methods. In all the tests conducted, results of case similarity measures using the Modified-Tversky method is greater than or equal to the calculations performed using the Jaccard dan Tversky methods. The test results also provide an average level of performance in processing text input is 89.3%. However, in future work can be developed text processing algorithms in order to improve the performance of text processing. Cases on the case-base should be added so that the system can handle the problems of mood disorders. Therefore, it takes a lot of supporting data relating to the case of mood disorders.

Refferences

[1] A. Aamold, "*Case-based reasoning: Foundation issues". AICOM* **7**, pp 39-59, 1994

[2] A. Aamold and E. Plaza, "Case-based Reasoning: foundation issues, methodological variation and System approach", *AI Communication 7(1), pp. 39-59,* 1994

[3] A. Nasir and A. Muhith, "The Fundamental of Psychiatric Nursing: Introduction and Theory", Salemba Medika, Jakarta, 2010

[4] M. Denis, and A. Jasmin, "Applying Case-based reasoning for mobile support in diagnosing infective diseases". International Conference on Signal Processing Systems, IEEE Computer Society, pp 779-783, 2009

[5] B. Chakraborty, I. Srinivas, P. Sood, V. Nabhi, D. Ghosh, " Case Based Reasoning Methodology for Diagnosis of Swine Flu", IEEE GCC Confererence and Exhibition, 19-22 Februari 2011, Dubai, United Arab Emirates, pp. 132-135, 2011

[6] P. P. S. Tomar, R. Singh, P. K. Saxena, and J. Sharma, "Case Based Medical Diagnosis of Occupational Chronic Lung Diseases From Their Symptoms and Signs", International Journal of Biometrics and Bioinformatics (IJBB), Volume (5) : Issue (4), pp. 216-224, 2011

[7] S. Pant, and S. R. Joshi, "Case-Based Reasoning In Neurological Domain", IEEE 978-1-4673-2590-5/12, 2012

[8] K. Ghorai, S. Saha, A. Bakshi, A. Mahanti, and P. Ray, "An mHealth Recommender for Smoking Cessation using Case Based Reasoning", 46th Hawaii International Conference on System Sciences, pp. 2695-2704, 2013

[9] B. J. Sadock and V. A. Sadock, "Mood disorder" In : *Kaplan & Sadock's Synopsis of Psychiatry: Behavioral Sciences/Clinical Psychiatry, 10th Edition.* Wolters Kluwer/Phiadelphia. Lippincott Williams & Wilkins, 2007

[10] E. Dalami, "Nursing Clients with Mental Disorder", Trans Info Media, Jakarta, 2009

[11] E. Vieta, "Managing Bipolar Disorder in Clinical Practice", second edition, London UK. Current Medicine Group Ltd., 2009

[12] S. Ferilli, F. Esposito, and D. Grieco, "Automatic Learning of Linguistic Resources for Stopword Removal and Stemming from Text", *Procedia Computer Science*, 38, 116–123, 2014

[13] O. Serban, G. Castellano, A. Pauchet, A. Rogozan, and J. P. Pecuchet, "Fusion of Smile, Valence and NGram Features for Automatic Affect Detection", *Humaine Association Conference on Affective Computing and Intelligent Interaction,* 2013

[14] G. Kowalski, "Information Retrieval Architechture and Algorithm", Springer, New York, USA. E-Book (e-ISBN 978-1-4419-7716-8), 2011

[15] S. K. Pal, and S. C. K. Shiu, "Foundation of Soft Case Based Reasoning", John Wiley & Sons, New Jersey, 2004

Estimation of Protein Content in Rice Crop and Nitrogen Content in Rice Leaves Through Regression Analysis with NDVI Derived from Camera Mounted Radio-Control Helicopter

Kohei Arai [1]
Graduate School of Science and Engineering
Saga University
Saga City, Japan

Osamu Shigetomi
Saga Prefectural Agricultural Research Institute
Saga Prefecture
Saga City, Japan

Masanori Sakashita
Department of Information Science
Saga University
Saga City, Japan

Yuko Miura
Saga Prefectural Agricultural Research Institute
Saga Prefecture
Saga City, Japan

***Abstract*—Estimation of protein content in rice crop and nitrogen content in rice leaves through regression analysis with Normalized Difference Vegetation Index: NDVI derived from camera mounted radio-control helicopter is proposed. Through experiments at rice paddy fields which is situated at Saga Prefectural Research Institute of Agriculture: SPRIA in Saga city, Japan, it is found that protein content in rice crops is highly correlated with NDVI which is acquired with visible and Near Infrared: NIR camera mounted on radio-control helicopter. It also is found that nitrogen content in rice leaves is correlated to NDVI as well. Protein content in rice crop is negatively proportional to rice taste. Therefore rice crop quality can be evaluated through NDVI observation of rice paddy field.**

Keywords—nitrogen content; NDVI; protein content; rice paddy field; remote sensing; regression analysis

I. INTRODUCTION

There are strong demands for saving human resources which are required for produce agricultural plants. In particular in Japan, now a day, the number of working peoples for agricultural fields is decreasing quite recently. Furthermore, the ages of the working peoples are getting old. Moreover, the agricultural fields are also getting wide through merging a plenty of relatively small scale of agricultural fields in order for maintain the fields in an efficient manner. Therefore, the working peoples have to maintain their fields in an efficient manner keeping the quality in mind.

Vitality monitoring of vegetation is attempted with photographic cameras [1]. Grow rate monitoring is also attempted with spectral reflectance measurements [2]. Bi-Directional Reflectance Distribution Function: BRDF is related to the grow rate for tealeaves [3]. Using such relation, sensor network system with visible and near infrared cameras is proposed [4]. It is applicable to estimate nitrogen content and fiber content in the tealeaves in concern [5]. Therefore, damage grade can be estimated with the proposed system for rice paddy fields [6]. This method is validated with Monte Carlo simulation [7]. Also Fractal model is applied to representation of shapes of tealeaves [8]. Thus the tealeaves can be asse3ssed with parameters of the fractal model. Vitality of tea trees are assessed with visible and near infrared camera data [9]. Rice paddy field monitoring with radio-control helicopter mounting visible and NIR camera is proposed [10] while the method for rice quality evaluation through nitrogen content in rice leaves is also proposed [11]. The method proposed here is to evaluate rice quality through protein content in rice crop with observation of NDVI which is acquired with visible and NIR camera mounted on radio-control helicopter.

The proposed method and system is described in the next section followed by experiments. The experimental results are validated in the following section followed by conclusion with some discussions.

II. PROPOSED METHOD AND SYSTEM

A. Radio Controlled Helicopter Based Near Infrared Cameras Utilizing Agricultural Field Monitoring System

The helicopter used for the proposed system is "GrassHOPPER" [1] manufactured by Information & Science Techno-Systems Co. Ltd. The major specification of the radio controlled helicopter used is shown in Table 1. Also, outlook of the helicopter is shown in Figure 1. Canon Powershot S100 [2] (focal length=24mm) is mounted on the GrassHOPPER. It

---

[1] http://www.ists.co.jp/?page_id=892

[2] http://cweb.canon.jp/camera/dcam/lineup/powershot/s110/index.html

allows acquire images with the following Instantaneous Field of View: IFOV at the certain altitudes, 1.1cm (Altitude=30m) 3.3cm (Altitude=100m) and 5.5cm (Altitude=150m).

Fig. 1. Outlook of the GrassHOPPER

TABLE I. MAJOR SPECIFICATION OF GRASSHOPPER

Weight	2kg (Helicopter only)
Size	80cm × 80cm × 30m
Payload	600g

Spectral response functions of filters attached to the camera used are shown in Figure 2.

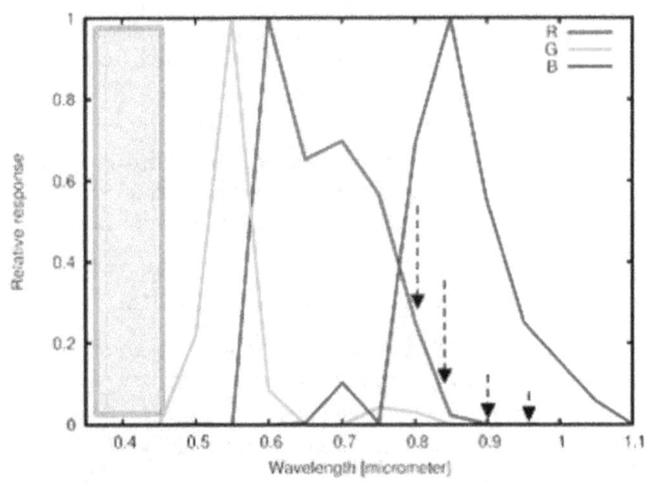

Fig. 2. Spectral Response of the Filter attached to Camera

In order to measure NIR reflectance, standard plaque whose reflectance is known is required. Spectralon[3] provided by Labsphere Co. Ltd. is well known as well qualified standard plaque. It is not so cheap that photo print papers are used for the proposed system. Therefore, comparative study is needed between Spectralon and the photo print papers.

The proposed system consist Helicopter, NIR camera, photo print paper. Namely, photo print paper is put on the agricultural plantations, tea trees in this case. Then farm areas are observed with helicopter mounted Visible and NIR camera. Nitrogen content in agricultural plants, rice crops in this case, is estimated with NIR reflectance.

B. *Regressive Analysis*

Linear regressive equation is expressed in equation (1).

$$N = a R + b \qquad (1)$$

where N, R denotes measured Nitrogen content in leaves as well as protein content in rice crops, and measured NDVI derived from visible and Near Infrared: NIR reflectance, respectively while a and b denotes regressive coefficients. There is well known relation between nitrogen content as well as protein content in rice crops and NDVI. Therefore, regressive analysis based on equation (1) is appropriate.

C. *Proposed Method for Rice Crop Quality Evaluation*

Rice crop quality can be represented with nitrogen content which are closely related to NDVI. Furthermore, it is well known that nitrogen content rich rice crops taste good while protein content rich rice crops taste bad. Therefore, rice crops quality can be evaluated with measured NDVI measured with camera data which is mounted on radio-control helicopter.

The proposed method and rice paddy field monitoring system with visible and NIR camera which is mounted on radio-control helicopter is based on the aforementioned scientific background.

D. *Rice Crop Field at Saga Prefectural Agricultural Research Institute: SPARI*

Specie of the rice crop is Hiyokumochi[4] which is one of the late growing types of rice species. Hiyokumochi is one of low amylase (and amylopectin rich) of rice species (Rice No.216).

Figure 3 and 4 shows layout of the test site of rice crop field at SPARI[5] which is situated at 33°13'11.5" North, 130°18'39.6"East, and the elevation of 52feet.

The paddy field C4-2 is for the investigation of water supply condition on rice crop quality. There are 14 of the paddy field subsections of which water supply conditions are different each other.

There are two types of water supply scheduling, short term and standard term. Water supply is stopped in the early stage of rice crop growing period for the short term water supply subsection fields while water supply is continued comparatively longer time period comparing to the short term water supply subsection fields.

---

3 https://www.google.co.jp/search?q=spectral+labsphere&hl=ja

4 http://ja.wikipedia.org/wiki/%E3%82%82%E3%81%A1%E7%B1%B3

5 http://www.pref.saga.lg.jp/web/shigoto/_1075/_32933/ns-nousisetu/nouse/n_seika_h23.html

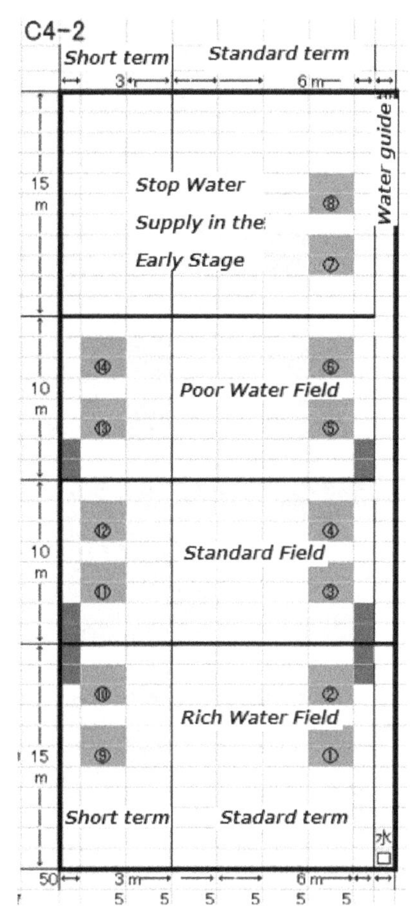

Fig. 3. Paddy field layout for investigation of water supply condition dependency on rice crop quality

Meanwhile, there are three types of water supply conditions, rich, standard, and poor water supply subsection fields.

On the other hand, test sites C4-3 and C4-4 are for investigation of nitrogen of chemical fertilizer dependency on rice crop quality. There are two types of paddy subsections, densely and sparsely planted paddy fields. Hiyokumochi rice leaves are planted 15 to 20 fluxes per m^2 on June 22 2012. Rice crop fields are divided into 10 different small fields depending on the amount of nutrition including nitrogen ranges from zero to 19 kg/10 a/nitrogen.

Nitrogen of chemical fertilizer is used to put into paddy fields for five times during from June to August. Although rice crops in the 10 different small fields are same species, the way for giving chemical fertilizer are different. Namely, the small field No.1 is defined as there is no chemical fertilizer at all for the field while 9, 11, and 13 kg/ 10 a/ nitrogen of after chemical fertilizer are given for No.2 to 4, respectively, no initial chemical fertilizer though. Meanwhile, 9, 11, 13 kg/10 a/nitrogen are given as after chemical fertilizer for the small field No.5, 6, and 7, respectively in addition to the 3 kg/10 a/nitrogen of initial chemical fertilizer. On the other hand, 12, 14, and 16 kg/10 a /nitrogen are given for the small fields No.5, 6, 7, respectively as after chemical fertilizer in addition to the initial chemical fertilizer of 3 kg/ 10 a/ nitrogen for the small field No. 15, 17, 19, respectively. Therefore, rice crop grow rate differs each other paddy fields depending on the amount of nitrogen of chemical fertilizer.

C4-3,4

50株/坪(甲)		70株/坪(乙)	
基肥 中追 穂肥I 穂肥II 実肥	基肥 中追 穂肥I 穂肥II 実肥	基肥 中追 穂肥I 穂肥II 実肥	基肥 中追 穂肥I 穂肥II 実肥
A2 ② 生育小 0 2 1.5 1.5	A8 ① 生育大 6 6 1.5 1.5	B8 ① 生育大 6 6 1.5 1.5	B2 ② 0 2 1.5 1.5
A3 ② 生育小 0 4 1.5 1.5	A7 ① 生育大 6 4 1.5 1.5	B7 ① 生育大 6 4 1.5 1.5	B3 ② 0 4 1.5 1.5
A4 ② 生育小 0 6 1.5 1.5	A6 ① 生育大 6 2 1.5 1.5	B6 ① 生育大 6 2 1.5 1.5	B4 ② 0 6 1.5 1.5
A5 ② 生育中 3 4 1.5 1.5	A5 ① 生育中 3 4 1.5 1.5	B5 ① 生育中 3 4 1.5 1.5	B5 ② 生育中 3 4 1.5 1.5
A6 ② 生育大 6 2 1.5 1.5	A4 ① 生育小 0 6 1.5 1.5	B4 ① 生育小 0 6 1.5 1.5	B6 ② 生育大 6 2 1.5 1.5
A7 ② 生育大 6 4 1.5 1.5	A3 ① 生育小 0 4 1.5 1.5	B3 ① 生育小 0 4 1.5 1.5	B7 ② 生育大 6 4 1.5 1.5
A8 ② 生育大 6 6 1.5 1.5	A2 ① 生育小 0 2 1.5 1.5	B2 ① 生育小 0 2 1.5 1.5	B8 ② 生育大 6 6 1.5 1.5
A1 ② 無肥料 0 0 0 0 0	A1 ① 無肥料 0 0 0 0 0	B1 ① 無肥料 0 0 0 0 0	B1 ② 無肥料 0 0 0 0 0

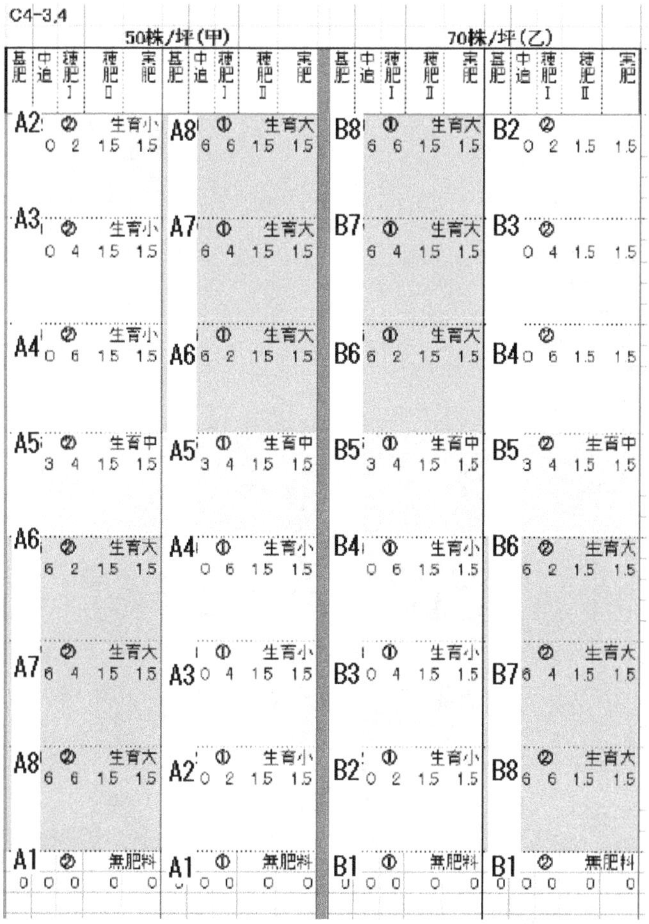

Fig. 4. Paddy filed layout for investigation of nitrogen of chemical fertilizer dependency on rice crop quality

III. Experiments

A. Acquired Near Infrared Camera Imagery Data

Radio wave controlled helicopter mounted near infrared camera imagery data is acquired at C4-2, C4-3, C4-4 in SPARI on 18 and 22 August 2013 with the different viewing angle from the different altitudes. Figure 4 shows an example of the acquired near infrared image. There is spectralon of standard plaque as a reference of the measured reflectance in between C4-3 and C4-4. Just before the data acquisition, some of rice crops and leaves are removed from the subsection of paddy fields for inspection of nitrogen content. Using the removed rice leaves, nitrogen content in the rice leaves is measured based on the Keldar method and Dumas method[6] (a kind of chemical method) with Sumigraph NC-220F[7] of instrument. The measured total nitrogen content in rice leaves and protein content in rice crops are compared to the NDVI.

---

[6] http://note.chiebukuro.yahoo.co.jp/detail/n92075

[7] http://www.scas.co.jp/service/apparatus/elemental_analyzer/sumigraph_nc-220F.html

Example of the acquired image is shown in Figure 5. Rice field name is annotated in the image. On the other hand, the acquired camera images on 18 August and 22 August are shown in Figure 6 (a) and (b), respectively. Meanwhile, these images have influences due to shadow and shade of rice leaves and water situated under the rice leaves as well as narrow roads between rice paddy fields. In order to eliminate the influences, thresholding process is applied to the acquired images.

Figure 7 shows the processed images of small portion of the images with the different threshold rages from 5 to 25. Through these trials, threshold of 25 is chosen for influences reduction. Geometric correction is applied to the acquired camera images after extraction of intensive study areas. Figure 8 (a) and (b) show the resultant corrected images acquired on 18 and 22 August 2013, respectively.

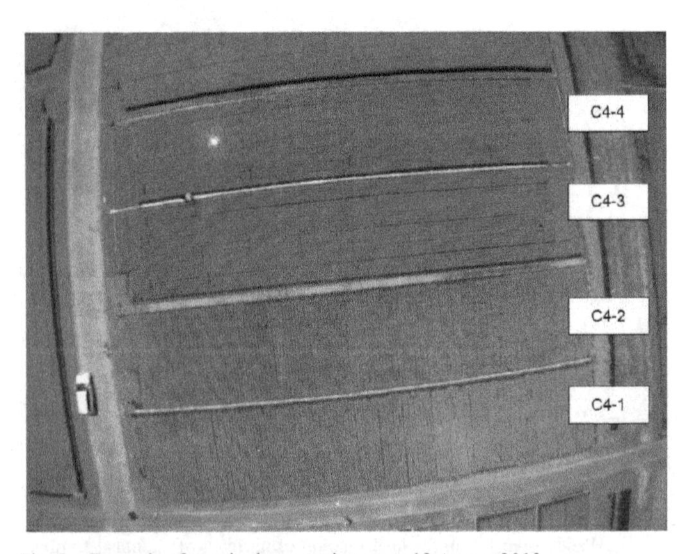

Fig. 5. Example of acquired camera image on 18 August 2013

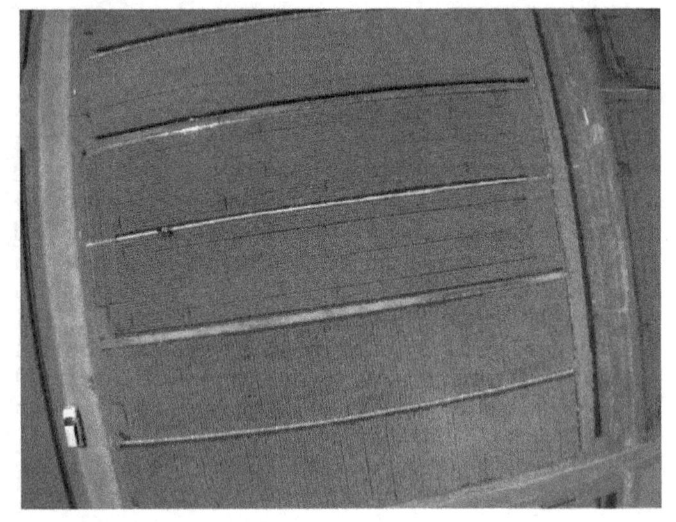

(a)August 18

(b)August 22

Fig. 6. Camera images acquired on 18 and 22 August 2013

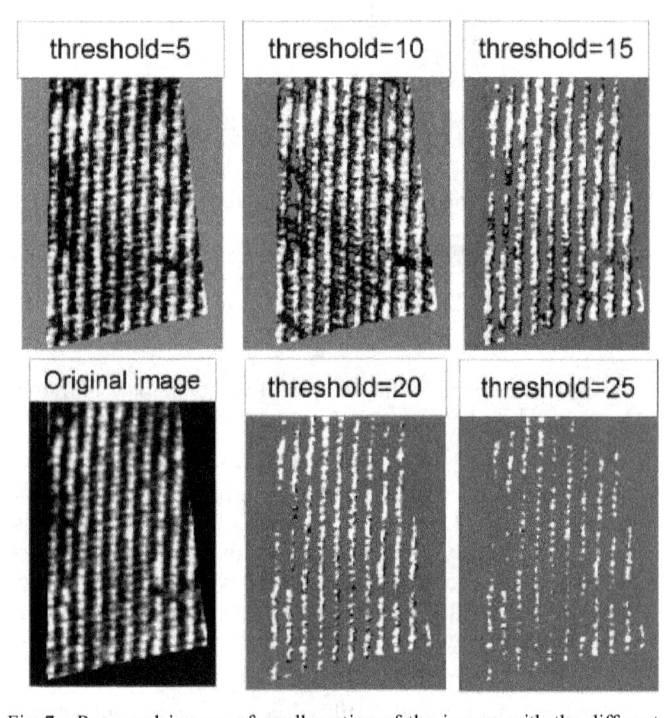

Fig. 7. Processed images of small portion of the images with the different threshold rages from 5 to 25.

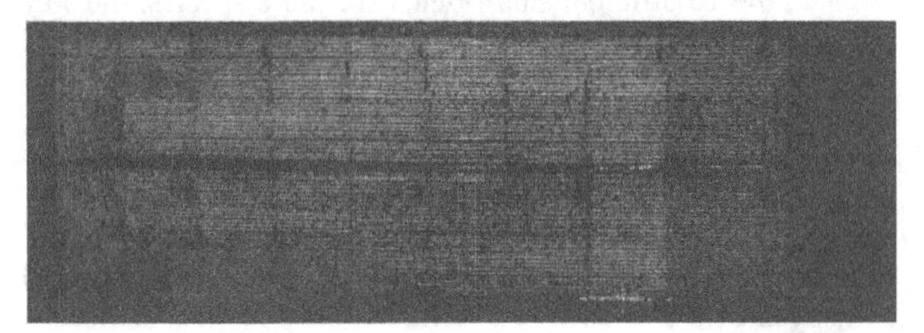

(a)August 18

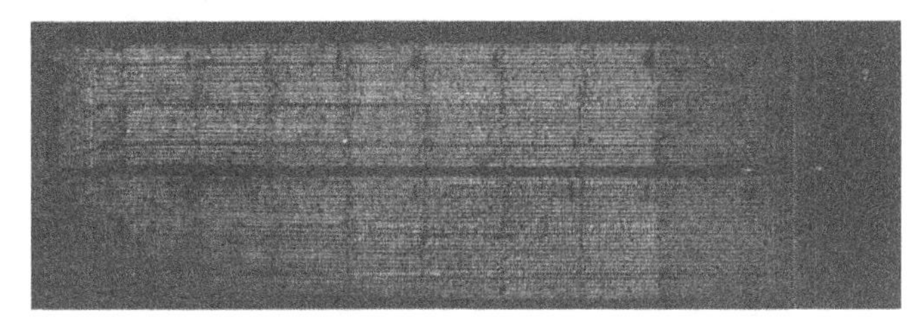

(b)August 22

Fig. 8. Resultant images of geometrically corrected which are acquired on 18 and 22 August 2013

On the other hand, measured nitrogen contents in rice leaves of rice paddy fields of partitioned A1 to A8 and B1 to B8 on 14 and 22 August 2013 are shown in Table 2 and 3, respectively. Nitrogen of chemical fertilizer, water management as well as plant density are different from each other partitioned rice paddy fields as aforementioned. Nitrogen content in the rice leaves seem to reflect the fact of chemical fertilizer of nitrogen, water supply management, and plantation density, obviously.

TABLE II. MEASURED NITROGEN CONTENT IN RICE LEAVES ON 14 AUGUST 2013

Farm Area	Nitrogen (%)
A1	2.61
A3	2.85
A5	2.84
A8	2.77
B1	2.82
B3	2.74
B5	3.16
B8	2.78

TABLE III. MEASURED NITROGEN CONTENT IN RICE LEAVES ON 22 AUGUST 2013

Farm Area	Nitrogen (%)
A1	2.46
A2	2.88
A4	2.97
A5	2.89
A6	2.67
A8	3.22
B1	2.33
B2	2.79
B4	2.84
B5	2.85
B6	2.96
B8	3.14

Relation between nitrogen as well as protein contents and NDVI for each paddy fields for August 18 and 22 2013 are shown in Figure 9 to 12.

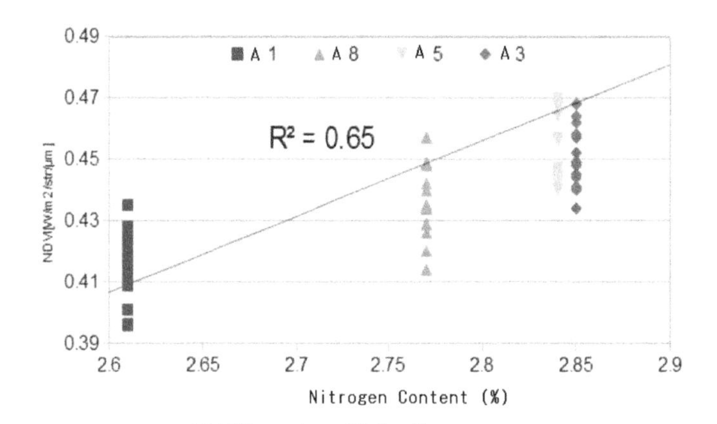

(a)A Site on August18 for all area

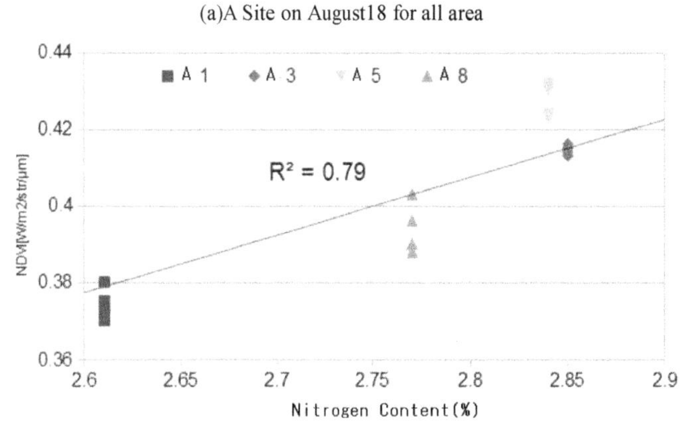

(b)A Site on August 18 for sampled areas

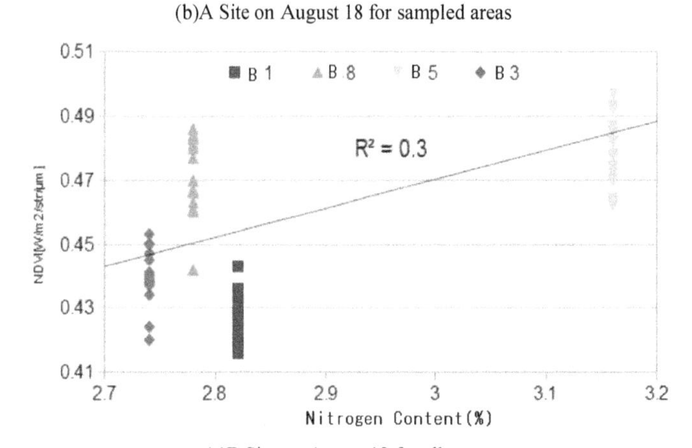

(c)B Site on August 18 for all area

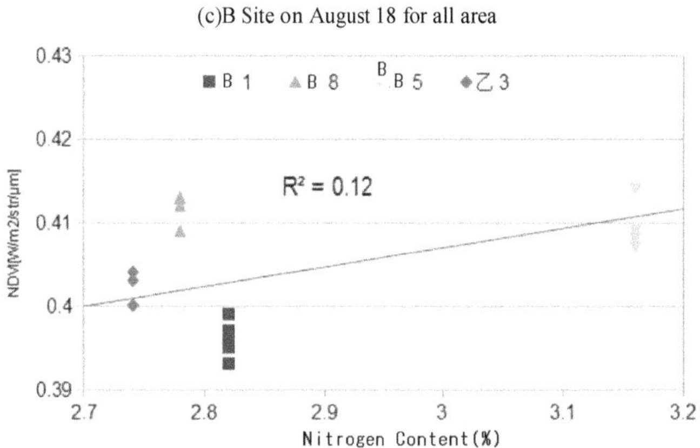

(d)B Site on August 18 for the sampled areas

Fig. 9. Relations between NDVI and the measured nitrogen content

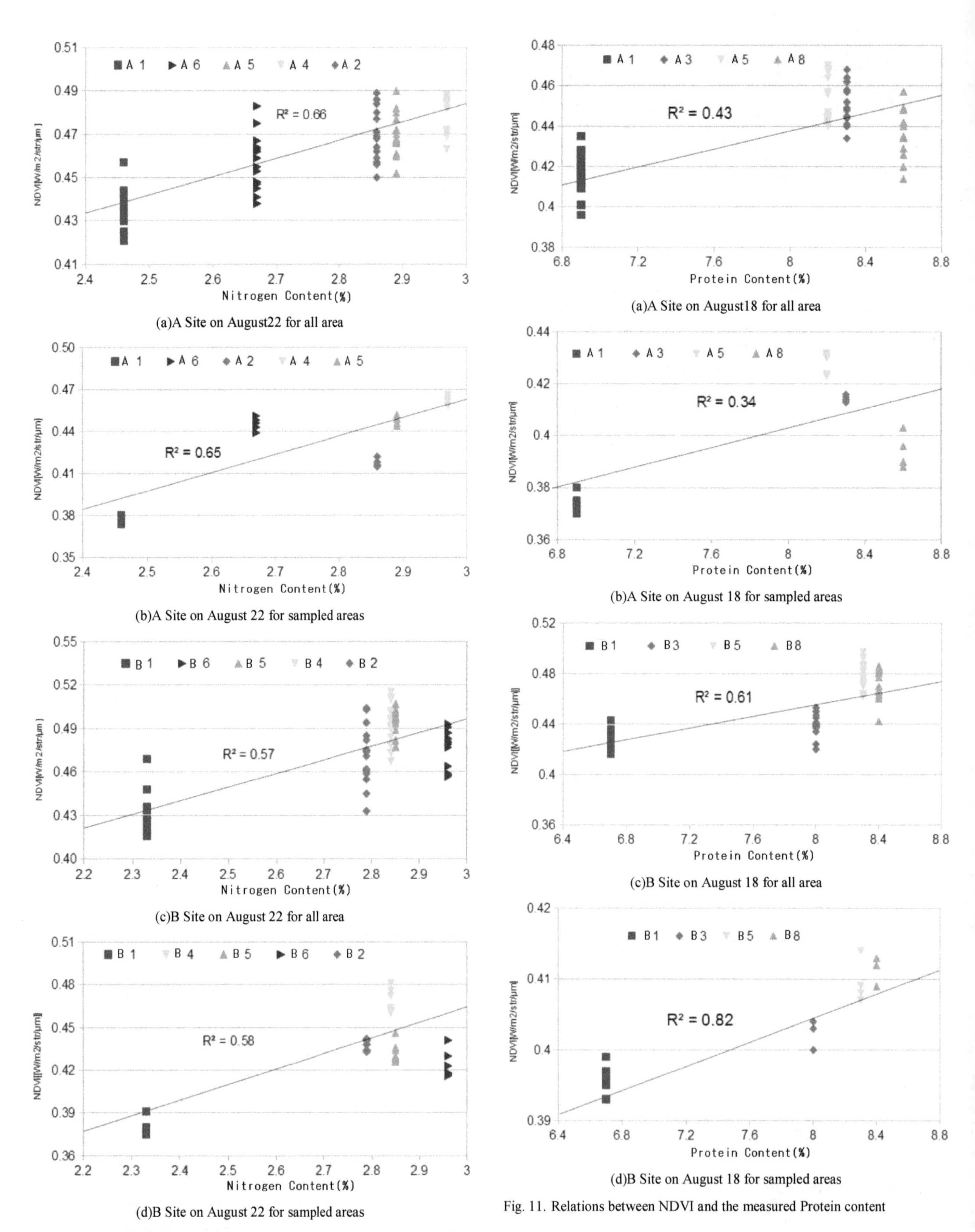

(a)A Site on August22 for all area

(b)A Site on August 22 for sampled areas

(c)B Site on August 22 for all area

(d)B Site on August 22 for sampled areas

Fig. 10. Relations between NDVI and the measured nitrogen content

(a)A Site on August18 for all area

(b)A Site on August 18 for sampled areas

(c)B Site on August 18 for all area

(d)B Site on August 18 for sampled areas

Fig. 11. Relations between NDVI and the measured Protein content

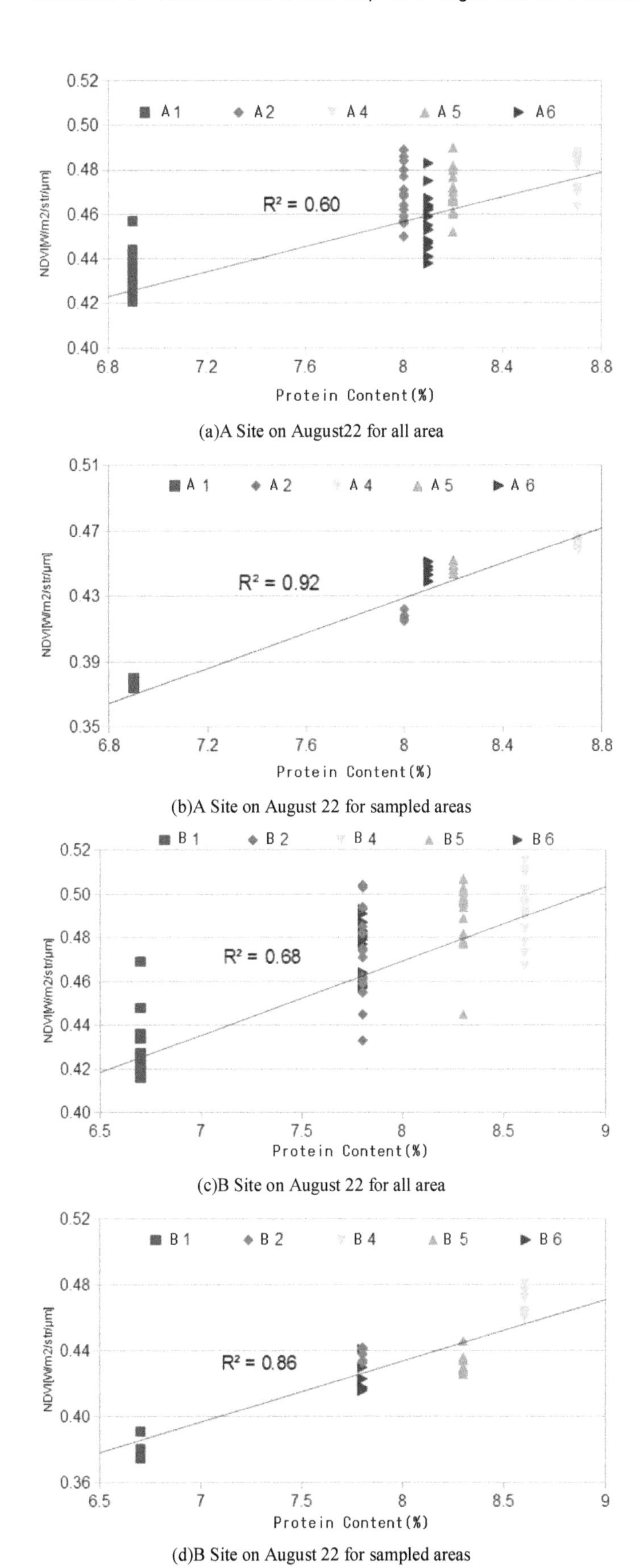

(a)A Site on August22 for all area

(b)A Site on August 22 for sampled areas

(c)B Site on August 22 for all area

(d)B Site on August 22 for sampled areas

Fig. 12. Relations between NDVI and the measured protein content

The sampled areas imply that the locations of surrounding areas at which rice leaves are picked up for measurement of nitrogen content. Meanwhile, all area implies whole area of the strip of the rice paddy field. The correlation coefficients for the sampled areas are much greater than those of all area. Also variances for the sampled areas are much smaller than those for all area. More importantly, R square values of protein content are greater than those of nitrogen content. This implies that rice crop quality which represented by protein content is much reflected by the nitrogen of chemical fertilizer, water supply management, and rice leaves density rather than nitrogen content in rice leaves directly.

IV. CONCLUSION

Estimation of protein content in rice crop and nitrogen content in rice leaves through regression analysis with Normalized Difference Vegetation Index: NDVI derived from camera mounted radio-control helicopter is proposed. Through experiments at rice paddy fields which is situated at Saga Prefectural Research Institute of Agriculture: SPRIA in Saga city, Japan, it is found that protein content in rice crops is highly correlated with NDVI which is acquired with visible and Near Infrared: NIR camera mounted on radio-control helicopter. It also is found that nitrogen content in rice leaves is correlated to NDVI as well. Protein content in rice crop is negatively proportional to rice taste. Therefore rice crop quality can be evaluated through NDVI observation of rice paddy field.

The correlation coefficients for the sampled areas are much greater than those of all area. Also variances for the sampled areas are much smaller than those for all area. More importantly, R square values of protein content are greater than those of nitrogen content. This implies that rice crop quality which represented by protein content is much reflected by the nitrogen of chemical fertilizer, water supply management, and rice leaves density rather than nitrogen content in rice leaves directly..

ACKNOWLEDGMENT

The author would like to thank Mr. Hideaki Munemoto of Saga University for his effort to conduct the experiments.

REFRENCES

[1] imating the growth and yield of rice, Journal of Crop Science, 58, 4, 673-683, 1989.

[2] Kohei Arai, Method for estimation of grow index of tealeaves based on Bi-Directional reflectance function:BRDF measurements with ground based netwrok cameras, International Journal of Applied Science, 2, 2, 52-62, 2011.

[3] Kohei Arai, Wireless sensor network for tea estate monitoring in complementally usage with Earth observation satellite imagery data based on Geographic Information System(GIS), International Journal of Ubiquitous Computing, 1, 2, 12-21, 2011.

[4] Kohei Arai, Method for estimation of total nitrogen and fiber contents in tealeaves with grond based network cameras, International Journal of Applied Science, 2, 2, 21-30, 2011.

[5] Kohei Arai, Method for estimation of damage grade and damaged paddy field areas sue to salt containing sea breeze with typhoon using remote sensing imagery data, International Journal of Applied Science,2,3,84-92, 2011.

[6] Kohei Arai, Monte Carlo ray tracing simulation for bi-directional reflectance distribution function and grow index of tealeaves estimation, International Journal of Research and Reviews on Computer Science, 2, 6, 1313-1318, 2011.

[7] K.Arai, Fractal model based tea tree and tealeaves model for estimation of well opened tealeaf ratio which is useful to determine tealeaf harvesting timing, International Journal of Research and Review on Computer Science, 3, 3, 1628-1632, 2012.

[8] K.Arai, H.Miyazaki, M.Akaishi, Determination of harvesting timing of tealeaves with visible and near infrared cameradata and its application to tea tree vitality assessment, Journal of Japanese Society of Photogrammetry and Remote Sensing, 51, 1, 38-45, 2012

[9] Kohei Arai, Osamu Shigetomi, Yuko Miura, Hideaki Munemoto, Rice crop field monitoring system with radio controlled helicopter based near infrared cameras through nitrogen content estimation and its distribution monitoring, International Journal of Advanced Research in Artificial Intelligence, 2, 3, 26-37, 201

7

Overview on the Using Rough Set Theory on GIS Spatial Relationships Constraint

Li Jing
[1]College of Mobile Telecommunications, Chongqing University of Posts and Telecommunications
Chongqing, China

Zhou Wenwen
[2]Chongqing University of Posts and Telecommunications
Chongqing, China

***Abstract*—To explore the constraint range of geographic video space, is the key points and difficulties to video GIS research. Reflecting by spatial constraints in the geographic range, sports entity and its space environment between complicated constraint and relationship of video play a significant role in semantic understanding. However, how to position this precision to meet the characteristic behavior extraction demand that becomes research this kind of problem in advance. Taking Rough set theory reference involved, that make measuring space constraint accuracy possible. And in the past, many GIS rough applications are based on the equivalence partition pawlak rough set. This paper analyzes the basic math in recent years in the research of rough set theory and related nature, discusses the GIS uncertainty covering approximation space, covering rough sets, analysis of it in the geographic space constraint the adjustment range.**

Keywords*—*space constraint; GIS; rough set; fuzzy geographic; spatial relationship

I. INTRODUCTION

GIS is a spatial information system that with the capacity of the collection, storage, management, analysis and description of the earth surface and space distribution [1, 2]. And the powerful spatial analysis ability provides advantages for understanding GIS semantics, besides it becomes the crucial premise of intelligent security monitoring prototype system by using geographical boundaries. The scale is an important concept in geographical spatial cognition, also geographic process and how to extract behavior features to satisfy movement elements have a high degree of dependence to scale, therefore, to study on such problems can be converted into discussion and explore on the positioning constraints space accuracy. But in the traditional processing method of GIS data can appear error or uncertainty, which causes the result is not entirely reliable, even error, will eventually lead to the decision-making mistakes or failure. However, rough Set as a new mathematical tool of knowledge as Deal with uncertainty and fuzzy information recently has widely put into use in GIS [3, 4]. Thus it becomes a new means to measure space constraints accuracy problem.

Studies on spatial data uncertainty at home and abroad mainly focused on space position data [5] and attribute data [6], and the unsureness of spatial relationships [7], etc. The point is the study of the uncertainty modeling, precision analysis, transmission method and visual expression [8-10]. Due to the geographical phenomenon of changing over time, the spatial data can only express one geographical phenomenon in one particular time, by that among of these studies, the considered space targets are mainly to explicitly recognize space entities (such as roads, rivers, etc.), meanwhile the uncertainty is mostly caused by measurement, digital data acquisition, and subsequent spatial analysis [11, 12]. In the approaches, to a large extent to use the methods of observational error processing in geodesy [13]. The vagueness refers to the thing can be clear described by the predefined attributes, but the boundary of the adjacent target is hard to differentiate clearly. Thus, the rough set theory has been widely used in the field of image processing, which includes system simplification, remote sensing image segmentation and remote sensing image recognition, etc. [14, 15].

This paper mainly probes into the uncertainty of this kind of spatial constraint, which includes introducing rough set into the geographical border uncertainty research, besides its basic concepts and essence (in part 2), the expression methods and rules of spatial relationship, and the applied analysis in GIS (in part 3), finally this paper discusses the current research progress, future development and difficulties (in part 4 and part 5).

II. ROUGH EXPRESSION AND REASONING OF SPATIAL RELATIONSHIP

A. Basic conception of rough set

Rough set theory is a mathematical tool proposed by professor Pawlak in 1982 which can quantitative analyze and deal with imprecise, inconsistent, incomplete information and knowledge [16]. Rough set theory, the initial prototype from the relatively simple information model, its basic idea is formed by classified concepts and rules of the relational database, through the classification of the equivalence relation and classification for the target approximate knowledge discovery. And the following are four kinds of rough set basic definitions methods [17, 18]:

1) Equivalence relation and the indiscernibility relationship

Set R is the equivalence relation limited on the U field (Meet the reflexivity, symmetry, and transitivity), recorded as $R \subseteq V \times U$. In U field, all $x \in v$ has the equivalent relation with the collection of R, recorded as $[x] R = \{y \in U| (x, y) \in R\}$. U/R indicates all the sets of R constitute of the equivalence classes that is quotient set.

And set R is as the equivalence relation clan, in which $P \subseteq R$, and $P \neq \emptyset$. All the intersection of equivalence relations in P is called the indistinguishable relationship of P, which can be recorded as ind(P), and that is $[x]ind(p) = \cap(R \in P)[X]R$.

2) The upper approximation, lower approximation and rough sets

The uncertainty of rough set theory is based on the concept of upper and lower approximation. Set R is the equivalence relation limited on the U field, considered to the set $X \in U$, coupling (RX, $\overline{R}X$) is called a rough approximation of X in the approximate space (U, R), and

$$RX = \{x \in U \| [x] R \subseteq X\}, \overline{R} = \{x \in U \| [x] R \cap X \neq \emptyset\} \quad (1)$$

In which RX, $\overline{R}X$ indicate the R upper and lower approximation of X respectively. Also the set $bnR(X) = RX-\overline{R}X$ refers to the R boundary region of X. The rough set defines that when $bnR(X) = \emptyset$, means $RX=\overline{R}X$, it refers to that X is R accurate sets and when $bnR(X) \neq \emptyset$, means $RX \neq \overline{R}X$, it calls X is R rough sets.

3) Variable precision rough set definition

The variable precision rough set definition is the extension of Pawlak rough set, it introduces β ($0 \leq \beta < 0.5$) to the basic rough set, therefore it should define the majority incorporate coefficient β, before the variable precision rough set.

Definition: Set X and Y are the non-void subset of the limited discourse domain U. If to each e ($e \in X$), there is $e \in Y$, and called it Y includes X that is recorded as $X \subseteq Y$. And also make

$$C(x, y)=\begin{cases}1 - |X \cap Y| / |x| & |x| > 0 \\ 0 & |x| = 0\end{cases} \quad (2)$$

In which $|x|$ is the cardinal number of set X and C (x, y) is the relative error resolution of set X regarded to set Y.

Definition: Based on the majority incorporate coefficient relationship, set (U, R) as approximation spaces, in which discourse domain U is non-empty limited set, and R is the equivalence relation on U, $U/R = \{E_1, E_2, \ldots, E_n\}$ is the set constituted of equivalence class or basic set of R. According to $X \subseteq U$ define the β lower approximation $\underline{R}_\beta$ and the β upper approximation $\overline{R}_\beta$ of X, and it approaches to variable precision rough set definition as:

$$\underline{R}_\beta X = \cup \{E \in U/R \mid c(E, X) \leq \beta\} \quad (3)$$

$$\overline{R}_\beta X = \cup \{E \in U/R \mid c(E, X) < 1 - \beta\} \quad (4)$$

4) Reduction

Set U as a discourse domain, Q and P are defined as two equivalence relation clusters on U besides $Q \subseteq P$, if Q is independent and ind(P)=ind(Q), then call Q is the absolutely reduction of the equivalence relation cluster P and recorded it as red(P). All the absolutely relationship sets in P are the core of equivalence relation cluster P, recorded as core (P).

The most significant difference of rough set theory and other theories of dealing with uncertainty and imprecise problems is about processing not provide any prior information except data collection, so the description of the uncertainty or processing can be said to be more objective [19], therefore it provides preferential conditions for the study of spatial direction relationship.

B. The Rough thoughts of spatial direction relationship

If it is necessary to integrate the rough set theory to the GIS, must start from the basic data model to establish uncertain, fuzzy geographic data model, so as to solve the two kinds of inaccuracy and fuzzy problems between the direction relationships of the objects, and then solve the uncertainty caused by fuzzy object fuzzy boundary problems. In the spatial relationships [20], there are fuzzy and precise objects in the space object, therefore, the spatial direction relation can be mainly divided into four types: fuzzy objects and the direction relationship between fuzzy objects, the fuzzy objects and the direction relationship between precise objects, the direction relationship between the fuzzy and precise objects, and also precise objects and the direction relationship between the precise objects. Because the space objects can approximately be expressed by the rough set, the spatial direction relationship between fuzzy and precise objects can be solved by researching on the relationships between its upper and lower rough approximations sets.

Set the fuzzy objects A and B, the upper and lower rough approximations sets of them are RA, $\overline{RA}$, RB, $\overline{RB}$ separately, and adopt 048 to represent the direction relationships of them. When the grading of the direction relationships is as 8, the direction relationship knowledge base is regarded as {N, NE, E, SE, S, W, SW, NW, O}, when the grading is four direction relationships, the knowledge base is regarded as {N, E, S, W, 0}. The rough expression of the spatial objects direction relationships is to represent the concept of complex direction relations as a collection of basic knowledge of the knowledge base. In which the classification knowledge of the discourse domain space is known, and the key is to confirm the relational functions between the concepts and the basic knowledge. There are two kinds of accuracy and fuzziness issues about direction relationships of the objects. The first one is caused by the fuzzy boundaries of the fuzzy objects, and the second one is by the improper methods which are caused by adopting the basic direction relationships of the knowledge base to represent the object direction relationships. The former is inherent, and the latter is the issue of methods and cognition.

The rough expression of spatial direction relationships. Set the extension cord of the outside rectangular of the object A divide the space region into O_i ($1 \leq i \leq n$), and n is as the resolution ratio of the direction. The membership function of objects B and Oi is as:

$$U(B \in O_i)=\begin{cases}1, & B \cap O_i \neq 0 \\ 0, & B \cap O_i = 0\end{cases} \quad (5)$$

The direction relation of objects A and B is $O_{AB} = \{O_i | u(B \in O_i) = 1\}$. Another definition is: the upper and lower rough approximations sets of the fuzzy objects direction relationships O_{AB} are $\underline{O}_{AB}$ and $\overline{O}_{AB}$, $\underline{O}_{AB} = \{O_{\underline{AB}}\}$, $\overline{O}_{AB} = \{O_{\overline{A}B}, O_{A\overline{B}}, O_{\overline{AB}}\}$.

Because the fuzzy and precise objects can be unified expressed by the upper and lower rough approximations sets, therefore by adopting the rough set is with the ability to unify the approximately express of the direction relationships of the fuzzy and precise objects in the frame, in order to process and reason. The upper and lower rough approximations sets of the fuzzy and precise objects direction relationships are equal, so the boundary is empty, it is consistent with the traditional expression method which eight direction relationship is based on projection, the approximation precision is 1. The boundary of the rough expression of spatial direction relationships is mainly caused by the boundary of the fuzzy objects, so the method can describe the fuzziness direction relationships of the which are caused by the fuzziness of the fuzzy objects boundary, it mainly is to approximately express the fuzzy direction relationship created by the fuzzy objects boundary, but it fail in solving the second kind of inaccuracy and fuzzy issues. [21]

III. Geography Spatial Relation Rules Extraction Based on Rough Set

A. *Rough set expression of spatial relationship*

To quantitatively express all kinds of the geological phenomenon spatial relationship, and then effectively converted into the format of the rough set data processing method, is the necessary conditions to use rough set rules to extract major spatial relationship of geological phenomena. Because of the reason that the rough set need to represent the data into the form of a two-dimensional table as processing the data, accordingly it requires to various kinds of geological phenomenon spatial relationship of two-dimensional form.

1) choose spatial relationship:

To aim at the specific issues of the geography, it chooses the specific spatial relationships of the geological phenomenon as the research objects according to the prior knowledge. Such as the respective features of different geological phenomenon: the water cycle, atmospheric circulation, ocean vortex, land usage and coverage, select the major effected spatial relationship factors such as the distance, topology and etc.

2) quantitative expression of spatial relationship:

To aim at the various spatial geological phenomenon, it adopts the appropriate description methods to quantitatively describe the spatial relationships, for example by employing the Euclidean distance to quantitatively describe the distance.

3) Construct spatial relationship decision table:

To convert quantitative description of geological phenomena spatial relationships into the form of a decision table. And the rows of the decision table say the research objects of geological phenomena, on the other side the columns of the decision table represent two parts: The former part known as condition attributes, on behalf of all kinds of geological phenomenon spatial relationships, the latter part of the decision table is decision attributes, the values of them are specific geological results. The values of each row are the quantitative descriptions of spatial relationships approached by various description methods of spatial relationship (except decision attributes). By using this two-dimensional table to express the spatial relationship of geological phenomena, we can employ the method of the rough set to analyze and extract the main spatial relationship rules of the geological phenomenon.

B. *The spatial relationship rules extraction*

Using the rough set method to process the geological phenomenon of intrinsic spatial relation rules extraction, it is mainly divided into the following steps:

1) The spatial relationship of rough sets expression: aim to the study of geological problems, by the method in III (A), do the processing of expressing the geological phenomena of spatial relationship to the data processing format of the rough set -- the form of a decision table.

2) Using the discretization method of the rough set theory to get the decision table then to discretize. As the rough set to process the decision table, it requests the values in the decision table expressed by discrete data (such as integer, string type, enumeration type), therefore, before processing the data it must do the decision table discretization.

3) Using the attribute reduction algorithm of the rough set to do the processing of spatial relationship reduction on the space relationship the discrete decision table of the geological phenomenon space relationship, and finally form the space relationship decision rule table. The spatial relationship decision table after reducing then become the space relationship decision rule table. Because the results of the space relationship reduction are not unique, and each reduction result of the space relationship decision table will become one space relationship decision rule table, so the finally, space relationship decision rule table is the "and" of all the space relationship decision rule tables that came from each reduction result. To the final spatial relationship rule, that asks for calculating the coverage and confidence of the spatial relationship decision rules.

IV. Rough Set Theory in the Application of GIS Spatial Relationships

A. *GIS data*

Data analysis is an important part of GIS data processing. Rough set theory has some unique opinions such as knowledge granularity, new membership, which makes rough set particularly suitable for data analysis, therefore, there are some successful applications by using the rough set theory in GIS data analysis, for example, to adopt Worboys to handle the inaccuracy caused by multi-space or multi-semantic resolution ratio [22, 23] models like Theresa based on rough set and Egg-Yolk model study on the fuzzy and uncertainty problems of spatial data [24]; Du introduces the rough set theory into the expression of direction relationships, and present the direction relationship rough expression method, variable precision rough representation methods and rough reasoning method of direction relationship, which leads to enhance the processing and analysis ability to handle accuracy and fuzziness, and also can unify the direction relationship between the fuzzy objects and the precise objects into a framework [25]; Shi has already discussed on the rough set theory in the application of GIS uncertainty problems, which shows the rough set theory is valuable in GIS uncertainty, but recent researches have not get deeply [26].

B. Spatial data mining

GIS is the main part of the spatial database development and contains a large number of spatial and attribute data, which has more rich and complex semantic information than the general database, and hides abundant information, all of these are very necessary for data mining. Spatial data mining means to extract the information users interested in which includes common relationships of spatial patterns and features, or spatial and non-spatial data, and some other general data characteristics hidden in the database data. Accordingly spatial data shows increasing important in the found and remake nature projects of people activity, the research and application of spatial data mining also increasingly aroused concerns, and the rough set theory is one of the important methods introduced to the data mining, in 1995 Theresa Beaubouef tried to describe a database model based on the original rough sets theory, and introduced some rough relational database models which include systems involving ambiguous, imprecise, or uncertain data [27], and Wang used GIS attribute mining as an example to analyze the application of rough set in GIS data mining [28].

C. Fuzzy geographical object modeling

The fuzzy object modeling have a wide range of meanings. The real world is complicated and full of all kinds of uncertainty, however in GIS, traditional geographic object modeling only consider the clear objects cannot reflect the complexity and uncertainty of the real world well. That causes the poor decision ability of GIS based on these kind models, which leads to hinder the development of GIS intellectualization. Using rough set to describe fuzzy object, is with the ability to fully represent the fuzziness of fuzzy objects, therefore abundant researches and applications of geographic object modeling based on the rough set theory have emerged, such as the research of Liao [29] is based on the rough set theory to transfer method to consider the polygon boundary of data fuzziness, and to employ the membership function to determine the uncertainty of the polygon boundary. Besides Du [30] combined the advantages of the rough sets and fuzzy sets dealing the fuzziness and uncertainty of the spatial data to express the fuzzy objects, and leads to expand the space data model expressing ability of fuzzy data.

D. The combination of rough set and other soft computing methods

The rough set theory is one kind of soft computing method, and the purpose of the soft computing method is to adapt to the inaccuracy of the real world around, to explore the tolerance to the accuracy, the uncertainty and partial real, and in order to achieve hand lability, robustness, and better contact with reality, whose function model is the mind of human. The main methods of calculating software are with rough sets and fuzzy sets, neural networks, genetic, and the theory of transport, etc. As solving practical problem, to adopt several computing technologies collaboratively rather than mutually exclusively has superiority compared with using one kind of computing technology. And also, it can combine the various sources of knowledge, technology and methods which solving complicated practical problems ask for. Due to the rough set has certain shortcomings as processing the data, it is necessary to combine the rough set method with other uncertain methods. At present, there are some applications of GIS data processing that combined the rough set with other soft computing methods, and the more commonly used is the combination of rough set and neural network or fuzzy sets [31, 32].

V. CONCLUSION

Rough set theory is a data analysis tool, which provides a powerful tool for the expression of GIS uncertainty information and processing, and offers favorable conditions to solve uncertain boundary space constraints. In which the fuzzy set and probability statistics method are also the commonly used methods dealing with uncertain information, but these methods need some additional information or data prior knowledge, such as fuzzy membership function and probability distribution, however sometimes it is not easy to get the information. On the other hand the rough set theory just use the information provided by the data itself, without any prior knowledge, at the same time has great advantage to reveal and express multi-level spatial knowledge.

To make a better use Rough set theory in GIS, there are still many problems to be solved. Mainly displays in: rough set can only be used for discrete space, and must be qualitative, therefore only apply to raster data, the application of vector data is difficult to determine; Rough set theory to study the expression of uncertainty in spatial analysis: recently rough set is used in attribute data, involved little in the location data uncertainty. Combining rough set and other uncertain methods, it although has made some achievements, but still there is a lot of unsolved problems ask for further research. With the further increasing of GIS data processing requirements, rough set theory is widely used to spatial data processing, at the same time, it will promote the development of the future GIS data processing technology, especially the spatial decision support system.

REFERENCES

[1] Zhang Xiao-Xiang,Yao Jing, Li Man-Chun, A Review of Fuzzy Sets on Spatial Data Handling[J], Remote Sensing Information, 2005(2).

[2] Robinson V B. Some implications of fuzzy set theory applied to geographic databases [J]. Computers Environment and Urban Systems, 1988, (12):89-97.doi:10.1016/0198-9715 (88) 90012-9.

[3] Liu Wenbao, DengMin, Analyzing Spatial Uncertainty of Geographical Region in GIS [J], Journal Of Remote Sensing, 2002, 6(1).

[4] LIAO Wei-hua, Method Study of GIS Data Transformation Based on Fuzzy Rough Set[J], Remote Sensing Technology And Application, 2007, 22(6)

[5] (Zhang Jingxiong Du Daosheng, Field-based Models for Positional and Attribute Uncertainty [J], Acta Geodaetica Et Cartographica Sinica, 1999, 28(3).

[6] Shi Wenzhong, Wang Shuliang, State of the Art of Research on the Attribute Uncertainty in GIS Data[J], Journal Of Image And Graphics, 2001, 6(9)

[7] CHENG Jicheng, JIN Jiangjun, The Uncertainty of Geographic Data[J], Geo-Information Science,2007,9(4)

[8] WANG Xiaoming, LIUYu, ZHANGJing, Geo -Spatial Cognition : An Overview [J], Geography and geographic information science,2005,21(6),pp.1-10.

[9] Roy AJ, Stell JG. Spatial relations between indeterminate regions. International Journal of Approximate Reasoning, 2001,27(3),pp.205-234.

[10] Leung Y, Ma J H , Goo dchild M F. A General Framework for Error Analysis in Measurement based GIS [C]. The 2nd International Symposium on Spatial Data Quality, Hong Kong, 2003.

[11] Hu Shengwu, Wang Hongtao, Representation and Properties Researches about Fuzzy Geographic Entities[J], Geomatics & Spatial Information Technology, 2006,29(2).

[12] Jonathan Lee, Introduction: Extending Fuzzy Theory to Object-Oriented Modeling[J], International journal of intelligent systems, 2001, 16(7).

[13] Cheng Tao, Deng Min, LI Zhilin, Representation Methods of Spatial Objects with Uncertainty and Their Application in GIS[J], Geomatics and Information Science of Wuhan University, 2007, 32(5), pp.389-393.

[14] Sun Lixin, Gao Wen, Selecting The Optimal Classification Bands Based On Rough Sets[J], Pattern recognition and artificial intelligence, 2000, 13(2).

[15] XU Yi, LI Longshu, Image Segmentation Based on Rough Entropy and K-Means Clustering Algorithm[J], Journal Of East China University Of Science And Technology(Natural Science Edition), 2007, 33(2).

[16] Pawlak Z. Rough set. International Journal of Computer and Information Sciences, 1982(11), pp.341-356.

[17] Han Zhenxiang, Zhang Qi, Wen Fushuan, A Survey on Rough Set Theory and Its Application[J], CONTROL THEORY & APPLICATIONS,1999, 16(2).

[18] Pawlak Z. Rough set-theoretical aspects of reasoning about data[M],.Dordrecht:Kluwer Academic Publishers,1991

[19] Wang Guoyin, Yao Yiyu, YuHong, A survey on rough set theory and applications[J], Chinese journal of computer, 2009, 32(7), pp. 1229-1246.

[20] Liao Weihua. GIS uncertainty analysis based on covering rough set [J], Science of Surveying and Mapping. 2012, 37(4), pp.154-156.

[21] Burrough P A, Frank A U. Geographic Objects with Indeterminate Boundaries[M].Basingstoke:Taylor and Francis,1996.

[22] WORBOYS M. computation with imprecise Geospatial Data[J].Computers, Environment and Urban Systems,1998.85-106.doi:10.1016/S0198-9715(98)00023-4.

[23] WORBOYS M. Imprecision in Finite Resolution Spatial Data [J].Geo Informatica,1998,(03):257-279.).

[24] THERESA B, FEDERICK E P. Vague regions and spatial relationships: A rough set approach [A].2001.pp.313-318.

[25] Du Shihong, WangQiao, Spatial Orientational Relations Rough Reasoning[J], Acta Geodaetica Et Cartographica Sinica, 2003, 32(4) : 334—338.

[26] Wang Shuliang, Li Deren, Theory and Application of Geo-rough Space[J], Geomatics And Information Science Of Wuhan University, 2002, 27(3), pp. 274—282.

[27] Theresa Beaubouef, Frederick E. Petry, Bill P. Buckles, Extension Of The Relational Database And Its Algebra With Rough Set Techniques[J], Computational Intelligence, 1995, 11(2),pp.233-245.

[28] Deng Xueqing, Dong Guangjun, Gis Attribute Data Mining Based On Rough Set Theory[J], SURVEYING AND MAPPING OF SICHUAN, 2003, 26(4) .

[29] LIAO Wei-hua, Method Study of GIS Data Transformation Based on Fuzzy Rough Set[J], Remote Sensing Technology And Application, 2007, 22(6)

[30] Du Shihong, WangQiao, The Reserch of Rough Expression of Fuzzy Objects and their Spatial Relations[J], Journal Of Remote Sensing, 2004, 8(1)

[31] Greco, S., Matarazzo, B.S., S lowi'nski, R., "Rough membership and Bayesian confirmation measures for parameterized rough sets", Rough sets, Fuzzy Sets, Data Mining, and Granular Computing, Proceedings of RSFDGrC'05, LANI 364 1, PP. 314 - 324, 2005

[32] Ali Azadeha, Morteza Saberi, An integrated Data Envelopment Analysis–Artificial Neural Network–Rough Set Algorithm for assessment of personnel efficiency[J], Expert Systems with Applications, 2011, 38(3), pp. 1364–1373.

8

Density Based Support Vector Machines for Classification

Zahra Nazari
Department of Information Engineering
University of the Ryukyus
Okinawa, Japan

Dongshik Kang
Department of Information Engineering
University of the Ryukyus
Okinawa, Japan

***Abstract*—Support Vector Machines (SVM) is the most successful algorithm for classification problems. SVM learns the decision boundary from two classes (for Binary Classification) of training points. However, sometimes there are some less meaningful samples amongst training points, which are corrupted by noises or misplaced in wrong side, called outliers. These outliers are affecting on margin and classification performance, and machine should better to discard them. SVM as a popular and widely used classification algorithm is very sensitive to these outliers and lacks the ability to discard them. Many research results prove this sensitivity which is a weak point for SVM. Different approaches are proposed to reduce the effect of outliers but no method is suitable for all types of data sets. In this paper, the new method of Density Based SVM (DBSVM) is introduced. Population Density is the basic concept which is used in this method for both linear and non-linear SVM to detect outliers. Experiments on artificial data sets, real high-dimensional benchmark data sets of Liver disorder and Heart disease, and data sets of new and fatigued banknotes' acoustic signals can prove the efficiency of this method on noisy data classification and the better generalization that it can provide compared to the standard SVM.**

Keywords—SVM; Density Based SVM; Classification; Pattern Recognition; Outlier removal

I. INTRODUCTION

Support Vector Machines is an important example of kernel methods, one of the key areas in machine learning. It is originated from the theoretical foundations of the Statistical Learning Theory and Structural Risk Minimization (SRM) [1, 2]. SVM was introduced by Vapnik and colleagues in 1970's, but its major developments were formed in 1990's. The main idea behind SVM is to find an optimal separating hyperplane with maximized margin. The maximum margin reduces the empirical risks (training errors) and causes a very good generalization performance. SVM became very famous because of its high ability in generalization and good performance in pattern recognition (digit recognition, computer vision, and text & speech categorization, etc.) and have found application in a wide variety of areas [2].

Classification with SVM is formulated as a quadratic programming which can be solved by using optimization algorithms. In binary classification problems the standard SVM can be used and data points will be classified without any misclassification. However in real world problems, sometimes there are many data points which are corrupted by noises or misplaced on the wrong side. These data points are called outliers and sensitivity of SVM to these outliers is a weak point for this algorithm. There are many approaches proposed to reduce this sensitivity; the Central SVM method (CSVM) which is using class center vectors [3], Adaptive Margin SVM for classification which propose a reformulation of the minimization problem [4], Mapping original input space to normalized feature space for increasing the stability to noise [5], Robust SVM for solving the over fitting problem [6], and Fuzzy SVM [7] are some examples of proposed approaches to reduce the effects of outliers and noises.

Fuzzy SVM is developed on the theory of the SVM and fuzzy membership for each data point shows the attitude of the corresponding point toward one class and also represents the importance of the data points to the decision boundary. The data points with a bigger fuzzy membership will be treated more important and will contribute more to the learning of decision boundary [7].

This paper is organized as follows. The theory of Support Vector Machines will be explained in section II. The Basic concept which is used to develop DBSVM will be explained in section III. Density Based SVM will be introduced in section IV. Experiments and comparison of standard SVM performance to DBSVM performance will be discussed in section V.

II. SUPPORT VECTOR MACHINES

Data classification process using SVM includes two stages: learning is the first stage, the aim of which is to analyze labeled data and learn a mapping from x to y where $y = \{1, \dots, C\}$ (with C being the number of classes) and to build a classifier. The second stage is predicting which is using the established model for predicting on novel inputs. SVM is one of the most successful classification algorithms and its important property is that the determination of the model parameters corresponds to a convex optimization problem, and so any local solution is also a global optimum [8]. The basis of the theory of SVM for classification problems will be reviewed in the following.

A. Hard Margin (Linear) SVM

The linearly separable case is the easiest classification problem which is rare in practice. In this case data pairs can be classified perfectly and the empirical risk can be set to zero. In linearly separable cases, among all the separating hyperplanes which minimize the empirical risk, the one with the largest margin is required. This can be expressed as the idea that a classifier with a smaller margin will have a higher expected risk [2]. Suppose that a set of 2-dimensional labeled training points $\{(x_1, y_1), (x_2, y_2), \dots, (x_n, y_n)\}$ is given and each of them has a class label $y \in \{-1,1\}$ which denotes the two classes separately. During the learning stage the machine finds parameters w and b of the decision function $f(x)$ given as:

$$f(x) = sgn(x.w^T + b) \tag{1}$$

where w is the weight vector and b is the bias. SVM, after learning by training points can produce an output for unknown data point according to above decision function (1). The linearly separable data points can be classified by solving the following quadratic program:

$$\begin{cases} \min \ \frac{1}{2}\|w\|^2 \\ y_i(x_i.w^T + b) \geq 1 \qquad i = 1, \dots, N \end{cases} \tag{2}$$

B. Soft Margin SVM

In previous section the training points were assumed that are linearly separable and the resulting support vector machine will give exact separation of the training points which is not very realistic. Sometimes in real-world problems the training points are overlapped (slightly nonlinear) and some samples cannot be classified correctly and the constraint in (2) will not be satisfied. Therefore classification violation must be allowed in the SVM. In practice the soft margin will be allowed. This approach allows some training points to be on the wrong side of the separating hyperplane, but with a penalty that increases with the distance from hyperplane [2, 9]. To do this, the nonnegative variable $\xi \geq 0$ will be used to measure the amount of this violation and (2) will be modified to (3):

$$\begin{cases} \min \ \frac{1}{2}\|w\|^2 + C\sum_{i=1}^{N}\xi_i \\ y_i(x_i.w^T + b) \geq 1 - \xi_i \qquad i = 1, \dots, N \\ \xi_i \geq 0 \qquad i = 1, \dots, N \end{cases} \tag{3}$$

where $C.\sum\xi_i$ is the distance of error samples to their correct places. Parameter C>0 (the only free parameter in SVM) controls the trade-off between slack variable penalty and the margin [8].

C. Non-linear SVM

In case of considerable class overlapping (seriously nonlinear) of the training points, soft margin SVM classifiers are unable to separate the samples into classes appropriately. Therefore SVM transforms samples x from original input space to a higher dimensional feature space by a non-linear vector mapping function $\Phi(x) = R^n \to F$. However the vector mapping function (Φ) leads to high computational expenses. Thus, this transformation can be performed by kernel function which allows more simplified representation of the data. Polynomial, Sigmoidal, and Gaussian (RBF) are some popular kernel functions for this kind of transformation [2, 8, 10, 11].

The different distribution in the feature space enables the fitting of a linear hypersurface in order to separate all samples into the classes. Classification is easier in higher dimensions, but computation is costly. The resulting separating hypersurface in feature space will be optimal in the sense of being a maximal margin classifier with respect to training points [2]. The vector $\Phi(x_i)$ in the feature space corresponds to vector x_i in the original space. The solution in the SVM does not depend directly to input vectors, rather to dot product between input vectors, and so the dot product of $\Phi(x_i).\Phi(x_j)$ is needed. It would be preferable to be able to define the dot

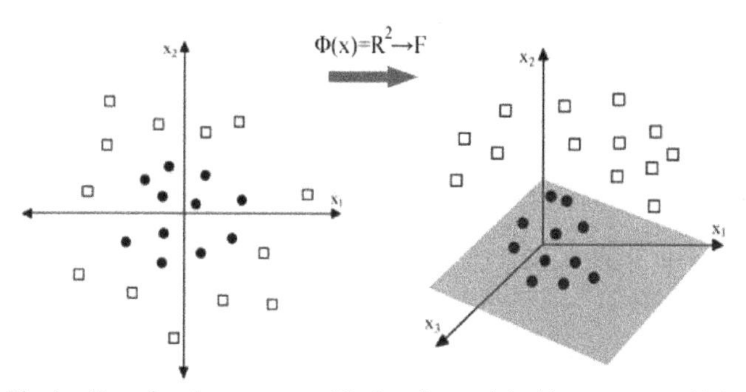

Fig. 1. Transforming non-separable data from original input space to higher dimensional feature space

product directly rather than defining the mapping Φ explicitly. The kernel function computes the dot product of training points in feature space and there will be no need to define Φ explicitly [12]. By using Lagrange multiplier and kernel method, the QP for nonlinear cases is as below:

$$\begin{cases} \max \ \sum_{i=1}^{N}\alpha_i - \frac{1}{2}\sum_{i=1}^{N}\sum_{j=1}^{N} y_i y_j \alpha_i \alpha_j K(x_i, x_j) \\ \sum_{i=1}^{N} y_i\alpha_i = 0 \\ 0 \leq \alpha_i \leq C \qquad i = 1, \dots, N \end{cases} \tag{4}$$

III. Basic Concept

Population Density is the basic concept which is used to develop Density Based SVM. Population density is the way of measuring the population per unit of area or volume. The term of population density was used by Henry Drury Hamess in 1937 for the first time, then widely used to measure the decrement and increment of densities and finally applied as an indicator to compare the area's population density. The concept of population density indicates the relationship between number of population and the occupied space by them.

$$Population\ density = \frac{\text{number of population}}{\text{area}} \tag{5}$$

By using this concept, the densely populated and less populated areas can be determined. Considering the training points as the population, those samples placed in less populated areas or the areas with low population densities can be treated as outliers. These outliers are not very important, but dramatically are affecting on performance of SVM algorithm.

Outliers are unusual data points that are inconsistent with other observations. In statistics an outlier is an observation with an abnormal distance from most other observations. Generally presence of an outlier may cause some sort of problems. An outlier may be due to gross measurement error, coding/recording error, and abnormal cases, but a frequent cause of outliers is a mixture of two distributions and they can be occurred by chance in any distribution [13, 14, 15]. There are two strategies to deal with outliers: first, outlier detection or removal as a part of preprocessing; second, developing a robust modeling method to be insensitive to outliers [14, 15]. Density Based SVM is based on the first strategy.

IV. Density Based SVM

The main goals of Density Based SVM is reducing effects of outliers, maximizing margin, providing better generalization, and adjusting the decision boundary according to the density of data sets. Meanwhile Density Based SVM reduces the number of support vectors which decreases computational complexity. It is noteworthy that in Density Based SVM, input vectors are those which are in highest-confidence area of data set and they are more informative than other input vectors.

Density Based SVM can detect outliers or data points which are out of the densely populated area. To detect these outliers, first the densely populated area of a data set should be determined. The data points which are located in the densely populated area will be considered as important (meaningful) points and other as less important (meaningless) which can be misclassified or ignored. Although the concept of population density is used to develop Density Based SVM, the formula is different with (5). In this method the distance (Euclidean & Mahalanobis) between data points of one data set plays the main role to determine the area with high population density.

A. Density Based SVM with Euclidean Distance

Euclidean distance measures the distance between two points by formula (6) in Euclidean space [16]. Suppose that a set of 2-dimensional data $\{(x_1,y_1),(x_2,y_2),\ldots,(x_n,y_n)\}$ is given. First the Euclidean distance between all data points of one class should be calculated. For example the Euclidean distance between point 1 and 2, 3, … , n and the Euclidean distance between point 2 to 1, 3, …, n and so on.

$$D(a,b) = \sqrt{(a_1-b_1)^2+(a_2-b_2)^2} \tag{6}$$

The next step is summing up all distances for each point. For example the total distance for point 1 is $d_1 = [\mathrm{D}(1,2) + \mathrm{D}(1,3) + \ldots + \mathrm{D}(1,n)]$ where n is the number of data points in one data set. The total distances for all data points of one data set is needed to calculate the average distance which will be used to determine data points which are inside and outside of densely populated area. The average distance can be calculated as follows:

$$Average\_d = \frac{\sum_{j=1}^{N}\sum_{i=1}^{N}\sqrt{(x_j-x_i)^2+(y_j-y_i)^2}}{n} \tag{7}$$

$$if\ d_i > Average\_d \rightarrow x_i = outlier$$

After calculating the $Average\_d$ by (7), those data points with ($d \le Average\_d$) should go to group 1 which is the new training set and others to group 2. The space which is occupied by group 1 is the area with high population density and data points inside this group will be considered as important data points. Those data points in group 2 will be considered as less important or outliers and they will not contribute in training phase [17].

Algorithm 1:

1- For each data point x_i:
   - Calculate the Euclidean distance between x_i and all other data points by (6)
   - Sum up all the distances calculated for one point as d
2- Sum up all d values as $total\_d$.
3- Divide $total\_d$ by number of data points of one set as $Average\_d$ by (7)
4- Set all data points with ($d \le Average\_d$) in one group
5- New group contains the most important data points and others will be considered as outliers.

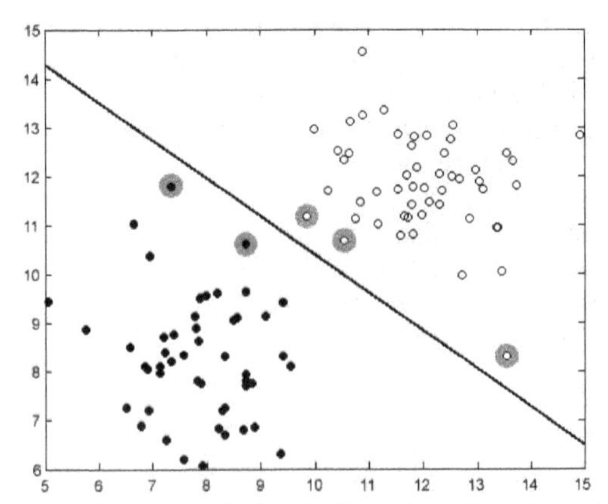

Fig. 2. Result of standard SVM; Outliers exist and margin is small

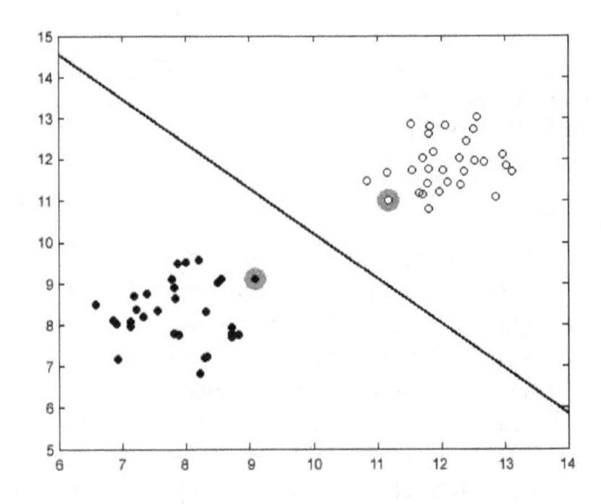

Fig. 3. Result of DBSVM; outliers are removed and margin is bigger

Applying algorithm 1 as preprocessing on both data sets of the classification problem in Fig .2, can help to detect outliers, reducing the number of support vectors, and maximizing the margin. The difference of margin with and without outliers is shown in Fig .2 and Fig. 3 respectively.

The described algorithm can be applied on both linear and non-linear SVM. In non-linear cases it can be done either in original input space or feature space and there will be no difference in result. In case of applying algorithm in input space, removing outliers from data set will also reduce the dimensionality of data points in feature space and there will be no need to change the algorithm and all should be done like as previous description. However in case of applying mentioned algorithm in feature space, there will be a small difference. Since the kernel matrix should be positive semi-definite and symmetric, after removing outliers it will become asymmetric. In this case by removing each data point, the corresponding column also should be removed. For example point x_3 in below matrix is an outlier, consequently in addition to the 3rd row, the 3rd column also should be removed from kernel matrix.

$$\begin{bmatrix} k(x_1,x_1) & k(x_1,x_2) & k(x_1,x_3) & \cdots & k(x_1,x_n) \\ k(x_2,x_1) & k(x_2,x_2) & k(x_2,x_3) & \cdots & k(x_2,x_n) \\ k(x_3,x_1) & k(x_3,x_2) & k(x_3,x_3) & \cdots & k(x_3,x_n) \\ & & \vdots & & \\ k(x_n,x_1) & k(x_n,x_2) & k(x_n,x_3) & \cdots & k(x_n,x_n) \end{bmatrix}$$

B. *Density Based SVM with Mahalanobis Distance*

In this section the Mahalanobis distance will be used instead of Euclidean distance. Euclidean distance measures the distance between two points by formula (6) in Euclidean space. The Mahalanobis distance is the distance from x to the quantity μ. This distance is based on the correlation between variables or the variance-covariance matrix. Mahalanobis distance is unit less and it takes into accounts the correlation of the data set and does not depend on the scale of measurement [16]. The Mahalanobis distance of point x to the mean of distribution can be calculated by formula (8) and Mahalanobis distance of point x to point y can be calculated by formula (9):

$$D_m(x) = \sqrt{(x-\mu)^T S^{-1}(x-\mu)} \quad (8)$$

$$D_m(x,y) = \sqrt{(x-y)^T\, S^{-1}(x-y)} \quad (9)$$

where μ is the mean of the distribution and S^{-1} is the inverse covariance matrix. Here, to determine the densely populated area, the Mahalanobis distance of each point to the mean μ of the data set is used. Same as the previous section, the average distance should be calculated and then, those data points with $(D_m) \leq Average\_D$ should go to group 1 as important points and others to group 2 as outliers.

$$\text{Average\_D} = \frac{\sum_{i=1}^{N} \sqrt{(x_i-\mu)^T S^{-1}(x_i-\mu)}}{n} \quad (10)$$

$$if\ D_m > Average\_D \rightarrow x_i = outlier$$

Algorithm 2:

1- For each data point x_i:
  - Calculate the Mahalanobis distance of x_i to the mean μ of data set as D by (8)

2- Sum up all D values as $total\_D$.

3- Divide $total\_D$ by number of data points of one set as $Average\_D$ by (10)

4- Set all data points with $D \leq Average\_D$ in one group

5- New group contains the most important data points and others will be considered as outliers.

C. *Density Based SVM for Special Cases*

So far, the considered data sets had one center and the distribution of data points were around that center. However sometimes data points are distributed very widely and it seems they have more than one center. To deal with this problem, before applying algorithm 1 or 2, the method of K-means clustering should be used to cluster data points and then algorithm 1 or 2 can be applied for each cluster separately.

K-means is one of the most popular clustering algorithms, and it is an iterative descent clustering method. K-means finds k clusters in a given data set and number of k should be defined by user. Each cluster is described by a single point called centroid. Centroid means it's at the center of all the data points in a cluster. K-means is a simple algorithm based on similarity and the measure of similarity plays an important role in the process of clustering [18, 19, 20].

The k-means algorithm works like this: First k randomly centroids will be placed, next, each point in the data set will be assigned to the nearest centroid by measuring the Euclidean distance between point and all centroids. After this step, the centroids will be updated by taking the mean value μ of all the points assigned to them. This process will be repeated until the assignments stop changing. The result of k-means depends to two factors: first the value of k; second the initial selection of centroids [21, 22, 23].

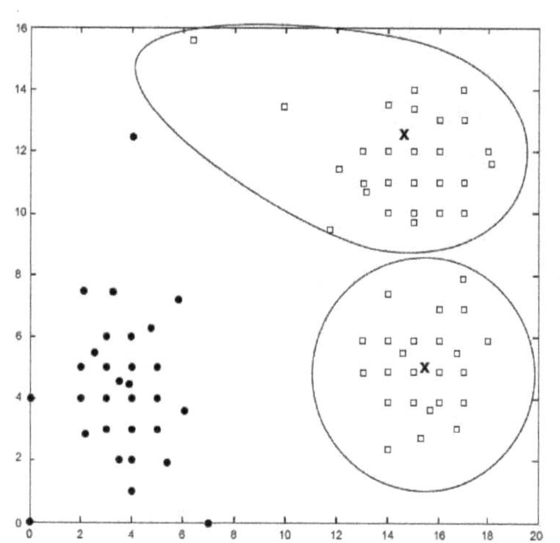

Fig. 4. Result of applying k-means clustering method on one data set

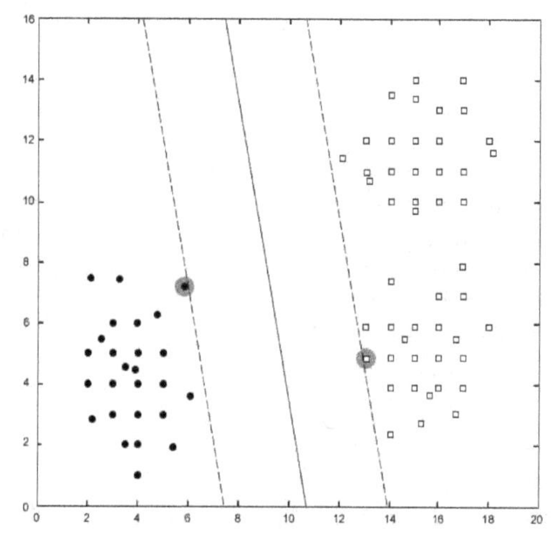

Fig. 5. Result of DBSVM after clustering the widely distributed data set

V. EXPERIMENTS & RESULTS

In order to validate the performance of Density Based SVM, two types of Experiments were performed with different data sets of binary classification problems. The first type of experiment was performed using 2 and 3 dimensional artificial data sets and the second type was performed using two high dimensional benchmark data sets and one set of high dimensional banknote data. The K-fold Cross Validation method is used for all data sets.

A. Artificial data classification

The artificial data sets which are used for this experiment are generated at random with normal distribution. These data sets are used with different standard deviations, and below tables represent the results of applying standard SVM and Density Based SVM on them by 2-fold cross validation.

TABLE I. CHARACTERISTICS OF ARTIFICIAL DATA SETS

Linearly separable data sets	**# of instances**
set1_a & set1_b (μ=8 & 12, SD=1)	tr=200, ts=200
set1_c & set1_d (μ =8 & 12, SD=2)	tr=200, ts=200
Non-linear data sets	**# of data**
set2_a & set2_b (μ=8 &12, SD=3)	tr=200, ts=200
set2_c & set2_d (μ=8 &12, SD=4)	tr=200, ts=200

TABLE II. LINEAR ARTIFICIAL DATA CLASSIFICATION

Data sets	SVM (Linear)	DBSVM (Linear+Euc)	DBSVM (Linear+Maha)
set1_a & set1_b	99%	99%	99%
#SV	35	30	30
set1_c & set1_d	80%	83%	83%
#SV	47	36	37

TABLE III. NON-LINEAR ARTIFICIAL DATA CLASSIFICATION

Data sets	SVM (RBF)	DBSVM (RBF+Euc)	DBSVM (RBF+Maha)
set2_a & set2_b	76%	82%	82%
#SV	70	45	48
set2_c & set2_d	62%	68%	70%
#SV	86	50	56

According to above results, Density Based SVM with Euclidean distance can perform better than with Mahalanobis distance on data sets which have smaller standard deviations and are linearly separable or slightly non-linear. However Density Based SVM with Mahalanobis distance performs better on data sets with bigger standard deviation values that are seriously nonlinear.

B. Benchmark data classification

Two benchmark data sets are used. They are medical data of Liver Disorder and Heart Disease which are obtained from real life and can be downloaded from the Repository of machine learning databases of the well-known University of California at Irvine (UCI) [25]. These data sets are used by 2-fold and 5-fold cross validations.

a) Liver Disorder Data

TABLE IV. CHARACTERISTICS OF LIVER DISORDER DATA SET

Data set characteristics:	Multivariate
Attribute characteristics:	Categorical, Integer, Real
Number of instances:	345
Number of attributes:	7

TABLE V. LIVER DISORDER CLASSIFICATION BY LINEAR KERNEL

Data sets	SVM (Linear)	DBSVM (Linear+Euc)	DBSVM (Linear+Maha)
Liver Disorder (2-fold)	66.9%	70.8%	69.7%
#SV	127	83	70
Liver Disorder (5-fold)	68.5%	70.5%	71.6%
#SV	200	125	122

TABLE VI. LIVER DISORDER CLASSIFICATION BY POLYNOMIAL KERNEL

Data sets	SVM (Poly)	DBSVM (Poly+Euc)	DBSVM (Poly+Maha)
Liver Disorder (2-fold)	57.5%	59.12%	57.7%
#SV	120	85	90
Liver Disorder (5-fold)	63.69 %	63.69%	67.73%
#SV	210	135	124

TABLE VII. LIVER DISORDER CLASSIFICATION BY RBF KERNEL

Data sets	SVM (RBF)	DBSVM (RBF+Euc)	DBSVM (RBF+Maha)
Liver Disorder (2-fold)	60.3%	60.3%	60.3%
#SV	172	131	115
Liver Disorder (5-fold)	57.5%	58%	58%
#SV	225	160	150

b) Heart Disease Data

TABLE VIII. CHARACTERISTICS OF HEART DISEASE DATA SET

Data set characteristics:	Multivariate
Attribute characteristics:	Categorical, Real
Number of instances:	270
Number of attributes:	13

TABLE IX. HEART DISEASE CLASSIFICATION BY POLYNOMIAL KERNEL

Data sets	SVM (Poly)	DBSVM (Poly+Euc)	DBSVM (Poly+Maha)
Heart Disease (2-fold)	72 %	71%	79.99%
#SV	86	50	30
Heart Disease (5-fold)	70.36 %	70.36%	79.99%
#SV	86	45	30

TABLE X. HEART DISEASE CLASSIFICATION BY RBF KERNEL

Data sets	SVM (RBF)	DBSVM (RBF+Euc)	DBSVM (RBF+Maha)
Heart Disease (2-fold)	60.6 %	60.6%	60.6%
#SV	135	85	75
Heart Disease (5-fold)	60%	60%	60%
#SV	96	50	58

C. New and Fatigued Banknote classification

To classify the new and fatigued banknotes, two sets of acoustic signals of new and fatigued U.S. one dollar banknote which are recorded by measurement system of acoustic signal are used for both training and testing. In this case the amplitude differences are considered as the characteristic value. The acoustic signal sets are mapped from very high-dimensional to four-dimensional data. Steps for converting data to four-dimensional are as follows [26]:

Step 1. Calculating the amplitude difference from the sample data of forward and backward (see Fig.6).

Step 2. Assigning each calculated data to the horizontal and vertical axes.

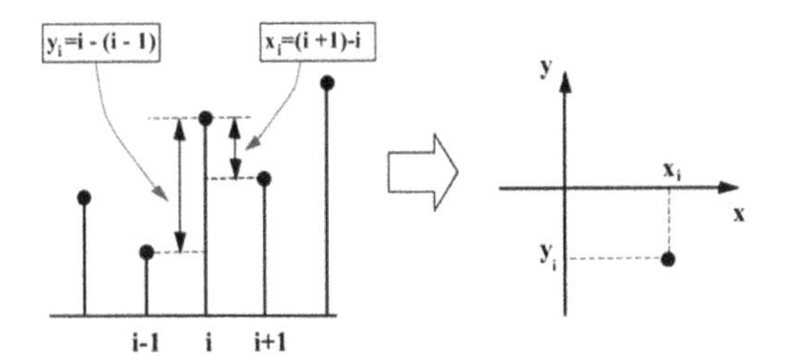

Fig. 6. Calculating amplitude difference from sample data of forward and backward

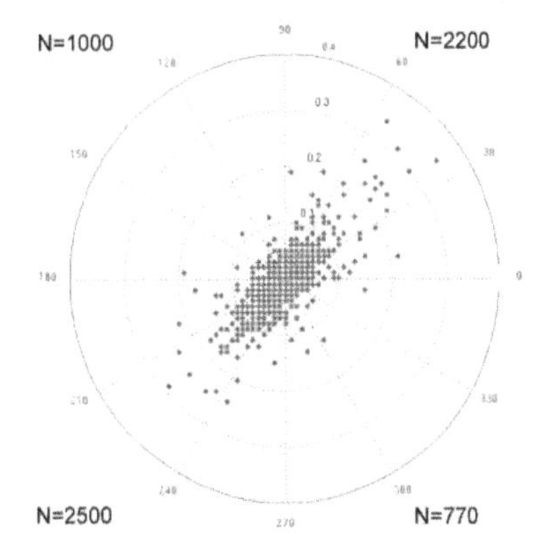

Fig. 7. Sample of data in polar coordinate 4-dim $x_i = (2200, 1000, 2500, 770)$

Step 3. Converting from Cartesian coordinate into the polar coordinate; all elements are divided by the fan-shaped domain.

Step 4. Number of elements which are distributed over each domain gives a four-dimensional data for each banknote acoustic signal (see Fig. 7).

The data set which is used for this experiment contains 48 four-dimensional data of new banknotes and 128 four-dimensional data of fatigued banknotes in four different levels. The data set is divided at random into two disjoint subset of the same size for 2-fold cross validation. Results of experiments with standard SVM with different kernels and Density Based SVM (Euclidean distance & Mahalanobis distance) are shown in following tables.

TABLE XI. CHARACTERISTICS OF BANKNOTE DATA SETS

Banknote Data Sets	# of instances	# of attributes
S00. New banknote	48	4
S01. Fatigued banknote (level 1)	32	4
S02. Fatigued banknote (level 2)	32	4
S03. Fatigued banknote (level 3)	32	4
S04. Fatigued banknote (level 4)	32	4

TABLE XII. BANKNOTE DATA CLASSIFICATION BY LINEAR KERNEL

Banknote Data	SVM (Linear)	DBSVM (Linear+Euc)	DBSVM (Linear+Maha)
Average Result of Binary Classification	66.77%	72.55%	69%
#SV	17	9	12
Multiclass Classification	39.56%	45%	42%
#SV	55	30	45

TABLE XIII. BANKNOTE DATA CLASSIFICATION BY POLYNOMIAL KERNEL

Banknote Data	SVM (Polynomial)	DBSVM (Poly+Euc)	DBSVM (Poly+Maha)
Average Result of Binary Classification	63.68%	68.85%	65%
#SV	13	8	10
Multiclass Classification	38%	43%	37.5%
#SV	50	32	36

TABLE XIV. BANKNOTE DATA CLASSIFICATION BY SPLINE KERNEL

Banknote Data	SVM (Spline)	DBSVM (Spline+Euc)	DBSVM (Spline+Maha)
Average Result of Binary Classification	62.24%	65%	63.83%
#SV	17	10	11

TABLE XV. BANKNOTE DATA CLASSIFICATION BY RBF KERNEL

Banknote Data	SVM (RBF)	DBSVM (RBF+Euc)	DBSVM (RBF+Maha)
Average Result of Binary Classification	69%	70.5%	69%
#SV	32	20	20
Multiclass Classification	35%	35%	35%
#SV	75	45	48

According to the results of different types of experiments on artificial data sets, benchmark data sets and banknote data sets, it can be claimed that Density Based SVM can provide better generalization ability, reduces the effects of outliers, and it can decrease the number of support vectors. Number of support vectors has a direct influence on the time required to evaluate the SVM decision function and also on the time required to train the SVM.

Considering the results presented in previous tables, algorithm 1 can be useful for linearly separable and slightly overlapping classes, and algorithm 2 can be useful for those classes with considerable overlapping (seriously nonlinear).

VI. CONCLUSION

In this paper, the new method of Density Based Support Vector Machines is introduced. Density Based SVM tries to decrease the effects of outliers on SVM performance. The basic concept which is used in this method is population density. By using this concept, the densely populated area of each data set can be found. Those data points which are inside this area are located in highest confidence area of data set and will be considered as most important points and others as outliers. To find this area, two algorithms are proposed; algorithm 1 uses Euclidean distance and algorithm 2 uses Mahalanobis distance.

SVM finds the optimal separating hyperplane under the effects of outliers, but this method first removes outliers as a preprocessing and adjusts the separating hyperplane/decision boundary according to the density of data sets. Support vectors in Density Based SVM are from high confidence area of data set. Although the main goal of Density Based SVM is removing outliers, it is also maximizing margin, reducing number of support vectors which results reducing computational complexity and gives better generalization ability. Different experiments on artificial data sets and real high dimensional data sets are performed to prove the validity of this method. Considering the results of experiments, the Density Based SVM can be useful on different types of noisy data sets. It increases the SVM performance and considerably reduces number of support vectors.

The future work to be done is to make some changes in this method to become more effective on RBF kernels. Because according to the results of experiments, Density Based SVM only reduces the number of support vectors and computational complexity, but does not increase the generalization ability while using RBF kernel.

ACKNOWLEDGMENT

The authors would like to thank Mr. Ikugo Mitsui, Mr. Nobuo Shoji, and Mr. Kenji Ozawa from Japan Cash Machine Co., Ltd, for their support and providing the measurement system of acoustic signals for experiment C.

REFERENCES

[1] V. N. Vapnik, The Nature of Statistical Learning Theory, Springer 2000.

[2] V. Kecman, Learning and Soft Computing, Support Vector Machines, Neural Networks, and Fuzzy Logic Models, The MIT Press 2001.

[3] X. Zhang, Using Class Center Vectors to Build Support Vector Machines, IEEE, pp.4-6, 1999.

[4] R. Herbrich, J. Watson, Adaptive Margin Support Vector Machines for Classification, Microsoft Research, pp. 2-4, 1999.

[5] AB. A. Graf, Classification in a Normalized Feature Space Using Support Vector Machines, IEEE, pp. 1-3, 2003.

[6] Q. Song, Robust Support Vector Machine with Bullet Hole Image Classification, IEEE, pp. 3-4, 2002.

[7] C. F. Lin, Fuzzy Support Vector Machines, IEEE, pp. 3-4, 2002.

[8] C. M. Bishop, Pattern Recognition and Machine Learning, Springer 2006.

[9] A. R. Webb and K. D. Copsey, Statistical Pattern Recognition, John Wiley & Sons 2011.

[10] V. N. Vapnik, Statistical Learning Theory, AT&T Research Laboratories, John Wiley & Sons, 1998.

[11] S. Abe, Support Vector Machines for Pattern Classification, Springer 2010.

[12] B. Scholkopf & A. J. Smola, Learning with Kernels, Support Vector Machines, Regularization, Optimization & Beyond, The MIT Press 2002.

[13] V. Cherkassky and F. Mulier, Learning from Data, Concepts, Theory and Methods, IEEE Press 2007.

[14] D. Ripley, Robust Statistics, M.Sc. in Applied Statistics MT2004.

[15] S. Theodoridis and K. Koutroumbas, Pattern Recognition, Academic Press 1999.

[16] Y. Dodge, The Concise Encyclopedia of Statistics, Springer 2008.

[17] Z. Nazari, D. Kang, and H. Endo, Density Based Support Vector Machines, The 29th International Technical Conference on Circuits/Systems, Computers and Communications (ITC-CSCC), pp.1-3, 2014.

[18] P. Cichosz, Data Mining Algorithms Explained Using R, John Wiley & Sons 2015.

[19] T. Hastie, R. Tibshirani, and J. Friedman, The Elements of Statistical Learning, data mining, inference and prediction, Springer 2009.

[20] P. Harrington, Machine Learning in Action, Manning Publications Co 2012.

[21] S. Marsland, Machine Learning an Algorithmic Perspective, CRC Press 2009.

[22] E. Alpaydin, Introduction to Machine Learning, The MIT Press 2010.

[23] J. Bell, Machine Learning Hands-on for Developers and Technical Professionals, John Wiley & Sons 2015.

[24] T. Segaran, Programming Collective Intelligence, O'REILLY Media. Inc 2007.

[25] Machine Learning Repository, available online at: (https://archive.ics.uci.edu/ml/datasets/ Liver+Disorders).

[26] M. Higa, D. Kang, H. Miagi, Classification of Fatigue Bill based on Acoustic Signals, pp. 1-4, 2012.

Students' Weakness Detective in Traditional Class

Artificial Intelligence

Fatimah Altuhaifa
College of Computer Engineer & Science
Prince Mohammed bin Fahd University
Alkhobar, Saudi Arabia

Abstract—**In Artificial Intelligent in Education in learning contexts and domains, the traditional classroom is tough to find students' weakness during lecture due to the student's number and because the instruction is busy with explaining the lesson. According to that, choosing teaching style that can improve student talent or skills to performs better in their classes or professional life would not be an easy task. This system is going to detect the average of students' weakness and find either a solution for this or instruct a style that can increase students' ability and skills by filtering the collected data and understanding the problem. After that, it provides a teaching style.**

Keywords—emotional learner prediction; voice identifier and verifier; weakness detecting; artificial intelligent in education

I. Introduction

Students' weakness is one of the most important factors that prevent students' improvement. For that, researchers built a lot of e-learning that detects student's difficulty and provides the students with suitable learning style. Not all students have the motivation to use e-learning while most of them care to attend their classes. This paper aims to find a solution for students' weakness in an environment such as a traditional classroom. Understanding students' character is the primary factor for detecting difficulty. Analyzing students' emotion and activities can obtain the nature of that student.

II. Select A Problem

This paper purposed to invent a solution for a system that helps in improving and enhancing student's skills and weakness. It will find the average of students' weakness in the class by collecting learners' emotion, reasoning the data and representing solution depending on the database or knowledge-based that the system has. The aim of this software will be reached by collecting information about each student at class and touching student's voice and analyzing it to find the suitable teaching style.

III. The Solution

As each student has his/her weakness that can affect his/her performance in the professional life, this system detects the average of students' vulnerability for every class. Then it provides a teaching style that helps in improving students' strengths.

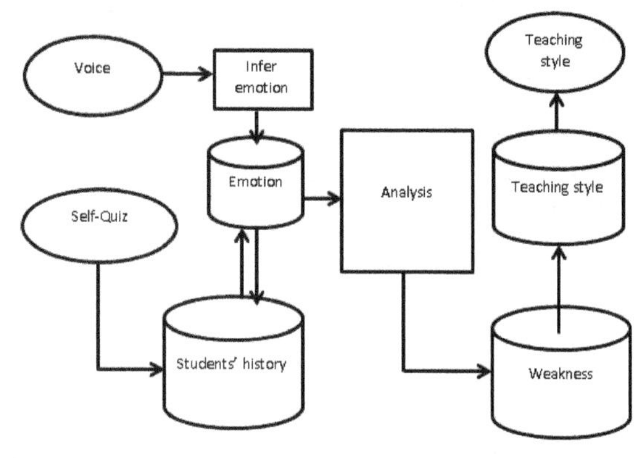

Fig. 1. Clarifying the solution

The system has one output and two inputs; one is students' voice, and the other is self-quiz for each student.

The system will touch the Students' voice for one week. Then the tone of speech is sent to software which will infer student's emotion. The system will have a database for storing students' emotion. Each student will have an ID, which is used to save assumed feelings with the unique voice. The ID with the unique voice will prevent the system from keeping the same emotion for the same student. Also, the system will add new colon if it is needed. Then the average for emotion will be taken and saved.

TABLE I. The Saved Emotions of Students

ID	Hash1	Hash2	Hash3	Hash4
01	Shy	Lack of self-confidence		
02	Sad	Fear	Anxiety	Despair
03	Angry	Anxiety		
04	Lack of self-confidence			
05	shy			

TABLE II. Shows the Average of Emotions

Average	Shy	Anxiety	Lack of self-confidence

The system has several emotions such as self-esteem, shy, fear, cooperative, sadness and exciting. And these feelings can be changed as the user want.

The second input is a self-quiz which each student is going to take at every beginning of the semester. This self-quiz will focus on student's character to enable the system inferring the weakness. Every student has to use his/her ID for accessing to this self-quiz, and this will prevent the student from doing the quiz more than once. After taking the information, the information will be sent to the database that has the history for each student. The most characters that system will focus on are self-confidence, self-awareness, work in the team and the ability for improving.

After the system stores the information in the students' history and stores the inferred emotions in the emotion database, the system will send the information and the inferred emotions sent to investigate block. The analyzing block will examine the self-quiz for each student, and it will put the result in the table, then it will take the average of students' characters. And because the result of self-quiz can be 50% for one character and 50% for another character, the system can record more than one or two characters for one ID.

TABLE III. DETECTING MORE THAN ONE CHARACTERISTIC FOR THE SAME PERSON

ID	Character1	Character2	Character3
01	Lack of Self-awareness	Lack of work in team	
02	self-confidence		
03	self-confidence		
04	Lack of improving		

TABLE IV. THE AVERAGE FROM THE FIRST TABLE

Average	self-confidence

Then the analysis block will relate the character and the emotion to each other to grip an appropriate weakness to the class. The system will send the gripped vulnerability to weakness the database which has a table for vulnerability and a proper teaching style for each weak point. After the system chooses the style of education, the system will send the result to the screen as a report which has the method of teaching.

Also, there is another problem that system can face. This problematic is one pair voice come from outside the class to the software. For solving this problem, the system will have an application that can be work on and work off by the instructor. This application is identifying students' voice at the beginning of the semester, or when any new student attends the class, who has done late registration for the course, the application can identify the students' voice when each student identify himself/herself to everyone at the first class of the semester.

1) How can the voice be touched?

2) How will the emotion extract from the tone of speech?

3) How can the system identify and verify the vocal sound?

4) What is the important of identifying and confirming the voice?

5) How are the data going to be analyzed?

6) Which is the type of teaching style going the system has?

7) How can the system collect the students' data?

8) How can the system deal with overlapping voice?

IV. METHODOLOGY AND ANSWERING SOME OF THE QUESTIONS

Most of the work will depend on how to handle the voice so the answer will be depending on understanding the natural of the sound and its signal. And more on the signal, because each type of speech has different vibration, frequency and signal which can help in understanding students' natural.

A. Catching the Voice

As known the voice travels through the medium (air, water, etc.) as a vibration. The system will have a Neumann microphone that one of the employees in the school has to install it on the ceiling. The mounted instrument will allow each sound to be easy to catch. This kind of tool is upright for using in studio, TV and room because it can detect the voice from a long distance.

B. Extracting Voice's Feature

When the microphone receives the speech, it will send it to the system. The system will extract at least one feature of the voice signal then it will determine the emotion that is related to the voice signal [1]. Each sound has different vibration, for example, the noise has high waves, and these waves are close while the beneath sound has vibrations that have small distant from each other.

Fig. 2. Neumann microphone [2]

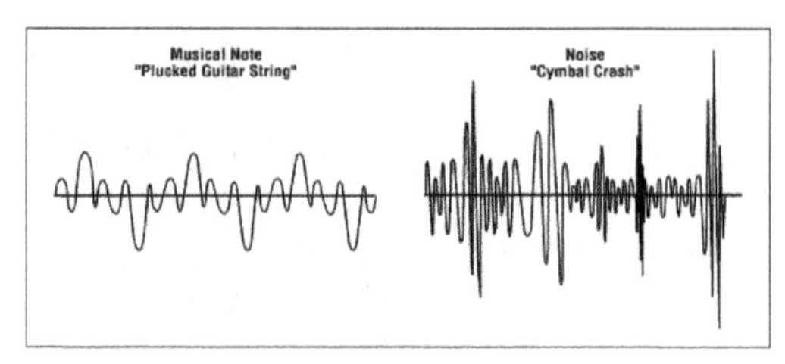

Fig. 3. Diversity in voice's frequency [1]

The Sound has three composed loudness, pitch, and timbre. Loudness is a strength or weakness of sound due to air pressure. The application can know the pitch by the frequency of the voice and rate repetition. Timber is the tone, color, or texture [1].

C. The Ability of Extraction Emotion From The Voice

A.Valery works in a system that can detect emotion in the voice signal. They claim that each emotion has different frequency and vibration. The system uses Object Oriented Programming (OOP). This programming helps in developing computer software and doing an analysis. This system receives the voice and does an analysis to it **[3].**

D. Identifying Verifying Voice

One of the inputs to the system is students' speech which will be touched in the class by the microphone for one week every month. Then, the vibrations of voice will be analyzed to infer the emotion of student. The system will need to check student's awareness every month to check students' improvement. Also, avoiding duplicate in emotion for the same student will be required for getting a perfect result. The system will need to recognize the voice and assigns it to specific ID, and it will need to verify the voice to ensure that voice has an ID to prevent the system from saving the duplicate voice of the same student. For this purpose, the system will have a database for storing voice's feature in it. The Hidden Markov Models with the Gaussian Mixtures (HMM-GM) is the used method [3]. With this approach, there are three ways for extraction voice's feature. These are filters, Speaker Dependent Frequency Filter Bank (SDFFB) and Speaker Dependent Frequency Cepstrum Coefficients (SDFCC). The filters work to find the domain of frequency by defining the discrete frequency domain. According to the different in vocal sound, the SDFFB's method is Linear Prediction Coefficients, while the system cannot use the SDFCC in some types of voice recognition because it "emphasizes the speaker influence too much" [4].

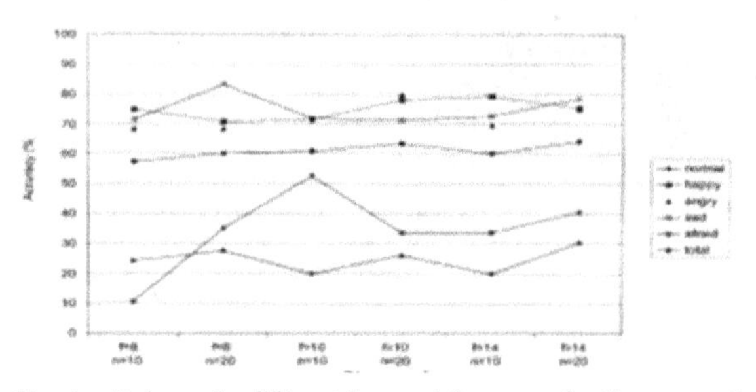

Fig. 4. It shows the different in sound frequency for diverse emotion [1]

E. Ability of Distinguishing Voices in Overlapping

T.WEI-OH and L.SHI-JIE wrote about a system that can distinguish between overlapping sounds and non-overlapping. And this system can determine who speak each part of overlapping sounds. The system will take the signal and checks if it is overlapping or not overlap then it will start analyzing the voice to determine who is speaking?[5].

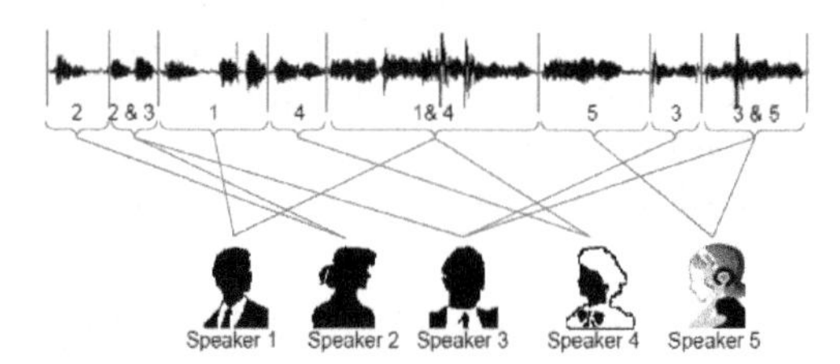

Fig. 5. Extracting the simulating voices [5]

F. Avoiding Duplicating Emotion For the Same Student

The system will need some algorithm or coding for achieving this goal. After verifying the voice and being sure if it is existing or not, then it will check the identification of the voice. It is not important to know exactly for which student this voice but it is important to assign each tone with unique ID. After recognition the voice, the system has to record the extracting emotion with its accent which means with its ID. But before recording the emotion, the system must ensure that this ID does not have this feeling. It the ID has the emotion the system will not record it while if the sense does not exist for that ID, the system has to record it.

```
If (newVoice == OldVoice){
If ( newEmotion == OldEmtoin)
Add emotion;
Else
Exit 0; }
Else{
If( saving == True )
Add newVoic;
Else
Exit 0;}
```

V. RELATED WORKS

This paper aimed to detect the student's knowledge state to help the system to find an appropriate learning plan for the student that improving the student and the learning effectively. It is a web-based educational system which provides the student with a quiz and observes his/her interaction on the web and his/her moving to the mouse and choosing the answer with observing the time for his/her action. Choosing appropriate learning plan depends on student diagnose. For achieving this purpose, the system will have student profile, student model, content model and learning plan. The application uses fuzzy logic supported modeling in analyzing student. The using of fuzzy logic supported modeling is depending on the student profile and the quiz **[6].** The way that the system is using for analyzing student and finding the appropriate plan for him/her is the most part that is related to evaluating students' weakness in the class. The reason that makes it the most important part is to diagnose student individually will be needed to discover the average of weakness in the class.

N.Aghaee and S.Ören discussed in their paper the process of finding the best solution for education style for the e-learning depending on the student's emotions. The primary inputs, at the System, are the student's personality and emotion while the output will be the teaching's style. The aim of this application is to inspire the student and enhance their ability. For reaching their goal, they used personality filter for software agent, emotion filter for software agent and MBTI indicator. The system has eight types of emotions which are fear, anger, sorrow, joy, disgust, acceptance, anticipation and surprise. The MBTI is the responsible for inference the suitable learning style, according to student's cognitive, emotion and character **[7]**. This paper related to our paper in the way in finding the appropriate learning style depending on the emotions and character but has different in the environment and how to collect the student's data. In this paper, the emotions will be inferred by facial expression while in our paper, the emotions will be inferred by using an agent that can cause analysis emotions from voice. In our paper, student's character will be assumed by student's history. In their paper, the environment is eLearning while our environment is a traditional classroom.

Another relative document to this article is A Model of the Student Behaviors in a Virtual Educational Environment, which is written by Moisil. This paper has a system that finds a suitable learning style on online for student depending on student's behavior, beliefs, and motivation. The system has different modeling in learning style for various kinds of students which help to determine the type of student performance. The system has four learning style. For achieving the most appropriate style, the designer provides the system with two types of questionnaire. One has 80 items, and the other has 40 items. The four learning style are Activists (Do), Reflectors (Review), Theorists (Conclude) and Pragmatist (Plan) **[8]**. This system has three agents that help in improving student's skills. These agents are a personal assistant, tutor, and the mediating agent. This system similar to our system in trying to find learning style depending on student behavior and beliefs while it differs in the environment. This system is e-learning while our system is in the real environment such as the classroom.

VI. Conclusion

The purpose of this application is to increase the students' skills and ability. For accomplishing that, the weakness of the student needs to be known which will enable the system to choose an appropriate style of education. This system depends on the analyzed students' emotion and history for getting the result.

This system will need to improve analyzing part in the feature. To see how the system can do the analysis and what is the proper technique for this purpose. Also, it needs work deeply in algorithm part or coding part for being sure that the analyzing done will. In addition to that, the system needs to mention the teaching style and how this refers to each weakness.

VII. Self-Quiz

1. **My need to take this course now:**
   - High. I need it straight away for a degree, job, or other important reason.
   - Moderate. I can take it on campus later or substitute another course.
   - Low. It's a personal interest that I can postpone it.
2. **Considering my professional and personal schedule, the amount of time I have to work on a course is:**
   - More than adequate for a campus class.
   - The alike as for a class on campus.
   - Less than adequate for a class on campus.
3. **I can classify myself as someone who:**
   - Often I can do things before the dead time.
   - Needs reminding to get things done on time.
   - Puts things off until the last minute.
4. **Feeling that I am one of a class is:**
   - Not particularly necessary for me.
   - Somewhat important to me.
   - Critical to me.
5. **As a reader, I would classify myself as:**
   - Good. I usually understand the text without help.
   - Average. I sometimes need help to understand the text.
   - Slower than average. **[5]**
6. **You have an assignment that you have to do it in group; you prefer to work with**
   - Friends.
   - No matter with whom.
   - Not with friends.
7. **You are going to present your work you prefer to perform in front of**
   - Only Friends.
   - No matter in front who.
   - Only the class and your instructor without other teachers.

Do you believe in yourself? Yes no

Are you happy in your major? Yes no

Is your major mostly what you talk about? Yes no

Do you feel guilty for doing the things that you want to do?Yes no **[9]**

References

[1] J. Beggs and D. Thede. *Designing Web Audi.* Chapter 2,The Science of Sound and Digital Audio. O'Reilly & Associates .2001.

[2] Boré, G., & Peus, S. (1999). *Microphones Methods of Operation and Type Examples.* Berlin: Druck-Centrum Fürst GmbH.

[3] A.Valery.(1999). System Method and Article of Manufacture for Detecting Emotion in Voice Signals through Analysis of a Plurality of Voice Signal Parameters. United States Patent **Publication**. G10L17/00; G10L17/00; (IPC1-7): G10L15/00.

[4] F.Orság. (2010). Speaker Dependent Coefficients for Speaker Recognition. *International Journal of Security & Its Applications*, 4(1), 31-47.

[5] T. WEI-HO, and L. SHIH-JIE. (2010). Speaker Identification in Overlapping Speech. *Journal of Information Science & Engineering,* 26(5), 1891-1903.

[6] D. Xu, H. Wang and.K. Su (2002). Intelligent Student Profiling with Fuzzy Models. *HICSS '02 Proceedings of the 35th Annual Hawaii International Conference on System Sciences (HICSS'02,*3(3), 0-7695-1435-9.

[7] *N.Aghaee and S.Ören (2008). Agents with Personality and Emotional Filters for an E-learning Environment. 2008 Spring Simulation Multiconference (SpringSim'08)- Poster Sessions (SCS-Poster sessions 2008).Ottawa, Canada, April 14 - 17, 2008*

[8] Moisil, (2008). A Model of the Student Behaviour in a Virtual Educational Environment. *International Journal of Computers, Communications & Control*, 3(3), 108-115.

[9] http://www.clt.odu.edu/oso/index.php?src=pe_isdlforme_quiz

10

New Cluster Validation with Input-Output Causality for Context-Based Gk Fuzzy Clustering

Keun-Chang Kwak
Dept. of Control and Instrumentation Engineering
Chosun University, 375 Seosuk-Dong
Gwangju, Korea

***Abstract*—In this paper, a cluster validity concept from an unsupervised to a supervised manner is presented. Most cluster validity criterions were established in an unsupervised manner, although many clustering methods performed in supervised and semi-supervised environments that used context information and performance results of the model. Context-based clustering methods can divide the input spaces using context-clustering information that generates an output space through an input-output causality. Furthermore, these methods generate and use the context membership function and partition matrix information. Additionally, supervised clustering learning can obtain superior performance results for clustering, such as in classification accuracy, and prediction error. A cluster validity concept that deals with the characteristics of cluster validities and performance results in a supervised manner is considered. To show the extended possibilities of the proposed concept, it demonstrates three simulations and results in a supervised manner and analyzes the characteristics.**

Keywords—Cluster Validation; Fuzzy clustering; Gustafson-Kessel clustering; Fuzzy covariance; Context based clustering; Input-output causality

I. Introduction

Intelligent systems that optimize using learning schemes without strict mathematical constraints are a very useful approach to construct modeling in complex environments[3][4]. A clustering approach [1-4][8][11-12] is one of the generic methods for determining the structure and parameters of an initial intelligent system. Once the initial structure and parameters are determined, the system can use various learning mechanisms for optimization. However, the method by which a system performs clustering is an interesting issue in itself [2][8][11]. Pattern recognition is one of the most interesting applications of intelligent systems, especially clustering method is useful approach of them. Clustering is a process in which groups of objects with high similarity, as compared to the members of other groups, are collected as clusters. The concept is highly similar to pattern classification or recognition. Generally, clustering methods perform well in an unsupervised manner to divide input spaces and extract useful information from data sets. This helps to construct intelligent systems [5]. [10] [11] such as neural networks and fuzzy systems that divide an input space into several local spaces, in turn allowing for ease of interpretation. In a clustering algorithm, selecting an appropriate number of clusters is a critical problem. A simple method to identify the proper number of clusters is to select the result that provides best performance. Another approach is to apply a cluster validation [6][7][14][17-19] using cluster parameters after the clustering algorithm is terminated. This method only needs clustering results and does not need any additional information such as performance results. Because of this property, many cluster validations have been proposed by researchers in the field of pattern recognition and widely used. In prior work, a semi-supervised clustering method [9][16] and a supervised clustering approach [10-12] have made use of output information. Additionally, context-based clustering methods [11-13] have used a context membership function, which was generated by a context term as output, and contained an input-output causality. This characteristic provides more quantitative information to perform the clustering. Conventional cluster validity methods induce a fixed value on the cluster validity. The cluster validity, including input-output causality such as the cluster validity of the output, has not yet been studied in a supervised manner. Any proposed cluster validity concept can obtain more flexible criterions when it uses the input-output causality or context information such as a context membership function. This means that when the cluster validity uses more than one cluster validity result, it can attempt to induce more flexible values for the cluster validity to adapt the input-output causality, or it can introduce a performance-dependent criterion. To achieve this, it proposes two combined cluster validity concepts that use the classification accuracy of a classification problem and a cluster validity of the context membership function. Among the cluster validity values, the proposed concept can choose a relative ratio to adjust the importance between the cluster validity of the input-output causalities, such as input/output CV, and performance accuracy. The proposed concept extends the cluster validity criterion to the supervised manner in the context-based clustering. The rest of this paper proceeds as follows. Section 2 describes related research, including clustering methods and cluster validity methods. In section 3, a new cluster validity concept that can be applied in a supervised manner is proposed. Section 4 then presents the results of experimental comparisons between our new cluster validation and previous approaches. In Section 5, the conclusion with a summary is given.

II. The Related Works

In this section, it briefly describes existing clustering methods and cluster validity methods. These methods based on new cluster validity. A context-based clustering method is introduced after our explanations of general clustering. Then, three cluster validity criterions will be used to briefly explicate cluster validities.

A. Unsupervised clustering methods

FCM [3][4] is a representative fuzzy clustering method that uses a partition matrix of the membership function between cluster centers and data sets. It measures similarity as follows:

$$\mu_{ik} = \frac{1}{\sum_{j=1}^{c}\left(\frac{d_{ik}}{d_{jk}}\right)^{\left(\frac{2}{m-1}\right)}} \quad (1)$$

where d_{ik} is the distance between a center c_i and kth data z_k. An m is a fuzzifier and the similarity μ_{ik} is the element of the partition matrix of the membership function. In the process, center c_i is updated by the similarity until a termination criterion is satisfied, as follows:

$$c_i = \frac{\sum_{k=1}^{N}(\mu_{ik})^m x_k}{\sum_{k=1}^{N}(\mu_{ik})^m} \quad (2)$$

Most cluster validity methods primarily use the partition matrix to evaluate the cluster validity.

Gustafson-Kessel (GK) [1][2] clustering uses the fuzzy covariance matrix to adapt elliptical shape cluster sets that use fuzzy covariance information, as shown in following equation:

$$F_i = \frac{\sum_{k=1}^{N}(\mu_{ik})^m (x_k - c_i)(x_k - c_i)^T}{\sum_{k=1}^{N}(\mu_{ik})^m} \quad (3)$$

The matrix A_i is combined by equation (4),

$$A_i = [\rho_i det(F_i)]^{1/n} F_i^{-1} \quad (4)$$

where ρ_i is a predefined constant to set to one. Then, the distance between center c_i and data x_k are measured by the following equation:

$$d_{ik}^{\ 2} = \left(x_k - c_i^{(l)}\right)^T A_i \left(x_k - c_i^{(l)}\right) \quad (5)$$

An updated GK cluster center is calculated as a weighted average by equation (2).

B. Supervised clustering methods

Context-based clustering [11] in a supervised manner uses a context membership function that regards input and output data as causally connected. When a context term, such as output space, can be grouped, connected input spaces are also meaningfully clustered. In the context term, the brief concept of context clusters is shown in Fig. 1. Different shapes are shown because of differences in measurement between simple Euclidean and fuzzy covariance metrics.

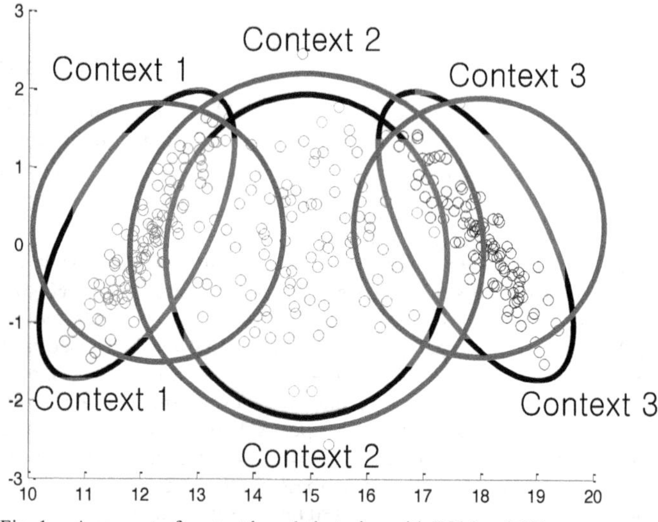

Fig. 1. A concept of context based clustering with FCM and GK

In the unsupervised manner, general similarity is calculated by equation (1). However, a similarity measure of the context clustering in the supervised manner is calculated by equation (6), adding context variable f_k which is induced by data x_k and context membership functions, as shown in Fig. 2.

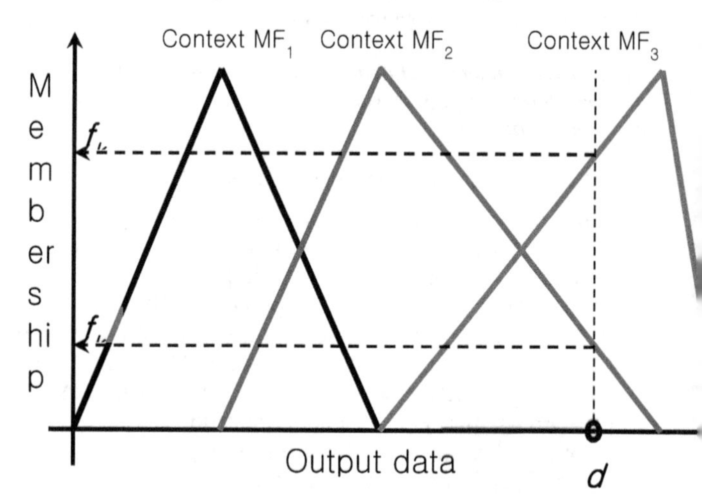

Fig. 2. The concept of context membership function

$$\mu_{ik} = \frac{f_k}{\sum_{j=1}^{c}\left(\frac{d_{ik}}{d_{jk}}\right)^{\left(\frac{2}{m-1}\right)}} \quad (6)$$

As shown Fig. 1, the f_k is induced by the context membership function when kth data is obtained by context membership functions two and three. Then, the equation (6) contains context information using f_k that assumes influencing input-output causality in the supervised manner.

C. Cluster validity

Cluster validity (CV) [6][7][14][18][19] is used to find the optimal number of clusters in a given data set. Bezdek proposed two CVs: the Partition Coefficient (V_{PC}), which minimizes an index value, and Partition Entropy (V_{PE}), which maximizes an index using a partition matrix as follows [6]:

$$V_{PC} = \frac{\sum_{j=1}^{n} \sum_{i=1}^{c} \mu_{ij}^{2}}{n} \tag{7}$$

$$V_{PE} = -\frac{1}{n} \sum_{j=1}^{n} \sum_{i=1}^{c} \mu_{ij} log_a (\mu_{ij}) \tag{8}$$

Xie and Beni [19] also proposed a CV index (VXB) that utilizes compactness and separation to find a minimized validity index, as follows:

$$V_{XB} = \frac{\sum_{i=1}^{c} \sum_{j=1}^{n} \mu_{ij}^{2} \|x_j - c_j\|^2}{n \left(\min_{i \neq k} \|c_i - c_k\|^2 \right)} \tag{9}$$

Kim [6] proposed a CV index (VK) for GK clustering that also finds a minimized validity index, as follows:

$$V_K = \frac{2}{c(c-1)} \sum_{p \neq q}^{c} \sum_{j=1}^{n} \left| c \left[\mu_{\widetilde{F_p}}(x_j) \cap \mu_{\widetilde{F_q}}(x_j) \right] h(x_j) \right| \tag{10}$$

Although there are many interesting extensions to the concept, a full explanation is not our present concern; thus, it limits the discussion to our extension of current CVs in a supervised manner using input-output causality.

III. The Proposed Cluster Validity Method

The proposed cluster validity (CV) concept, which it calls context-based cluster validity (CCV), uses more than two CV considerations, such as a CV of the input space clustering and performance results, or a CV of the context clustering. This means that it extends the conventional CV concept in the unsupervised manner to a supervised CV concept. In the clustering process, it assumes that the output information of the data is already known because clustering based on supervised learning uses the output data, as recognized by the context term.

Throughout the causality, the output is causally correlated with the input. To construct the input clusters, context-based clustering serves advanced information of the causality using f_k that includes a causality degree of input and output clusters, as shown in Fig. 3. There are two criterions of the CV that exist in the model as an input and an output side, respectively.

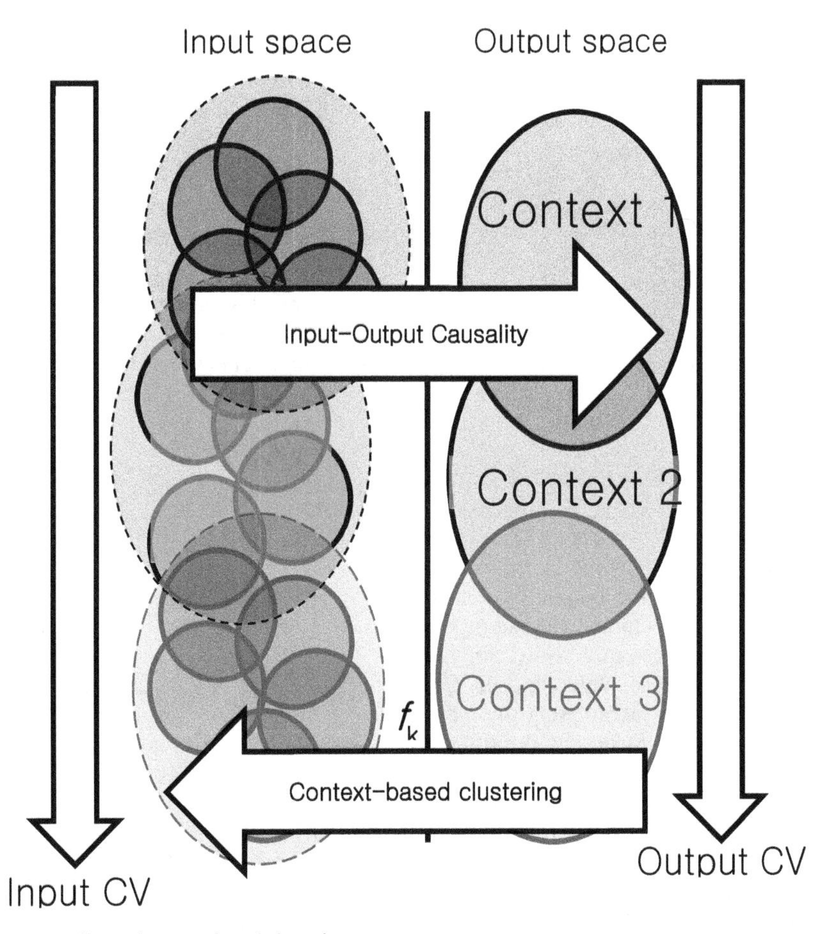

Fig. 3. The concept of input-output causality and context based clustering

The two types of context information are presented. The first is the accuracy (error) of the classification problems. The second is the CV of the partition matrix of the context membership in equation (12). In the classification problems, the context-based clustering method often does not obtain a context membership degree between zero and one. It only includes zero or one. Therefore, it cannot directly obtain the CV of the context membership function and then replace a classification error for adapting the causality. However, the classification error can be estimated easily by comparing the clustering results and the output data, such as class labels in the supervised manner. In case of very small values less than one, it amplifies the error to affect the CV result, with amplification ratio manually decided by minimum error value. This amplification helps to ensure an observed change in the CV curves. Eq. (11-1) contends that an induced new CV includes the CV of the input spaces and the classification error results in the context term. This CV concept influences the new CV result with the error. Despite getting a good input CV result, the proposed concept can have a bad CV value when classification error increases on the context term. In addition, Eq. (11-2) is the form of applying influence parameter α. It can influence an effect ratio of the context term such as error.

$$Proposal = CV\ of\ input \times (1 - error) \times (Amplification) \tag{11-1}$$

$$new\ CV = \alpha \times (CV\ of\ input) + (1 - \alpha)(proposal) \tag{11-2}$$

$$ew\ CV = \alpha \times (CV\ of\ input) + (1 - \alpha)(CV\ of\ context) \tag{12}$$

In Eq. (12), a new cluster validity concept that uses the CV of the context term and adjusts the relative ratio using the variable α is proposed. The parameter α can adjust the influence ratio of the input-output relativity emphasis. Conventional CVs generally calculate a criterion to induce a value that has no possibility of adjustment. In this paper, the variable α is important as it allows us to adjust the influence of the context information. It extends the CV concept from a fixed value of the CV to a choice preference in the scope of the input-output relativity emphasis. When the output data have continuous values and do not have a label index, generating the CV of the context membership function easily allows for the application of the causality. In this case, the proposed CV concept can apply an extended CV evaluation using the input and output CV. In the context-based clustering during the supervised learning, the clustering algorithm generally optimizes the input clusters using an advanced similarity metric with input-output causality. Then, the cluster validity also needs to extend the validity criterions at that environment. It specifies that the first characteristic is input-output causality in supervised settings. The input characteristic is already in existence as the CV. When the context-based clustering algorithm cannot obtain the context membership degree, such as in classification problems that do or do not only belong to the class, it assumes that classification error can replace the context membership function to represent the input-output causality. To apply the context CV, the classification accuracy is used to estimate the context CV of the classification problem. However, when it can obtain the context CV, the proposed concept easily adapts the criterion through an Eq. (12) such that a regression problem is used by the context membership degree, alongside other information to influence the final result.

IV. Experimental Results

In this Section, it used two computer simulations to show the characteristics of the proposed concept. The simulations using MATLAB 12, which was run on a Windows 7 machine with an i7 2.80 GHz CPU and 16 GB of DDR3 RAM is performed. The three simulation data sets, including two synthetic classification problems and one real data set are used. The two synthetic data sets were generated by a random selection method that intentionally forced shapes to obtain the elliptical geometric structure. The outputs were composed of three and five class labels. The real data set was downloaded from the UCI machine learning repository. This data set has 506 instances and fourteen attribute numbers, including an output that comprises the median value of owner-occupied homes in $1000. Here it used two input attributes: the weighted distance to five Boston employment centers, and the lower status of the population. The synthetic data distribution is shown in Fig. 4. It has five groups with various shapes, distributions, and densities. The three class problem is also from the same data set where two central classes are merged into a new class and two-sided small classes are also merged into a new class.

A. Cluster validity Cluster validity in classification problems

The index values of five and three (5, 3 classes) to represent the cluster validity of the input space and the classification error between the inferred cluster label and the real output label are used.

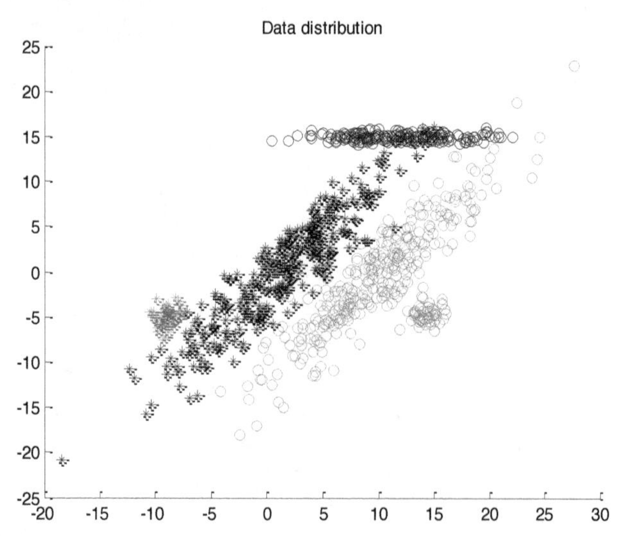

Fig. 4. Synthetic data distribution

To compare the change of the CV, all performance and CV results are normalized in Fig. 5 when the FCM algorithm is performed. The thick black line is a classification result that increases the classification performance when the number of clusters is increased. The thin red line is the cluster validity result of [19]. The dotted red line is the result of Eq. (11-1). The thick red line is a result of Eq. (11-2), which applies the input CV results and classification result with an influence parameter α of 0.5. The blue lines are similar to the CV of the [6]. Regarding the blue lines, the CV of the input and applied CV is a different curve. This means that if it knows the classification error then it can change the number of the clusters to fit the performance.

Figs. 5 and 6 show the CV results when FCM and GK clustering are performed. The cluster number scope is two to fifteen. In the three class problem, the Vk and our proposed concept are more different when the cluster number is increased. It is also possible to see the black line of the classification accuracy that influenced the proposed CV curve. In the five class problem, the cluster number is started from five to twenty. Figs. 7 and 8 show the CV results when FCM and GK clustering are performed.

B. Cluster validity in a regression problem

The CV results of the Boston housing regression [15] problem at the CFCM are shown in Fig. 9. The thick blue line is an input CV and the other lines are influenced by a CV of the context term as output and the influence parameter α in equation (12). The figure shows different results when influence parameter α is changed. As shown in Fig. 9, when the influence parameter α is already 0.5, a criterion value of the proposed concept is less than the input CV value. This means that the final determination including the CCV can change the optimal cluster number.

As illustrated in Fig. 10, it shows the result of the GK clustering when the influence parameter α is changed. It seems to have little effect compared with the FCM.

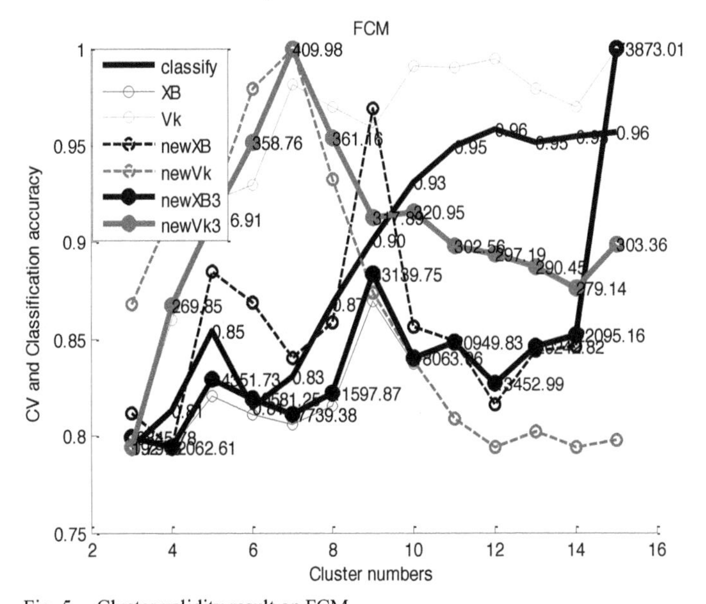

Fig. 5. Cluster validity result on FCM

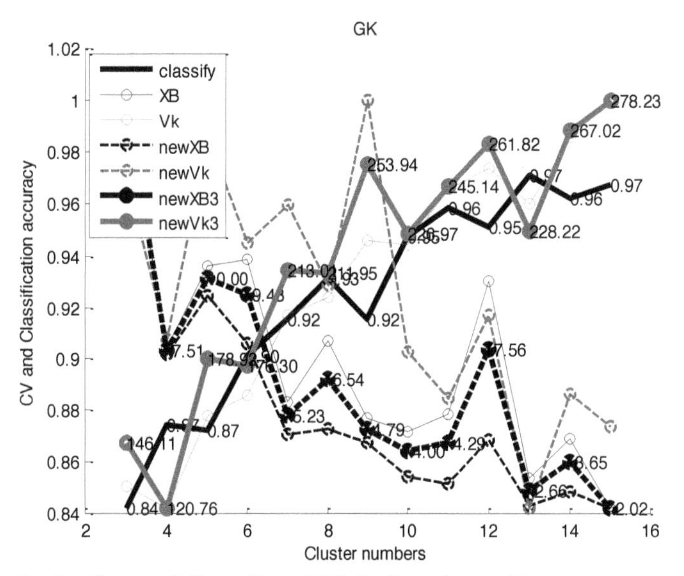

Fig. 6. Cluster validity results on GK in the three class problem

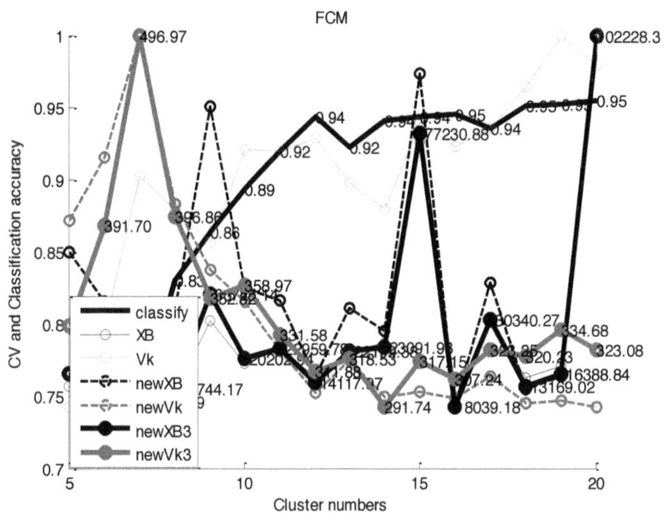

Fig. 7. Cluster validity result on FCM in the five class problem

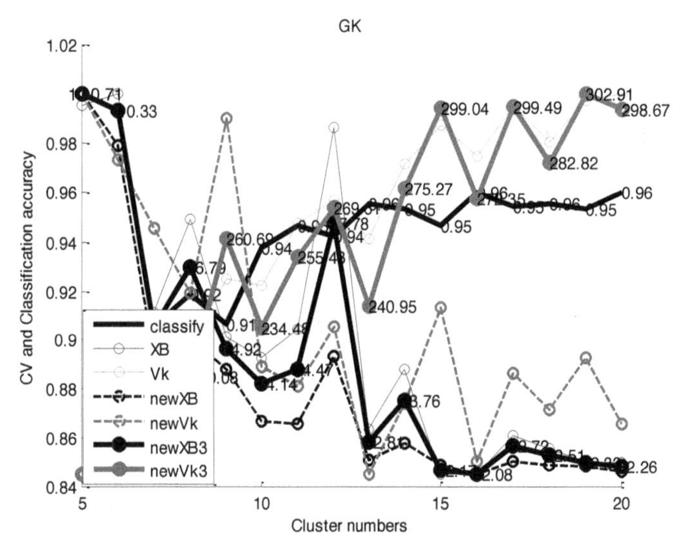

Fig. 8. Cluster validity result on GK in the five class problem

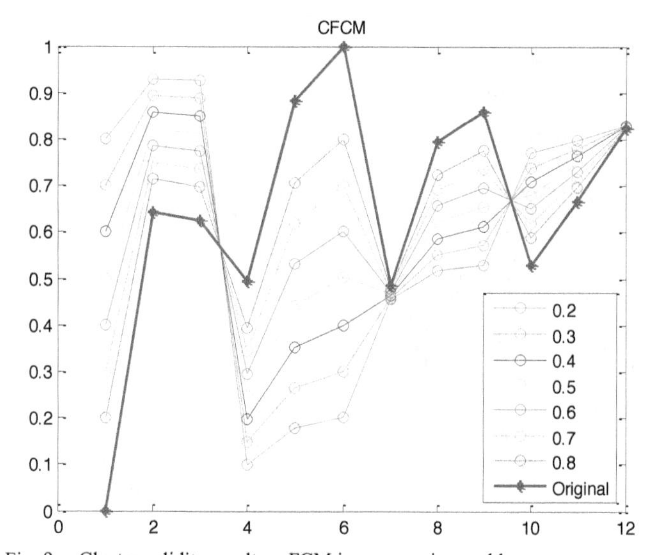

Fig. 9. Cluster validity result on FCM in a regression problem

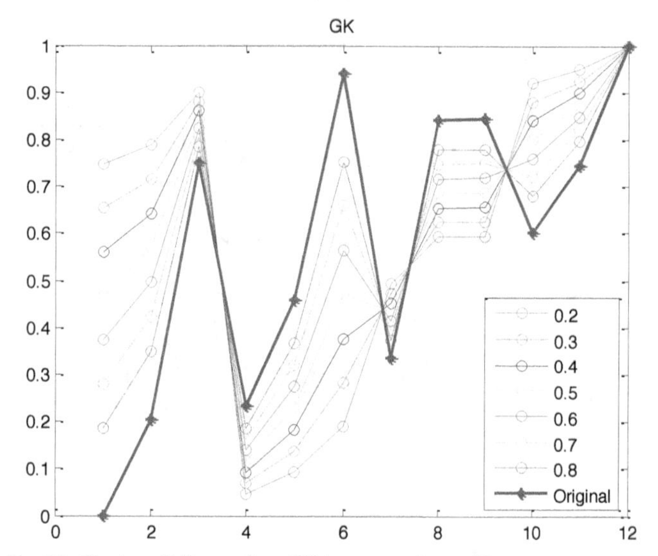

Fig. 10. Cluster validity result on GK in a regression problem

TABLE I. COMPARISON RESULTS OF CV

Case	Context number	Cluster number in a context	Cluster number	Input CV	Proposed CV
1	2	2	4	0	0.4
2	2	3	6	0.6425	0.7855
3	2	4	8	0.6235	0.7741
4	3	2	6	0.4921	**0.2952**
5	3	3	9	0.8825	0.5295
6	3	4	12	1.00	0.6001
7	4	2	8	**0.4849**	0.4698
8	4	3	12	0.7943	0.6554
9	4	4	16	0.8574	0.6933
10	5	2	10	0.5286	0.6492
11	5	3	15	0.6641	0.7304
12	5	4	20	0.8242	0.8266

Comparison of the values in Table 1 indicates that the best optimal cluster number is eight when only the input CV is used. However, in our concept, the best optimal cluster number is six at three context clusters. It has two cases of six clusters with different CV values at cases two and four.

As indicated by the CV results, it attempts to show the difference between conventional CV approaches and our proposed concept. Our approach has two advanced characteristics. First, it extends the cluster validity concept from the unsupervised to the supervised setting. In addition, introducing influence parameter α provides a more varied range of possible extensions.

V. CONCLUSIONS

In this paper, a new cluster validation method for context-based clustering in a supervised manner has developed. By adding more information to the context term, the cluster validation concept extends the possible application from unsupervised to supervised settings. Applying an input-output causality and an influence parameter provide wider choice in the cluster validity. This approach easily adapts to the context-based clustering. Conventional cluster validity values tend to have fixed values or constants and do not consider the input-output causality. Our proposed cluster validity extends this constancy to offer greater flexibility by using various elements and adjustments, such as α. Instead of constancy in the unsupervised settings, the proposed concept has sufficient scope to determine the most suitable number of clusters. In the instruction of an intelligent system using clustering, our approach can provide more marginal choice to determine the best overall parameters. Context-based clustering can adapt various context membership functions to improve performance. Thus, applying various membership functions in context terms and, later, analyzing the results of cluster validity will be very interesting opportunities for further research. Future work should also include applying the semi-supervised clustering and related works.

ACKNOWLEDGEMENTS

This research was supported by Basic Science Research Program through the National Research Foundation of Korea (NRF) funded by the Ministry of Science, ICT and Future Planning (NRF-2013R1A1A2012127)

REFERENCES

[1] I. Gath, A. B. Geva, "Unsupervised optimal fuzzy clustering", *IEEE Trans on Pattern Analysis and Machine Intelligence* Vol. 11, No. 7, pp. 778-780, 1989.

[2] D. E. Gustafson, W. C. Kessel, "Fuzzy clustering with a fuzzy covariance matrix", *IEEE Conference on Decision and Control including the 17th Symposium on Adaptive Processes*, Vol. 17, pp. 761-766, 1978.

[3] S. Haykin, Neural Networks: A Comprehensive Foundation 2nd. Prentice Hall, 1999.

[4] J. S. R. Jang, C. T. Sun, and E. Mizutani, Neuro-Fuzzy and Soft Computing: A Computational Approach to Learning and Machine Intelligence, Prentice Hall, 1997.

[5] S. S. Kim, H. J. Choi, K. C. Kwak, "Knowledge extraction and representation using quantum mechanics and intelligent models", *Expert System with Applications*, Vol. 39, No. 3, pp. 3572-3581, 2012.

[6] Y. I. Kim, D. W. Kim, D. H. Lee, K. H. Lee, "A cluster validation index for GK cluster analysis based on relative degree of sharing", *Information Sciences*, Vol. 168, No. 4, pp. 225-242, 2004.

[7] S. H. Kwon, "Cluster validity index for fuzzy clustering", *Electronics Letters*, Vol. 34, No. 22, pp. 2176-2177, 2002.

[8] R. Krishnapuram, J. Kim, "A Note on the Gustafson-Kessel and Adaptive Fuzzy Clustering Algorithms", *IEEE Trans on. Fuzzy Systems*, Vol. 7, No. 4, pp. 453-461, 1999.

[9] M. H. C. Law, A. Topchy, A. K. Jain, "Clustering with Soft Group Constraints. Structural, Syntactic, and Statistical Pattern Recognition", *Lecture Notes in Computer Science*, Vol. 3138, pp. 662-670, 2004.

[10] W. Lu, W. Pedrycz, X. Liu, J. Yang, P. Li, "The modeling of time series based on fuzzy information granules", *Expert Systems with Applications*, Vol. 41, No. 8, 3799-3808, 2014.

[11] W. Pedrycz, "Conditional fuzzy C-Means", *Pattern Recognition Letters*, Vol. 17, pp. 625-632, 1996.

[12] W. Pedrycz, "Conditional fuzzy clustering in the design of radial basis function neural networks", *IEEE Trans. on Neural Networks*, Vol. 9, No. 4, pp.745-757, 1999.

[13] W. Pedrycz, K. C. Kwak, "Linguistic models as a framework of user-centric system modeling", *IEEE Trans. on Systems, Man, and Cybernetics-Part A,* Vol. 36, No. 4, pp.727-745, 2006.

[14] B. Rezaee, "A cluster validity index for fuzzy clustering.", *Fuzzy Sets and Systems*, Vol. 161, No. 23, pp. 3014-3025, 2010

[15] D. A. Belsley, E. Kuh, R. E. Welsh, *Regression Diagnostics: Identifying Influential Data and Source of Collinearity*, John Wiley & Sons, Inc, 1980.

[16] K. Wagstaff, C. Cardie, S. Rogers, S. Schroedl, "Constrained K-means Clustering with Background Knwledge", *Proceeding of the Eighteenth International Conference on Machine Learning*, pp.577-584. 2001.

[17] W. Wang, Y. Zhang, "On fuzzy cluster validity indices", *Fuzzy Sets and Systems* , Vol. 158, No. 19, pp. 2095-2117, 2007.

[18] K. L. Wu, M. S. Yang, "A cluster validity index for fuzzy clustering", *Pattern Recognition Letters*, Vol. 26, No. 9, pp. 1275-1291, 2005.

[19] X. L. Xie, G. Beni, "A validity measure for fuzzy clustering", *IEEE Trans on Pattern Analysis and Machine Intelligence*, Vol. 13, No. 8, pp. 841-847, 1991.

Proposal of Tabu Search Algorithm Based on Cuckoo Search

Ahmed T. Sadiq Al-Obaidi
Department of Computer Sciences
University of Technology
Baghdad, Iraq

Ahmed Badre Al-Deen Majeed
Quality Assurance Department
University of Baghdad
Baghdad, Iraq

Abstract—**This paper presents a new version of Tabu Search (TS) based on Cuckoo Search (CS) called (Tabu-Cuckoo Search TCS) to reduce the effect of the TS problems. The proposed algorithm provides a more diversity to candidate solutions of TS. Two case studies have been solved using the proposed algorithm, 4-Color Map and Traveling Salesman Problem. The proposed algorithm gives a good result compare with the original, the iteration numbers are less and the local minimum or non-optimal solutions are less.**

Keywords—Tabu Search; Cuckoo Search; Heuristic Search; Neighborhood Search; Optimization; 4-Color Map; TSP

I. Introduction

A huge collection of optimization techniques have been suggested by a crowd of researchers of different fields; an infinity of refinements have made these techniques work on specific types of applications. All these procedures based on some common ideas and are furthermore characterized by a few additional specific features. Among the optimization procedures, the iterative techniques play an important role; for most optimization problems no procedure is known in general to get directly an "optimal" solution [1].

The general steps of an iterative procedure consists in constructing from a current solution i to the next solution j and in checking whether one should stop there or perform another step. Neighborhood search methods are iterative procedures in which a neighborhood $N(i)$ is defined for each feasible solution i, and the next solution j is searched among the solutions in $N(i)$ [2,3,4].

The origin of the Tabu Search (TS) went back to the 1970s and the modern form of TS was derived independently by Glover and Hansen [4,5]. The hybrids of the TS have improved the quality of solutions in numerous areas such as scheduling, transportation, telecommunication, resource allocation, investment planning. The success of the TS method for solving optimization problems was due to its flexible memory structures which allowed the search to escape the trap of local optima and permitted to search the forbidden regions and explored regions thoroughly [2].

Cuckoo search was inspired by the obligate brood parasitism of some cuckoo species by laying their eggs in the nests of other host birds (of other species). Some host birds can engage direct conflict with the intruding cuckoos. For example, if a host bird discovers the eggs are not their own, it will either throw these alien eggs away or simply abandon its nest and build a new nest elsewhere [7].

The objective of this paper is to improve the tabu search using the nature-inspired algorithm which cuckoo search. The outline of this paper is as follows. Section 2 describes the concepts of Tabu Search method with two basic algorithms. Section 3 includes the concepts of Cuckoo Search. Section 4 deals with proposal of Tabu-Cuckoo Search (TCS) algorithm. Section 4 presents 2 case studies which are solved by TCS and TS with experimental results of each one. Section 5 includes the conclusions of this paper.

II. Tabu Search

Tabu Search (TS) is a meta-heuristic search which is designed to cross the boundaries of feasibility and search beyond the space of local optimality. The use of flexible memory based structures is the center strategy of the TS method [7]. While most exploration methods keep in memory essentially the value $f(i^*)$ of the best solution i^* visited so far, TS will also keep information on the itinerary through the last solution visited. Such information will be used to guide the move from i to next solution j to be chosen in $N(i)$. The role f the memory will be to restrict the choice of some subset of $N(i)$ by forbidding for instance moves to some neighbor solutions [8]. It would therefore be more appropriate to include TS in a class of procedures called dynamic neighborhood search techniques [7].

Formally let us consider an optimization problem in the following way : given a set S of feasible solutions and a function $f : S \rightarrow \Re$, find some solution i^* in S such that $f(i^*)$ is acceptable with respect to some criterion (or criteria). Generally a criterion of acceptability for a solution i^* would be to have $f(i^*) \leq f(i)$ for every i in S. In such situation TS would be an exact minimization algorithm provided the exploration process would guarantee that after a finite number of steps such an i^* would be reached [5,7].

In most contexts however no guarantee can be given that such an i^* will be obtained; therefore TS could simply be viewed as an extremely general heuristic procedure. Since TS will in fact include in its own operating rules some heuristic techniques, it would be more appropriate to characterize TS as a *metaheuristic*. Its role will often be to guide and to orient the search of another (more local) search procedure [8].

As a first step towards the description of TS, the classical descent method will be illustrated [1]:

Step 1: Choose an initial solution *i* in *S*.
Step 2: Generate a subset V^* of solution in *N(i)*.
Step 3: Find a best *j* in V^* (i.e. such that $f(i) \le f(k)$ for any *k* in V^*) and set *i* to *j*.
Step 4: If $f(j) \ge f(i)$ Then stop, Else go to Step 2.

In a straightforward descent method, we would generally take $V^*=N(i)$. However this may often be too time-consuming: an appropriate choice of V^* may often be a substantial improvement.

Except for some special cases of convexity, the use of descent procedures is generally frustrating since the researchers are likely to be trapped in a local minimum which may be far (with respect to the value of *f*) from a global minimum [1,2].

As soon as non-improving moves are possible, the risk visiting again is a solution and more generally of cycling is presented. This is the point where the use of memory is helpful to forbid moves which might lead to recently visited solutions. If such memory is introduced we may consider that the structure of *N(i)* depend upon the itinerary and hence upon the iteration k; so we may refer to *N(i,k)* instead of *N(i)*. With these modifications in mind we may attempt to formalize an improvement of the descent algorithm in a way which will bring it closer to the general TS procedure. It could be stated as follows (i^* is the best solution found so far and *k* the iteration counter) [1,2]:

Step 1: Choose an initial solution *i* in *S*. Set $i^*=i$ and k=0.
Step 2: Set $k=k+1$ and generate a subset V^* of solution in *N(i,k)*.
Step 3: Choose a best *j* in V^* (with respect to *f* or to some modified function f') and set $i = j$.
Step 4: If $f(i) < f(i^*)$ Then set $i^*=i$.
Step 5: If a stopping condition is met Then stop, Else go to Step 2.

Observe that the classical descent procedure is included in this formulation (the stopping rule would simply be $f(i) \ge f(i^*)$ and i^* would always be the last solution).

In TS some immediate stopping conditions could be the following [1, 2, 9]:

- $N(i,k+1)=\varnothing$.
- k is larger than the maximum number of iterations that allowed.
- the number of iterations since the last improvement of i* is larger than a specified number.
- evidence can be given than an optimum solution has been obtained.
- tabu list is full.
- no improved solutions.

While these stopping rules may have some influence on the search procedure and on its results, it is important to realize that the definition of *N(i,k)* at each iteration *k* and the choice of V^* are crucial [2].

The definition N(i,k) implies that some recently visited solutions are removed from N(i); they are considered as tabu solutions which should be avoided in the next iteration. Such memory based on recent will partially prevent cycling. For instance keeping at iteration k a list T (tabu list) of the last |T| solutions visited will prevent cycles of size at most |T|. In such case N(i,k)=N(i)-T will be taken. However this list T may be extremely impractical in use; therefore the exploration process in S in terms of moves from one solution to the next [1,2]. In addition to, there are other versions of TS algorithms, but the above is the classical.

III. Cuckoo Search

CS is a heuristic search algorithm which has been proposed recently by Yang and Deb [10]. The algorithm is inspired by the reproduction strategy of cuckoos. At the most basic level, cuckoos lay their eggs in the nests of other host birds, which may be of different species. The host bird may discover that the eggs are not its own and either destroy the egg or abandon the nest all together. This has resulted in the evolution of cuckoo eggs which mimic the eggs of local host birds. To apply this as an optimization tool, Yang and Deb used three ideal rules [10, 11]:

1) Each cuckoo lays one egg, which represents a set of solution co-ordinates, at a time and dumps it in a random nest;

2) A fraction of the nests containing the best eggs, or solutions, will carry over to the next generation;

3) The number of nests is fixed and there is a probability that a host can discover an alien egg. If this happens, the host can either discard the egg or the nest and this result in building a new nest in a new location. Based on these three rules, the basic steps of the Cuckoo Search (CS) can be summarized as the pseudo code shown as below [10, 11, 12].

Cuckoo Search via Levy Flight Algorithm

Input: Population of the problem;
Output: The best of solutions;

```
Objective function f(x), x = (x1, x2, ...xd)^T
Generate initial population of n host nests xi
                                  (i = 1, 2, ..., n)
While (t <Max Generation) or (stop criterion)
    Get a cuckoo randomly by Levy flight
    Evaluate its quality/fitness Fi
    Choose a nest among n(say,j)randomly
    If (Fi > Fj) replace j by the new solution;
    A fraction(pa) of worse nests are abandoned and new
     ones are built;
    Keep the best solutions (or nests with quality
     solutions);
    Rank the solutions and find the current best;
    Pass the current best solutions to the next generation;
End While
```

When generating new solution $x^{(t+1)}$ for, say cuckoo *i*, a Levy flight is performed

$$x^{(t+1)}_i = x(t)_i + \alpha \oplus Levy(\beta) \ \ldots\ldots \ (1)$$

where $\alpha > 0$ is the step size which should be related to the scales of the problem of interests. In most cases, we can use α = 1. The product $\oplus$ means entry-wise walk while multiplications. Levy flights essentially provide a random walk while their random steps are drawn from a Levy Distribution for large steps

$$Levy \sim u = t^{-1-\beta} \quad (0 < \beta \leq 2) \ldots\ldots\ldots (2)$$

this has an infinite variance with an infinite mean. Here the consecutive jumps/steps of a cuckoo essentially form a random walk process which obeys a power-law step-length distribution with a heavy tail. In addition, a fraction *pa* of the worst nests can be abandoned so that new nests can be built at new locations by random walks and mixing. The mixing of the eggs/solutions can be performed by random permutation according to the similarity/difference to the host eggs.

IV. Proposal of Tabu Search Algorithm Based on Cuckoo Search

Generally, in the most heuristic search algorithms, the guarantee of finding the optimal solutions is the big problem. Also, local minimum (or maximum) represent the second big problem. Therefore, the heuristic search algorithms still in continuous developing. In this work, an attempt to improve the performance of TS using CS which is provides more diversity to candidate solutions of TS. CS will call in the TS when there are no more good solutions in TS. Initially, CS will be work with best solutions list (*B*) and replace the old solutions of tabu list by the CS solutions to provide a good diversity to TS candidate solutions. In other words, any iterative exploration process should in some instance accept also non-improving moves from i to j in V^* (i.e. $f(j) > f(i)$) if one would like to escape from local minimum, CS does this. Therefore the proposed version of TS will be more heuristic and robust to find the optimal solution or at least reduce the local minimum problem. The suggested TCS as following:

Step 1: Choose an initial solution i in S. Set $i^*=i$ and $k=0$.
Step 2: Set $k=k+1$ and generate a subset V^* of solution in $N(i,k)$.
Step 3: Choose a best j in V^* and set $i = j$.
Step 4: Select best subset from $N(i,k)$ add in B.
Step 5: If there is no best solution Then call the Cuckoo Search with best subset from Tabu List.
Step 6: Select the best solutions from Cuckoo Search output to add in the Tabu List.
Step 7: If a stopping condition is met Then stop, Else go to Step 2.

where B represent the currently best solutions list which is contain the best neighbors of V^*, so the algorithm can recover the best previous states when the route of behavior far of the goal. The update step of B means delete the used neighbors and rearrange the others. In the next section illustrates the performance of TCS algorithm compare with others TS algorithms.

V. Case Studies and Experimental Results

Two standard optimization problems were used to test the proposal algorithm and to compare their performances with the original algorithm.

A. 4-Color Map Problem

The celebrated 4 Color Map Theorem states that any map in the plane or on the sphere can be colored with only four colors such that no two neighboring countries are of the same color. The problem has a long history and inspired many people (including many non-mathematicians and in particular countless high school students) to attempt a solution [13].

The proof of the four color theorem by Haken and Appel [14] was so involved it required computational support to complete. It is well known that determining if a graph can be colored by a certain number of colors is NP-complete, but it is also known that even approximating the chromatic number of a graph is NP-hard [15]. There exist two main categories of algorithms: *successive augmentation algorithms* [16], which color a graph one vertex at a time, disallowing vertices from being re-colored and *iterative improvement algorithms*, which allow backtracking and re-coloring. Leighton's [17] RLF algorithm is an example of the first and Tabu searches and genetic algorithms are examples of the second [18].

In 4-color map problem there is a vector (N), where N is the number of cities in the map. An adjacency array of dimension NxN is used to identify the neighborhood of adjacent cities. The neighborhood search operator used is simply swapping two randomly chosen points.

B. Traveling Salesman Problem TSP

TSP is one of the major success stories for optimization because of its simplicity and applicability (or perhaps simply because of its intriguing name), the TSP has for decades served as an initial proving ground for new ideas related to both these alternatives. These new ideas make the TSP an ideal subject for a case study [19].

The origins of the Traveling Salesman Problem (TSP) are somewhat mysterious. It is a classical combinatorial optimization problem and can be described as follows: a salesman, who has to visit clients in different cities, wants to find the shortest path starting from his home city, visiting every city exactly once and ending back at the starting point. More formally [19]:

Given a set of n nodes and costs associated with each pair of nodes, find a closed tour of minimal total cost that contains every node exactly once.

In other words, a set $\{c_1, c_2, \ldots, c_N\}$ of *cities* is given and for each pair $\{c_i, c_j\}$ of distinct cities a *distance* $d(c_i, c_j)$. The goal is to find an ordering Π of the cities that minimizes the quantity

$$\sum_{i=1}^{N-1} d(c_{\Pi(i)}, c_{\Pi(i+1)}) + d(c_{\Pi(N)}, c_{\Pi(1)})$$

This quantity is referred to as the *tour length*, since it is the length of the tour a salesman would make when visiting the cities in the order specified by the permutation, returning at the end to the initial city. The concentrated in this paper would be on the *symmetric* TSP, in which the distances satisfy [19]:

$$d(c_i, c_j) = d(c_j, c_i) \text{ for } 1 \leq i,\ j \leq N$$

In computing terms the problem can be represented by a graph where all the nodes correspond to cities and the edges between nodes correspond to direct roads between cities [19].

In 4-color map problem there is a vector (N), where N is the number of cities in the tour. An adjacency array of dimension NxN is used to identify the neighborhood of adjacent cities. The neighborhood search operator used is simply swapping two randomly chosen points.

C. Results

The researchers of TS have been proposed several modifications and hybrids algorithms with other techniques, one of these are Simulated Annealing Tabu Search (SATS) [20]. In this paper the proposed TCS will be compared with standard TS and SATS to illustrate the performance of each one.

In this paper, results of average 10 independent runs for all of these algorithms have proved that all of these algorithms are good technique capable of finding solutions close to the optimum, but a local minimum problem occur in very special cases. Results indicate that the proposal algorithm TCS have a faster convergence than the original TS and SATS.

Figure 1 illustrates the curve of number of iteration with number of cities in 4-color map problem in only solved cases using TS, SATS and TCS. Figure 2 illustrates the number of local minimum non-optimal solutions occur with number of cities in 4-color map problem using TS, SATS and TCS. Figure 3 illustrates the curve of number of iteration with number of cities in TSP in only solved cases using TS, SATS and TCS. Figure 4 illustrates the number of local minimum and non-optimal solutions occur with number of cities in TSP using TS, SATS and TCS.

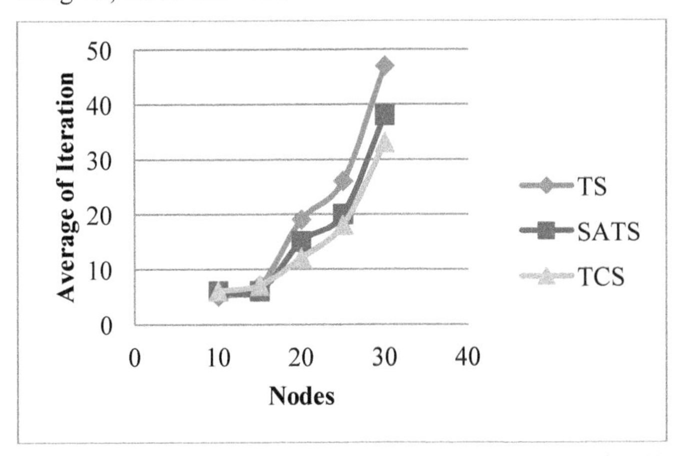

Fig. 1. Average of No. of Iterations for 4-Color Map Problem Using TS, SATS and TCS

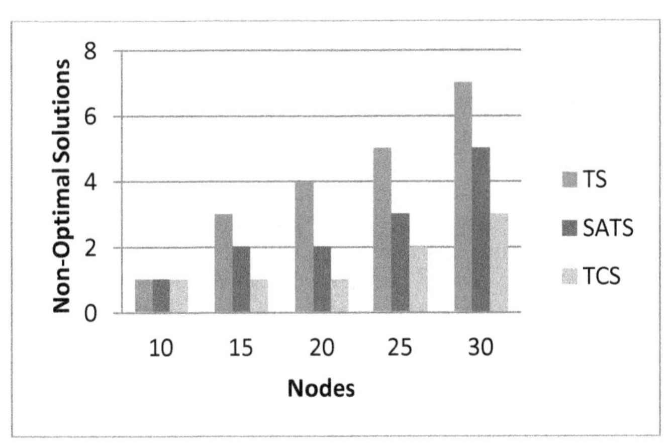

Fig. 2. Average of Non-Optimal Solutions for 4-Color Map Problem Using TS, SATS and TCS

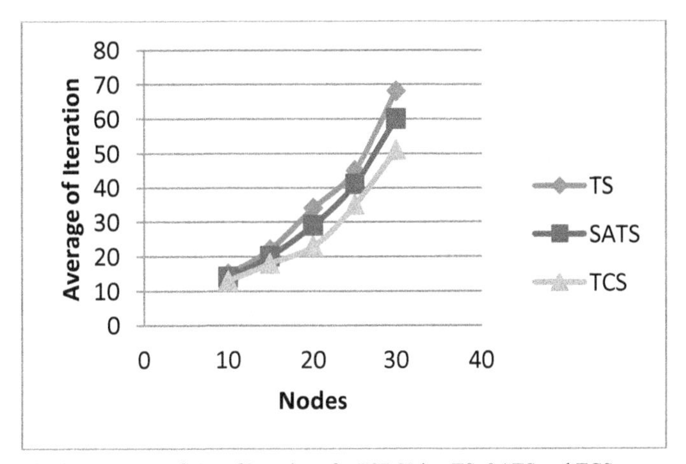

Fig. 3. Average of No. of Iterations for TSP Using TS, SATS and TCS

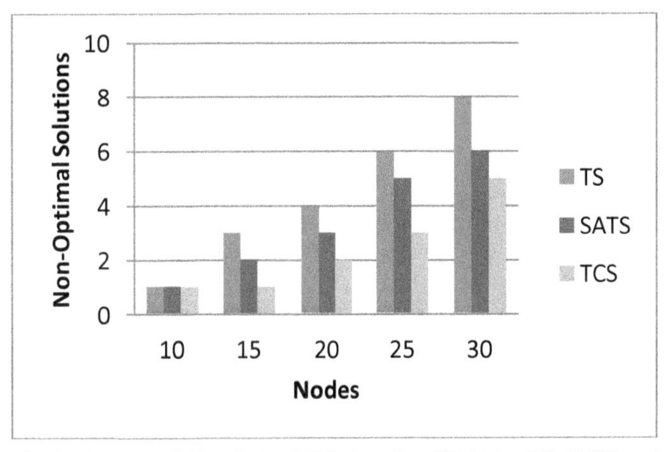

Fig. 4. Average of Non-Optimal Solutions for TSP Using TS, SATS and TCS

VI. Conclusions

The presented approach TCS is an important version of TS. TCS can increase the performance of optimal solutions finding, also, it can reduce the non-optimal solutions and local minimum problem. TCS depends on storing the best neighbors in the currently best solutions list to use these solutions in the CS to for improving whenever the algorithm in local minimum or cannot find the new best neighbor. The suggested approach achieves two important features of methods' searching which are called intensification and diversification. TCS gives less iteration numbers compare with TS and SATS. Also it has been reduced the non-optimal solutions and local minimum problem.

References

[1] A. Hertz, E. Taillard and D. de Werra, "*A Tutorial on Tabu Search*", EPFL, 1995.

[2] F. Glover, M. Laguna, A. Hertz, E. Taillard and D. de Werra, "*Tabu Search*", Annals of Operation Research, Vol. 41, 1993.

[3] F. Glover, "*Tabu Search, Part 2*", ORSA Journal on Computing 2, pp. 4-32, 1990.

[4] P. Hansen, and N. Mladenović, N., "*Variable Neighborhood Search: Principles and Applications*", European Journal of Operational Research, 130, pp. 449-467, 2001.

[5] F. Glover, "*Tabu Search, Part 1*", ORSA Journal on Computing 1, pp. 190-206, 1989.

[6] R. B. Payne, M. D. Sorenson, and K. Klitz, "*The Cuckoos*", Oxford University Press, (2005).

[7] F. Glover and M. Leguna, "*Tabu Search*", Kluwer Academic Publisher, 1997.

[8] A. Hertz and D. de Werra, "*The Tabu Search Metaheuristic : how we used it*", Annals of Mathematics and Artificial Intelligence 1, pp. 111-121, 1990.

[9] D. Deng, J. Ma and H. Shen, "*A Simple and Efficient Tabu Search Heuristics for Kirkman Schoolgirl Problem*", Technical Report, University of Turku, Finland, 2005.

[10] X. S. Yang and S. Deb, "*Cuckoo Search via Lévy Flights*". World Congress on Nature & Biologically Inspired Computing (NaBIC 2009). IEEE Publications. pp. 210–214, December, 2009.

[11] H. Zheng and Y. Zhou, "*A Novel Cuckoo Search Optimization Algorithm Base on Gauss Distribution*", Journal of Computational Information Systems 8: 10, 4193–4200, 2012.

[12] Xin-She Yang, "*Cuckoo Search and Firefly Algorithm*", Springer Press, 2014.

[13] Peter, Alfeld. "*Bivariate Splines and the Four Color Map Problem*", http://www.math.utah.edu/~alfeld/talks/S13/4CMP.html

[14] Wilson. Robin. "*Four Colors Suffice*", Princeton University Press, 2000.

[15] Garey. Johnson, D. S., "*Computers and Intractability: A Guide to the Theory of NP-Completeness*", San Francisco: Freeman, 1977.

[16] Lewandowski, Gary. Condon, Anne. "*Experiments with Parallel Graph Coloring Heuristics and Applications of Graph Coloring*". DIMACS Series in Discrete Mathematics, DIMACS ,1994.

[17] Leighton, F T. "*A Graph Coloring Algorithm for Large Scheduling Problems*", Journal of Research of the National Bureau of Standards, Vol. 84, No. 6, pp 489-506, 1979.

[18] Palmer, Daniel. Kirschenbaum, Marc. Shifflet, Jason. Seiter, Linda. "*Swarm Reasoning*", www.jcu.edu/math/swarm/papers/SIS2005.pdf

[19] Gaertner Dorian. "*Natural Algorithms for Optimisation Problems*". M.Sc. Thesis, Imperial College, 2004.

[20] A. Lim, B. Bodrigus and J. Zhang, "*Tabu Search Embedded Simulated Annealing for Shortest Route Cut and Fill Problem*", Journal of Operations Research Society, Vol. 56, No. 7, pp. 816-824, July 2005.

12

Thresholding Based Method for Rainy Cloud Detection with NOAA/AVHRR Data by Means of Jacobi Iteration Method

Kohei Arai[1]
Graduate School of Science and Engineering
Saga University
Saga City, Japan

***Abstract*—Thresholding based method for rainy cloud detection with NOAA/AVHRR data by means of Jacobi iteration method is proposed. Attempts of the proposed method are made through comparisons to truth data which are provided by Japanese Meteorological Agency: JMA which is derived from radar data. Although the experimental results show not so good regressive performance, new trials give some knowledge and are informative. Therefore, the proposed method suggests for creation of new method for rainfall area detection with visible and thermal infrared imagery data.**

***Keywords*—Jacobi itteration method; Multi-Variiate Regressive Analysis; AVHRR; Rainfall area detection; Rain Radar**

I. INTRODUCTION

Rainfall area detection with satellite based visible and thermal infrared sensor data is tough issue because the visible and thermal infrared sensor data represent just cloud surface reflectance and temperature. In general, the rainy clouds which cause rainfall can be divided into two kinds, nimbostratus and cumulonimbus. The reflectance and the temperature at the top of the cumulonimbus are relatively high and extremely cold, respectively because the height of the cumulonimbus is quite high. On the other hand, the reflectance and the temperature of the nimbostratus are comparatively low and relatively warm, respectively because the height of the nimbostratus is comparatively low. Meanwhile, no rainy cloud types show very similar characteristics in terms of cloud top reflectance and temperature. Therefore, it is extremely difficult to discriminate between rain and no rain clouds.

There are 10 types of clouds which include four types of cumulus, cumulonimbus, stratus, stratocumulus in the lower cloud, three types of nimbostratus high-rise clouds and high cumulus clouds in the middle clouds, in the high-rise clouds three types of cirrus, cirrocumulus, and cirrostratus. Moreover, these clouds are overlapped sometime. Therefore, it is tough to discriminate rainy clouds by using only reflectance and temperature at the top of the clouds.

Microwave radiometer data represent cloud liquid in rainy cloud. Therefore, some methods for detecting rainy clouds with microwave radiometer data have been proposed already [1]-[9]. On the other hand, limb sounding data also represent some extent of rainy clouds information. Therefore, some methods for rainy cloud detection based on limb sounding data have also been proposed so far [10]-[16].

The rainy cloud detection method proposed here is based on thresholding of visible and thermal infrared radiometer data by means of Jacobi iteration method. The proposed method is to be compared to the multiple linear regressive analyses by only using visible and thermal infrared data observed from space. In the method, Probability Density Function: PDF of the visible and thermal infrared data are calculated. Then the PDF is approximated with the best fit ideal normal distribution. After that, the visible and thermal infrared data are binarized (0 denotes no rain, and 1 means rain) with the most appropriate threshold determined by means of Jacobi iteration method.

The proposed method is described followed by experiments. The experimental results are validated with the posterior created weather maps by using rainfall radar data and the other meteorological data in the following section followed by conclusion with some discussions.

II. PROPOSED METHOD

A. Discrimination of Cloud Types

Within 10 types of clouds, cumulonimbus and nimbostratus clouds are major concern because I intend to discriminate between rainy clouds and the clouds without rainfall. The cumulonimbus clouds are situated in the lower layer of the atmosphere while the nimbostratus clouds are situated in the middle layer of the atmosphere. Therefore, relatively low cloud top temperature and comparatively low cloud top reflectance of clouds have to be found with visible and thermal infrared radiometer data. By using visible and thermal infrared data, appropriate threshold which allows discriminate nimbostratus / cumulonimbus and the other clouds has to be determined.

B. Jacobi Iteration Method

The proposed method uses Jacobi iteration method. The Jacobi iteration method is expressed as follows,

$$g \;= g \quad {}_1 + \alpha \quad F' \tag{1}$$

$$\alpha \;=\; \tfrac{1}{2} \tag{2}$$

$$F_k = \frac{1}{\sqrt{2\pi\sigma_{S1}^2}} exp\left(\frac{-(g_{k-1}-m_{S1})^2}{2\sigma_{S1}^2}\right) + \frac{1}{\sqrt{2\pi\sigma_{S2}^2}} exp\left(\frac{-(g_{k-1}-m_{S2})^2}{2\sigma_{S2}^2}\right) \quad (3)$$

where k denotes iteration number while F_k denotes summation of the PDF functions of the approximated normal distribution of S_i (i=1 denotes rainy cloud while i=2 denotes non rainy clouds of visible and thermal infrared radiometer data. Namely, the PDF functions of visible and thermal infrared data are firstly created then the most appropriate approximation normal distribution functions are calculated. After that, cross point between two approximated normal distribution is determined by using Jacobi iteration method. This cross point is used for threshold for discrimination between rainy and non-rainy clouds.

C. Process Flow

Fig.1 shows the process flow of the proposed method for discrimination of rainy and non-rainy clouds with visible and thermal infrared radiometer data.

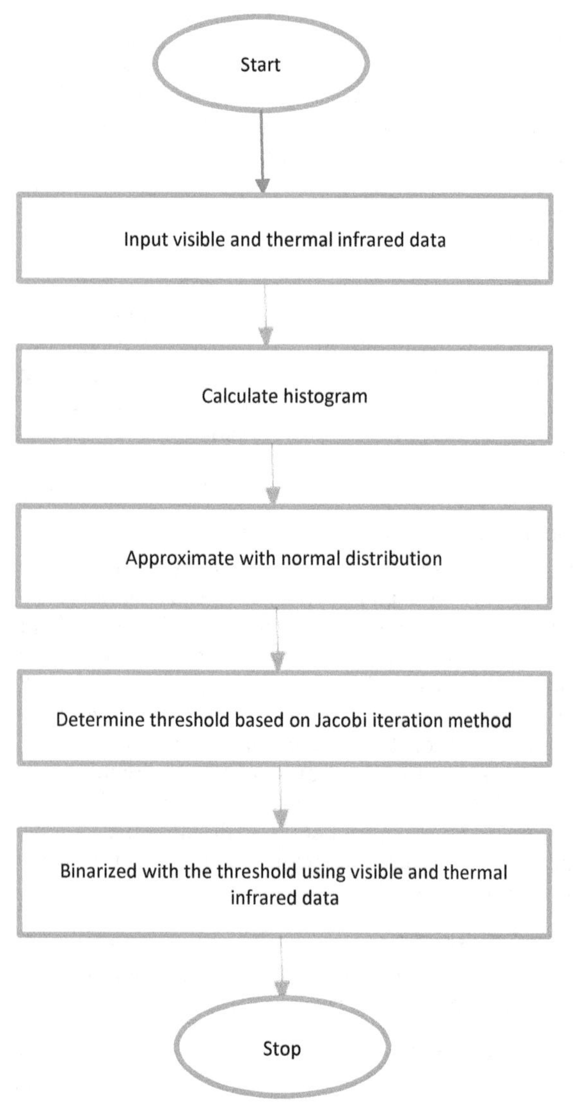

Fig. 1. Process flow of the proposed method for discrimination between rainy and non-rainy clouds with visible and thermal infrared data

III. EXPERIMENTS

A. Visible and Thermal Infrared Imagery Data Used

NOAA/AVHRR (National Oceanic and Atmospheric Administration / Advanced Very High Resolution of Radiometer) of visible and thermal infrared data of Tohoku, Japan which is acquired on February 12 1997 is used. Visible channel which covers the wavelength ranges from 0.73 to 1.10 μm is used while thermal infrared channel which covers the wavelength ranges from 10.3 to 11.3μm is used.

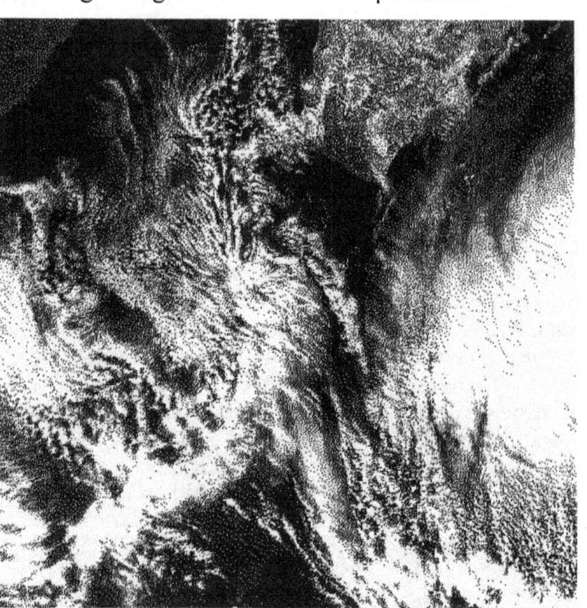

(a)Visible

(b)Thermal Infrared

Fig. 2. NOAA/AVHRR of visible and thermal infrared data of Tohoku, Japan which is acquired on February 12 1997 is used for the experiments

Fig.2 shows the visible and the thermal infrared imagery data used for the experiments. The images consists of 512 by 512 pixels (the pixel represent 2.2km by 2.2km ground surface areas). Radiometric resolution of visible channel is 0.1% of albedo while that of thermal infrared channel is 0.2degree C. These data are represented by 8 bits (256 levels) while minimum and maximum physical values correspond to 0 to 35% for visible channel while 243K to 294K for thermal infrared channel.

B. Truth Data Used

As a truth data of rainfall areas, radar data derived rainfall areas which is provided by Japanese Meteorological Agency: JMA is used. Fig.3 shows the radar data derived rainfall areras of image which consists of 500km by 500km (the pixel consists 2.6km by 2.6km). Black areas show the rainfall areas with 1 to 4mm/hr of rainfall rate while hatched areas shows the rainfall areas with less than 1mm/hr of rainfall rate.

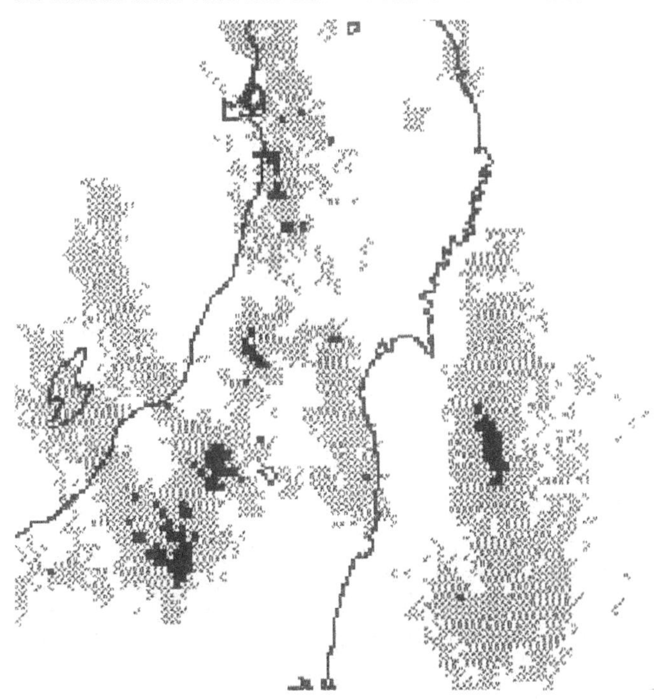

Fig. 3. Radar data derived rainfall areas which is observed at 10:00 in the morning on February 12 1997

Also, Fig.4 shows the imagery data which are used for multiple linear regressive analysis (Radar data on the right, Visible channel of imagery data in the middle, and Thermal infrared imagery data on the left, respectively) which are acquired on February 12 1997.

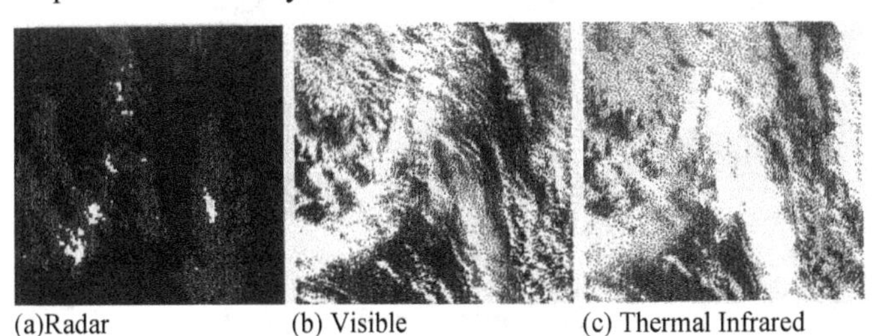

(a)Radar (b) Visible (c) Thermal Infrared

Fig. 4. Imagery data which are used for multiple linear regressive analysis (Radar data on the right, Visible channel of imagery data in the middle, and Thermal infrared imagery data on the left, respectively) which are acquired on February 12 1997

These visible and thermal infrared imagery data are extracted from the NOAA/AVHRR imagery data. Also, the cloud analysis information image which is acquired at 18:00 on that day provided by JMA and weather map at 12:00, noon on that day is shown in Fig.5 as reference data for rainfall areas. ① to ④ in Fig.5 (a) denotes cumulus clouds while ⑤ and ⑥ areas denote non rainy areas. The eastern portion of Japanese island, in particular, Hokkaido and Tohoku, there are relatively large cloudy areas. Particularly, ① area shows rainfall areas.

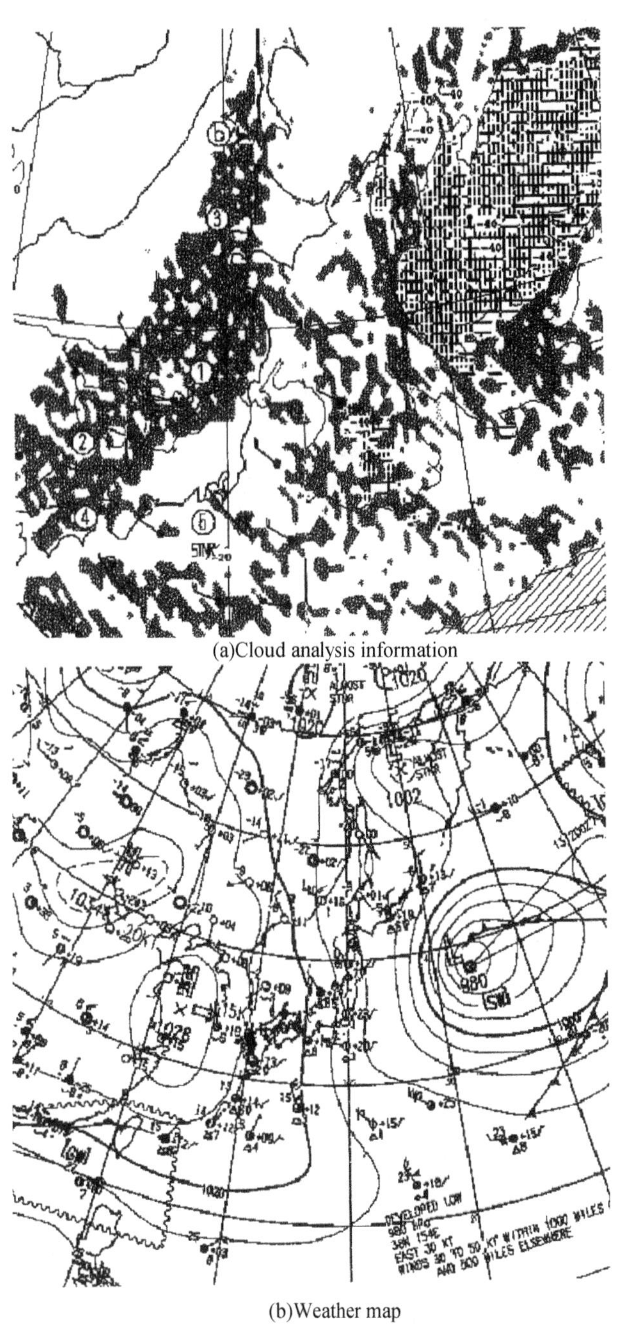

(a)Cloud analysis information

(b)Weather map

Fig. 5. Cloud analysis information image which is acquired at 18:00 on that day and weather map at 12:00, noon on that day are shown in Fig.5 as reference data for rainfall areas

C. *Experimental Results*

Fig.6 (a) shows histograms of rainy (S1) and non-rainy (S2) cloud areas of visible channel of imagery data and the approximate PDF functions for the histograms for both rainy and non-rainy cloud areas. Meanwhile, Fig.6 (b) shows histograms of rainy (S1) and non-rainy (S2) cloud areas of thermal infrared channel of imagery data and the approximate PDF functions for the histograms for both rainy and non-rainy cloud areas. The mean and variance of S1 of the visible channel of data are 168.874 and 877.135, respectively while those for S2 of visible channel of data are 238.292 and 748.170, respectively.

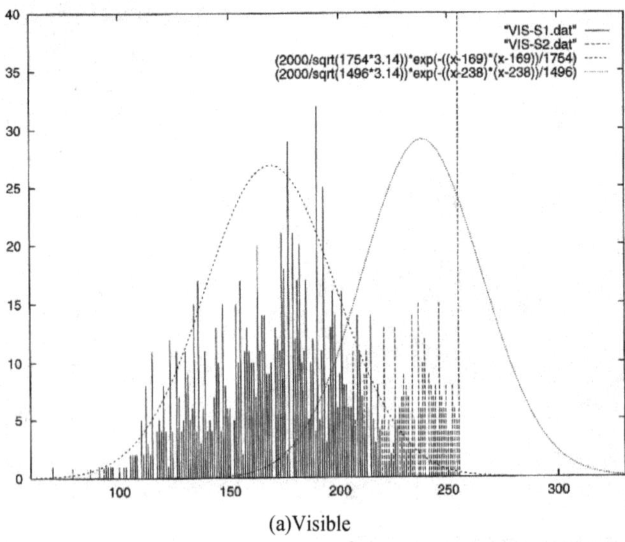

(a)Visible

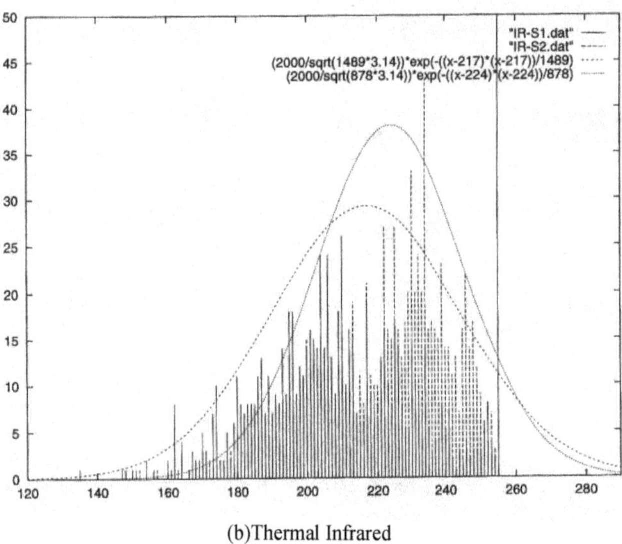

(b)Thermal Infrared

Fig. 6. Histograms of rainy (S1) and non-rainy (S2) cloud areas of thermal infrared channel of imagery data and the approximate PDF functions for the histograms for both rainy and non-rainy cloud areas

It is found that histograms of rainy and non-rainy clouds are very close for the thermal channel of data while those are relatively distinguishable for the visible channel of data.

Fig.7 (a) shows the binarized image of the visible channel of data while Fig.7 (b) shows that of the thermal infrared channel of data with the determined thresholds by the Jacobi iteration method, respectively.

D. *Validation of the Proposed Method*

Fig.8 (a) shows raw image of Rainfall radar while Fig.8 (b) shows the rainfall rate extracted image with rain fall radar data. On the other hand, Fig.9 shows the extracted rainfall areas with NOAA/AVHRR of visible and thermal infrared imagery data. White square box in the Fig.9 shows the corresponding area of interest with the rain radar derived rainfall areas.

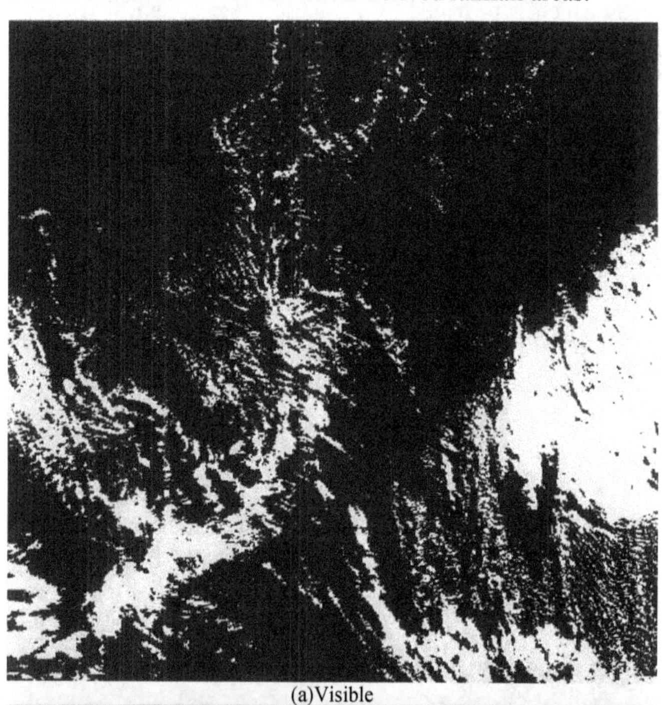

(a)Visible

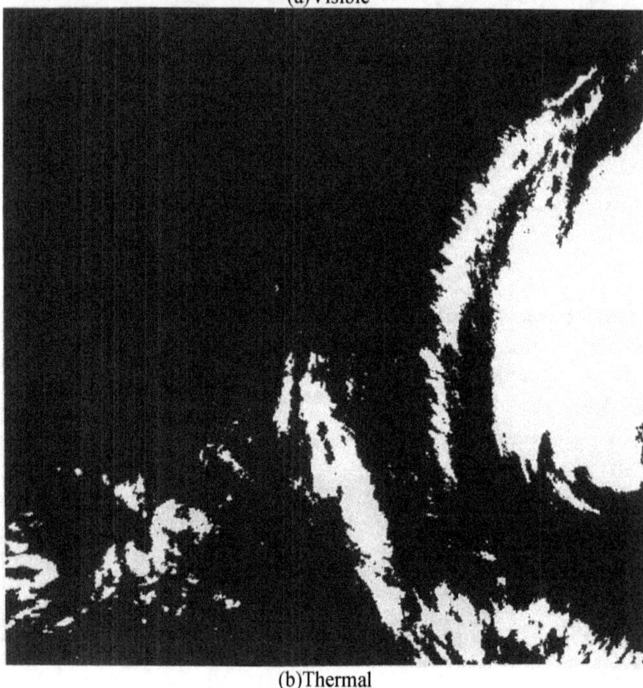

(b)Thermal

Fig. 7. Binarized images of the visible and the thermal infrared channels of data which are acquired on February 12 1997

Through comparisons between rainfall radar data derived rainfall rate image and the extracted rainfall areas with NOAA/AVHRR data based on the proposed method, it is found that the extracted rainfall areas with NOAA/AVHRR data based on the proposed method shows relatively heavily rainfall areas. The extracted rainfall areas with NOAA/AVHRR data based on the proposed method is corresponding to the rainfall areas with rainfall rate of 2 to 4 mm/hr.

(a)Rainfall radar

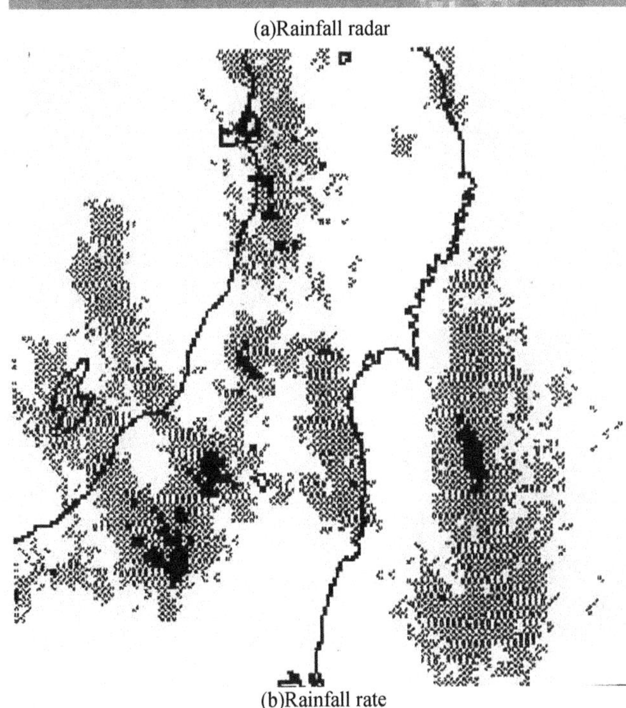

(b)Rainfall rate

Fig. 8. Rainfall radar image and the rainfall rate extracted image with rain fall radar data

Fig.10 (a) shows the binarized image of rainfall radar derived rainfall rate while Fig.10 (b) shows the binarized image of NOAA/AVHRR of visible and thermal infrared data derived rainfall areas based on the proposed method. Both images show marginal coincidence in terms of rainfall areas. Root Mean Square Difference: RMSD between the aforementioned two binarized images is 13.727. Therefore, it is marginal accuracy of rainfall area detection.

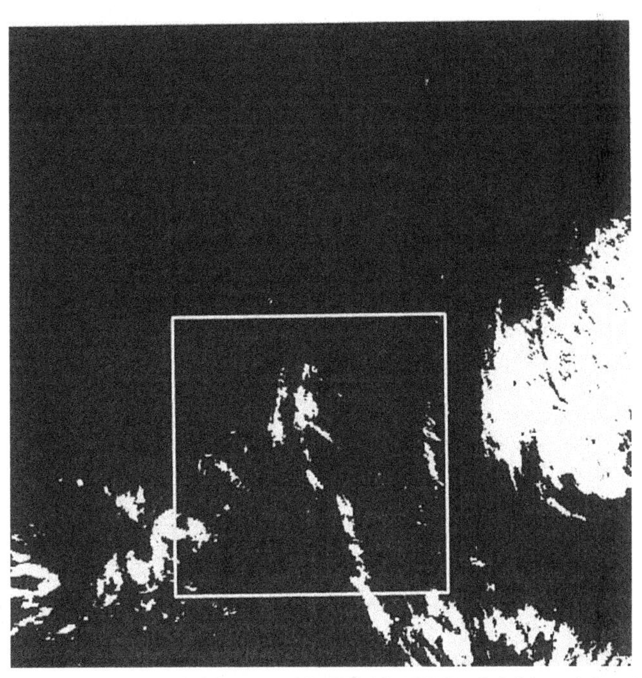

Fig. 9. Extracted rainfall areas with NOAA/AVHRR of visible and thermal infrared imagery data (White square box shows the corresponding area of interest with the rain radar derived rainfall areas)

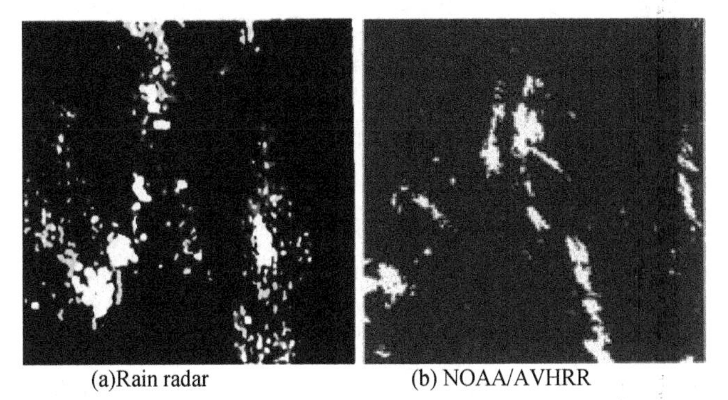

(a)Rain radar (b) NOAA/AVHRR

Fig. 10. Binarized images of rainfall radar derived rainfall rate and NOAA/AVHRR of visible and thermal infrared data derived rainfall areas based on the proposed method

E. Alternative Method for Rainfall Area Detection

As described before, there is the alternative method of rainfall area detection, multiple linear regressive analyses: MLRA based method. Namely, rainfall radar data derived rainfall rate is approximated with the NOAA/AVHRR of visible and thermal infrared radiometer data through the MLRA. Fig.11 (a) shows the scatter plots of the rainfall rate

(AME) and the visible and the thermal infrared channels of NOAA/AVHRR data while Fig.11 (b) shows MLRA equation which is the result from the MLRA expressing the rainfall rate as the functions of the visible channel and the thermal channel of NOAA/AVHRR data.

MLRA equation is expressed in equation (4).

$$Z=0.0155x+0.00941y+1.731 \quad (4)$$

where Z denotes rainfall rate while x and y denotes the visible and the thermal infrared channels of NOAA/AVHRR data.

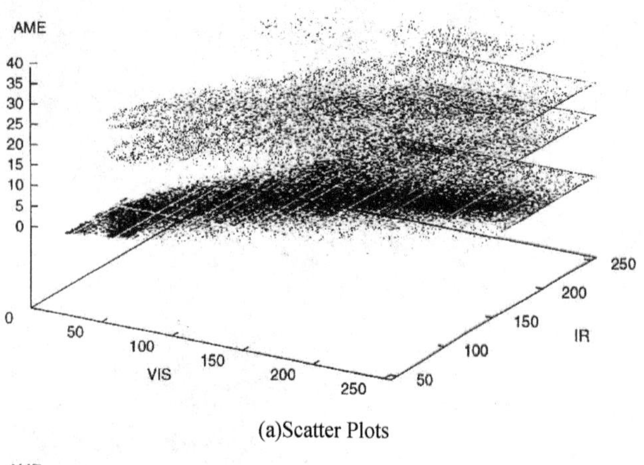

(a)Scatter Plots

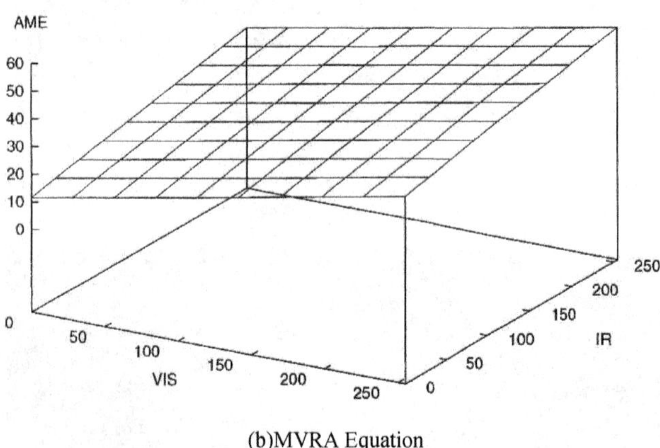

(b)MVRA Equation

Fig. 11. Scatter plots of the rainfall rate (AME) and the visible and the thermal infrared channels of NOAA/AVHRR data and MLRA equation which is the result from the MLRA expressing the rainfall rate as the functions of the visible channel and the thermal channel of NOAA/AVHRR data

The coefficient of determination of the MLRA is 0.020, and multiple correlation coeffient is 0.142. Also, degree of freedum corrected coefficient of determination is 0.020 while degree of freedum corrected multiple correlation coefficient is 0.141. Therefore, not so good correlation is found between rainfall rate and the visible and the thermal infrared channels of NOAA/AVHRR data. Consequently, the proposed method is superior to the MLRA based approach.

IV. CONCLUSION

Thresholding based method for rainy cloud detection with NOAA/AVHRR data by means of Jacobi iteration method is proposed. Attempts of the proposed method are made through comparisons to truth data which are provided by Japanese Meteorological Agency: JMA which is derived from radar data. Although the experimental results show not so good regressive performance, new trials give some knowledge and are informative. Root Mean Square Difference: RMSD between two binarized images of the rainfall radar derived rainfall rate and the NOAA/AVHRR derived rainfall area detected resultant image is 13.727. Therefore, it is concluded that the proposed method has a marginal accuracy of rainfall area detection. Therefore, the proposed method suggests for creation of new method for rainfall area detection with visible and thermal infrared imagery data.

Through the comparison between the proposed method and the multiple linear regressive analyses, it is concluded that the proposed method is superior to the Multiple Linear Regressive Analysis: MLRA based approach.

Further investigations are required for new additional information such as collocated microwave radiometer data and limb sounding data.

ACKNOWLEDGMENT

The author would like to thank Mr. Hirokazu Taniguchi of former student of Saga University for his effort to conduct the experiments.

REFERENCES

[1] Chuang, C. and K. V. Beard, A numerical model for the equilib- rium shape of electrified raindrops, J. Atmos. Sci., 47, 1374- 1389, 1990.

[2] Crewell, S., H. Czekala, U. LShnert, and C. Simmer, MICCY- a 22 channel ground-based microwave radiometer for atmopsh- eric research, submitted to Radio Science, 2000a.

[3] Crewell, S., U. LShnert, A. van Lammeren, and C. Simmer, Cloud remote sensing by combining synergetic sensor information, Phys. Chem. Earth (B), 25, No. 10-12, 1043-1048, 2000b.

[4] Czekala, H., S. Crewell, A. Hornbostel, A. Schroth, C. Simmer, and A. Thiele, Validation of microwave radiative transfer cal- culations for nonspherical liquid hydrometeors with ground- based measurements, J. Appl. Meteorol., 2000, (submitted for publication).

[5] Czekala, H. and C. Simmer, Microwave radiative transfer with non-spherical precipitating hydrometeors, J. Quant. $pectros. Radiat. Transfer, 60, 365-374, 1998.

[6] Fox, N., and A. J. Illingworth, The potential of spaceborne cloud radar for the detection of stratocumulus clouds, J. Appl. Me- teorol., 36, 676-687, 1997.

[7] Gfildner, J. and D. Spiinkuch, Results of year-round remotely sensed integrated water vapour by ground-based microwave radiometry, J. Appl. Meteorol., 38, 981-988, 1999.

[8] Mishchenko, M. I., Calculation of the amplitude matrix for a nonspherical particle in a fixed orientation, Appl. Opt., 39, 1026-1031, 2000.

[9] Solheim, F., J. Godwin, E. R. Westwater, Y. Han, S. Keihm, K. Marsh, and R. Ware, Radiometric profiling of temperature, water vapor and cloud liquid water using various inversion methods, Radio Science, 33, 393-404, 1998.

[10] Ameur, Z., Ameur, S., Adane, A., Sauvageot, H., Bara, K., (2004) Cloud classification using the textural features of Meteosat images. International Journal of Remote Sensing, 25, 21, 4491-4503

[11] Feidas H., Giannakos A., (2010) Identifying precipitating clouds in Greece using multispectral infrared Meteosat Second Generation satellite data. Theor Appl Climatol DOI: 10.1007/s00704-010-0316-5

[12] Haralick, R., Shanmugan, K., Dinstein, I., (1973) Texture features for image classification. Transactions on Systems, Man, and Cybernetics, 3, 6, 611-621

[13] Kwon, E., Sohn, B., Schmetz, J., Watts, P., (2010) Intercomparison of height assignment methods for opaque clouds over the tropics. J. Atmos. Sci., 46, 1, 11-19

[14] Strabala, K., Ackerman, S., (1993) Cloud Properties Inferred from 8-12μm Data. J. Appl. Meteor, 33, 212229

[15] Thies, B, Nauss T, Bendix J (2008b) A new technique for detecting precipitation at mid-latitudes during daytime using Meteosat Second Generation SEVIRI. 2008 EUMETSAT Meteorological Satellite Conference, Darmstadt, Germany

[16] Thies, B., Nauss T, Bendix J., (2008a) Discriminating raining from non-raining cloud areas at mid-latitudes using meteosat second generation SEVIRI night-time data. J. Appl. Meteor, 15, 219-230

A Method of Multi-License Plate Location in Road Bayonet Image

Ying Qian
The lab of Graphics and Multimedia
Chongqing University of Posts and Telecommunications
Chongqing, China

Zhi Li
The lab of Graphics and Multimedia
Chongqing University of Posts and Telecommunications
Chongqing, China

***Abstract*—To solve the problem of multi-license plate location in road bayonet image, a novel approach was presented, which utilized plate's color features, geometry characteristics and gray feature. Firstly, the RGB color image was converted to HSV color model and calculates the distance according to the plate's color information in the color space. Secondly, the license plate candidate regions were segmented by binary and morphological processing. Finally, based on the plate's geometry characteristics and gray feature, the license plate regions were segmented by and validated. In a certain degree, the method wasn't limited the plate's type, size, number, the location of the car and the background in the picture. It was tested using the road bayonet image.(*Abstract*)**

Keywords—multi-license plate location; color features; geometry characteristics; gray feature

I. INTRODUCTION

At present, the intelligent transportation system commonly used HD intelligent traffic cameras, which has a wide range of monitoring. The system can capture two or three vehicle lanes by using HD traffic cameras, which has significantly improved the efficiency. At the same time, the equipment cost and maintenance cost has been saved. The license plate recognition system is mainly composed of license plate location, character segmentation and character recognition. Among them, the license plate location is the premise and foundation of license plate character segmentation and character recognition.

There are many kinds of method for license plate location, but most methods aiming at the single license plate or the plate in the semi-structure environment, such as charging stations, small import and export. Which has constrained in the application, such as requiring the plate size and position in the image varies in a certain range. Jie Guo et al [1] convert the image from RGB color space to HSV color space and then segment the regions which satisfied the color feature of plate by calculating the distance and similarity in color space. To the segmented image, the texture and structural features are analyzed to locate the license plate correctly. De-hua Ren [2] proposed a color classification method based on the distance between different colors according to the Chinese car license plate color features and then segmented the license-plate's background color regions by scanning lines of picture and analyzing the line segments. Finally, these regions were translated into binary image in which license-plate's background color was dark and license-plate's foreground color was white, and validated by the license-plate's gray features. The two methods could adapt to license-plate's type, size, number and weren't limited to the location of the car and the background in the image. But in the application of the multi-license plate location in road bayonet image, as a result of multiple lanes, the disturbance of the trees, billboards and the reason of license plate dirty, wear hardly, the above methods can't locate all regions of license plate in the image.

This paper aimed at the problem of multi-license plate location in road bayonet image, firstly calculates the distance according to the plate's color features in HSV color space and then segments the regions of interest by binary and morphological processing. Finally, based on the plate's geometry characteristics and gray feature, the license plate regions were segmented and validated. The method combines the license-plate's color features, geometry characteristics and gray feature, which can locate all regions of license plate in road bayonet image.

II. COLOR IMAGE PREPROCESSING

A. Color models

According to different applications, color representation in different ways. RGB color model is used for display, TV and scanner device, use three basic colors of red, blue and green configures most of the color which human eye can see. The HSV color model is widely used in video and television broadcasting. H,S,V respectively represent the Hue, Saturation and Value, which corresponding with the color features of the human eye can perceive. This color model represented by Munsell three-dimensional space coordinate system, because of the psychological perception of independence between the coordinates, it can independently perceive the change of color components, and because of the color is linear scalability, which suitable for user judgment with the naked eye. At the same time as the HSV model corresponds to the painter of color model, which can reflect the human perception and discrimination to the color and suitable for similarity comparison of color image [1], so this paper used the HSV color space to segment the color image.

Because of the image generally use the RGB model, so the first thing is conversion. The relationship of each component between the HSV color model and the RGB color model is as follows [3]:

$$ax = \max\{r, g, b\}, min = \min\{r, g, b\}$$

$$H=\begin{cases}0^{\circ} & max=min\\ 60^{\circ}\times\frac{g-b}{max-min}+0^{\circ} & max=r,g\geq b\\ 60^{\circ}\times\frac{g-b}{max-min}+360^{\circ} & max=r,g<b\\ 60^{\circ}\times\frac{b-r}{max-min}+120^{\circ} & max=g\\ 60^{\circ}\times\frac{r-g}{max-min}+240^{\circ} & max=b\end{cases} \quad (1)$$

$$S=\begin{cases}0 & max=0\\ \frac{max-min}{max} & max\neq 0\end{cases} \quad (2)$$

$$V=max \quad (3)$$

Among them, Hue with the metric, range from 0° to 360°, Saturation values range from 0 to 1, Value range from 0 to 1.

B. The distance in color space

License plate background and character color of Chinese has a fixed collocation, mainly contain blue background with white characters, yellow background with black characters, white background with black or red characters and black background with white characters. The three components of R,G,B in values equal to 0 and 255 consisting of eight kinds of basic color [3], this paper selects four base color, which is associated with the plate background were blue, yellow, white and black. The RGB values of four base colors and the corresponding HSV values is shown in Table 1.

In HSV color space, the distance between $C_1=(h_1,s_1,v_1)$ and $C_2=(h_2,s_2,v_2)$ is as follows:

$$d(C_1,C_2)=[(v_1-v_2)^2+(s_1\cdot\cos h_1-s_2\cdot\cos h_2)^2+(s_1\cdot\sin h_1-s_2\cdot\sin h_2)^2]^{\frac{1}{2}} \quad (4)$$

Respectively calculated the distance from each pixel in HSV image to four base colors can obtain the feature image based on color features. The smaller the value of the pixel in feature image is, the color of the pixel in RGB space is more close to the base color.

TABLE I. RGB AND HSV VALUES OF FOUR BASE COLORS

Color space	Four base colors			
	Blue	*Yellow*	*White*	*Black*
RGB	0,0,255	255,255,0	255,255,255	0,0,0
HSV	240,1,1	60,1,1	0,0,1	0,0,0

C. Binarization of feature image

In order to further separate the license plate from complex background, also need convert the feature image into binary image. Before the binary processing should find the appropriate threshold *T*, usually the OTSU is the ideal method to obtain the threshold, but in this paper, due to the license plate region occupied small proportion in the image and presence of large area of blue or yellow interference region, the binary image obtained from OTSU method usually contains lots of independent information.

Found in the course of the experiment, the total area of the license plate region in the image in a certain range and the value of license plate area in feature image is small. Decreased the value of the threshold obtained from OTSU method, the number of white pixels reduced present certain rules in binary image.

The following describes specific steps to obtain appropriate threshold *T*:

1) Calculate the number of the white pixels in binary image through the OTSU method.

2) Decrease the threshold obtained from OTSU method, then calculate the reduce number of white pixels and the total number of white pixels in binary image which processed by the reduced threshold.

3) If the reduce number or the total number satisfied the rules obtained by experiment debugging, T is equal to the reduced threshold. Otherwise returns step 2).

The binary image converted from feature image by appropriate threshold *T* is as shown in Fig.1(b).

(a) The original image

(b) The binary image

Fig. 1. The original image and the binary image

D. Morphology operation

Although the binary processing has filter out most background information of the image, there are still some noise and vehicle information. Therefore, mathematic morphologic close and open operation are used to make the possible license plate region into the rectangular connected region.

The selection of structure element associated with the size of the license plate in road bayonet image, so set the close operation structure element in the image of three lanes is 5×10 and the structure element in four lanes image is 3×7. In order to remove the isolated points and smooth edge, set the open structure element is 2×2. The morphological image is as shown in Fig.2.

Fig. 2. The morphological image

E. Connected componet labeling

After morphology operation we can obtain the independent connected domain by 8- connected component labeling, save the results in $L(x, y)$ and the number as *N*. the process is as follows:

$$L(x,y) = \begin{cases} i & 1 \leq i \leq N \\ 0 & other \end{cases} \quad (5)$$

The connected domain is expressed as the candidate license plate region.

III. Location of License Plate

A. Geometry characteristics of license plate

License plate has obvious geometry characteristics, the width and height of plate are fixed and the ratio of width and height is in a certain range. In China, the ratio of small car license plate is 3.14 and the ratio of large car is 2. Considering the road bayonet image obtained from fixed traffic camera, the license plate size in the image is in a certain range and related to the size of the image. So calculate three characteristic values of the candidate license plate region, which are size, ratio and filling.

B. Location of candidate license plate

The connected domain is expressed as the candidate license plate region, so the size of the region is the area of the connected domain and the ratio is the ratio of the ratio of the minimum enclosing rectangle. The filling defined as follows:

$$P = \frac{area\ of\ connected\ domain}{area\ of\ minimum\ enclosing\ rectangle} \quad (6)$$

The following describes specific steps to locate license plate:

1) Set size range from S_{min} to S_{max} obtained by experiment, the connected domain which satisfied the range for the next step.

2) Conserding the reason of shooting angle and tilt, set the ratio range from 2 to 6.

3) The P is closer to 1, it means that the connected domain is closer to the license plate. Considering the deletion of part plate information in binary and morphology operation, set the P range from 0.6 to 1.

4) Segment the region satisfied all of the conditions mentioned above in the original image and eliminate satisfied regions and small regions in the morphological image.

C. Secondary location of candidate license plate

The regions exist in morphological image after location include the large area regions (include license plate or not) and the region disturbed by solitary lines or vehicle information. So use the secondary location of candidate regions to locate the rest of license plate in the image. The following describes specific steps:

1) Obtained the coordinates of the current connected domain and then segmented the region in the feature image.

2) Decrease the threshold T, obtained the candidate region through binary processing and morphology operation.

3) Segment the regions which satisfied the conditions mentioned in location of candidate license plate in the original image.

The result of the location of candidate license plate is shown in Fig.3 and the license plate segmented from the original image is shown in Fig.4. There are two false results.

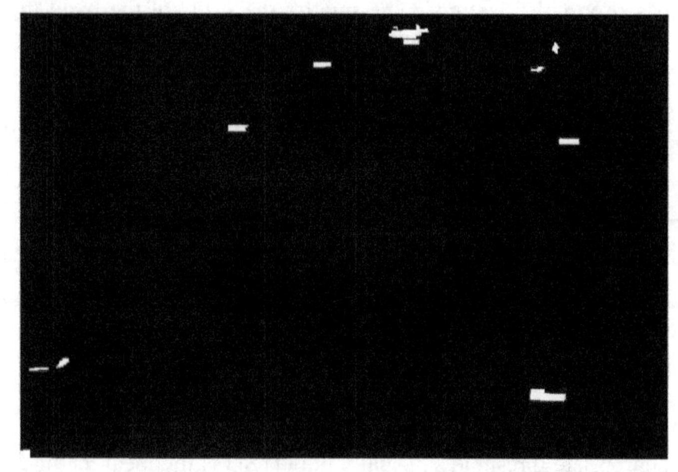

Fig. 3. The results of location in morpological image(red line marking the license plate region)

Fig. 4. The results of segmentation in the original image

D. Validation of license plate

The result of the license plate segmented from the original image may include some false license plate, so we validated the result based on the gray feature of license plate. The following describes specific steps:

1) Obtained the binary image of the color license plate segmented form original image by OTSU method.

2) Get the middle 80% of the binary image to remove the interference of the border in vertical projection.

3) Count the changing times of the character and the background, remove the result which unsatisfied with the gary feature.

The process of validation of license plate is shown in Fig.5.

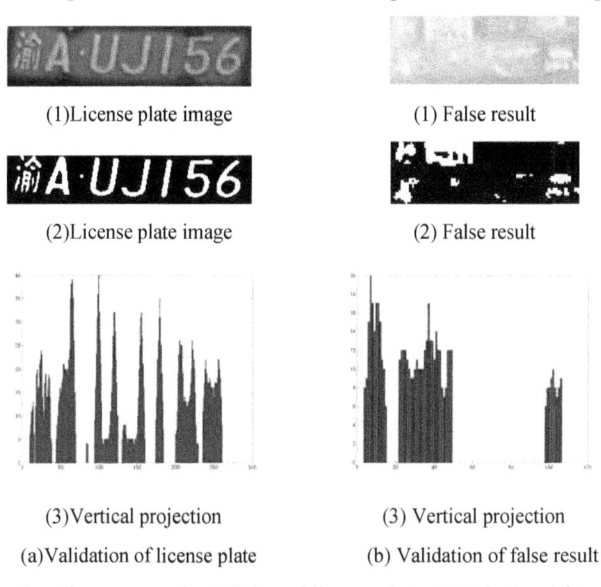

(1)License plate image (1) False result

(2)License plate image (2) False result

(3)Vertical projection (3) Vertical projection

(a)Validation of license plate (b) Validation of false result

Fig. 5. The process of validation of license plate (a)(1) is the right color license plate image, (a)(2) is the binary image of the right license plate, (a)(3) is the vertical projection image of license plate. (b)(1) is the false plate image, (b)(2) is the binary image of false plate image, (b)(3) is the vertical projection image of false result

IV. Experiment Results

Considering the size of road bayonet image is *4912×3264*, so at first we compress the size of image into *1228×816* to reduce the calculation time. The experiment shows that the method could locate all license plates in the road bayonet image. The part of experiment results are shown in Fig.6.

(1)The original image (2)The results of segmentation

(a)The location of four lanes, different size and multi-license plate

(1)The original image (2)The results of segmentation

(b)The location of three lanes and blue plate with blue car

(1)The original image (2)The results of segmentation

(c)The location of three lanes, different type, multi-license plate

Fig. 6. Part of experiment results

V. Conclusions

This paper aimed at the problem of multi-license plate location in road bayonet image, proposed a method combines color features, geometry characteristics and gray feature of license plate. Firstly based on the color features of license plate, filter the most background information by calculating the distance in HSV color space and binary processing. Obtained the candidate license plate regions through the morphology operation and then based on the geometry characteristics of size, ratio and filling, locate all license plate in the image by secondary location. Finally validate the results of segmentation based on the gray feature of license plate and remove the false results. This method can locate the license plate in different position, size, number and direction in the road bayonet image, which is a method of adaptability.

References

[1] GuoJie, Shi Peng-fei. Color and texture analysis based vehicle license plate location [J]. Journal of Image and Graphics, 2002,7(5):472-476.

[2] Ren De-hua. Multi-license plate extraction based on color features in nature complex environment [J]. Journal of Image and Graphics, 2009,14(12):2517-2526.

[3] Makoto Miyahara, Yasuhiro Yoshida. Mathematical transform of (R,G,B) color data to Munsell(H,V,C) color data [A]. In: Proceeding of SPIE Conference on Visual Communications and Image Processing [C]. Cambridge, MA, USA, 1988: 650-657.

[4] ZhengChengyong. A novel license plate location method on RGB color space [J]. Journal of Image and Graphics, 2010,15(11):1623-1628.

[5] GuoTianshu. A car plate location method based on itself's structural features [J]. Computer & Information Technology, 2008,10:51-57.

[6] Tan Siting, Hu Zhikun. An effective integration method for license plate location based on HSV color space [J]. Computers and Applied Chemistry, 2011,28(7):903-906.

[7] Li Wen-Ju, Liang De-qun, Zhang Qi, Fan Xin. A novel approach for vehicle license plate location based on edge-color pair [J]. Chinese Journal of Computers, 2004,27(2):204-208.

[8] Gan Ling, Sun Bo. Multiple license plate location based on separation projective and morphology operation [J]. Application Research of Computers, 2010, 29(7) : 2730-2732.

Optimum Band and Band Combination for Retrieving Total Nitrogen, Water, Fiber Content in Tealeaves Through Remote Sensing Based on Regressive Analysis

Kohei Arai [1]
Graduate School of Science and Engineering
Saga University
Saga City, Japan

Abstract—**Optimum band and band combination for retrieving total nitrogen, water and fiber content in tealeaves with remote sensing data is investigated based on regressive analysis. Based on actual measured data of total nitrogen, fiber and water content in tealeaves as well as remotely sensed visible to near infrared reflectance data with 5nm of wavelength steps and ASTER/VNIR onboard Terra satellite, regressive analysis is conducted. As the results, it is found that 1045nm is the best wavelength for retrieving total nitrogen content while 945nm is the best wavelength for fiber content retrieval. Also it is found that 545nm is the best wavelength for water content. On the other hand, it is found that 350 and 750nm wavelength combination is the best for estimation of total nitrogen content while 535 and 720 wavelength combination is the best for fiber content estimation. It also found that 545 and 760nm wavelength combination is the best for water content retrieval.**

Keywords—regressive analysis; total nitrogen content; tealeaves; fiber content;; water content

I. Introduction

It is highly desired to monitor vitality of crops in agricultural areas automatically with appropriate measuring instruments in order to manage agricultural area in an efficient manner. It is also required to monitor not only quality but also quantity of vegetations in the farmlands. Vegetation monitoring is attempted with red and photographic cameras [1]. Grow rate monitoring is also attempted with spectral observation [2].

This paper deals with automatic monitoring of a quality of tealeaves with earth observation satellite, network cameras together with a method that allows estimation of total nitrogen and fiber contents in tealeaves as an example. Also this paper describes a method and system for estimation of quantity of crop products by using not only Vegetation Cover: VC and Normalized Difference Vegetation Index: NDVI but also Bi-directional Reflectance Distribution Function: BRDF because the VC and NDVI represent vegetated area while BRDF represents vegetation mass, or layered leaves.

Total nitrogen content corresponds to amid acid which is highly correlated to Theanine: 2-Amino-4-(ethylcarbamoyl) butyric acid for tealeaves so that total nitrogen is highly correlated to tea taste. Meanwhile fiber content in tealeaves has a negative correlation to tea taste. Near Infrared: NIR camera data shows a good correlation to total nitrogen and fiber contents in tealeaves so that tealeaves quality can be monitored with network NIR cameras. It is also possible to estimate total nitrogen and fiber contents in leaves with remote sensing satellite data, in particular, Visible and near infrared: VNIR radiometer data. Moreover, VC, NDVI, BRDF of tealeaves have a good correlation to grow index of tealeaves so that it is possible to monitor expected harvest amount and quality of tealeaves with network cameras together with remote sensing satellite data. BRDF monitoring is well known as a method for vegetation growth [3],[4]. On the other hand, degree of polarization of vegetation is attempted to use for vegetation monitoring [5], in particular, Leaf Area Index: LAI together with new tealeaves growth monitoring with BRDF measurements [6].

It is not well known that the most preferable wavelength bands for observation of vegetation. Vitality of vegetation can be expressed with nitrogen, fiber and water contents in the leaves. Therefore, it is better to determine appropriate wavelength for retrieving theses parameters. In order to determine appropriate wavelength bands for estimation of total nitrogen, fiber and water contents in tealeaves, regressive analysis is conducted. Through regressive analysis, it is clarified that appropriate single wavelength and double wavelength for the retrievals with respect to the actual truth data sets of the parameters and hyperspectral data of reflective radiance from the tealeaves.

In the following section, research background is described followed by method for determination of appropriate single and double wavelength for retrievals. The regressive analysis results are summarized followed by conclusion and some discussions.

II. Research Background

A. *Vegetation Area Monitoring and Agricultural, in Paticular, Tea Farm Area Monitoring System*

The proposed tea estate monitoring system is illustrated in Figure 1. Visible and NIR network cameras are equipped on the pole in order to look down with 10-80 degrees of incident angle (these angles allow BRDF measurements). The pole is used for avoid frosty damage to the tealeaves using fan mounted on the pole (for convection of boundary layer air). With these network cameras, reflectance in the wavelength region of 550nm (red color) and 870nm (NIR) are measured together with BRDF assuming that vegetated areas are homogeneous and flat. BRDF is used for estimation of Grow Index (GI) and BRDF correction from the measured reflectance of the tealeaves.

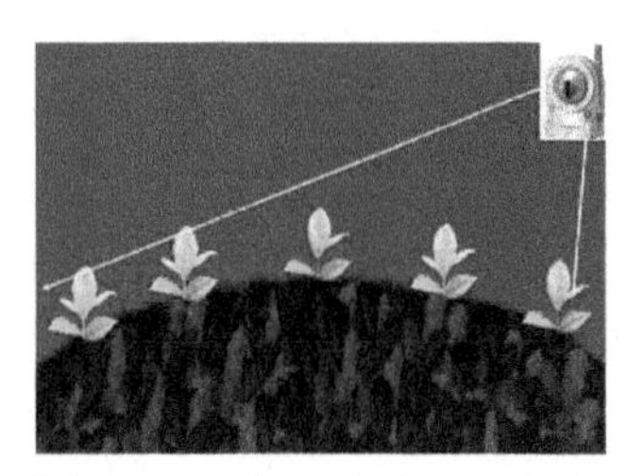

Fig. 1. Illustrative view of the proposed vegetation monitoring system with two network cameras, visible and NIR

These are controlled through Internet terminals. Visible Pan-Tilt-Zoom: PTZ network camera and NIR filter (IR840) attached network camera is equipped on the pole. PTZ cameras are controlled by mobile phone as well with "mobile2PC" or Internet terminal with "LogMeIn" of VNC services [7] through wireless LAN connected to Internet. Acquired camera data are used for estimation of total nitrogen and fiber contents as well as BRDF for monitoring grow index. An example of visible camera image acquired in daytime is shown in Figure 2 (a) while that for NIR camera image acquired in nighttime is shown in Figure 2 (b).

The cameras are connected to the Internet through the network card of W05K that is provided by AU/KDDI. Through http://119.107.81.166:8080, the acquired image data are accessible so that it is easy to access the data from Internet terminals. Panasonic BB-HCM371 cameras are used for the experiments. Solar panel of G-500 (12V, 500mA, 8.5W) with battery of SG-1000 is used together with Xpower75 (60W) of inverter.

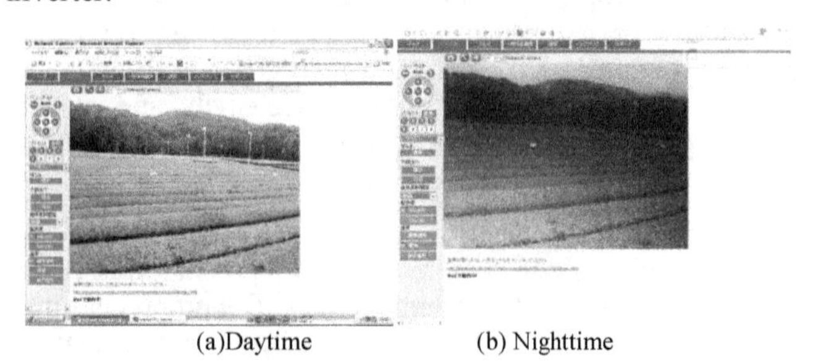

(a)Daytime (b) Nighttime

Fig. 2. Examples of farmland monitored visible camera images.

On the other hand, weather station data can be accessible from the URL of http://katy.jp/mapstation/ of data server provider through wireless LAN connection from the weather station to the Internet terminal. Figure 3 shows examples of the images displayed onto mobile phone. Not only camera imagery data, but also weather station data can be monitored with mobile phone. Figure 4 (a) and (b) shows overall weather station data of atmospheric pressure, solar direct and diffuse irradiance, leaf wetness, soil moisture, etc. and time duration of air-temperature and relative humidity of the tea estate while Figure 4 (c) shows web camera imagery data.

(a) New tealeaves appears partially (b) New tealeaves covers all over the surface

Fig. 3. Typical photos of new tealeaves grow process taken with network camera at tea estate of the prefectural tea research institute of Saga in the begging of April (a) and the late of April (b).

(a) Overall weather station data

(b) Air-temperature and relative humidity

(c) Camera image data

Fig. 4. Data displayed onto mobile phone

B. Tea Farm Area Monitoring with HyperSpectrometer

Other than these, hyper-spectral sensor can be equipped at the tea farm areas. Due to the fact that two bands of visible and near infrared cameras are not good enough in terms of estimation accuracy of nitrogen, fiber and water contents of tealeaves. Therefore, single and double wavelength bands for getting better accuracy of nitrogen, fiber and water contents have to be determined.

C. Dataset for Determination of Approporiate bands for Nitrogen, Fiber and Water Content Estimat5ion

Intensive study area is situated at the Saga Prefectural Tea Institute in Ureshino-city, Saga, Japan. ASTER/VNIR image of the site is shown in Figure 5.

Fig. 5. Terra/ASTER/VNIR images of Saga acquired on May 16 in 2008 (False color representation: Blue Band #1, Green Band #2, Red Band #3).

Figure 6 shows enlarged image of ASTER/VNIR image of Saga Prefectural Tea Institute: SPTI. In particular, nitrogen content in tealeaves is shown in Figure 6 (b). Red circles shows four tea farm areas which are situated in East, West, South and North direction of Saga Prefectural Tea Institute.

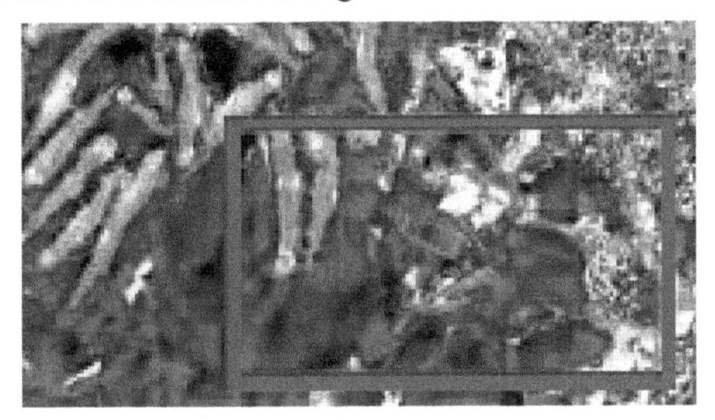

(a)Portion of ASTER/VNIR image of Ureshino, Saga

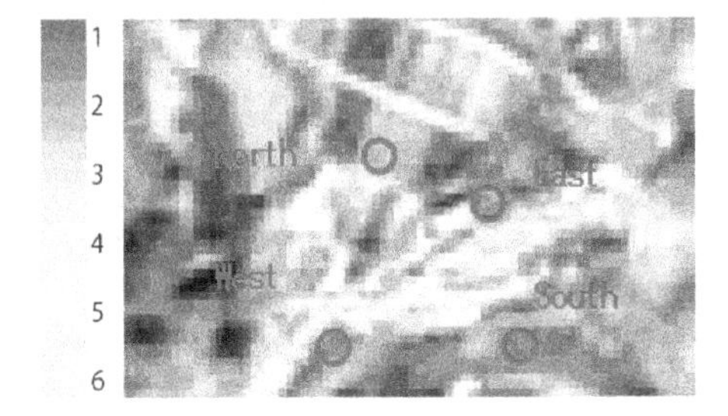

(b)Enlarged image of Saga Prefectural Tea Institute

Fig. 6. Enlarges ASTER/VNIR image and total nitrogen contents in tealeaves at the tea estate (Red circles shows tea estates. Grayscale shows TN% of nitrogen contents in tealeaves derived from equation (1) of TN=22.474 Ref (Band#3)-10.177).

SPTI is situated at (33:07'2.9"N, 129:59'42.5"E,elevation: 130m) at the center location. In terms of species of the tea farm areas, East tea field has Yabukita tea farm area while North tea field has Yabukita tea farm and Okumidori.

Meanwhile, West tea field has Benifuki tea farm while South tea field has Ohiwase tea farm. Just before the harvesting tealeaves, in May 2008, spectral reflectance is measured. Figure 7 shows the reflectance. Meantime, total nitrogen, fiber and water content in the tealeaves are also measured. Thus, correlation can be calculated with these dataset through correlation analysis. Figure 8 shows the calculated correlations

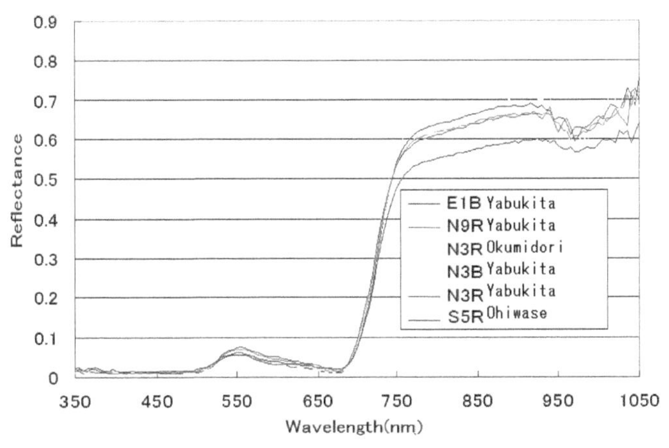

Fig. 7. Spectral reflectance measured at East, North and South tea farm areas situated at SPTI on 5 May 2008.

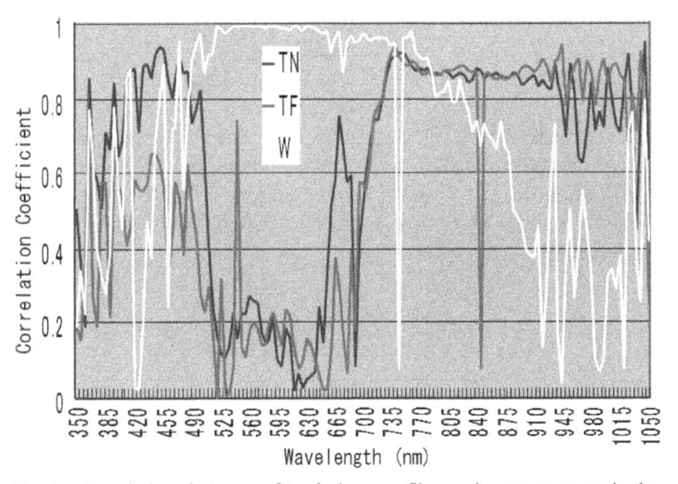

Fig. 8. Correlations between of total nitrogen, fiber and water contents in the tealeaves and the measured spectral reflectance

III. EXPERIEMNTS

A. Single Spectral Band for Estimation of TN, Fiber, and Water Contentsin TealeavesSlope Effect

Using the correlations between TN, fiber, and water content in tealeaves and spectral reflectance measured at SPTI, Saga Japan on May 5 2008, just before the harvesting tealeaves, regressive analysis is conducted. Through regressive analysis with single band with 5nm band width, the most appropriate spectral bands for estimation of TN, Fiber, and Water contents in tealeaves are estimated. Table 1, 2, and 3 show the results from the regressive analysis for TN, Fiber, and Water contents in tealeaves, respectively.

TABLE I. RESULT FROM REGRESSION FOR TN CONTENT ESTIMATION WITH SINGLE SPECTRAL BAND WITH 5NM OF BAND WIDTH

Nitrogen	1045nm
R	0.9502
R^2	0.9029
StDev	0.3375
No.	6

TABLE II. RESULT FROM REGRESSION FOR FIBER CONTENT ESTIMATION WITH SINGLE SPECTRAL BAND WITH 5NM OF BAND WIDTH

Fiber	945nm
R	0.9502
R^2	0.9029
StDev	0.3375
No.	6

TABLE III. RESULT FROM REGRESSION FOR WATER CONTENT ESTIMATION WITH SINGLE SPECTRAL BAND WITH 5NM OF BAND WIDTH

Water	545nm
R	0.9999
R^2	0.9997
StDev	0.0004
No.	6

As the results, it is found that the most appropriate spectral bands for estimation of TN, Fiber and Water contents in tealeaves are 1045, 945, and 545 nm. The regressive analysis is conducted based on Pearson's correlation with 95 % of confidence level. The regressive errors of TN, Fiber, and Water contents in tealeaves are shown in Table 4, 5, and 6, respectively.

TABLE IV. REGRESSION ERROR FOR TN CONTENT ESTIMATION WITH SINGLE SPECTRAL BAND WITH 5NM OF BAND WIDTH

Field Name	Species	TN(%)	Est.TN	Reg.Error
E1B	Yabukita	4.7	4.168	0.283
N9R	Yabukita	4.6	4.938	0.114
N3R1	Okumidori	4.9	4.914	0.000197
N3R2	Yabukita	5	4.889	0.0123
N3B	Yabukita	5	5.099	0.00983
S5R	Ohiwase	2.5	2.691	0.0367

TABLE V. REGRESSONE ERROR FOR TN CONTENT ESTIMATION WITH SINGLE SPECTRAL BAND WITH 5NM OF BAND WIDTH

Field Name	Species	Fiber(%)	Est(Fiber)	Reg.Error
E1B	Yabukita	20.4	14.99	29.2681
N9R	Yabukita	17.3	16.66	0.4096
N3R1	Okumidori	1.5	4.98	12.1104
N3R2	Yabukita	19.6	21.66	4.2436
N3B	Yabukita	17.2	16.45	0.5625
S5R	Ohiwase	32.2	33.46	1.5876

TABLE VI. REGRESSION ERROR FOR TN CONTENT ESTIMATION WITH SINGLE SPECTRAL BAND WITH 5NM OF BAND WIDTH

Field Name	Species	Water(%)	Est(Water)	Reg.Error
E1B	Yabukita	0.7623	0.7625	4E-08
N9R	Yabukita	0.7318	0.7314	1.6E-07
N3R2	Yabukita	0.7555	0.7754	0.000396
N3B	Yabukita	0.7271	0.7424	0.000234

As the result, it is found that fiber content in tealeaves is the most difficult followed by TN content and water content There are not available data of water content of truth data for the test sites of N3R1, S5R.

B. Double Spectral Band for Estimation of TN, Fiber, and Water Contents in Tealeaves

The most appropriate two spectral bands with 5 nm of band width for estimation of TN, Fiber, and Water contents in tealeaves are determined through regressive analysis using the aforementioned correlation data between truth data and estimated data. The results from the regressive analysis are shown in Table 7, 8, and 9, respectively.

TABLE VII. REGRESSION RESULT FOR TN CONTENT ESTIMATION WITH SINGLE SPECTRAL BAND WITH 5 NM OF BAND WIDTH

Nitrogen	350&750nm
R	0.9906
R^2	0.9812
StDev	0.1713
No.	6

TABLE VIII. REGRESSION RESULT FOR FIBER CONTENT ESTIMATION WITH SINGLE SPECTRAL BAND WITH 5 NM OF BAND WIDTH

Fiber	535&720nm
R	0.9798
R^2	0.96
StDev	2.538
No.	6

TABLE IX. REGRESSION RESULT ERROR FOR WATER CONTENT ESTIMATION WITH SINGLE SPECTRAL BAND WITH 5 NM OF BAND WIDTH

Water	545&760nm
R	0.9999
R^2	0.9999
StDev	0.0003
No.	6

As the results, it is found that the most appropriate band combination for estimation of TN, Fiber, and Water contents in tealeaves are 350 and 750 nm, 535 and 720 nm, 545 and 760 nm, respectively.

C. *Comparison of Estimation Accuracy Among Single, Double Spectral Band, and ASTER/VNIR Spectral Bands for Estimation of TN, Fiber, and Water Contents in Tealeaves*

Estimation accuracy for TN, Fiber, and Water contents in tealeaves is evaluated with ASTER/VNIR spectral bands and is compared to the aforementioned estimation accuracy with the most appropriate single, and double spectral bands with 5 nm band width. The results from the comparisons are shown in Figure 3, 4, and 5, for TN, Fiber, and Water contents in tealeaves.

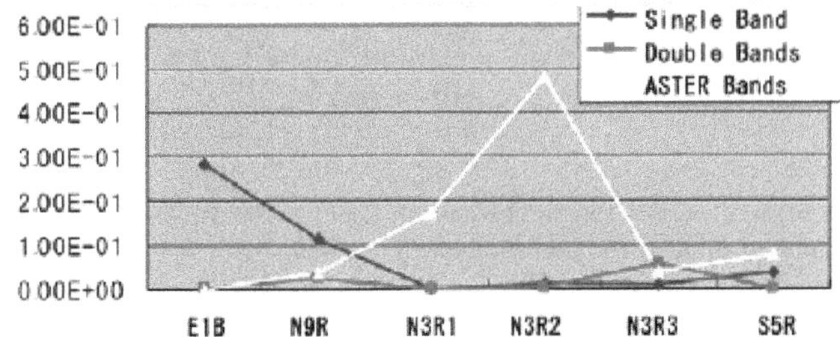

Figure 9 Comparison among single, double, and ASTER/VNIR spectral bands for estimation of TN content in tealeaves

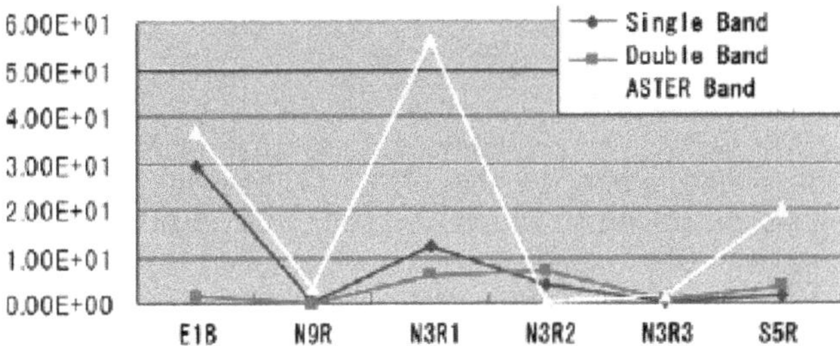

Figure 10 Comparison among single, double, and ASTER/VNIR spectral bands for estimation of Fiber content in tealeaves

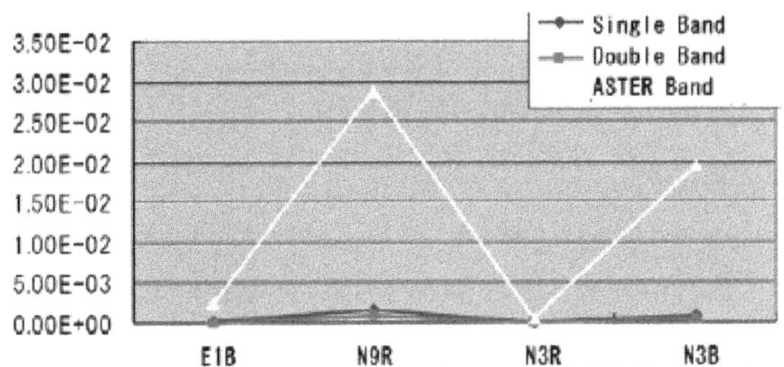

Figure 11 Comparison among single, double, and ASTER/VNIR spectral bands for estimation of water content in tealeaves

The results show that two spectral bands case (double) shows the best estimation accuracy followed by single spectral band, and ASTER/VNIR. Due to the fact that ASTER/VNIR spectral bands are broad in comparison to the single and double spectral bands with 5 nm of band width, ASTER/VNIR spectral bands case shows the worst estimation accuracy. Also it is found that estimation accuracy depends on the tea farm areas of intensive study areas.

IV. Conclusion

Optimum band and band combination for retrieving total nitrogen, water and fiber content in tealeaves with remote sensing data is investigated based on regressive analysis. Based on actual measured data of total nitrogen, fiber and water content in tealeaves as well as remotely sensed visible to near infrared reflectance data with 5nm of wavelength steps and ASTER/VNIR onboard Terra satellite, regressive analysis is conducted.

As the results, it is found that 1045nm is the best wavelength for retrieving total nitrogen content while 945nm is the best wavelength for fiber content retrieval. Also it is found that 545nm is the best wavelength for water content. On the other hand, it is found that 350 and 750nm wavelength combination is the best for estimation of total nitrogen content while 535 and 720 wavelength combination is the best for fiber content estimation. It is also found that 545 and 760nm wavelength combination is the best for water content retrieval. The results show that two spectral bands case (double) shows the best estimation accuracy followed by single spectral band, and ASTER/VNIR. Due to the fact that ASTER/VNIR spectral bands are broad in comparison to the single and double spectral bands with 5 nm of band width, ASTER/VNIR spectral bands case shows the worst estimation accuracy. Also it is found that estimation accuracy depends on the tea farm areas of intensive study areas.

Acknowledgment

The author would like to thank Mr. Shin-ichi Motomura for his efforts through experiments and simulations.

References

[1] Arai, K, Lecture Notes on Remote Sensing, Morikita-Shuppan, Co.Ltd., 2005

[2] C.C.Borel and S.A.Gerst, Nonlinear spectral mixing models for vegetative and soils surface, Remote Sensing of the Environment, 47, 2, 403-416, 1994.

[3] R.N.Clark and T.I.Roush, Reflectance spectroscopy: Quantitative analysis techniques for remote sensing applications, Journal of Geophysical Research, 89, B7, 6329-6340, 1984.

[4] B.Hapke, Bidirection reflectance spectroscopy, I. Theory, Journal of Geophysical Research, 86, 3039-3054, 1981.

[5] Mersenne Twister (MT), http://www.math.sci.hiroshima-u.ac.jp/~m-mat/MT/mt.html

[6] B.Nash and J.Conel, Spectral reflectance systematic for mixtures of powered hypersthenes, labradoride and ilmenite, Journal of Geophysical Research, 79, 1615-1621, 1974.

[7] R.Singer, Near infrared spectral reflectance of mineral mixtures: Systematic combinations of pyroxenes olivine and iron oxides, Journal of Geophysical Research, 86, 7967-7982, 1974.

[8] R.Singer and T.B.McCord, Mars; Large scale mixing of bright and dark surface materials and implications for analysis of spectral reflectance, Proc., 10th Lunar and Planetary Sci., Conf., 1835-1848, 1979.

Instruments and Criteria for Research and Analysis of the Internet Visibility of Bulgarian Judicial Institutions WEB-Space*

Nayden Valkov Nenkov
Faculty of Mathematics and Informatics
University of Shumen "Episkop Konstantin Preslavsky"
Shumen, Bulgaria

Mariana Mateeva Petrova
Faculty of Mathematics and Informatics
St. Cyril and St. Methodius University of Veliko Turnovo,
Bulgaria

***Abstract*—e-Justice has been under discussion at European level since 2007. The article describes some tools and displays objective criteria for evaluating the WEB-pages of judicial institutions in Bulgaria. A methodology is offered in order to improve the organization and functioning of the judicial institutions. It is used to conduct experimental tests for analysis and assessment of the main characteristics of the Bulgaria courts' WEB-sites. The results provide grounds for findings and recommendations leading to improved communication and the presence of these institutions in the WEB.**

***Keywords*—judicial institution; WEB-page; SEO (search engine optimization); evaluation criteria; court**

I. Introduction

The evaluation of the judicial WEB-sites is an important task in the context of the radical reform made in this area. This evaluation must be consistent with the overall vision and the project to build an e-government with the EC requirements and the standards which exist for e-administration and e-services to the population in Bulgaria [1, 3].

On 18 December 2008 The European Parliament adopted a Resolution on e-Justice, on 22 October 2013 it adopted a Resolution on e-Justice calling for the use of electronic applications, the electronic provision of documents, the use of videoconferencing and the interconnection of judicial and administrative registers to be increased, in order to further reduce the cost of judicial and out-of-court proceedings [4, 7].

The existing European roadmap covers the objectives for the European projects in the field of e-Justice up until 2013. Some of the existing projects will only bear visible results after that period since the development of European-wide IT projects of preliminary groundwork.

Along with the many administrative, organizational, social and aesthetic requirements [6, 10, 11], there are also technical ones. They have been largely set in a number of tools to evaluate the WEB-content of the sites. The existence of a web-site is not enough. The questions, related to the Internet visibility of the web-site, are also of importance.

The official list of all courts is published on the website of the Supreme Judicial Council. Links to the websites of all courts are available on the website of the Supreme Cassation Court in the section "Useful links".

The first stage of the study was implemented in January 2013 between 07 and 14. The survey covered all courts by type: 7 Appellate courts; 28 District courts; 113 Regional courts; 5 Military courts; 28 Administrative courts and the Specialized Criminal Court [5, 6].

II. Exposition

The subject of the study is to improve the presence of judicial institutions on the WEB in order to provide the necessary services citizens and the transparency of their activities. Part of the reviewed tools is used for SEO-optimization, but it is not the focus here and goes beyond the scope of issues discussed.

Before evaluating a WEB-site of an administrative unit of the judicial system, we should establish the criteria and their weight in the overall assessment. The analyzed sources [2, 3, 8, 12] offer various criteria that show different quantitative and qualitative characteristics of the sites.

There are numerous tools that facilitate both WEB-designers and experts in the creation of this type of software, as well as the experts and managers responsible for them.

Therefore are need to validate the WEB-sites and their codes according to the standard of WWW [2, 12] and evaluate to various characteristics such as site rating, and more.

In order to meet the set requirements and criteria for accessibility level, set by the European Commission about the websites of public administration, and also to meet the requirements of the current Internet technologies, the Web-sites of the institutions must adhere to the standards of WCAG 2.0 and the level of compliance "Double-A". Web sites of the institutions should cover the accessibility level Double-A according to the latest standards of the World Wide Web Consortium - Web Content Accessibility Guidelines 2.0 (WCAG 2.0) by the using best practices and techniques.

III. Tools

Here are discussed some of the most commonly used tools in practice, which give an idea of the types of tests and WEB-sites' inspection procedures.

* The research is financed by project № 08-306/12.03.2015, Research on intelligent methods and applications of simulators for neural networks and optimal methods of learning process of University of Shumen

A. *The validator of W3C - Markup Validation Service*

Markup Validation Service http://validator.w3.org/ - checks the validity of WEB-documents by using the scripting languages: HTML, XHTML, SMIL, MathML and checking whether the site carries out the ISO / IEC standard 15445: 2000 Information technology - Document description and processing languages - HyperText Markup Language (HTML) [2, 11]. For the validation a specific content such as RSS / Atom feeds or CSS styles, MobileOK content, or to find broken links is which there are other varieties of that validator are used for.

B. *SEO-optimization Tool*

"SEO-optimization Tool" [11, 12] also deserves attention, but only some of its functions can be used for free: check site ranking, loading speed and coding URL, while others like the automatic SEO analysis and analysis of external links are to be paid for.

C. *Open Site Explorer Tool*

The tool „Open Site Explorer"(fig.1) is powerful and multi-functional [9], gives an option for complex optimization. It allows to evaluate the rating of the domain, the WEB-site, the link metrics, the social metrics (not available in trial version), and the quantitative assessment of the sanctioned spam and inbound links to the site.

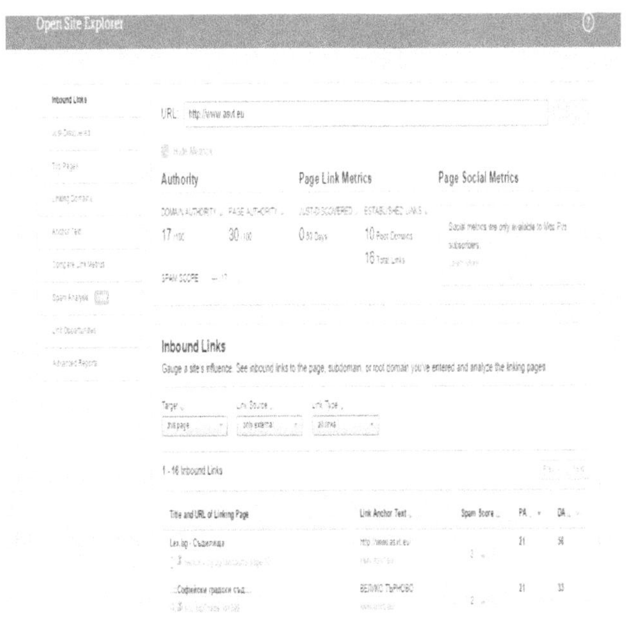

Fig. 1. Tool for complex SEO-optimization and testing of WEB-sites

D. *PINGDOM TOOLS*

PINGDOM TOOLS provide some excellent possibilities for precise monitoring of WEB-sites (Fig. 2). Pingdom performs a global monitoring of WEB-sites and WEB-applications [8]. It has the following functions:

Uptime MONITORING – tests and verifies the presence of sites in the WEB-space every minute automatically from over 60 selected global locations

- REAL USER MONITORING – accumulates and stores valuable performance data of the WEB-site, based on actual visits from users in order to improve its performance.
- TRANSACTION MONITORING - shows if important interactions and operations with the site such as registration, search or downloading files are slow or crashed.
- DevOps - integrates with other mostly cloud applications to correlate site data with given indicators in real-time, in order to improve productivity.
- RELIABLE – allows any problems encountered to be checked by a second opinion to achieve filtering of false alarms through this double check.
- ROOT CAUSE ANALYSIS - determines the causes of errors and interruptions in the WEB-site or server to solve problems and prevent their reoccurrence.

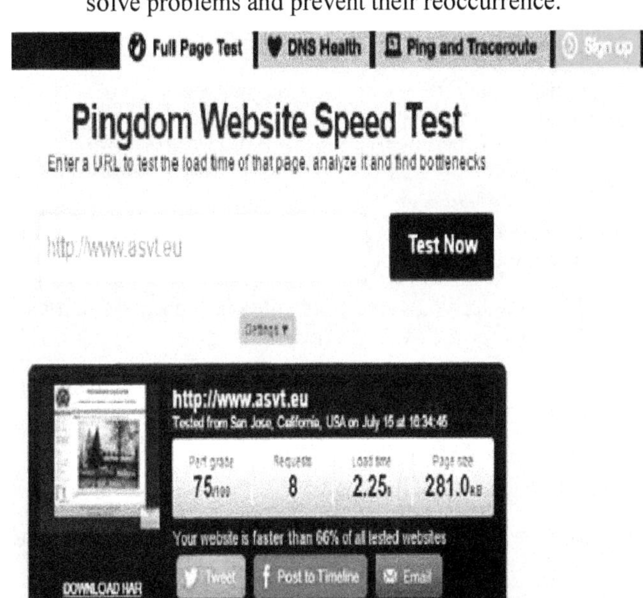

Fig. 2. Pingdom tool for monitoring WEB-sites and WEB-applications

IV. CRITERIA

The evaluation ***E*** of WEB-pages from sites of the judicial system can be defined by the multitude:

$$E = \{K_{ij}, O_j, S_j\} \quad (1),$$

where $K_{ij} = V_{jk}$; $i = 1,..,m$; $j = 1,..,n$; $k = 1,...,b$; V_{jk} – numeric evaluations of the criteria, m – number of evaluation criteria (received by the tools for evaluating WEB-sites),

n — number of evaluated judicial institutions, b – extreme values of the scale of the evaluated values, complex criteria O_j *and* S_j have values in the interval + or –[0,…1].

The calculation of the criterion for evaluating the judicial WEB-sites is one of the most important stages in the process. You can define the following rules for calculation.

1) The value of the first criteria K1 is derived from the check with the validator described above (fig. 1) and is defined:

- If the site cannot be validated $\mathbf{K_1}$ is not evaluated and is marked with **NO** in Table 1, part of which is shown below;
- If it is validated with errors, the derived value $\mathbf{V_1}$ is recorded in the same table and $\mathbf{K_1} = \mathbf{V_1}$ for the evaluated object;
- If there are no errors $\mathbf{V_1 = 0}$ and so $\mathbf{K_1 = 0}$.

2) The criterion reflecting a site's ranking in productivity $\mathbf{K_2}$ is defined like that:

- If $\mathbf{K_2 = 0}$ then $\mathbf{K_2}$ is not evaluated and we do not record anything in Table 1 (leave an empty cell);
- Else the derived value $\mathbf{V_2}$ is filled in the same table and $\mathbf{K_2 = V_2}$ for the evaluated object.

The site is ranked, taking into account the results of tests passed until its conducting.

3) The number of requests necessary for loading the site V_3 define the criteria $\mathbf{K_3}$

- If $\mathbf{K_3 = 0}$ then $\mathbf{K_3}$ is not evaluated and we do not record anything in Table 1 (leave an empty cell)
- Else the derived value $\mathbf{V_3}$ is filled in the same table and $\mathbf{K_3 = V_3}$ for the evaluated object.

This value describes the number of required elements of the site that need to be loaded for its proper operation.

4) Site loading time – V_4 defines the criterion $\mathbf{K_4}$
It is defined in the following way:

- If $\mathbf{K_4 = 0}$ then $\mathbf{K_4}$ is not evaluated and we do not record anything in Table 1 (leave an empty cell)
- Else the derived value $\mathbf{V_4}$ is filled in the same table and $\mathbf{K_4 = V_4}$ for the evaluated object in seconds.

The largest and most authoritative company in this business GOOGLE ranks well a site only if it loads quickly. If the site is slow, it cannot optimize well.

The rating is good at loading speed from 0 to 1 sec, 2 to 3 seconds is average and the owners should work to improve it, and more than 4 seconds means that the owners must definitely optimize it.

5) Site total size - $\mathbf{K_5}$
This criteria is defined like that:

- If $\mathbf{K_5=0}$ then $\mathbf{K_5}$ is not evaluated and we do not record anything in Table 1 (leave an empty cell)
- Else the derived value $\mathbf{V_5}$ is filled in the same table and $\mathbf{K_5 = V_5}$ for the evaluated object in seconds.

The size of the WEB-site should be optimal, depending on its purpose. The table is filled in with values in kilobytes (Kb). It is preferable the site to have a minimum size.

The results of the experiment conducted in May 2015 on 182 courts in Bulgaria are shown in Table 1.

TABLE I. PART OF THE EVALUATIONS OF THE JUDICIARY WEB-SITES IN BULGARIA

№	Name	WEB-site 2015	Valid. of WEB-sites (errors) V_1	Site rank by prod. (out of 100) V_2	Num. requests V_3	Load time (s) V_4	Site size (Kb) V_5
1	2	4	5	6	7	8	9
1	Burgas Court of Appeal	www.bgbas.org	21	87	29	0,829	131,4
2	Varna Court of Appeal	www.appealcourt-varna.org	1	85	20	1,39	250,5
5	Sofia Court of Appeal	acs.court-bg.org	no				
14	Administrative Court Burgas	http://www.admcourt-bs.org	19	82	29	1,97	250,5
…	…	…	…	…	…	…	…
182	Regional Court - Yambol	http://yambol.court-bg.org	no				

In order for the WEB-site to be evaluated, it is necessary to consider the impact of all criteria on its functionality. This is accomplished by making the following steps, which are used in the methodology [5, 6]:

A. Remove the results of experimental data beyond borders

It is believed they are due to errors in the reporting of the primary data obtained or other non-specific events during the experiment.

For each criteria, the mean square deviation is calculated and it is determined whether there are values out of range (- 3σ, + 3σ). If there are such values they are brought to value of the nearest border. WEB-sites that cannot be validated are excluded from calculations of further steps!

B. Normalization of criteria for evaluated objects - Web-sites of the courts

$$K_{ij} = \frac{v_{ij} - v_{javg}}{\sigma} * 100 \quad (2)$$

Where K_{ij} – i criteria for the j site i= 1,…, m; j=1,….n.

Performed through transformation that takes into account the averages and deviations from them.

The obtained results are Table 2 where each of the rows contains a vector with values of the criteria K_1 to K_5 (column 2 to 6) for a WEB-site. The values which are negative are below average importance, and those with a positive sign are above average importance.

TABLE II. VALUES OF CRITERIA AND RANKING OF THE SITES OF THE JUDICIARY

№	K_1	K_2	K_3	K_4	K_5	O_{ij}	S_{ij}	R.
1.	2.	3.	4.	5.	6.	7.	8.	9.
160	-0,000963726	-0,3635488	-0,02933	-0,00128	-0,00102	0,760647	0,15212946	1
12	0,014442872	0,360612219	-0,01798	-0,00142	-0,00086	0,729849	0,14596982	2
74	-0,005407937	0,360612219	0,01891	-0,00149	-0,00076	0,72707	0,14541391	3
7	0,007924696	0,360612219	-0,03784	-0,00082	-0,0012	0,697213	0,13944259	4
27	-0,001852569	-0,323317632	-0,04351	-0,00138	-0,00076	0,695997	0,13919947	5
144	-0,002148849	0,360612219	-0,035	-0,00155	0,000831	0,681209	0,13624183	6
57	-0,004222814	0,360612219	-0,02933	-0,00145	-0,00091	0,681092	0,13621841	7
...	...	...	...	...	...	...	...	...
104	0,003184	-0,001468	-0,000953	-0,000894	-0,000346	0,001239	0,000248	**137**

R. – rating.

C. The integrated evaluation of each site is determined by the expression:

$$O_i = \sum_{j=1}^{n} w_j \, v_{ij} \quad (3)$$

where i =1,...m, j =1,...,n, w_j -weight coefficient indicating the importance of each criterion, v_{ij} - numeric value of criteria j for site i.

This evaluation O_{ij} weighs the different criteria in the final evaluation. In this case the weight of the first and the second criteria is $w_{1,2}$ =2, which means that they are basic and have a two times greater effect than the other three which are of weight $w_{3,4,5}$ =1. They are shown in column 7 of Table 2.

D. Complex evaluation of sites

$$S_i = \frac{\sum_{j=1}^{n} O_{ij}}{n} \quad (4)$$

where i =1,...m, m-number of the evaluated sites (does not include non-validating sites), j =1,...,n, ,n=5, O_{ij} - integrated evaluation j for site i.

The complex evaluation of WEB-sites of the judicial system is calculated as the average of the evaluation of other criteria and their weighted influence. Thus the qualities of the development and functioning of the sites are considered. The results are in column 9 of Table 2.

V. DISCUSSION

The results show that 45 (24.73%) of the sites cannot be validated and are excluded from the evaluation. This leads to a violation of the standard and the urgent need to take action to resolve the issue.

The remaining 137 sites are validated with different number of errors in the code, adversely affecting their quality. There are 16 sites in which no errors were made and complied with standard WWW [2, 12].

The integrated O_i and the complex evaluation S_i are illustrated in Fig. 3.

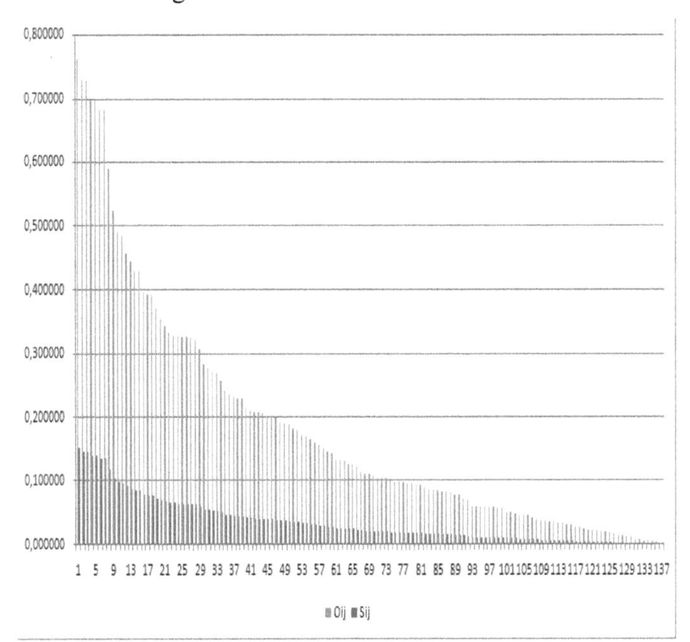

Fig. 3. Integrated and comprehensive assessment of websites

VI. CONCLUSION

The existing WEB-pages of the courts are evaluated by appropriately selected criteria, tools and methodology. The proposed conclusions and recommendations were obtained after analyzing the experimentally determined values of several criteria. They cover the important features of this type of communication and their visual expression in the WEB-space.

A large number (45) of the pages of judicial institutions are not validated and cannot be evaluated. They need to be adjusted and adapted to the standard for validating the content.

The remaining (137) WEB-pages have a large number of errors, which are subject to correction and improvement to reach a good level of maturity. This will ensure better results in the service of citizens and employees who work for these institutions.

REFERENCES

[1] G. Dimitrov, S. Manolov , L. Blagoev and A. Kuyumdzhieva, "State of the E-government in Bulgaria", Analysis, Law and Internet Foundation, Sofia, pp. 34-56, 2006.

[2] E. Castro, "HTML, XHTML & CSS", Visual Quickstart Guides, Sixth Edition, Peachpit Press, pp. 345-346, 2007.

[3] M. Petrova, "Informatization of public in Bulgaria", pp. 24-47, FABER, Veliko Tarnovo, 2014.

[4] M. Petrova and G. Dimitrov, "E-justice - the guarantee of an effective judicial system. Initiatives in Bulgaria", REVISTA NAŢIONALĂ DE DREPT (Publicaţie periodică ştiinţifico-practică), Universitatea de Stat

din Moldova, Chisinau, Moldova, nr.10-11 (133-134), pp. 169-173, 2011.

[5] M. Petrova, "Internet-presence and internet-exposure of the courts in Bulgaria", VI International Scientific Conference "Innovations in technology and education", KuzGTU, Belovo, May, 2013, v. 2, pp.159-165.

[6] M. Petrova, "Research of web-sites in the district courts of Bulgaria regulatory compliance", VI International Scientific Conference "Innovations in technology and education", KuzGTU, Belovo, May, 2013, v.2, pp. 159-165.

[7] Official Journal of the European Union, C 182, pp 2-13, Volume 57, June, 2014.

[8] "Optimize your Pingdom alerts", PINGDOM tutorial Tools, http://royal.pingdom.com/tag/tutorial/ Retrieved, July, 2015

[9] R. Fishkin and Moz Staff , "The beginer's Guide to SEO", http://d2eeipcrcdle6.cloudfront.net/guides/M-The-Beginners-Guide-To-SEO.pdf, Retrieved, August, 2015.

[10] M. Stefanova, T. Stefanov and O. Asenov, "Access to E-government' Services through Vein-code Biometric Identification", 7th Iberian Conference on Information Systems and Technologies, Madrid, Spain, June 2012, Vol. II, pp.174-176.

[11] Stefanov, T. "Methods for Assessing Information Sites", XLVII International Scientific Conference on Information, Communication and Energy Systems and Technologies ICEST'12, Bulgaria, Veliko Tarnovo, June 2012, Vol.2, pp. 455 – 458.

[12] Web Content Accessibility Guidelines (WCAG) Overview. World Wide Web Consortium, Retrieved, April, 2009.

Semantic Image Retrieval: An Ontology Based Approach

Umar Manzoor[1], Mohammed A. Balubaid[2]
[1] Faculty of Computing and Information Technology
[2] Industrial Engineering Department, Engineering Faculty,
King Abdulaziz University,
Jeddah, Saudi Arabia

Bassam Zafar[1], Hafsa Umar[3], M. Shoaib Khan[3]
[1]Faculty of Computing and Information Technology
King Abdulaziz University, Jeddah, Saudi Arabia
[3]National University of Computer and Emerging Sciences,
Islamabad, Pakistan

***Abstract*—Images / Videos are major source of content on the internet and the content is increasing rapidly due to the advancement in this area. Image analysis and retrieval is one of the active research field and researchers from the last decade have proposed many efficient approaches for the same. Semantic technologies like ontology offers promising approach to image retrieval as it tries to map the low level image features to high level ontology concepts. In this paper, we have proposed Semantic Image Retrieval: An Ontology based Approach which uses domain specific ontology for image retrieval relevant to the user query. The user can give concept / keyword as text input or can input the image itself. Semantic Image Retrieval is based on hybrid approach and uses shape, color and texture based approaches for classification purpose. Mammals domain is used as a test case and its ontology is developed. The proposed system is trained on Mammals dataset and tested on large number of test cases related to this domain. Experimental results show the efficiency / accuracy of the proposed system and support the implementation of the same.**

Keywords—Image Retrieval; Ontology; Semantic Image; Image Understanding; Semantic Retrieval

I. INTRODUCTION

Images / Videos are major source of content on the internet and the content is increasing rapidly due to the advancement in this area [10, 12, 13]. Digital Image processing / retrieval is one of the hottest research field and researchers from the last decade have proposed many efficient approaches for image analysis such as [6, 7, 14, 15] and retrieval [9, 11, 16, 17]. Image retrieval systems are usually based on keywords or text meta-data based [4, 18, 19] where the retrieval is done based on the textual description of the images. The description about the image is usually provided by the user. Most common search engines such as Google and Bing used keyword based search techniques; this approach is fast and effective; however it still has some disadvantages. In this approach, the image is described by a set of keywords or text-metadata and usually this information is provided by the user.

The keyword based image retrieval system matches user text query to the textual description of the images and return all the images whose description is the possible match. However, it is quite possible that the results returned contain irrelevant images. For example, you may find a dog picture while you are searching for human. This usually happens because the description of the irrelevant image contains that specific keyword. So, the major disadvantage of text-based image retrieval system is that it may return redundant or irrelevant images in the result [13, 4].

The accuracy of keyword based image retrieval systems is far from perfect because of the following reasons:

1) If the user made spell mistake while describing the image, this image will never be listed in the result because of this mistake.

2) Sometimes the user has to specify the image description / keywords in natural language which makes it difficult to describe the image as the user has little knowledge about the natural language.

3) It is very difficult to find appropriate keywords for image description (i.e. synonym plays important role in image retrieval).

In conclusion, keyword approach ignores the image features which sometimes results in irrelevant image retrieval [23, 24].

Content based Image retrieval (CBIR) has been studied for many years which focuses on extracting and comparing features from the images [20, 21, 22]. Image Features are usually extracted using dominant color, dominant texture, or shape (i.e. this technique focuses on the visual features of the image). Researchers in the last decade have demonstrated the efficiency and accuracy of CBIR based techniques, however, CBIR still lacks to understand the semantic analysis of the image. For example, if the user wants to search "Loin" images, CBIR system will not be able to map human concept into image feature (i.e. creating a semantic gap between the low-level image features and high-level human understandable concepts). Therefore, semantic analysis needs to be incorporated in content based image retrieval to reduce this gap.

Semantic technologies like ontology offers promising approach to image retrieval as it tries to map the low level image features to high level ontology concepts. Compared to the existing approaches (i.e. text / keyword based and content based image retrieval), Ontology based image retrieval focuses more on capturing semantic content (i.e. mapping image features to concepts), because this can help in satisfying user requirements in much better way.In this paper, we have proposed Semantic Image Retrieval: An Ontology based Approach which uses domain specific ontology for image retrieval relevant to the user query.

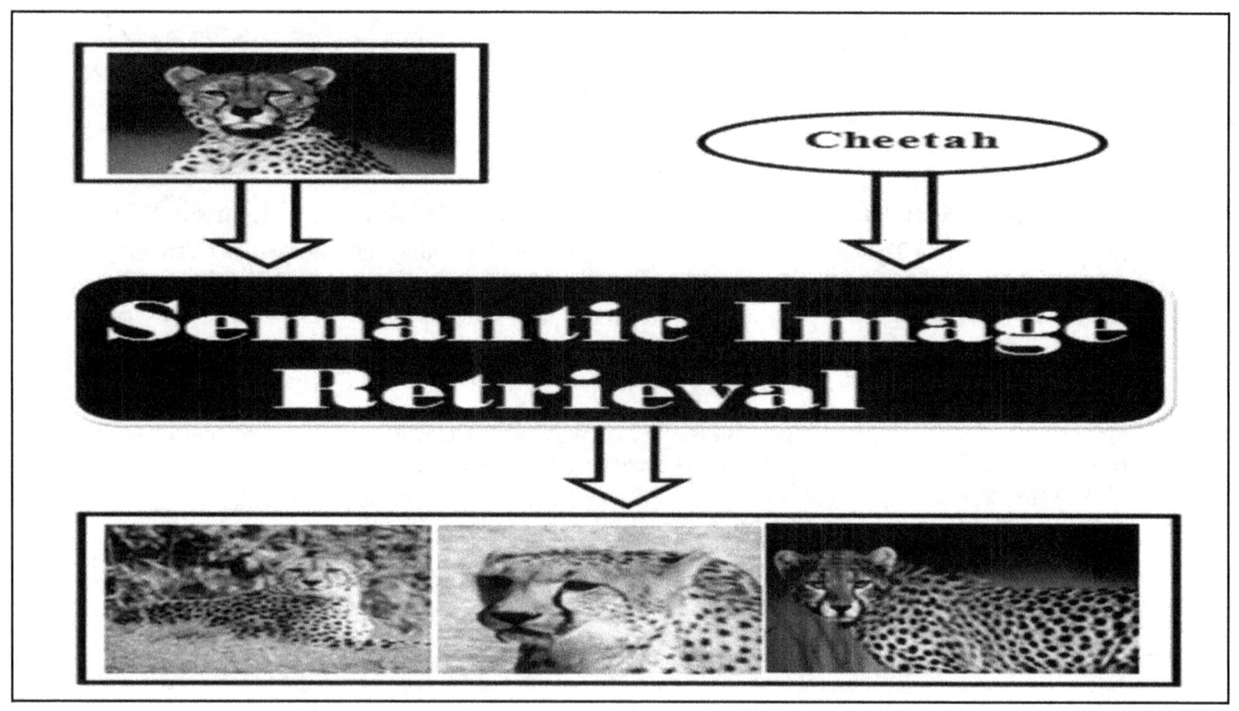

Fig. 1. Zero Level Architecture of Semantic Image Retrieval: An Ontology based Approach

The user can give concept / keyword as text input or can input the image itself. Mammals domain is used as a test case and its ontology is developed. The proposed system is trained on Mammals dataset and tested on large number of test cases related to this domain. Experimental results show the efficiency and accuracy of the proposed system.

The remainder of this paper is organized as follows. In Section 2, we present brief overview of ontology and image analysis, this section is followed by the discussion of literature survey. In Section 4, the proposed Semantic Image Retrieval: An Ontology based Approach architecture and classification mechanism is discussed. In Section 5, the experimental analysis of proposed solution is presented. Finally, the conclusion is drawn in Section 6.

II. Ontology and Image Analysis

The word ontology refers to the science of metaphysics which defines the nature with its properties and relations [8]. In Computer Science, ontology is a systematic arrangement of concepts, their properties and relations which exist in domain [25]. Common components of ontology includes Individuals, Classes, Attributes, Relations, Function terms, Restrictions, Rules, and Axioms; for more details related to these concepts please see [3, 5]. Ontology can be domain-specific or generic; the former means ontology concepts are defined with reference to the specific domain whereas the later means the concepts are defined in general (i.e. the meaning / relationship of these concepts are already defined by English language) [26].

The implementation of ontology is generally a hierarchal representation defining concepts and their relationships. Three kind of relationships namely is-a, instance-of and part-of are generally used in the ontology; for more information please see [27, 28]. Ontology are usually develop to share common understanding of information among entities or softwares where each node in the ontology is a concept containing set of attributes and relationships.

In the last decade, Ontologies have been widely used for knowledge representation and sharing. Ontology-based systems have been used in diverse areas such as software maintenance, Business Process Management, Biomedical Informatics, Knowledge Sharing, Knowledge Integration, Semantic Web, Fuzzy Systems, Supply chain management, Healthcare, Text Classification, Medical Domain, Robotics, Autonomic Computing, System Modelling, etc.

The idea of using the ontologies in Image processing for content used retrieval is not new; in the last decade, researchers have proposed many efficient solutions using Ontologies for content based Image processing and retrieval such as [29-34]. The existing approaches can broadly be categorized into three types namely 1) Color based techniques 2) Shape based technique and 3) Texture based technique. The color based approaches proposed calculate the color histogram of the image and use the same for classification, shape based approaches identify the shape(s) in the image and use it for classification whereas the texture based approaches identify the texture in the image and use it for classification purpose.

Each of the approaches discussed above have some limitation, for example the color based technique will work effectively on the color-dominant image dataset whereas it will be outperformed by other technique on non-color-dominant image dataset. Similarly shape detection in complex images are hard and texture based approaches will be outperformed on non-texture-based image dataset. In this paper, we have proposed a hybrid technique which uses color, shape and texture feature of the image and use these features for classification.

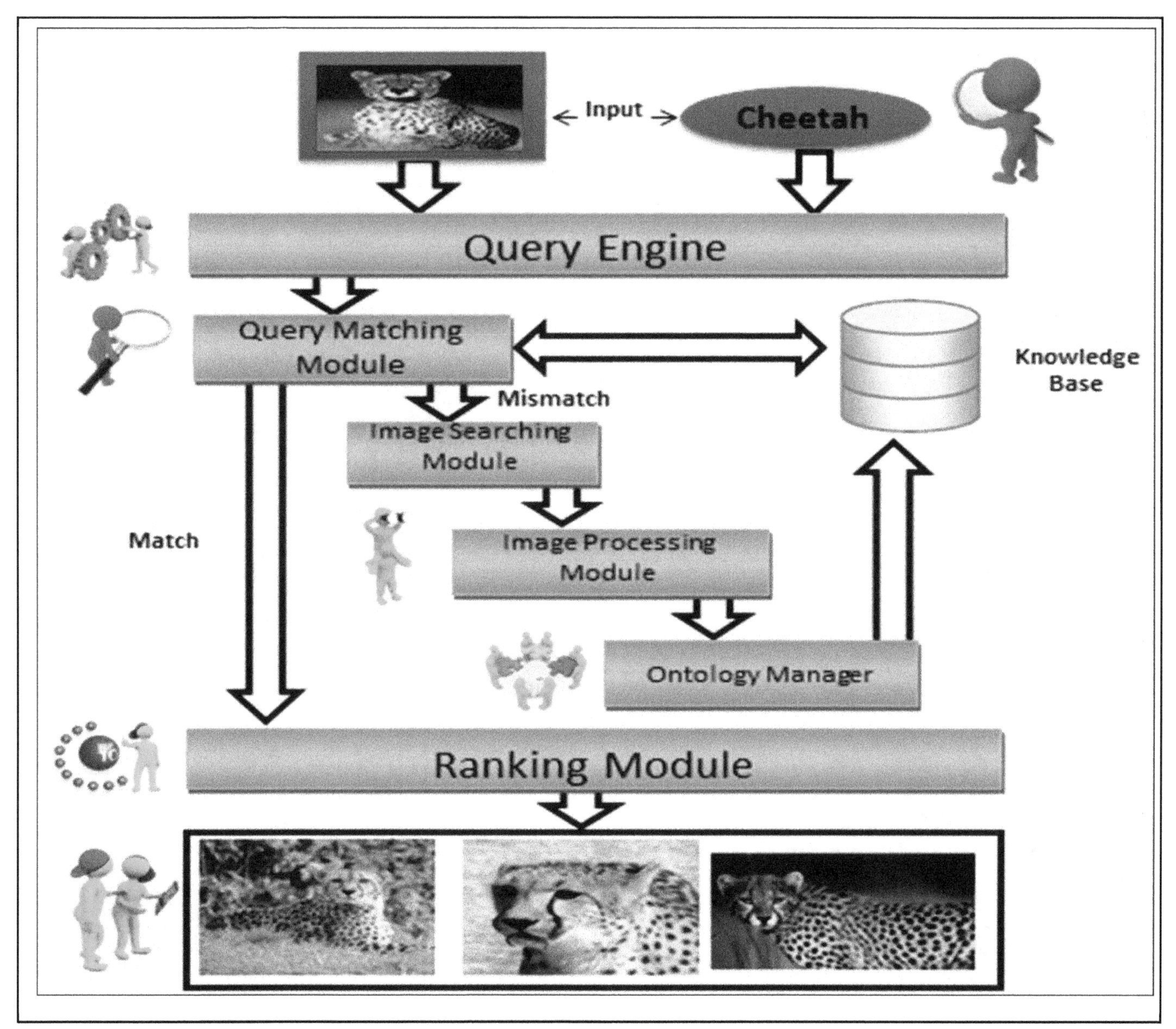

Fig. 2. System Architecture of Semantic Image Retrieval: An Ontology based Approach

III. Literature Review

A lot of research has been conducted on Image Retrieval (IR) on the basis of content similarity. Many techniques have been used to enhance the results of image search. These approaches include hierarchical knowledge-based systems for Image Retrieval as researched by Kurtz, Camille, et al [40] in 2014. The semantic gap between the low-level image features and their high level semantics has always ruined the retrieval quality. So to cope up with this problem, Fernández Miriam et al. [36] used an ontology based approach for the enhancements of the image semantics. This research aimed to solve the restriction of the keyword based searching to support the semantic based Image Retrieval. The concept of semantic indexing has also been studied in the field of ontology based retrieval systems. The literature review on Image Retreival based on semantic concepts by Riad Alaa et al. [38] had a great impact on the Image Retrieval field as it was very helpful for improving the semantic image retrieval systems accuracy. In this research various image search techniques are described for reduction of semantic gap. Furthermore, based on existing methods and application requirements author have suggested few future assessments. Another important survey was conducted by Liu Ying et al. [39] in 2007 about the recent technical achievements on semantic based Image Retrieval; majority of the recent publications were included as the test data for the survey covering diverse amount of aspects in this area. Similar work has also been conducted on medical images by Xu J et al. in [41], the authors focused on the key features of the image (e.g., shape, texture) in this research. The authors concluded that the performance of most CBIR systems is forced by these features because they cannot efficiently model the expectations of the user. All of existing studies helped in improving the results of content based images retrieval and lowering down the semantic gap between the user requirements and the search results.

IV. System Architecture

Semantic Image Retrieval (SIR): An Ontology based Approach system architecture describes the working of the various components / modules of the system and their interaction with each other. Figure 2 shows the detail system architecture of SIR and consists of the following modules:

- Query Engine
- Matching Module
- Ontology Manager

A. *Query Engine*

Query Engine is responsible to take input from the user using the web interface; the input contains the content which the user wants to search. The input can be provided in two ways by the user.

*1) **Text Input:*** The first method of providing the input to the SIR is text based. In this approach the user is required to enter the text containing the information about the thing that he / she wants to search. This approach is commonly used in the current search engines, e.g. Google, Bing, AltaVista etc. The main focus of incorporating this approach in SIR is to provide ease to the users as they do not have to learn the new way of interacting with the SIR. The user has to simply write down the text query (e.g. Cheetah, Elephant, Horse etc and the same is passed to Text based Query module.

*2) **Image Input:*** The second method of providing the input to the SIR is image based. In this approach the user is required to provide the image of the object(s) which he/she wants to search. The input image can contain a single object or multiple objects. The user is also provided some options (optional) to describe the input image. This approach is feasible when the user wants to search related objects / images similar to the one he / she has. Furthermore, this method provides flexibility in the input method, as it gives new dimension to the searching. After taking input from the user, Query Engine built the query for the input. As Ontology based Knowledge base is used, the query is built in SPARQL language. The query building process consists of the following two components.

a) Text based Query: This module is responsible for building the query for the text based input. In Step 1, all standard stop-list / stemmer words like ("is", "the", "on", "and"…) are removed from the input text. In Step 2, SPARQL query is generated with all possible "AND" and "OR". The generated query is then passed to the Matching Module for the further processing.

b) Image based Query: This module is responsible for building the query for image based input. In Step 1, object(s) in the image are detected using shape based feature extraction as described in [2]. After object detection, two sub-steps are performed: In the first step, the detected objects are passed to

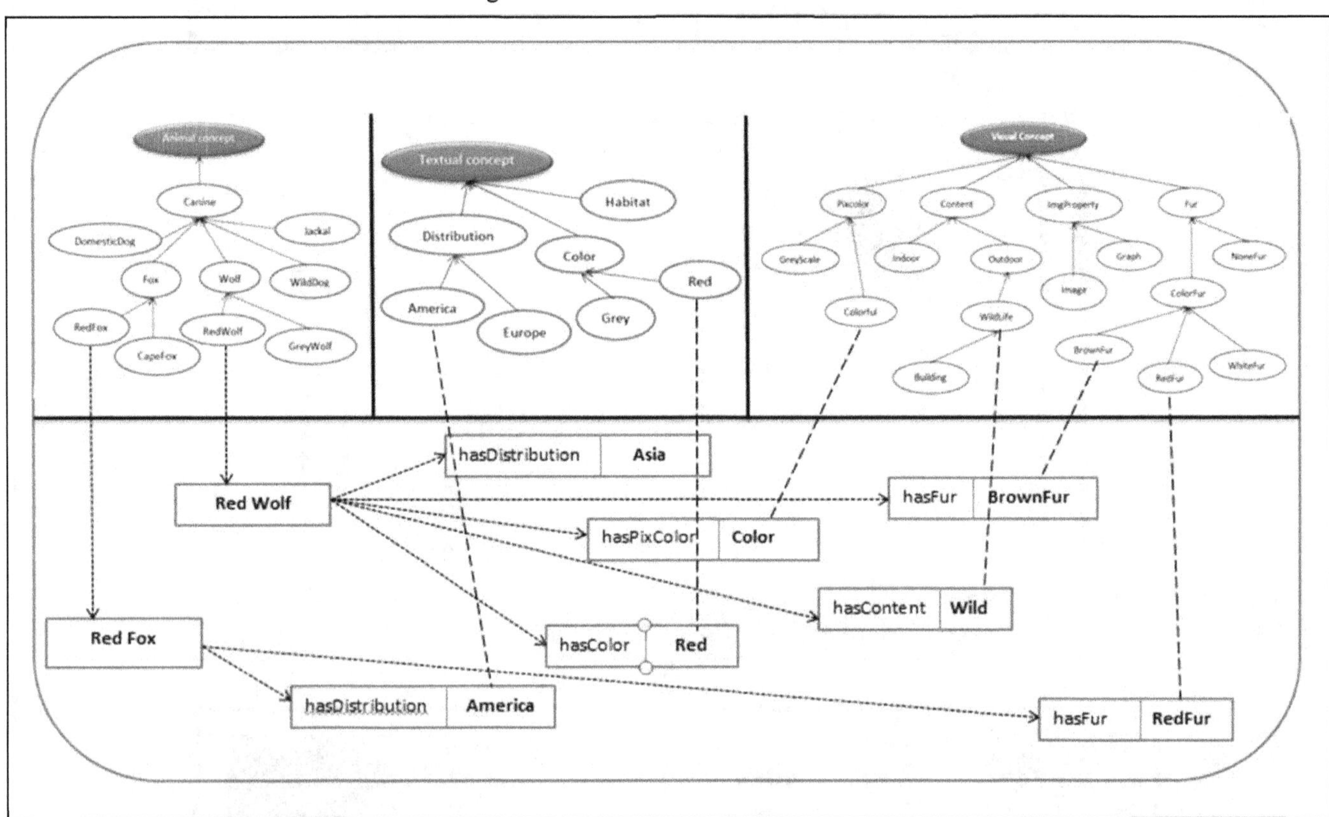

Fig. 3. Partial Ontology Knowledgebase

Color based Feature Extraction technique which uses MTH algorithm proposed by Guang-Hai Liu et al in [1] to calculate the color value and pixel color of the objects; In the second step, the detected objects are passed to texture classification technique proposed by Mohsen Zand et al in [35] to identify texture / pattern (if any) in the detected objects. In Step 3, the low level features extracted using the previous two steps are converted into high level ontology concepts; the image description if provided in search by the user are also converted into ontology concepts, after completing this step SPARQL query is generated using these parameters.

B. Matching Module

Matching Module takes SPARQL query as input from the Query Engine and executes the same on the Ontology Knowledge Base to retrieve the most related images. If the query results in successful search, the output images are passed to ranking module for result ranking. If the search is unsuccessful (i.e. relevant images are not found in our knowledge base), matching module performs the following three steps:

Image Search: Matching Module searches the internet for relevant images by querying existing search engine (i.e. Google or Bing). The results returned by search engine are passed to Image Processing Module for content verification.

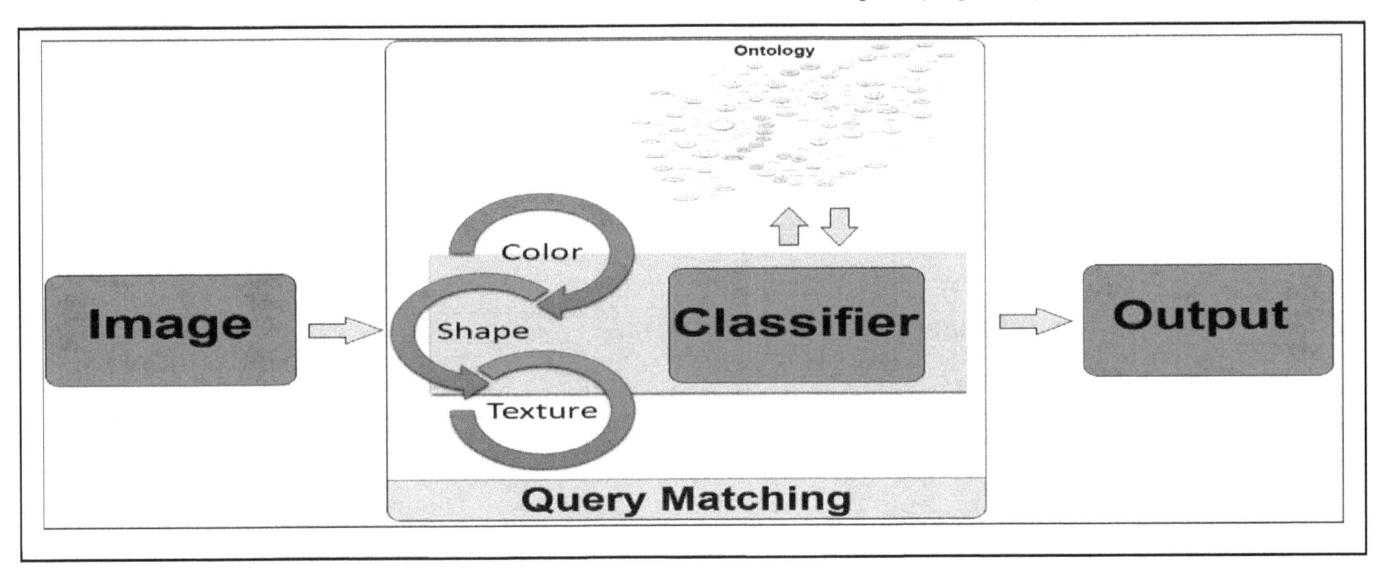

Fig. 4. Query Matching

1) ***Image Processing:*** The images returned by search engine may not be relevant to the user query; therefore the content of each image needs to be verified. This module is responsible to check the images for the compliance with the input query. The objects in each image are detected using shape based feature extraction and these objects are passed to 1) Color based Feature Extraction technique which uses MTH algorithm proposed by Guang-Hai Liu et al in [1] to calculate the pixel color and color value of the objects and 2) texture classification technique proposed by Mohsen Zand et al in [35] to identify texture / pattern (if any) in the objects. In the next step, the low level features extracted in the previous step are converted into high level ontology concepts; afterwards SPARQL query is generated using these concepts and executed on the ontology knowledgebase. If the result class(es) matches user search query, the image is included in the resultant set otherwise it is discarded. As a result only the related images remains and the non-relevant images are discarded in this step.

2) ***Ontology Manager:*** Ontology Manager is responsible to insert the new relevant images features and concepts (gathered from the web and filtered in the previous step) in the ontology knowledge base.

C. Ranking Module

Ranking module is responsible to rank the images according to relevance with the user query. The resultant image set passed by Query Matching Module contains image and matching value (which is calculated as a sum of matched ontology concepts with reference to user query); the result set is sorted in descending order according to the matching value. After sorting, top ten images are displayed to the user (i.e. most matched images are showed first) and the remaining are displayed on user request in the decreasing order.

V. SIMULATION

Initially for the experimentation, we trained Semantic Image Retrieval (SIR) and built the ontology concepts using 900 images which contain pictures of 20 different mammals. Partial training dataset is shown in figure 5. We have evaluated SIR on large number of test cases; results were promising and showed the efficiency of the proposed system. In this section, few of the test cases are presented and discussed in detail.

Fig. 5. Partial Training set

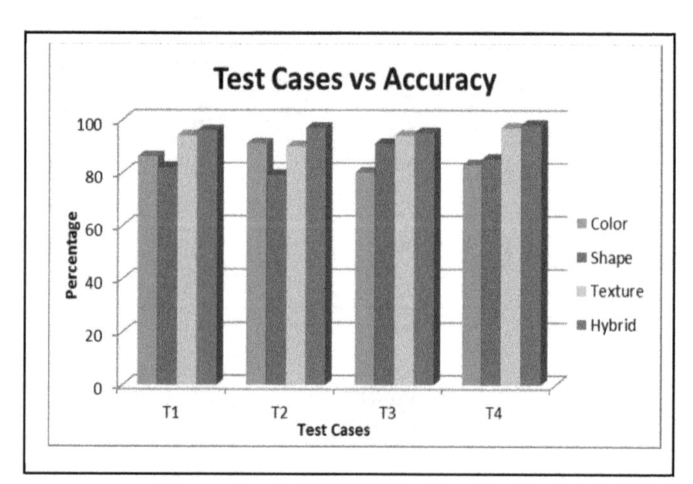

Fig. 6. Test Cases vs Accuracy

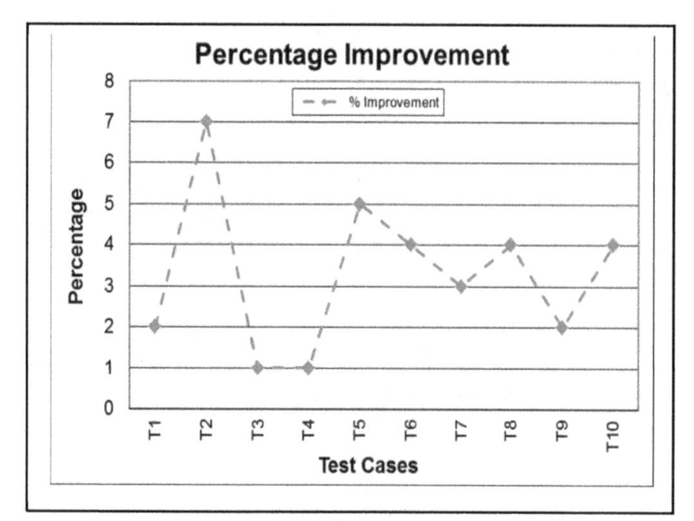

Fig. 7. Percentage Improvement vs Test Cases

Figure 6 shows the accuracy comparison of color based shape based, texture based and our proposed approach with reference to four different test cases. As depicted by figure 6, our proposed hybrid approach outperforms these approaches with reference to accuracy.

Figure 7 shows the percentage improvement of proposed hybrid technique over number of test cases; as shown in figure 7 the proposed solution improvement percentage varies over number of test cases; this is because the content of images present in each test case plays an important role.

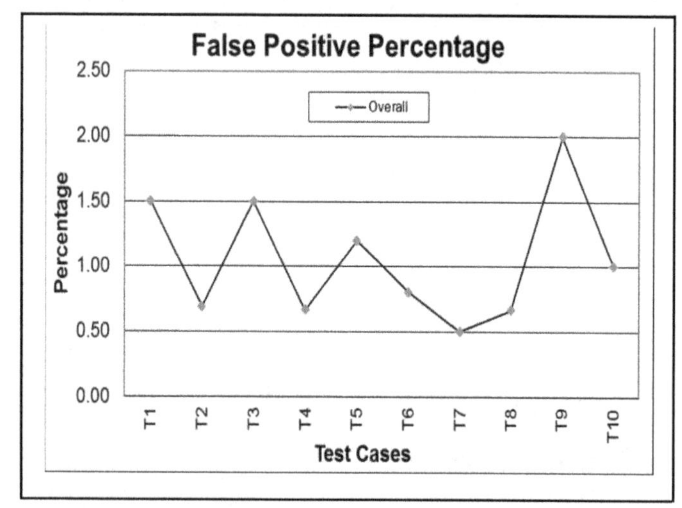

Fig. 8. False Positive Percentage vs Test Cases

Figure 8 shows false positive percentage over number of test cases, the proposed solution false positive percentage ranges from 0.60 to 2 percent in the test cases which shows the result accuracy of the proposed solution.

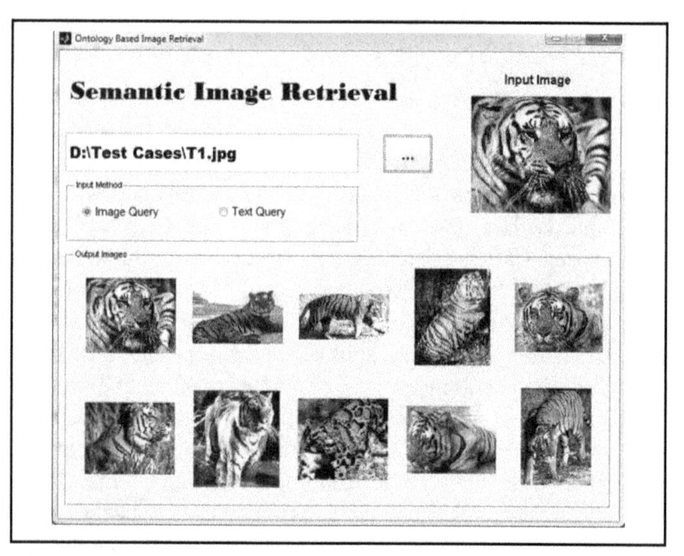

Fig. 9. Test Case 1

In figure 9, the user used cheetah image as input; Query Engine generates the query for the same and executes it on ontology knowledge base. The resultant images are found in the knowledge base, therefore web image search, image filtration and ontology updation steps are skipped in this test case.

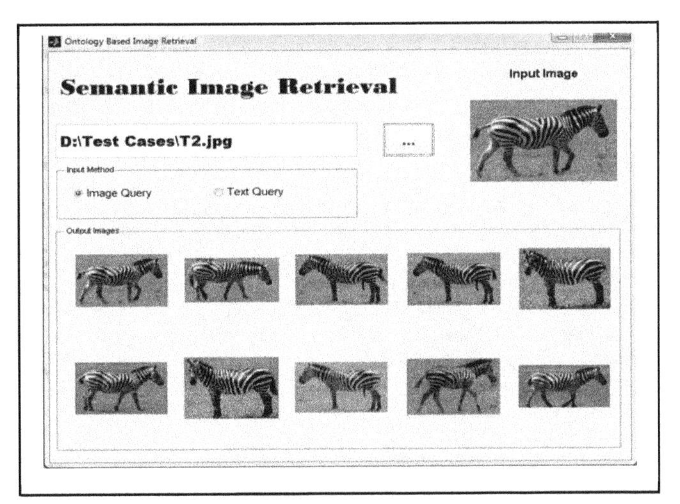

Fig. 10. Test Case 2

The results are passed to Ranking Module which ranks the results and displayed it to the user as shown in Figure 9. Figure 10 and 11 are similar to the first test case (figure 9) where the user enters an image and relevant images are returned to the user.

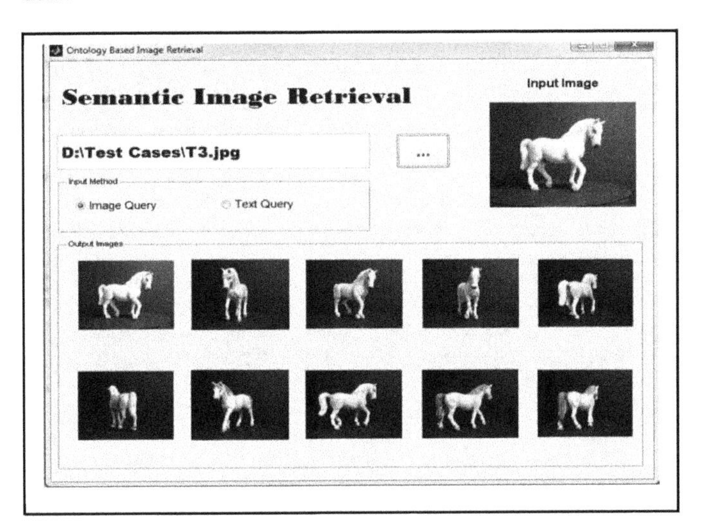

Fig. 11. Test Case 3

In test case 4, the user used text input feature of the SIR system and provided the input as text. SIR generates the corresponding query for the same and executes it on the knowledge base. The related images are displayed to the user as shown in figure 12.

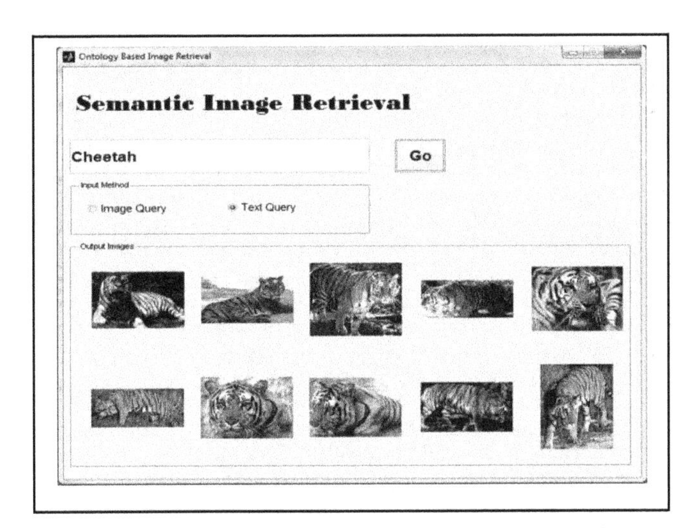

Fig. 12. Test Case 4

VI. CONCLUSION

Image retrieval systems are usually based on keywords or text meta-data based. Most common search engines such as Google and Bing are based on keyword based search techniques. This approach is fast and effective; however it still has some disadvantages. Content based Image retrieval (CBIR) has been studied for many years which focuses on extracting and comparing features from the images. Researchers in the last decade have demonstrated the efficiency and accuracy of CBIR based techniques, however, CBIR still lacks to understand the semantic analysis of the image. Semantic technologies like ontology offers promising approach to image retrieval as it tries to map the low level image features to high level ontology concepts. In this paper, we have proposed Semantic Image Retrieval: An Ontology based Approach which uses domain specific ontology for image retrieval relevant to the user query. The proposed system has been tested on large number of test cases; experimental results shows the efficiency and effectiveness of the proposed technique.

REFERENCES

[1] Guang-Hai Liu, Lei Zhang, Ying-Kun Hou, Zuo-Yong Li, Jing-Yu Yang, Image retrieval based on multi-texton histogram, Pattern Recognition (2010), Volume 43, Pages 2380–2389.

[2] Alexander Toshev, Ben Taskar and Kostas Daniilidis, Shape-Based Object Detection via Boundary Structure Segmentation, International Journal of Computer Vision (2012), Volume 99, Number 2, Pages 123-146.

[3] Umar Manzoor, Samia Nefti, Yacine Rezgui, "Categorization of malicious behaviors using ontology-based cognitive agents", Data & Knowledge Engineering (2013), Volume 85, May 2013, Pages 40–56.

[4] Y. Liu,D. Zhang,G. Lu,W.-Y. Ma,A survey of content-based image retrieval with high-level semantics, Pattern Recognition (2007), Volume 40, Issue 11, Pages 262–282.

[5] Umar Manzoor, Samia Nefti, "iDetect: Content Based Monitoring of Complex Networks using Mobile Agents", Applied Soft Computing, Volume 12, Issue 5, May 2012, Pages 1607-1619.

[6] Shotton J, Winn J, Rother C, Criminisi A (2009) Texton boost for image understanding: Multi-class object recognition and segmentation by jointly modeling texture, layout, and context. International Journal of Computer Vision, 81(1), 2009.

[7] Shotton J, Blake A, Chipolla R, Contour-based learning for object detection. In Proceeding of International Conference on Computer Vision (2005).

[8] Awatef Al Azemi, Samia Nefti, Umar Manzoor, Yacine Rezgui "Building a Bilingual Bio-Ontology Platform for Knowledge Discovery", International Journal of Innovative Computing, Information and Control, Volume 7, Number 12, Dec 2011, Pages 7067-7075.

[9] T. Quack, U. Monich,L. Thiele, B.S. Manjunath, Cortina: a system for large-scale, content-based web image retrieval, in:Proceedings of the 12th annual ACM international conference on Multimedia, 2004.

[10] Naveed Ejaz, Umar Manzoor, Samia Nefti, Sung Wook Baik "A Collaborative Multi-Agent Framework for Abnormal Activity Detection in Crowded Areas", International Journal of Innovative Computing, Information and Control, Volume 8, Number 6, June 2012, Pages 4219-4234.

[11] N. Alajlan, M.S.Kamel, G.H.Freeman, Geometry-based image retrieval in binary image databases,IEEE Transactions on Pattern Analysis and Machine Intelligence (2008), Volume 30, Issue 6, Pages 11003–11013.

[12] Kamran Manzoor, Atique Ahmed, Sohail Ahmad, Umar Manzoor, Samia Nefti "V-NIP Ceaser: Video Stabilization System", Communications in Computer and Information Science Volume 111, 2010, pp 350-356.

[13] Umar Manzoor, Naveed Ejaz, Nadeem Akhtar, Muhammad Umar, M Shoaib Khan, Hafsa Umar "Ontology based image retrieval", IEEE The 7th International Conference for Internet Technology And Secured Transactions (ICITST-2012), pp. 288-293, 2012.

[14] Peter Veelaert, Kristof Teelen, "Adaptive and optimal difference operators in image processing", Pattern Recognition, Volume 42, Issue 10, October 2009, Pages 2317-2326.

[15] Dacheng Tao, Dianhui Wang, Fionn Murtagh, "Machine learning in intelligent image processing", Signal Processing, Volume 93, Issue 6, June 2013, Pages 1399-1400.

[16] Xiang-Yang Wang, Hong-Ying Yang, Yong-Wei Li, Wei-Yi Li, Jing-Wei Chen "A new SVM-based active feedback scheme for image retrieval" Engineering Applications of Artificial Intelligence, Volume 37, January 2015, Pages 43-53

[17] Ming Zhang, Ke Zhang, Qinghe Feng, Jianzhong Wang, Jun Kong, Yinghua Lu "A novel image retrieval method based on hybrid information descriptors" Journal of Visual Communication and Image Representation, Volume 25, Issue 7, October 2014, Pages 1574-1587.

[18] Ren-Jie Wang, Ya-Ting Yang, Pao-Chi Chang "Content-based image retrieval using H.264 intra coding features", Journal of Visual Communication and Image Representation, Volume 25, Issue 5, July 2014, Pages 963-969.

[19] Subrahmanyam Murala, Q.M. Jonathan Wu, "Expert content-based image retrieval system using robust local patterns" Journal of Visual Communication and Image Representation, Volume 25, Issue 6, August 2014, Pages 1324-1334.

[20] Malay Kumar Kundu, Manish Chowdhury, Samuel Rota Bul "A graph-based relevance feedback mechanism in content-based image retrieval", Knowledge-Based Systems, Volume 73, January 2015, Pages 254-264

[21] Daniel Carlos Guimarães Pedronette, Jurandy Almeida, Ricardo da S. Torres "A scalable re-ranking method for content-based image retrieval", Information Sciences, Volume 265, 1 May 2014, Pages 91-104

[22] Hong-Ying Yang, Yong-Wei Li, Wei-Yi Li, Xiang-Yang Wang, Fang-Yu Yang "Content-based image retrieval using local visual attention feature", Journal of Visual Communication and Image Representation, Volume 25, Issue 6, August 2014, Pages 1308-1323.

[23] Ying Liua, Dengsheng Zhanga, Guojun Lua, Wei-Ying Mab "A survey of content-based image retrieval with high-level semantics" Pattern Recognition, Volume 40, Issue 1, January 2007, Pages 262–282.

[24] Ryszard S. Choraś, "Content-Based Image Retrieval — A Survey" Biometrics, Computer Security Systems and Artificial Intelligence Applications, 2006, pp 31-44.

[25] Umar Manzoor, Samia Nefti, Yacine Rezgui "Autonomous Malicious Activity Inspector – AMAI" Natural Language Processing and Information Systems, Lecture Notes in Computer Science Volume 6177, 2010, pp 204-215.

[26] Umar Manzoor, Bassam Zafar "Multi-Agent Modeling Toolkit – MAMT" Simulation Modelling Practice and Theory, Volume 49, December 2014, Pages 215–227

[27] Francesco Rea, Samia Nefti-Meziani, Umar Manzoor, Steve Davis "Ontology enhancing process for a situated and curiosity-driven robot" Robotics and Autonomous Systems, Volume 62, Issue 12, December 2014, Pages 1837–1847.

[28] Umar Manzoor, Mati Ullah, Arshad Ali, Janita Irfan, Muhammad Murtaza "A Tool for Agent Based Modeling – A Land Market Case Study" Information Systems, E-learning, and Knowledge Management Research, Communications in Computer and Information Science Volume 278, 2013, pp 467-472.

[29] Stefan Poslad, Kraisak Kesorn "A Multi-Modal Incompleteness Ontology model (MMIO) to enhance information fusion for image retrieval", Information Fusion, Volume 20, November 2014, Pages 225-241.

[30] Camille Kurtz, Adrien Depeursinge, Sandy Napel, Christopher F. Beaulieu, Daniel L. Rubin "On combining image-based and ontological semantic dissimilarities for medical image retrieval applications" Medical Image Analysis, Volume 18, Issue 7, October 2014, Pages 1082-1100.

[31] Mohsen Sardari Zarchi, Amirhasan Monadjemi, Kamal Jamshidi "A semantic model for general purpose content-based image retrieval systems" Computers & Electrical Engineering, Volume 40, Issue 7, October 2014, Pages 2062-2071.

[32] Gowri Allampalli-Nagaraj, Isabelle Bichindaritz "Automatic semantic indexing of medical images using a web ontology language for case-based image retrieval" Engineering Applications of Artificial Intelligence, Volume 22, Issue 1, February 2009, Pages 18-25.

[33] Nicolas Eric Maillot, Monique Thonnat "Ontology based complex object recognition" Image and Vision Computing, Volume 26, Issue 1, 1 January 2008, Pages 102-113.

[34] Enamul Hoque, Orland Hoeber, Minglun Gong "CIDER: Concept-based image diversification, exploration, and retrieval" Information Processing & Management, Volume 49, Issue 5, September 2013, Pages 1122-1138.

[35] Mohsen Zand, Shyamala Doraisamy, Alfian Abdul Halin, Mas Rina Mustaffa "Texture classification and discrimination for region-based image retrieval" Journal of Visual Communication and Image Representation (2014), doi:10.1016/j.jvcir.2014.10.005.

[36] Fernández, Miriam, et al. "Semantically enhanced Information Retrieval: an ontology-based approach." Web Semantics: Science, Services and Agents on the World Wide Web 9.4 (2011): 434-452.

[37] Kara, Soner, et al. "An ontology-based retrieval system using semantic indexing." Information Systems 37.4 (2012): 294-305.

[38] Riad, Alaa M., Hamdy K. Elminir, and SamehAbd-Elghany. "A Literature Review of Image Retrieval based On Semantic Concept." International Journal of Computer Applications 40.11 (2012): 12-19.

[39] Liu, Ying, et al. "A survey of content-based image retrieval with high-level semantics." Pattern Recognition 40.1 (2007): 262-282.

[40] Kurtz, Camille, et al. "A hierarchical knowledge-based approach for retrieving similar medical images described with semantic annotations." Journal of biomedical informatics (2014).

[41] Xu J, Faruque J, Beaulieu CF, Rubin DL, Napel S. A comprehensive descriptor of shape: method and application to content-based retrieval of similar appearing lesions in medical images. J Digit Imaging 2012;25:121–8.

17

A Study of Routing Path Decision Method Using Mobile Robot Based on Distance Between Sensor Nodes

Yuta Koike
Department of Information and Communication Engineering, Tokyo Denki University
Tokyo, Japan

Kei Sawai
Department of Information and Communication Engineering, Tokyo Denki University
Tokyo, Japan

Tsuyoshi Suzuki
Department of Information and Communication Engineering, Tokyo Denki University
Tokyo, Japan

***Abstract*—We propose Robot Wireless Sensor Networks (RWSNs) management method for maintaining wireless communication connectivity for a mobile robot teleoperation with considering a distance between sensor nodes. Recent studies for reducing disaster damage focus on a disaster area information gathering in underground spaces. Since information gathering activities in such post disaster underground spaces present a high risk of personal injury by secondary disasters, a lot of rescue workers were injured or killed in the past. On basis of this background, gathering information by utilizing the mobile robot is discussed in wide area. However, maintaining wireless communication infrastructures for teleoperation of a mobile rescue robot in the post-disaster underground space by various reasons. Therefore we have been discussing the wireless communication infrastructures construction method for teleoperation of the rescue robot by utilizing the RWSN. In this paper, we evaluated the proposed method for changing routing path by utilizing the RWSN in field operation test in order to confirm the availability of performance of communication connectivity and the throughputs between End-to-End communications via constructed network.**

Keywords—Wireless Sensor Networks; Moblie Robot Tele-Operation; Maintaining Throughput; Routing Path

I. Introduction

Gathering information in disaster areas is very important for assessing the situation, avoiding secondary disasters and managing disaster reduction [1]–[7]. However, if a disaster occurs in a congested city, the rescue team cannot gather information because of the complicated urban structure. In general, gathering information from a bird's eye view with an unmanned air vehicle (UAV) is a useful method in a disaster area. However, in an urban area with many underground spaces where information gathering by using a UAV is difficult, checking on the extent of the damage, which is important for avoiding secondary disasters, is difficult [8]. Also, rescue teams cannot organize a rescue plan for underground spaces. In this situation, the rescue team has to gather damage information by entering into the underground spaces directly and share them. However, when the communication infrastructure is broken, rescue teams cannot cooperate because of disconnect between above-ground and underground spaces.

Therefore, rescue workers face secondary disaster risks increasing by a sudden situation changes. For example, in the underground disasters in Korea in 2003, many casualties occurred among rescue workers because of smoke damage. The rescue workers could not expect the smoke damage because they could not gather enough disaster area information beforehand. This is a typical case of underground secondary disaster that occurred due to the rescue team having entered into underground space without adequate information. In the future, disaster area information gathering in a closed area such as an underground spaces is important to contribute the reducing secondary disaster risks and formulating an appropriate rescue plan. On the basis of above background, recent researches have focused on disaster area information gathering method using wireless sensor networks (WSNs) in closed areas. Thus we have proposed robot wireless sensor networks (RWSNs) that include the WSN and a mobile rescue robot.

The WSN consists of spatially distributed sensor nodes (SNs) to cooperatively monitor physical or environmental conditions, such as temperature, sound, vibration, pressure, motion and so on. Then, the WSNs can provide a wireless communication infrastructure in place without established communication infrastructures. Therefore, the WSN is discussed as one of methods to construct the communication infrastructure and gather information in disaster area. However, existing construction method of the WSN is difficult to deploy SN in underground space such as post-disaster environment where rescue worker cannot enter.

The RWSN enables construction of wireless communication infrastructure by using the WSN and the mobile rescue robot in post-disaster underground space (Fig. 1). In our WSN deployment method for constructing the RWSN, we adopt a method that the mobile rescue robot deploys SNs which are wirelessly connected by defined routing path in advance. The mobile rescue robot moves into the underground space, and deploys SNs onto the own movement path. Deployed SN is connected to adjacent SNs wirelessly one by one, and then the WSN is expanded in underground space. An operator remotely controls the mobile rescue robot via constructed WSN communication infrastructure. In the network topology of this WSN, each SN is linearly connected to prevent the error of routing control. Generally, the WSN is able to decide the routing path of data transfer automatically by utilizing the RSSI between SNs, the throughput of End-to-End communication or the rate of packet loss. The routing path of the WSN is reconstructed by changes of these communication qualities. However, the routing path is repeatedly reconstructed in the situation that the communication connection between SNs is disconnected frequently. This situation is a problem for the system of the mobile robots teleoperation. An abeyance of the wireless communication connection degrades an operability of the mobile robot teleoperation and the performance of the disaster area information gathering. Then the communication qualities are often changed in disaster area by the disaster damages, the routing path is repeatedly reconstructed in the WSN. Therefore, in order to prevent the decline of the mobile robot activity, we adopt the network topology that routing path of the WSN is linearly connected. For such system, this paper describes a strategy of a routing path decision method using a mobile rescue robot for maintenance of the communication quality between SNs.

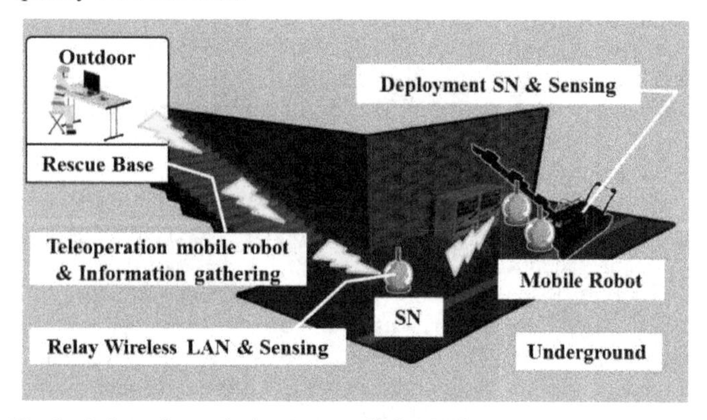

Fig. 1. Information-gathering system utilizing RWSN

II. RELATED WORKS

WSN construction methods have been discussed in the many research fields. In these researches, deployment methods have been proposed based on considering evaluation scales of factors such as packet routing, communication connectivity, energy efficiency, and coverage area[9]-[14].

Kumer et al. proposed a scheme for homogeneous distribution of randomly deployed mobile sensor networks to achieve maximum coverage while maintaining connectivity [15]. In this scheme, to achieve the maximum coverage and better connectivity, mobile nodes deploy themselves at the point which is divided other node's communication range into six. Wang et al. proposed a shortest moving path algorithm for the mobile robot to deploy a linear WSN [16]. This algorithm considers that the mobile robot has to return starting point to reload sensors and impact of different deployment strategies. Rizzo et al. proposed the deployment method of a robot team in fading environments [17]. In this method, to maintain constant connectivity and high signal quality in the communication network formed by the robots and the base station, the robot deployment is driven by RSSI measurements.

However, these methods don't consider an End-to-End network connectivity or a communication quality such as throughput. Many researches are premised on communication link being maintained automatically. Moreover, it is possible that the routing path reconstruction is difficult to maintain the communication quality, when the communication connection is disconnected by multi path fading in underground spaces. Therefore, considering routing path and communication quality in the WSN are important to maintain the End-to-End network connectivity.

III. MANAGEMAENT METHOD OF MAINTAINING COMMUNICATION CONNECTIVITY UTILIZING MOBILE ROBOT

A. Prior Conditions

In our proposed system, the SNs are deployed by the mobile rescue robot to construct the WSN. An assumed environment where the WSN is constructed has entrance stairs and the first basement floor.

The entrance stairs in under-ground is required to set up at intervals 30 [m], and passage way is built in line in Japanese building standard low. Therefore, our proposed system gathers this area's information. Then we discussed the WSN construction method by utilizing the mobile robot in this area. In the wireless communication of the RWSN, IEEE 802.11 series are adopted for wireless communication between SNs including the mobile rescue robot, which has been used as proven communication protocol in many studies of mobile robots and WSNs [18]-[22]. Then in our proposed method, we treat a mobile robot as a SN in the WSN. Heterogeneous networks includingsome SNs and various mobile robots are difficult to manage the system control. Especially, maintaining the stability of the system control is not easy under disaster situation in underground spaces. In this environment, to simplify the network structure is necessary to construct the stable communication connection. Therefore we simplified the network structure by treating a mobile robot as a SN. From here onwards, the IEEE802.11 series is also adopted for communication system of the mobile robot in the RWSN.

In our SN deployment method for constructing the WSN, the mobile rescue robot delivers the previously wireless connected SNs. The mobile robot deploys them onto the own movement path to construct the WSN. The operator can control the mobile rescue robot via constructed WSN communication infrastructure. In the network topology of this RWSN, each SN is linearly connected for prevention of the routing control error. As mentioned before, we adopt the network topology that the SNs are linearly connected, and the routing path is defined beforehand to prevent the decline of the mobile rescue robot activity when constructing the WSN.

B. Requested Specification

IEEE 802.11 series are necessary to keep the throughput that is more than 1.0 [Mbps] between the operator and the mobile robot (End-to-End communication) [23]. The throughput is defined as the number of packet transferred per unit time in a communication network. In the WSN construction by utilizing the mobile robot, the throughput between End-to-End communications has to be maintained for comfortable mobile robot teleoperation. The construction length of the WSN is required 50 [m] by concerning the distance of first basement floor 30 [m] and entrance stairs 20

[m]. The communication connectivity by IEEE802.11 series, however, is characterized by decreasing in turn area covered with concrete material such as the underground space. Thus, for WSN construction, our proposed system considers this communication characteristic to avoid a network disconnection risks.

The communication system among the RWSN uses IEEE802.11b standard that is tolerant to communication disruption from obstacles. The theoretical values of throughput by using IEEE802.11b are 11.0 [Mbps], and then the actual measurement values are lowered around 7.0 [Mbps] by the efficiency of the various factors in the real environment. This wireless LAN protocol can connect to SNs within 100 [m] in 1-hop communication on the straight line. The throughput is required 1.0 [Mbps] or more for the mobile robot teleoperation with keeping high communication connectivity.

In the ad-hoc network constructed on IEEE802.11b communication link, the maximum number of SNs to maintain 1.0 [Mbps] or more is five, and then throughput between each SN should be kept 6.0 [Mbps] or more. In the ad-hoc networks constructing the WSN, the delay of data transfer increases with increasing the number of hops between the source and the destination SNs. In order to expand a communication distance with maintaining the throughput in the WSN construction, the necessary number of SNs is decided beforehand, and then the routing path of these SNs is connected linearly. This method provides the high connectivity in an area covered with concrete material such as the underground space by deploying SNs as communication relay nodes to construct the WSN. In this system, the RWSN is constructed by using a source SN for the operator, three SNs for communication relay deployed by the mobile robot and the mobile robot regarded as one SN.

C. SN Deployment method for maintaining communication quality

In our proposed method, the mobile robot simultaneously measures communication quality while moving in the environment, and decides the deployment position of a SN. Therefore, the SN position can be determined flexibly against a change of radio wave condition, the mobile robot can cover with the whole target passageway. We proposed and evaluated the availability of several our deployment method in field operation test in the past [24]-[26].

To keep the throughput of 1.0 [Mbps] in End-to-End communication, two communication qualities between adjacent SNs need to be maintained. One is the RSSI that values over -86 [dBm] between two adjacent SNs (1-hop) communication are required. A wireless LAN module controlling the throughput speed constantly refers the RSSI for stability of the network connection. If the RSSI value is below -86 [dBm], the wireless LAN module controls the upper limit of throughput speed to 5.5 [Mbps]. Our proposed method also measures the RSSI value to predict the throughput speed control of the wireless LAN module. The other is the throughput that values over 6.0 [Mbps] between adjacent SNs (each 1-hop) are required for maintaining the throughput over 1.0 [Mbps] between End-to-End communications. Moreover, for the decision of SN deployment position, measurement of the throughput between the operator and the mobile robot as End-to-End communication is required to evaluate the communication quality. Therefore, both the throughput and RSSI values must be measured to satisfy the required communication performance for mobile robot teleoperation, and the robot measures both the throughput and RSSI value between each SN accordingly. The robot moves continuously to the destination while maintaining the 1.0 Mbps of throughput required for end-to-end communication and an RSSI value of -86 dBm between each SN(Fig. 2).

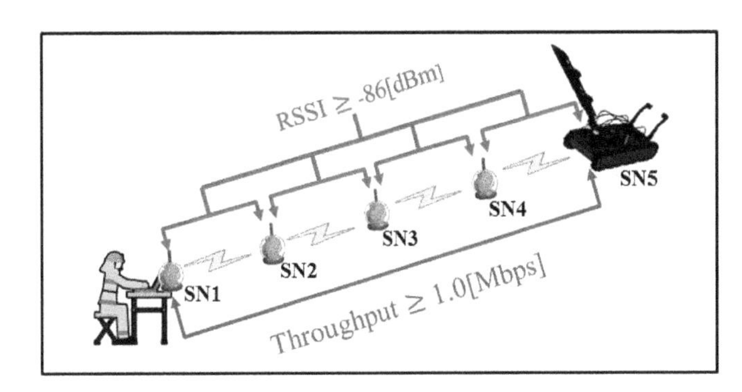

Fig. 2. SN Deployment method

D. Routing Path Decision Method Based on Communication Connectivity Between Adjacent SNs

As mentioned above, the proposed routing path decision method is necessary to prevent the communication disconnection and to maintain the throughput of over 1.0 [Mbps] between End-to-End communications.

In the network topology of the WSN constructed by the mobile robot, SNs are linearly connected and deployed. Then, the mobile robot moves on a straight line along deployed SN. At that time, the mobile robot communicates with a last deployed SN by 1-hop. Then, the distance between the mobile robot and this SN changes by moving the mobile robot. Generally, the throughput decays by degrading the RSSI and increasing the noise signal with the extension of distance between SNs. From this reason, it is necessary to decide the routing path between End-to-End communications flexibly against change of the distance between the SN and the mobile robot.

In order to maintain stability of the communication quality between each SN in the WSN constructed by our SN deployment method, we propose a decision method of a routing path between End-to-End communications based on relative positions of the mobile robot and SNs. When the positions of the mobile robot and a SN are the same, the mobile robot selects the SN of high connection priority, and then switches the routing path. A connection priority is different from the moving direction of the mobile robot. Each SN stores the routing path to the last deployed SN. Hence, using these routing path policies, the mobile robot can decide the routing path between End-to-End communications. Then, the throughput between End-to-End communications is increased or decreased depending on the number of hops by updating the routing path. Though, the proposed method maintains the throughput between End-to-End communications and prevents communication disconnection because the throughput and the RSSI value between SNs are stable.

Figure 3 to 5 show the workflow of this routing path decision method to be applied to the mobile robot (SN5). In the workflow, X is parameter of deployed SN ID, and Y is parameter of the high primary connection SN ID (hpc-SN ID).

Also, we define the source direction is that the mobile robot moves back to the passage, and the destination direction is that the mobile robot moves forward to the passage.

Moreover, Table 1 and 2 show a pattern of routing path. The workflow is outlined below.

1) The mobile robot records the SN deployment position with referring to own odometry and assigns an ID to the SN deployed one by one(X = 1, 2, 3, 4).

2) The mobile robot moves along the constructed WSN. Then the mobile robot measures movement distance by referring to the odometry.

3) The mobile robot compares self-position and the SN deployment position. If their positions match, the mobile robot refers to SN ID (choice SN ID, X). Otherwise, the workflow backs to the process (2).

4) The mobile robot identifies the forward direction and in case the mobile robot is moving the source direction (point of SN1: 0 [m]), the workflow progresses to the process (5). Otherwise (the destination direction: the mobile robot is facing the SN2, SN3, or SN4), the workflow progresses to the process (6).

5) For the movement to the source SN (SN1) direction, the mobile robot selects the SN (ID = X − 1) which has high connection priority, and multicast the packet to run the command changing routing path. Then, the SN which received the packet updates the routing table and connects to the mobile robot (Fig. 4).

6) For the movement to the destination SN (SN4) direction, the mobile robot selects the SN that SN (ID = X) which has high connection priority, and multicast the packet to run the command changing routing path. Then, the SN which received the packet updates the routing table and connects to the mobile robot (Fig. 5).

7) The mobile robot send the packet to the SN1 which is the source SN for the operator, and then each SN transfers received packet to the SN1 while updating the routing table. By repeating this, the mobile robot decides the routing path between End-to-End communications.

TABLE I. ROUTING PATH MOVEMENT TO SOURCE DIRECTION

Mobile Robot Position	Routing Path
SN2	SN1-SN5
SN3	SN1-SN2-SN5
SN4	SN1-SN2-SN3-SN5

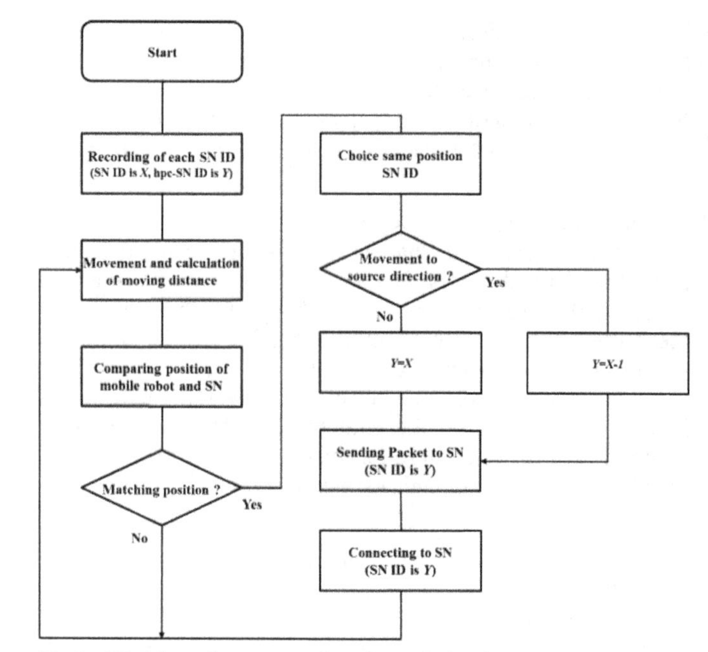

Fig. 3. Workflow of our proposed routing path decision method

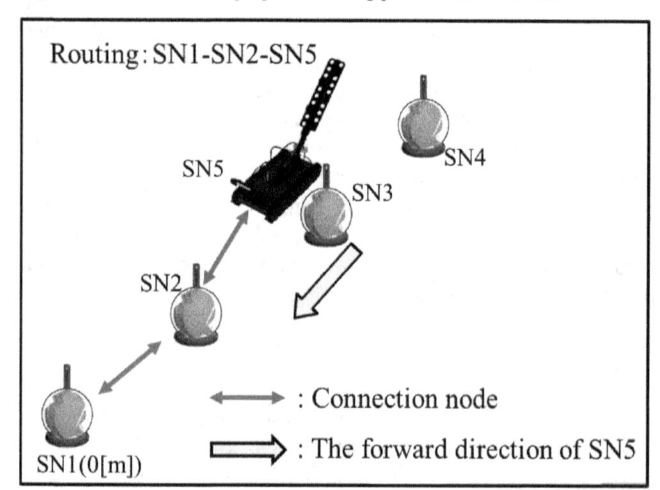

Fig. 4. Decision of routing path at source direction (e.g. mobile robot reaches point of SN3)

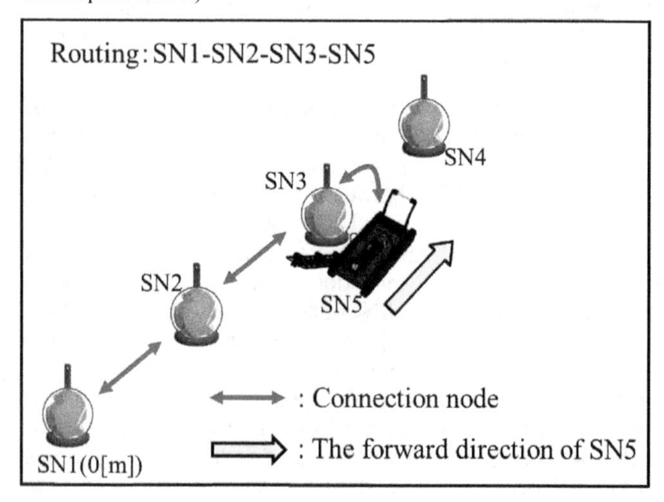

Fig. 5. Decision of routing path at destination direction (e.g. mobile robot reaches point of SN3)

TABLE II. ROUTING PATH MOVEMENT TO DESTINATION DIRECTION

Mobile Robot Position	Routing Path
SN2	SN1-SN2-SN5
SN3	SN1-SN2-SN3-SN5
SN4	SN1-SN2-SN3-SN4-SN5

IV. PERFORMANCE EVALUATION OF PROPOSED ROUTING METHOD

A. Experimental Conditions

This section describes experimental conditions for performance evaluation of proposed routing method. Pseudo failure SN disconnects WSN between the operator and the mobile robot, and then proposed routing method reconstructs the network in this experiment. The experiment verifies maintaining connectivity by measuring throughput between End-to-End communications. Stability of the throughput indicates success of change of the routing pass, and then WSN reconstruction prevents the disconnection of teleoperation between the operator and the mobile robot. In this experiment, we performed a cooperative evaluation of a fixed linear routing method and the proposed routing method.

The experiment was performed in the passageway with a length of 300 [m] or more in Tokyo Denki University, and the WSN was constructed by utilizing RWSN in this environment (Fig. 6). The SN configuring the WSN mounts the CPU board, memory device, CompactFlash disc, IEEE 802.11b/g wireless LAN module, and a battery. These devices are controlled by Linux OS (Debian) installed to the CPU board. By utilizing the "AODV-uu", Ad-Hoc networks can be constructed on the WSN. Table 3 shows the specification of the SN. The CPU board is "Armadillo-300" (@techno Inc.) in which Linux OS is implemented (Fig. 7 (a)). In RSSI value measurement, we used "wlanconfig" command included in the Linux wireless tools.

The throughput was measured by using "utest" provided by NTTPC Communications, Inc. The "utest" is a tool measuring throughput by transferring 10000 times of the packet of 1500 [byte] to communication partner. The throughput is calculated by received data per unit time. The crawler-type mobile robot, "S-90LWX" (TOPY INDUSTRIES, LIMITED) equipped with a SN deployment mechanism shown in Fig. 7 (b), is used as the mobile robot in this experiment.

The mobile robot constructed a WSN by utilizing our proposed SN deployment strategy prior to the evaluation. Each SN were deployed at SN1: 0 [m], SN2: 80 [m], SN3: 128 [m], SN4: 136 [m], and SN5: 196 [m]. After that, we measured 10 times of RSSI value and throughput between 1-hop connections in each 4 [m] interval in two situations of the linear routing network and the proposed routing method while the mobile robot (SN5) moved from the point of 196 [m] to the point of 0 [m]. The operator controlled the mobile robot as SN5 connected to SN1 via the WSN. The WSN provided communication infrastructure to the operator for wireless teleoperation of the mobile robot in this experiment.

Fig. 6. Experimental environment

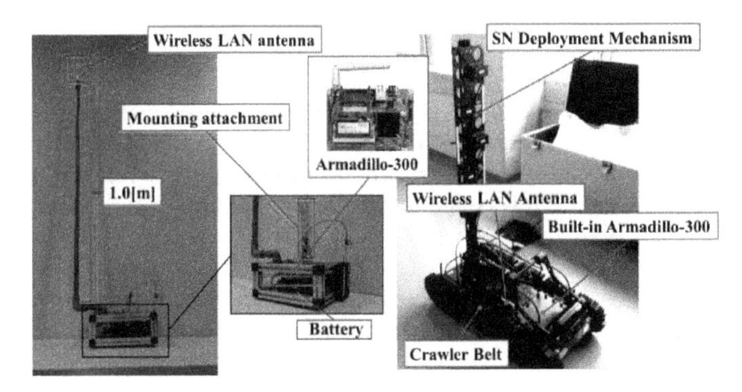

(a) Sensor Node (b) S-90LWX

Fig. 7. Configuration devices on WSN

TABLE III. SPECIFICATION OF WIRELESS SENSOR NODE

Sensor Node	
Operating system	Linux Kernel 2.6 (Debian)
CPU board	Armadillo-300 (ARM 200[MHz])
Weight	1.5 [kg]
Height×Width×Length	225 [mm] × 180 [mm] × 380 [mm]
Battery No. 1	Output : 5 [V], 1.8 [A]
Battery No. 2	Output : 12 [V], 2.1 [A]
Operating time	3 [hour]

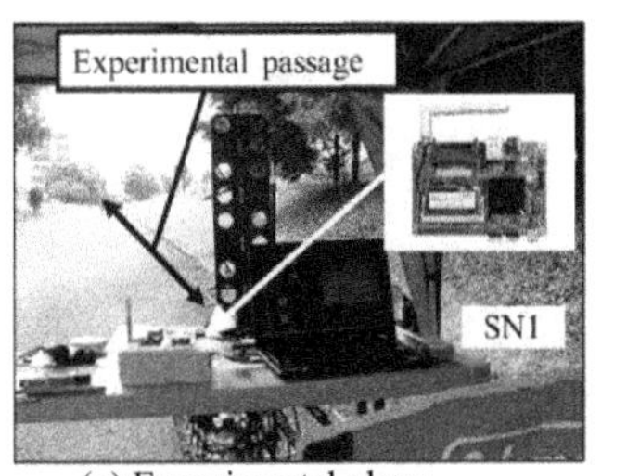

(a) Experimental place

(b) S-90LWX with SN

Fig. 8. Overview of experimental environment

B. Experimental Results

Figure 9 and 10 show experimental results that are the mean value of RSSI and throughput in fixed linear routing network. Figure 11 and 12 show the values in proposed routing network. In fixed linear routing network, measurement intervals of RSSI and throughput were between SN4 and SN5.

In the proposed routing network, SN4 connects SN5 between 196 [m] and 136 [m], SN3 connects SN5 between 132 [m] and 128 [m], SN2 connects SN5 between 124 [m] and 80 [m], and SN1 connects SN5 between 76 [m] and 0 [m] (Fig. 11 and 12). Thus measurement intervals of RSSI and throughput were above intervals in proposed routing network.

We confirmed that the mobile robot as SN5 was able to connect to WSN in all measurement intervals of the fixed linear routing network and the proposed routing network. Then the operator could control the mobile robot by connecting WSN in both routing network without communication failures such as network disconnection, multipath fading of radio wave, various environmental noise and so on.

V. Discussion

In the case of using the fixed linear routing network in the performance evaluation, the throughput between End-to-End communications was lower than 1.0 [Mbps] at the point of 80 [m] (Fig. 9). It shows maintaining the throughput of 6.0 [Mbps] between SN4 and SN5 was difficult with the decrease of RSSI because the distance between SN4 and SN5 was expanded (Fig. 10). Also, the throughput between SN4 and SN5 was unstable from the point of over 80 [m] where the upper limit value of the throughput was adjusted frequently. Then, the operability of the mobile robot might decrease.

Against that, the case of using our proposed routing method, the throughput between End-to-End communication increased from 1.4 [Mbps] to 7.0 [Mbps] because the number of hops between SN1 and SN5 decreased (Fig. 11). Therefore, we confirmed that the throughput of 1.0 [Mbps] between End-to-End communications and 7.0 [Mbps] between the mobile robot and the SN connected by 1-hop were maintained at all SN deployed positions (Fig. 12). These results show the availability of our proposed method in this field test.

VI. Conclusion

This paper proposed the WSN management method of changing routing path considering communication disconnection by the effect of multi path fading. The proposed method maintained communication conditions that throughput

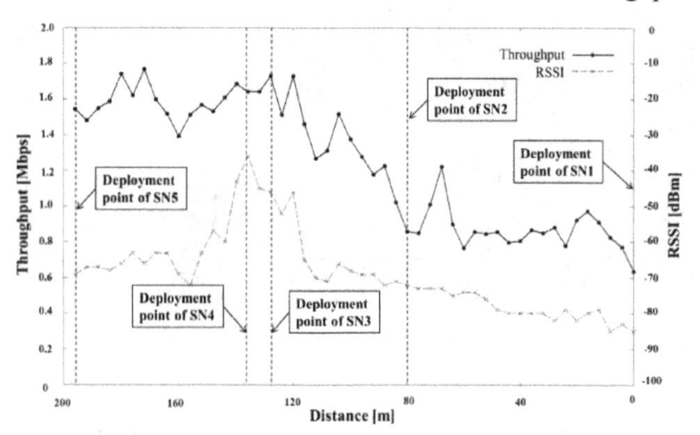

Fig. 9. Experimental result of RSSI and End-to-End throughput in fixed linear routing network

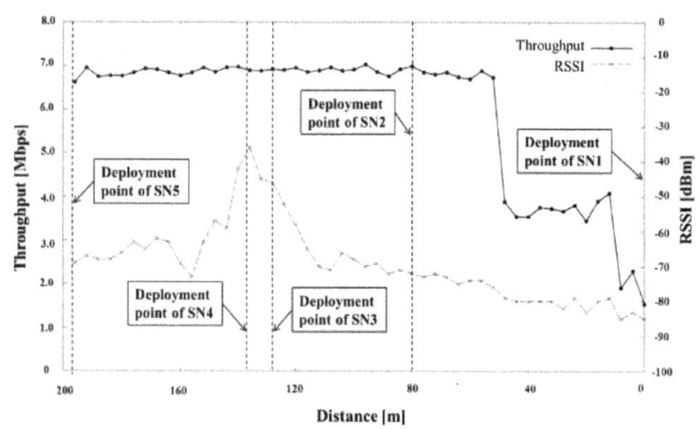

Fig. 10. Experimental result of RSSI and 1-hop throughput in fixed linear routing network

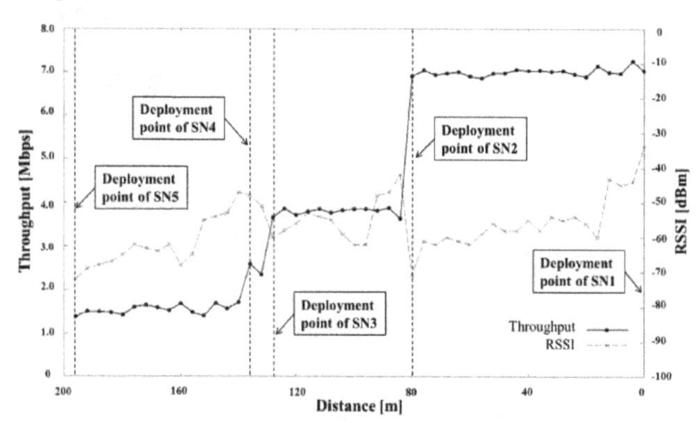

Fig. 11. Experimental result of RSSI and End-to-End throughput utilizing our proposed routing network

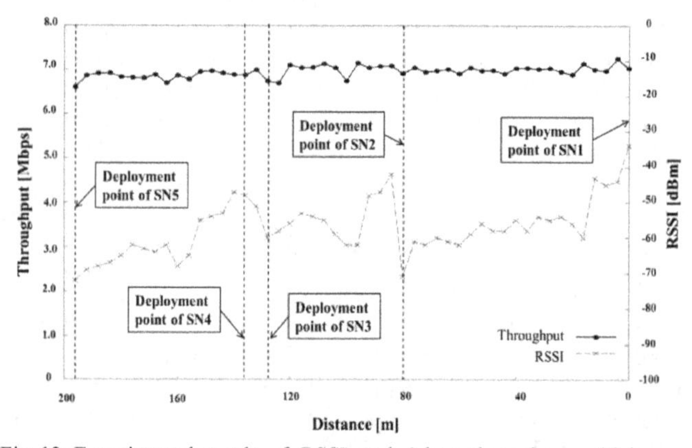

Fig. 12. Experimental result of RSSI and 1-hop throughput utilizing our proposed routing network

between End-to-End communications in the WSN enables smooth teleoperation of the mobile robot in a post-disaster underground space. Experimental results showed the effectiveness of the proposed method that enables to manage the WSN in the field test. The rapid implementation of actions to reduce secondary disasters in disaster areas requires the stable referral of disaster information. Therefore, this WSN managing method that maintains the throughput stable by utilizing the mobile robot is effective for gathering disaster area information in actual disaster scenarios.

In future work, we should consider the communication disconnection on WSN due to SN failure. In the underground space, the SN may break down by the effect of secondary disasters. It means that information gathering by the RWSN becomes impossible. We will consider the improved SN re-deployment method to repair the communication disconnection by integrating our proposed routing path decision methods.

ACKNOWLEDGMENT

This work was partially supported by the Research Institute for Science and Technology of Tokyo Denki University, Grant Number Za10-01 / Japan.

REFERENCES

[1] CHI Hao-yuan, LIU Xu, XU Xiao-dong, "A Framework for Earthquake Disaster Mitigation System," Proceedings of 2011 China located International Conference on Information Systems for Crisis Response and Management (ISCRAM), pp.490-495, 2011.

[2] Huang AN, "China's Emergency Management Mechanisms for Disaster Prevention and Mitigation," Proceedings of International Conference on E-Business and E-Government (ICEBEG), pp.2403-2407, 2010.

[3] Yoshiaki KANAEDA, Kazushige MAGATANI, "Development of the device to detect SPO2 in the Field,"31st Annual International Conference of the IEEE EMBS, pp.412-415, September 2009.

[4] O. Mizuno, A. Takashi, S. Yamamoto, and K. Asatani, "Sustainable operation technologies for the mitigation information network in urban area", In Humanitarian Technology Conference (R10-HTC) 2013 IEEE Region 10, pp. 255-260, 2013.

[5] A. P. Tang, and A. P. Zhao, "A Decision Supporting System for Earthquake Disaster Mitigation", Intelligent System Design and Engineering Application (ISDEA), 2012 Second International Conference on, IEEE, pp.748-751, 2012.

[6] M. Asif, T. Nitin, K. Ullah, and M.S. Sarfraz, "A Web-based Disaster Management-Mitigation Framework Using Information and Communication Technologies and Open Source Software", JU Journal of Information Technology (JIT), Vol.1, pp.15-18, 2012.

[7] Helge-Bjorn Kuntze, Christian W. Frey, Igor Tchouchenkov, Barbara Staehle, Erich Rome, Kai Pfeiffer, Andreas Wenzel and Jurgen Wollenstein, "SENEKA - Sensor Network with Mobile Robots for Disaster Management," Homeland Security (HST), pp.406-410, 2012.

[8] Abishek T K, Chithra K R and Maneesha V. Ramesh, "ADEN:Adaptive Energy Efficient Network of Flying Robots Monitoring over Disaster Hit Area," Proceedings of 8th IEEE International Conference on Distributed Computing in Sensor Systems (IEEE DCOSS), pp.306-310, 2012.

[9] E. Budianto, M.S. Alvissalim, A. Hafidh, A. Wibowo, W. Jatmiko, B. Hardian, P. Mursanto and A. Muis, "Telecommunication Networks Coverage Area Expansion in Disaster Area using Autonomous Mobile Robots : Hardware and Software Implementation," Proceedings of International Conference on Advanced Computer Science and Information Systems (ICACSIS), pp.113-118, 2011.

[10] Andrew Chiou, and Carol Wynn, "Urban Search and Rescue Robots in Test Arenas: Scaled Modeling of Disasters to Test Intelligent Robot Prototyping," Proceedings of International Conference on Autonomic and Trusted Computing (ATC), pp.200-205, 2009.

[11] R.C. Luo, O. Chen, "Mobile sensor node deployment and asynchronous power management for wireless sensor networks." Industrial Electronics, IEEE Transactions on, Vol.59, Issue.5, pp.2377-2385, 2012.

[12] Wing-Yue Geoffrey Louie, and Goldie Nejat, "A victim identification methodology for rescue robots operating in cluttered USAR environments," Advanced Robotics, vol. 27, issue. 5, pp. 373-384, 2013.

[13] Andrew Markham and Niki Trigoni, "Magneto-Inductive NEtworked Rescue System (MINERS):Taking Sensor Networks Underground," Proceedings of the 11th international conference on Information Processing in Sensor Networks (IPSN '12), pp. 317-328, 2012.

[14] Josh D. Freeman, Vinu Omanan, and Maneesha V. Ramesh, "Wireless Integrated Robots for Effective Search and Guidance of Rescue Teams," Proceedings of 8th International Conference on Wireless and Optical Communications Networks (WOCN 2011), pp. 1-5, 2011.

[15] A. Kumar, V. Sharma, and D. Prasad, "Distributed Deployment Scheme for Homogeneous Distribution of Randomly Deployed Mobile Sensor Nodes in Wireless Sensor Network." International Journal of Advanced Computer Science and Applications (IJACSA), The Science and Information organization, Vol.4, No.4, pp.139-146, 2013.

[16] Z. Wang, X. Zhao, and X. Qian, "Carrier-based sensor deployment by a mobile robot for wireless sensor networks", Control Automation Robotics & Vision (ICARCV), 2012 12th International Conference on. IEEE, pp.1663-1668, 2012.

[17] C. Rizzo, D. Tardioli, D. Sicignano, L. Riazuelo, J. L. Villarroel, and L. Montano, "Signal-based deployment planning for robot teams in tunnel-like fading environments", The International Journal of Robotics Research, Vol.32, No.12, pp.1381-1397, 2013.

[18] H. Sato, K. Kawabata and T. Suzuki, "Information Gathering by wireless camera node with Passive Pendulum Mechanism," International Conference on Control, Automation and Systems 2008 (ICCAS2008), pp.137-140, 2008.

[19] T. Yoshida, K. Nagatani, E. Koyanagi, Y. Hada, K. Ohno, S. Maeyama, H. Akiyama, K. Yoshida and S. Tadokoro, "Field Experiment on Multiple Mobile Robots Conducted in an Underground Mall," Field and Service Robotics Springer Tracts in Advanced Robotics, vol. 62, pp365-375, 2010.

[20] H. Jiang, J. Qian, and W. Peng, "Energy Efficient Sensor Placement for Tunnel Wireless Sensor Network in Underground Mine," Proceedings of 2nd International Conference on Power Electronics and Intelligent Transportation System (PEITS 2009), pp. 219-222, 2009.

[21] J. Xu, S. Duan and M. Li, "The Research of New Type Emergency Rescue Communication System in Mine Based on Wi-Fi Technology," Proceedings of IEEE 3rd International Conference on Communication Software and Networks (ICCSN), pp. 8-11, 2011.

[22] K. Nagatani, S. Kiribayashi, Y. Okada, K. Otake, K. Yoshida, S. Tadokoro, T. Nishimura, T. Yoshida, E. Koyanagi, M. Fukushima and S. Kawatsuma, "Emergency Response to the Nuclear Accident at the Fukushima Daiichi Nuclear Power Plants using Mobile Rescue Robots," Journal of Field Robotics, vol. 30, no. 1, pp. 44-63, 2013.

[23] J. Yamashita, K. Sawai, Y. Kimitsuka, T. Suzuki,Y. Tobe, "The design of direct deployment method of sensor nodes by utilizing a rescue robots in disaster areas," SICE Annual Conference 2008, pp183, 2B3-4, 2008.

[24] Tsuyoshi Suzuki, Kei Sawai, Hitoshi Kono and Shigeaki Tanabe, "Sensor Network Deployment by Dropping and Throwing Sensor Node to Gather Information Underground Spaces in a Post-Disaster Environment," Descrete Event Robot, iConcept PRESS, in Press. 2012.

[25] K. Sawai, H. Kono, S. Tanabe, K. Kawabata, T. Suzuki, "Design and Development of Impact Resistance Sensor Node for Launch Deployment into Closed Area," In international journal of sensing for industry(Sensor Review), Emerald Group Publishing Ltd., Vol. 32, pp.318 – 326, 2012.

[26] S. Tanabe, K. Sawai and T. Suzuki, "Sensor Node Deployment Strategy for Maintaining Wireless Sensor Network Communication Connectivity," International Journal of Advanced Computer Science and Applications (IJACSA), The Science and Information organization, Vol.2, No. 12, pp.140 – 146, 2011.

Highly Accurate Prediction of Jobs Runtime Classes

Anat Reiner-Benaim
Department of Statistics
University of Haifa
Haifa, Israel

Anna Grabarnick
Department of Statistics
University of Haifa
Haifa, Israel

Edi Shmueli
Intel Corporation
Haifa,
Israel

***Abstract*—Separating the short jobs from the long is a known technique to improve scheduling performance. This paper describes a method developed for accurately predicting the runtimes classes of the jobs to enable the separation. Our method uses the fact that the runtimes can be represented as a mixture of overlapping Gaussian distributions, in order to train a CART classifier to provide the prediction. The threshold that separates the short jobs from the long jobs is determined during the evaluation of the classifier to maximize prediction accuracy. The results indicate overall accuracy of 90% for the data set used in the study, with sensitivity and specificity both above 90%.**

Keywords—Runtime Prediction; Job Scheduler; Server Farms; Classifier; Mixture Distribution

I. Introduction

Supplying job schedulers with information on how long the jobs are expected to run enabled the development of the backfilling algorithms, which leverage the information to pack the jobs more efficiently and improve system utilization [1]. The backfilling algorithms, however, were designed for parallel systems, in which the jobs require many processors in order to execute, and processor fragmentation (idleness) is a big concern. Thus in parallel system environments the scheduler needs to know the actual runtimes of the jobs (use numeric predictions) to be able to optimize the schedule and improve performance [10].

Our work targets systems in which most jobs are serial, like server farms that are used for software testing. In serial system environments sophisticated scheduling algorithms are not required, and in order to improve performance it is enough to simply separate the short jobs from the long ones, and assign them to different queues in the system [12]. The separation reduces the likelihood that short jobs will be delayed after long ones, improves the average turn-around times of the jobs and overall system throughput (Figure 1).

Respectively, to implement such a system it is enough to only predict the runtime classes of the jobs – whether they will be short or long, in order to assign them to the right queue. On the other hand, any misclassification of the jobs can severely impact performance. For example, mistakenly assigning long jobs to the short jobs queue will cause many of the short jobs to be delayed, average turnaround time to increase, and the overall throughput to decrease as a result.

Motivated by the later usage model (server farms), a method that allows predicting the runtime classes of the jobs with high accuracy was developed.

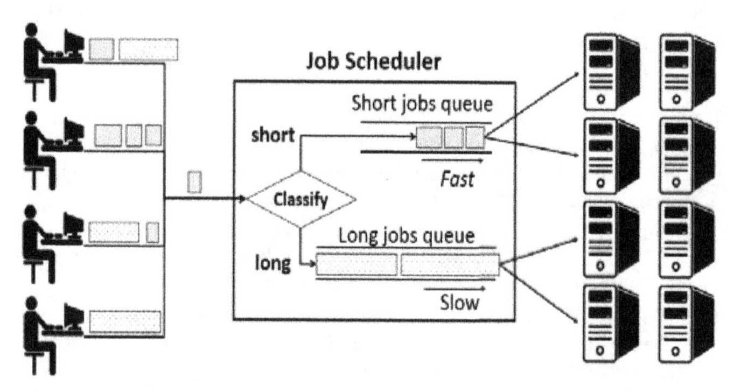

Fig. 1. Separating the short jobs from the long reduces the likelihood that short jobs will be delayed after long ones and improves system performance

The method is based on applying a log transformation on the runtimes of the jobs (historical records), revealing a mixture of two overlapping Gaussian distributions that represent the short and long jobs. We use the mixture model to determine the distribution parameters and to set the initial separation threshold between the short and long runtime populations.

A key design aspect for the proposed method is to be able to predict the classes with high accuracy. In order to achieve high accuracy, the threshold that separates the short jobs from the long is not determined in advance (which can lead to an eventual high misclassification rate). Instead, the threshold is determined as part of the evaluation of the classifier: a subset of the data that is close to the means of the distributions is used for training the classifier, and then the full dataset is used to select the threshold that optimizes a desired target function.

For class prediction for newly incoming jobs, the CART classifier is used [18]. CART is suitable for binary classification and can account for both continuous and categorical classifying variables, and is based on a tree optimizing algorithm that minimizes classification error while reducing overfitting by branch pruning.

The proposed method was applied on a job trace obtained from one of Intel's server farms, and which contained more than one million job records. Setting the target on achieving the best trade-off between misclassifications of short jobs and misclassifications of long jobs resulted in prediction accuracy of 90% (total misclassification rate of 10%) on the independent validation set. The predictions were based on estimated distribution means of 140 and 3,500 seconds for the "short" and "long" classes, respectively, and a separating threshold of 608 seconds.

This paper is organized as follows. Section 2 describes the data that is used for training, testing and validating the model. Section 3 describes the initial class labeling based on the mixture model analysis. Section 4 reviews the CART model. Section 5 describes the learning algorithm along with the optimal threshold determination procedure for best accuracy. Section 6 describes the results of the study. Section 7 surveys related work and Section 8 concludes the paper.

II. THE DATA

Our data is based on two traces obtained from one of Intel's server farms. The first trace, which was used to train the model, contained a sample of around one million job records that executed in the farm during a period of ten consecutive days. The second trace, which was used to validate the model, contained a sample of additional 755,000 records (approximately) of jobs that executed during a period of seven consecutive days. The validation on independent data is important for establishing the robustness of the model obtained in the training stage.

Each record in the traces contained 13 fields pertaining to a particular job. The continuous variables: "Submitime", "Starttime" and "Finishtime" indicated when the job was submitted, when the job started, and when the job finished executing, respectively. In order not to reveal any information about the workload, the traces did not contain any descriptive information about the jobs. Instead, the values in the fields were transformed into discrete values (categorical variables) that can be used for the analysis. In addition, the names were also transformed in order not to reveal information about the possible meanings of the values.

Table I groups the 9 categorical variables and roughly explains the meaning of each group. Table II outlines basic statistics on each of the variables.

TABLE I. ROUGH GROUPING OF THE 9 CATEGORICAL VARIABLES

Group	tab	Relates to	Example
A	3	Scheduling information	Resources requested by the job
B	2	Execution-specific information	Command line and arguments
C	4	Association information	Project and component

TABLE II. STATISTICS REGARDING THE CATEGORICAL VARIABLES

Variable	# of categories	# of missing (in training data)
A1	9	0
A2	7	0
A3	5	0
B1	44	173
B2	22	184
C1	2	0
C2	5	239
C3	6	184
C4	32	0

In addition to the above, two additional categorical variables were defined, day and hour, based on the three continuous variables in the trace. These variables indicate the day of the week (1 for Sunday to 7 for Saturday) and the hour of the day (0 to 23), the job was submitted, started, and finished executing, respectively. Figure 2 shows the distribution of the respective temporal categorical variables along the timeline axis. As can be seen, during weekdays longer jobs are typically submitted during the morning hours, with occasional peaks in runtime during evening hours. During weekends, peaks in runtime also occur during the afternoon and evening hours.

Figure 3 shows the distribution of the jobs runtime "as-is", and after applying a log base 2 transformation on the runtime. As can be seen in Figure 3a, the vast majority of the jobs are short (the shortest job ran 3.5 seconds), and there are few long ones (the longest job ran for nearly 9 days). This well corresponds to previous observations made on the runtime, describing a phenomenon that characterizes many production workloads [11].

Transforming the runtime to the log scale (Figure 3b) reveals a mixture pattern of two main Gaussian-like distributions (some-times referred to as a "hyper lognormal distribution), with a stretching right tail.

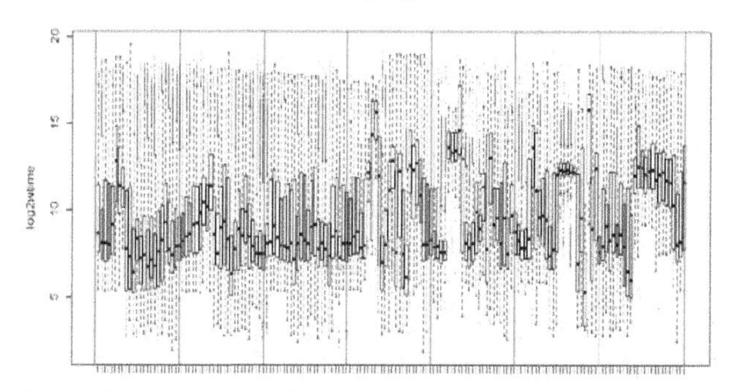

Fig. 2. Runtime boxplots (in log base 2 scale) along time (Sunday through Saturday). The days are separated by vertical red lines. Each tick mark along the time axis marks an hour of the day

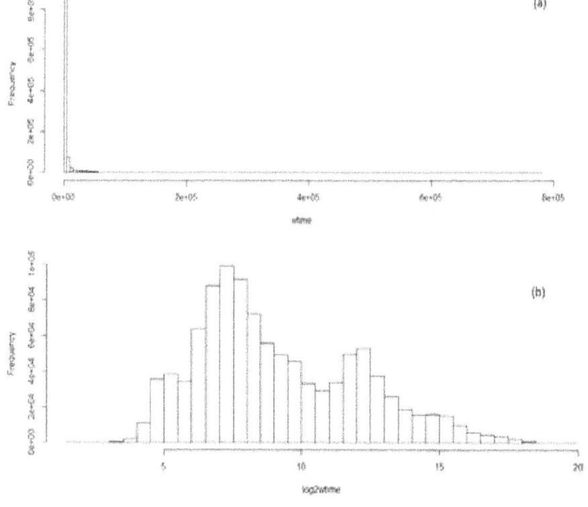

Fig. 3. Histogram of runtime. (a) Raw data (b) After log base 2 transformation

III. Class Construction by Mixture Distribution Analysis

The first component in the proposed analysis sets the base for defining the two runtime classes by estimating the mixture distribution parameters, and then labeling each job as "short" or "long". Once the predictor variables are selected through a training-and-testing algorithm (see Section 5.1), the model is optimized by selecting the mixture threshold which provides the best performance, namely minimizes the prediction error or approaches the desired sensitivity-specificity combination (see Section 5.2).

The Gaussian (normal) mixture model has the form

$$f(x) = \sum_{m=1}^{M} \alpha_m \phi(x;\ \mu_m, \Sigma_m)$$

with mixing proportions α_m, $\sum_m \alpha_m = 1$, and each Gaussian density has a mean μ_m and covariance matrix Σ_m. The parameters are usually fit by the maximum likelihood approach using the EM algorithm, which is a popular tool for simplifying difficult maximum likelihood problems.

Based on the mixture observed in Figure 2, one to four mixture components may be defined. However, for the purpose of the current study it was decided to focus only on two classes, namely "short jobs" and "long jobs". Thus the runtime Y is modeled as a mixture of the two normal variables

$$Y_1 \sim N(\mu_1, \sigma_1^2), \qquad Y_2 \sim N(\mu_2, \sigma_2^2).$$

Y can be defined by

$$Y = (1 - \Delta) \cdot Y_1 + \Delta \cdot Y_2,$$

where $\Delta \in \{0,1\}$ with $\mathbb{P}(\Delta = 1) = \pi$. This generative representation is explicit: generate a $\Delta \in \{0,1\}$ with probability π, and then depending on the outcome, deliver either Y_1 or Y_2. Let $\phi_\theta(x)$ denote the normal density with parameters $\theta = (\mu, \sigma^2)$. Then the density of Y is

$$g_Y(y) = (1 - \pi)\phi_{\theta_1}(y) + \pi\phi_{\theta_2}(y).$$

Suppose the model is fit to the data by maximum likelihood. The parameters are

$$\theta = (\pi, \theta_1, \theta_2) = (\pi, \mu_1, \sigma_1^2, \mu_2, \sigma_2^2).$$

The log-likelihood based on N training cases is

$$l(\theta;\ \mathrm{Z}) = \sum_{i=1}^{N} \log[(1 - \pi)\phi_{\theta_1}(y_i) + \pi\phi_{\theta_2}(y_i)].$$

Direct maximization of $l(\theta;\ \mathrm{Z})$ is quite difficult numerically due to the sum of terms inside the logarithm. There is, however, a simpler approach. We consider unobserved latent variables Δ_i taking values 0 or 1 as earlier: if $\Delta_i = 1$ then Y_i comes from distribution 2, otherwise Y_i comes from distribution 1. Suppose the values of the Δ_i's are known. Then the log-likelihood would be

$$l(\theta;\ \mathrm{Z}, \Delta) = \sum_{i=1}^{N} [(1 - \Delta_i) \log \phi_{\theta_1}(y_i) + \Delta_i \log \phi_{\theta_2}(y_i)] + \sum_{i=1}^{N} [(1 - \Delta_i) \log \pi + \Delta_i \log(1 - \pi)]$$

and the maximum likelihood estimates of μ_1 and σ_1^2 would be the sample mean and the sample variance of the observations with $\Delta_i = 0$. Similarly, the estimates for μ_2 and σ_2^2 would be the sample mean and the sample variance of the observations with $\Delta_i = 1$.

Since the Δ_i values are actually unknown, the procedure continues in an iterative fashion, substituting for each Δ_i in the previous equation its expected value

$$\gamma_i(\theta) = \mathbb{E}(\Delta_i | \theta, \mathrm{Z}) = \mathbb{P}(\Delta_i = 1 | \theta, \mathrm{Z}),$$

which is also called the responsibility of model 2 for observation i.

We use the following procedure, known as the EM algorithm, for the two-component Gaussian mixture:

1) Take initial guesses for the parameters $\hat{\pi}, \hat{\mu}_1, \hat{\sigma}_1^2, \hat{\mu}_2, \hat{\sigma}_2^2$ (see below).

2) Expectation step: compute the responsibilities

$$\hat{\gamma}_i = \frac{\hat{\pi}\phi_{\hat{\theta}_2}(y_i)}{(1 - \hat{\pi})\phi_{\hat{\theta}_1}(y_i) + \hat{\pi}\phi_{\hat{\theta}_2}(y_i)}, \tag{1}$$

$$i = 1, 2, \ldots, N.$$

3) Maximization step: compute the weighted means and variances,

$$\hat{\mu}_1 = \frac{\sum_{i=1}^{N}(1 - \hat{\gamma}_i) y_i}{\sum_{i=1}^{N}(1 - \hat{\gamma}_i)}, \qquad \hat{\sigma}_1^2 = \frac{\sum_{i=1}^{N}(1 - \hat{\gamma}_i)(y_i - \hat{\mu}_1)^2}{\sum_{i=1}^{N}(1 - \hat{\gamma}_i)},$$

$$\hat{\mu}_2 = \frac{\sum_{i=1}^{N} \hat{\gamma}_i y_i}{\sum_{i=1}^{N} \hat{\gamma}_i}, \qquad \hat{\sigma}_2^2 = \frac{\sum_{i=1}^{N} \hat{\gamma}_i (y_i - \hat{\mu}_1)^2}{\sum_{i=1}^{N} \hat{\gamma}_i},$$

and the mixing probability,

$$\hat{\pi} = \frac{\sum_{i=1}^{N} \hat{\gamma}_i}{N}.$$

4) Iterate steps 2 and 3 until convergence.

In the expectation step, a "soft" assignment of each observation to each model is done: the current estimates of the parameters are used to assign responsibilities according to the relative density of the training points under each model. In the maximization step, the responsibilities are used within weighted maximum-likelihood fits to update the estimates of the parameters.

A simple choice for initial guesses for $\hat{\mu}_1$ and $\hat{\mu}_2$ is two randomly selected observations y_i. The overall sample variance $\sum_{i=1}^{N} \frac{(y_i - \bar{y})^2}{N}$ can be used as an initial guess for both $\hat{\sigma}_1^2$ and $\hat{\sigma}_2^2$. The initial mixing proportion $\hat{\pi}$ can be set to 0.5.

The "mixtools" R package [15, 16] was used for the mixture analysis, with the function "normalmixEM" for parameter and posterior probability (responsibility) estimation.

IV. The CART Model

The CART (Classification and Regression Trees) model, also named the decision tree model, is an approach for making either quantitative or class prediction. The CART model is non-parametric, thus no assumptions are made regarding the underlying distribution of the predictor variables, enabling CART to handle numerical data that are skewed or multi-modal. Both continuous and categorical predictors can be considered, including ordinal ones.

CART identifies classifying, or "splitting", variables based on an exhaustive search of all classifying possibilities with the available variables. Useful CART trees can be generated even when there are missing values for some variables, by using "surrogate" variables, which contain information similar to the missing variables.

CART analysis consists of the following steps:

- Tree building, during which a tree is built using recursive splitting of nodes. This process stops when a maximal tree has been produced. The higher the splitter variable in the tree, the higher its importance in the prediction process.
- Tree "pruning", which is a simplification of the tree by cutting nodes off from the maximal tree.
- Optimal tree selection, which selects one tree from the set of pruned trees with the least evidence of over fit.

Each path from the root of a decision tree to one of its leaves can be transformed into a rule. Less complex decision trees are preferred, since they are easier for interpretation and may be more accurate.

V. MODEL LEARNING AND OPTIMIZATION PROCEDURE

Once labeled data is obtained, a supervised learning technique is used for the purpose of generating a classification rule. In the first stage a set of variables that will be included in the model are selected, while evaluating the performance of each model, and in the second stage a final model is obtained by using the selected variables on the full dataset and evaluate the model based on the performance target function.

This stage of the analysis is done on a subset of the training data (containing the ten days period), which is extracted as follows. Since the two observed runtime distributions overlap, observations that are within 0.5 standard deviations off the two means are selected, such that they will be distant from the overlapping region and will belong to the corresponding classes with high certainty (Figure 4). A total of 257,467 observations labeled short=1 (belonging to the Gaussian population with the smaller mean) and 192,205 observations with short=0 (belonging to the Gaussian population with the larger mean) were selected. Together they made around 43% of the data.

A five-fold cross-validation procedure is then performed in order to select variables and evaluate each model, by iteratively dividing the data in random into a training set (80% of the learning data) and an evaluation set (20% of the learning data) and implementing the CART model with the mixture threshold of 0.5. The importance measure, which considers how high the splitting variable is in the tree, is averaged across all iterations for each variable, and the variable having an importance score above the baseline level is selected.

Once a set of variables is selected for each model type, a model is fit to the full training data. We account for the two types of misclassification error, the false "positive" classification and the false "negative" classification. Defining classification into "short" as "positive", the former refers to the erroneous classification of a long job into the "short" class, and the latter refers to the erroneous classification of a short job into the "long" class.

In the job runtime context, sensitivity is defined as the proportion of short jobs classified as short, while the specificity is defined as the proportion of long jobs classified as long. Subtracting the specificity from 1 will give the proportion of long jobs erroneously classified as short. For the CART classifier, the mixture threshold that yields the best tradeoff between the two errors is chosen. The full set of sensitivity-specificity combinations can be summarized in a pseudo-ROC (Receiver Operating Characteristic) curve, in which the sensitivity is plotted against 1-specificity, for each threshold of the probability obtained in the mixture model (the final value obtained for equation 1).

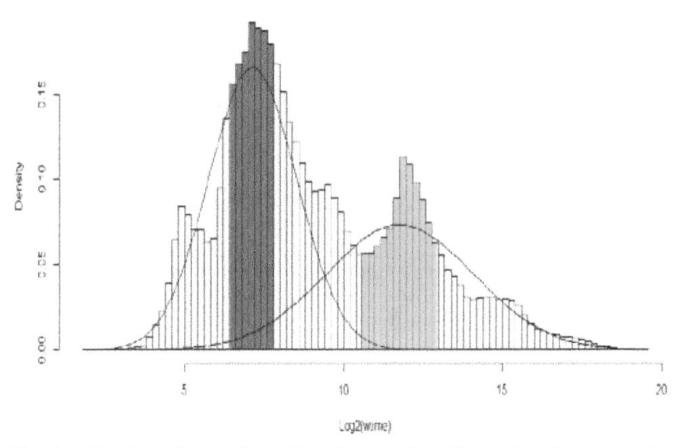

Fig. 4. Runtime (in log base 2 scale) density. The red and green colored regions mark the observations selected for the learning process

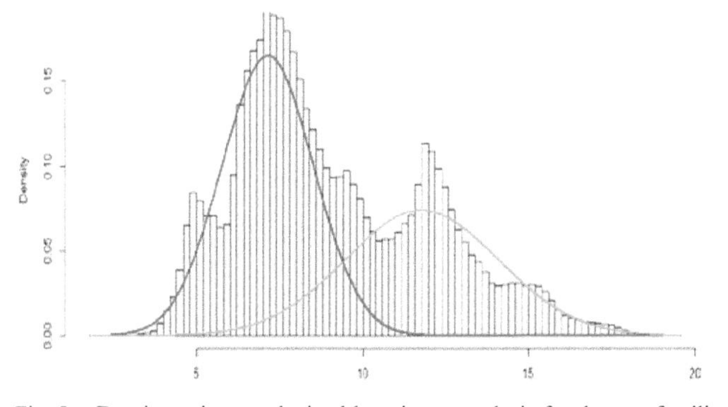

Fig. 5. Density estimates obtained by mixture analysis for the two families underlying the runtime distribution (on the log base 2 scale). The red line marks the estimated density for the "short" class, while the green line marks the estimated density for the "long" class

A model performing a perfect discrimination has an ROC curve that passes through the upper left corner (100% sensitivity, 100% specificity). Therefore the closer the ROC curve to the upper left corner, the higher the overall accuracy of the test. Yet, the consequences, or costs, of each type of error may vary among applications and among policy makers. Thus the optimal threshold may allow higher weight to one of the errors on account of the other.

VI. RESULTS

A. Class definition

Implementing the mixture model clustering approach, two Gaussian families underlying the runtime distribution (on the log base 2 scale) were defined. The density estimates are super imposed on the runtime density in Figure 5. The parameters for each family and the mixing proportions are detailed in Table III.

The mixture analysis also yields the posterior probability, as defined by equation (1), for each observation to belong to the "short" class. For a probability threshold of 0.5, 631,059 observations (nearly 60%) are classified into the "short" class, while 411,053 observations are classified into the "long" class. Once a classifier is found (see the next subsection), the threshold is refined to optimize the sensitivity-specificity tradeoff.

B. Classifying by CART

Applying the CART classifier on the training data through the cross-validation procedure, six variables obtained high importance scores (Figure 6). The classifier achieved a total misclassification error of 3.5%.

TABLE III. DENSITY PARAMETER ESTIMATES OBTAINED BY THE MIXTURE ANALYSIS

	First Gaussian Family		Second Gaussian Family	
Mixing Proportion	0.57		0.43	
Mean	7.13	$2^{7.13}$	11.78	$2^{11.78}$
Standard Deviation	1.38	$2^{1.38} = 2.60\ sec$	2.32	$2^{2.32} = 4.99\ sec$

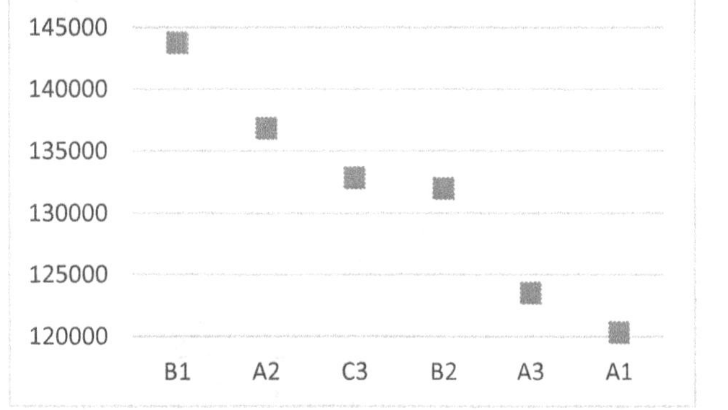

Fig. 6. Top ranking importance scores for the CART model, averaged across 150 cross-validation iterations

A model containing the six selected variables was fit to the full dataset for a series of class mixture threshold. The pseudo-ROC in Figure 7 presents the sensitivity-specificity combinations obtained for a set of threshold probability used for the mixture distribution. The best performing model is the one using the threshold of 0.45, which achieves sensitivity of 92.5% and specificity of 91.1%, with a total misclassification error of 8.9%. This threshold corresponds to a runtime of 9.25 on the log scale, or 608 seconds on the original scale.

The selected model yielded a tree containing four of the six variables that were tried (Figure 8). The total misclassification rate was 8.08%. Implementing the obtained tree on the validation data resulted in a total misclassification error of 9.17%, with specificity of 91.5% and sensitivity slightly beyond 90%.

VII. RELATED WORK

Supplying the scheduler with information on how long the jobs are expected to run has always been a challenging task. In general, two approaches were used estimating runtime. The first is to ask the users to supply the information, and the other is to try and predict the runtimes automatically using historical data on jobs that have already completed.

Asking the users to estimate the runtimes has been shown to be highly inaccurate, as users tend of overestimate the runtimes in order to prevent the scheduler from killing their jobs [1]. Furthermore, Tsafrir et al. [2] has observed that the users further tend to "round" the estimates, thereby limiting the scheduler's ability to optimize the schedule. Bailey et al. [7] have shown that users are quite confident of their estimates, and that most likely they will not be able to improve much the accuracy of their estimates.

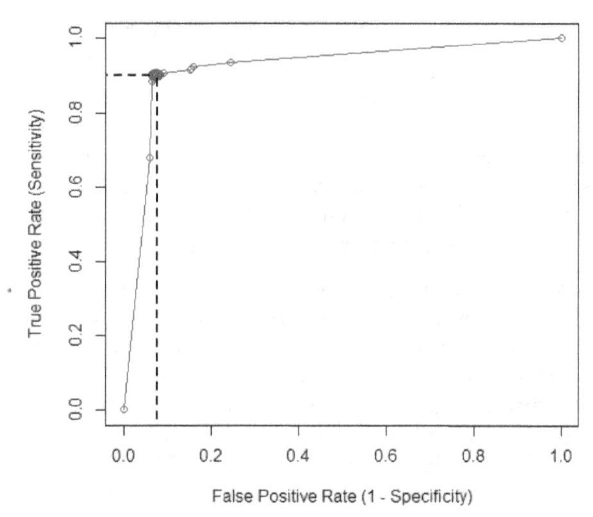

Fig. 7. The pseudo-ROC curve obtained by the CART classifier for the full training data. The blue circle marks the optimal tradeoff between sensitivity and specificity (enhanced by the dashed lines), obtained for mixture probability threshold of 0.45

Predicting the runtimes automatically is therefore the default alternative. This is usually composed of two steps:

1) Identifying classes of similar jobs within the historical jobs records, and

2) Using the aforementioned classes to predict the runtimes for newly submitted jobs.

Gibbons [4] and Downey [5] classified the jobs using a statically defined set of attributes, e.g. user, executable, queue, etc. For newly submitted jobs, Gibbons used the 95th percentile of the runtimes in the respective class, while Downey used a statistical model that was based on a log-uniform distribution of the runtimes in order to provide the prediction.

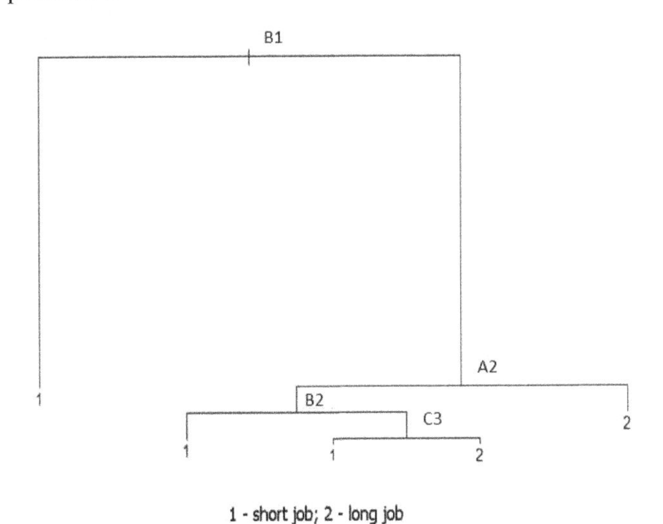

Fig. 8. The final tree obtained on the full data

Smith et al. [6] suggested the use of genetic algorithms to refine the selection of attributes used for the classification, and achieved up to 60% improvement in accuracy compared to the static approaches. Respectively, Kapadia et al. [8] used instance-based learning and Krishnaswamy et al. [9] applied rough-set theory. Finally, Tsafrir et al. [10] showed that complicated prediction techniques may not be required if the scheduling algorithm itself can be modified, and suggested to average the last two jobs by the same user in the history.

These techniques, however, were mainly designed for parallel systems, in which the scheduler needs to the actual runtimes of the jobs (use numeric predictions) to be able to optimize the schedule. For the server farm usage model which is targeted in the current work, sophisticated scheduling algorithms are not required, and the paper shows that it is enough to simply separate the jobs into short and long in order to improve performance [12]. Our work provides the facility to perform the separation and hence forms the basis for enabling such systems.

VIII. Discussion and Conclusions

Predicting the runtimes of jobs using actual numeric values is of high importance for parallel systems. Here, fragmentation is a big concern, and in order to minimize it (namely to fill the holes in the schedule), it is important for the scheduler to know the exact runtime of the jobs. For other types of systems like server farms used for software testing, it is enough to only predict the runtime classes of the jobs e.g., short or long, in order to send the jobs to the right queue and improve performance.

Motivated by the later usage model, a method that facilitates highly accurate prediction of the job runtime class was developed. The method leverages the fact that the runtimes may be represented as a mixture of two or more distributions, in order to train a classifier that will be used to predict the runtime classes of newly incoming jobs. In order to achieve high accuracy, the threshold that separates the short jobs from the long is determined during the evaluation of the classifier. In a real system the threshold can be periodically communicated to the scheduler to help deciding on the right allocation of resources for the different job classes.

The work presented is based on a single data set that was obtained, and included validation on data independent of the training data. Yet, in spite of its promising results, additional testing is required on more data sets in order to establish complete confidence in the robustness of the proposed method. However, due to the size of the data (over one million jobs), and the fact that the mixture distribution is known to be evident in the many real-world workloads, there is a strong reason to believe that with small adjustments e.g., to the number of classes, the proposed method can be tuned to sustain the workloads as well.

A uniqueness of the proposed approach is in conducting an initial step that is aimed to define runtime classes based on the empirical runtime distribution. While other existing approaches for predicting job runtime do not used categorization into classes but rather estimate the runtime numerically, the end result of all approaches is in an overall system efficiency. If efficiency can be defined and measured reliably, it can serve to compare the ultimate performance of all methods. Such assessment is under design and is planned to be conducted as a next step of the presented study.

Acknowledgment

We thank Prof. Dror G. Feitelson of the Hebrew University, Israel, for the useful comments, Evgeni Korchatov from Intel for helping to obtain and prepare the data and Eran Smadar from Intel for supporting this work.

References

[1] Feitelson, D.G., Mu'alem Weil, A.: Utilization and predictability in scheduling the IBM SP2 with backfilling. In 12th IEEE Int'l Parallel Processing Symp. (IPPS), 542–546 (1998)

[2] Tsafrir, D., Etsion, Y., Feitelson, D.G.: Modeling User Runtime Estimates, Job Scheduling Strategies for Parallel Processing (JSSPP), 1-35 (2005)

[3] Lawson, B.G., Smirni, E., Puiu, D.: Self-Adapting Backfilling Scheduling for Parallel Systems. Proceedings of the 2002 International Conference on Parallel Processing (ICPP), 583–592 (2002)

[4] Gibbons, R.: A Historical Application Profiler for Use by Parallel Schedulers. Job Scheduling Strategies for Parallel Processing (JSSPP), 58-77 (1997)

[5] Downey, A.B.: Predicting Queue Times on Space-Sharing Parallel Computers. Proc. 11th IEEE Int'l Parallel Processing Symp. (IPPS), 209-218 (1997)

[6] Smith, W., Foster, I., Taylor, V.: Predicting application run times using historical information. In the 4th Workshop on Job Scheduling Strategies for Parallel Processing (JSSPP), 122–142, Lect. Notes Comput. Sci. vol. 1459 (1998)

[7] Lee, C.B., Schwartzman, Y., Hardy, J., Snavely, A.: Are user runtime estimates inherently inaccurate? Job Scheduling Strategies for Parallel Processing (JSSPP), 253–263 (2004)

[8] Kapadia, N.H., Fortes, J.A.B., Brodley, C.E.: Predictive Application-Performance Modeling in a Computational Grid Environment. Proc. IEEE Int'l Symp. High Performance Distributed Computing (HPDC), 6 (1999)

[9] Krishnaswamy, S., Loke, S.W., Zaslavsky, A.: Estimating Computation Times of Data-Intensive Applications. IEEE Distributed Systems Online, vol. 5, no. 4 (2004)

[10] Tsafrir, D., Etsion, Y., Feitelson, D.G.: Backfilling using system-generated predictions rather than user runtime estimates. IEEE Trans. Parallel & Distributed Syst. 18(6), 789-803 (2007)

[11] Feitelson, D.G.: Metrics for mass-count disparity. In 14th Conf. Modeling, Analysis, and Simulation of Comput. and Telecomm. Syst., 61-68 (2006)

[12] Harchol-Balter, M., Crovella, M., Murta, C.: On Choosing a Task Assignment Policy for a Distributed Server System. IEEE Journal of Parallel and Distributed Computing (JPDC), vol. 59, no. 2, 204-228 (1999)

[13] Hastie, T., Tibshirani R., Friedman, J.: The Elements of Statistical Learning: Data Mining, Inference, and Prediction, Springer (2001)

[14] Maimon, O., Rokach, L.: The Data Mining and Knowledge Discovery Handbook, Springer, XXXVI (2005)

[15] Benaglia, T., Chauveau, D., Hunter, D.R., Young, D.S.: Mixtools: An R Package for Analyzing Finite Mixture Models, Journal of Statistical Software, Vol. 32, No. 6, 1-29 (2006)

[16] Benaglia, T., Chauveau, D., Hunter, D.R., Young, D.S, Elmore, R., Hettmansperger, T., Thomas, H., Xuan, F.: Package 'Mixtools' - Tools for Analyzing Finite Mixture Models, Repository CRAN (2014)

[17] Dudoit, S., Fridlyand, J., Speed, T.P.: Comparison of Discrimination Methods for the Classification of Tumors Using Gene Expression Data. Journal of the American Statistical Association Vol. 97, No. 457, 77-87 (2002)

[18] Lewis, R.J.: An Introduction to Classification and Regression Tree (CART) Analysis. Presented at the 2000 Annual Meeting of the Society for Academic Emergency Medicine in San Francisco, California (2000)

Lung Cancer Detection on CT Scan Images: A Review on the Analysis Techniques

H. Mahersia[1], M. Zaroug[1]
[1]Department of Computer Science, College of science and arts of Baljurashi, Albaha University, Albaha, Kingdom of Saudi Arabia

L. Gabralla[2]
[2]Faculty of Computer Science & Information Technology University of Science &Technology, Khartoum, Sudan

Abstract—**Lung nodules are potential manifestations of lung cancer, and their early detection facilitates early treatment and improves patient's chances for survival. For this reason, CAD systems for lung cancer have been proposed in several studies. All these works involved mainly three steps to detect the pulmonary nodule: preprocessing, segmentation of the lung and classification of the nodule candidates. This paper overviews the current state-of-the-art regarding all the approaches and techniques that have been investigated in the literature. It also provides a comparison of the performance of the existing approaches.**

Keywords—Classification; Computed Tomography; Lung cancer; Nodules; Segmentation

I. Introduction

The Lung cancer (LC) is the second most common cancer in both men and women in Europe and in the United States and represents a major economic issue for health care systems, accounting for about 12.7% of all new cancer cases per year and 18.2% of cancer deaths. In particular, each year there are approximately 1,095,000 new cancer cases and 951,000 cancer-related deaths in men and 514,000 new cases and 427,000 deaths in women [3].

Lung cancer is caused by uncontrollable irregular growth of cells in lung tissue. These lung tissue abnormalities are often called Lung nodules. They are small and roughly spherical masses of tissue, usually about 5 millimeters to 30 millimeters in size. In general, They can be categorized into 4 groups [78][94][59] including: juxta-vascular, well-circumscribed, pleural tail, and juxta-pleural. Figure 1 shows some examples of these categories. Pulmonary nodules are the characterization of the early stage of the lung cancer.

Investigations have shown that the curability of this deadly cancer is nearly 75%, if it is recognized early enough because it is easier to treat and with fewer risks. Therefore, the early diagnosis of malignant nodules is a crucial issue for reducing morbidity and mortality.

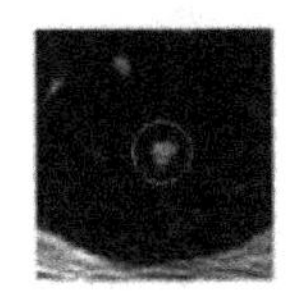
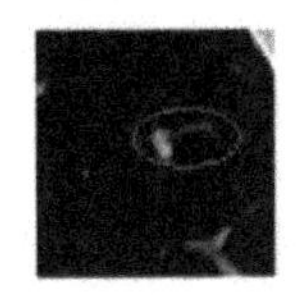
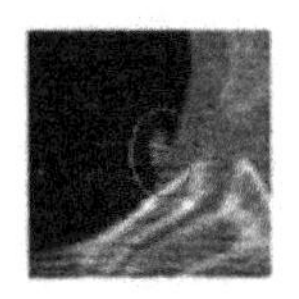
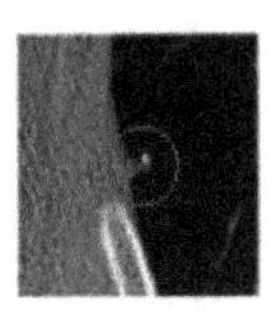

Fig. 1. Nodule's classification from [78][94]: Respectively from left to right, Well-circumscribed, vascularized, juxta-pleural and pleural-tail

Computer-aided diagnosis (CAD) systems are efficient schemes that have been developed for the detection and characterization of various lesions in the field of the diagnosis of lung cancer. The main objective of such systems is to assist the radiologist in the different analysis steps and to offer him a second opinion to the final decisions.

Thus, Researchers are becoming more and more concerned with the elaboration of automated CAD systems for lung cancer. Many publications proposed different automated nodule recognition systems using image processing, and including, different techniques for segmentation, feature extraction and classification.

II. Review of Existing Nodule Detection Methods

In literature, authors proposed several methods for automated and semi-automated detection of pulmonary nodules [59]. However, all these works involved four steps to detect the pulmonary nodule: pre-processing, extraction of nodule candidates, reduction of false positives and classification. Figure 2 shows these steps in details.

The next part focuses on the different studies involving these steps.

A. Pre-processing

Computed Tomography (CT) is considered as one of the best methods to diagnose the pulmonary nodules [76]. It uses x-rays to obtain structural and functional information about the human body. However, the CT image quality is influenced a lot by the radiation dose. The quality of image increases with the significant amount of radiation dose [15], but in the same time, this increases the quantity of x-rays being absorbed by the lungs. To prevent the human body from all kind of risk, radiologists are obliged to reduce the radiation dose, which affects the quality of image and is responsible for noises in lung CT images.

Pre-processing step aims to reduce the noises in these images. Different filtering techniques were proposed in literature to remove these noises, such as median filtering [11][49][7][41][62][63][82], wiener filtering [27][76][77], Gaussian filter [39][72][89][73][47], bilateral filtering [84] and a specific high-pass filter [32]. Many others works combine median filters with Laplacian filters by a differential technique, which subtracts a nodule suppressed image (through a median filter) from a signal enhanced image (through a Laplacian matched filter with a spherical profile) [34][35][16]. A difference image, containing nodule enhanced signal, is then obtained and used for the next stages.

In [84], the authors compare different pre-processing methods with various filters and suggest that bilateral filter provides better performances for pre-processing medical images. In addition, Bae et al. used a morphological filter to enhance the image region [8], whereas, Ochs et al. [68] and Paik et al. [71] applied in their studies, a spherical enhancement filter to enhance the nodule like structure in CT images.

In [7], the authors affirm that an Adaptive Median filtering is required to correct the poor contrast caused by poor lighting conditions during image acquisition. They generated a low frequency image by replacing each the pixel value with a median pixel value computed over a square area of 5x5 pixels. Then, a contrast limited adaptive histogram (CLAHE) equalization technique is used to improve the contrast of the CT pre-processed image.

In other hand, Farag et al. insist in [27][28] that the filtering approach to use must preserve object boundaries and detailed structures, Sharpen the discontinuities to enhance morphological structures and efficiently remove noise in homogeneous physical regions. In their work, the authors used both the Wiener and anisotropic diffusion filters.

Recently, other filters have been developed to enhance lung structures in 3-D images. Many researchers employed filters based on eigenvalues of the Hessian matrix [42][51][75][60]. Frangi et al. [31] further developed this approach by defining a 3D multi-scale structure enhancement filter based on the eigenvalues of the Hessian matrix and applying it to the enhancement of vessels. More later, Rikxoort et al. was the first to propose a supervised enhancement approach based on single phase and multi-phase methods [74]. In [92], the authors applied a set of 3D morphologic filters to separate the nodule from other surroundings structures, such as vessels and bronchi.

B. Segmentation

Segmentation of the lung regions is the second stage of the methods processing scheme. It refers to the process of partitioning the pre-processed CT image into multiple regions to separate the pixels or voxels corresponding to lung tissue from the surrounding anatomy. Various approaches have been used for lung segmentation and they can be categorized into two main groups: 2D approaches and 3D approaches.

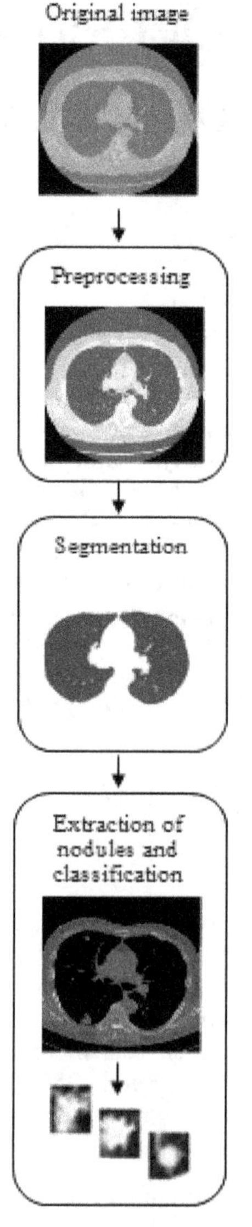

Fig. 2. The general scheme of lung nodule detection system

1) 2D-based approaches

In this section, we systematically review the state-of-the-art of the segmentation methods for lung CT images. Due to the large number of segmentation methods, we have categorized these methods into five intuitive groups for easier comprehension: thresholding-based, stochastic, region-based, contour-based, and learning-based methods, as shown in Figure 3.

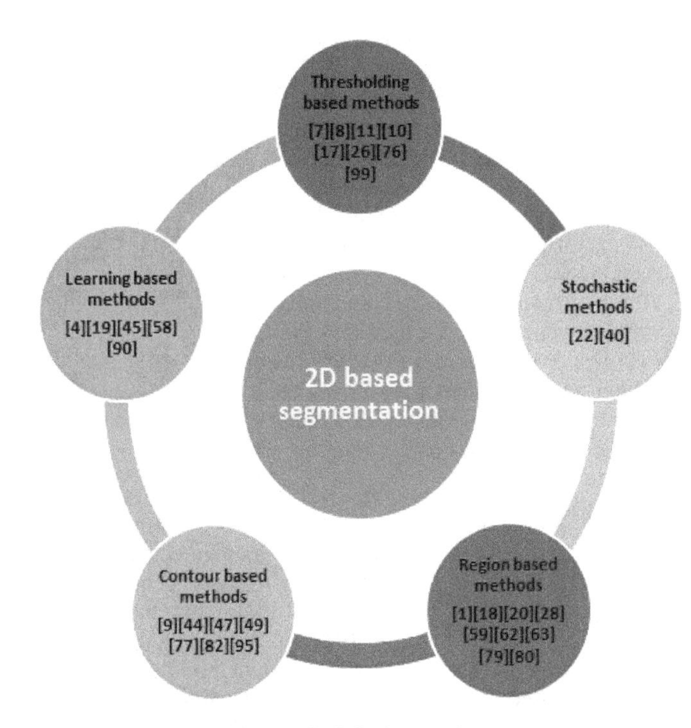

Fig. 3. 2D-based segmentation methods for lung CT images

Thresholding is a simple segmentation technique that converts a gray-level image into a binary image by defining all pixels greater than some value to be foreground and all other pixels are considered as background [7][11][10][17][26][8][76][99]. In [17], the authors separate nodule candidates from CT images using mathematical morphology and grey level thresholding. In [7], image histogram is used to find two value of threshold and then a multilevel thresholding and a connected-component labeling step is applied to the image in order to segment candidate nodule regions. Furthermore, simple thresholding that exploit the intensity characteristics of lung CT scans was presented in Farag et al. [27], El-Baz et al. [22], and Giger et al. [34] for separation of the nodule candidates from the background image. Bae et al. performs in [8] thresholding and seeded segmentation to isolate the juxta-pleural nodule from other structures.

In the same context, Shao et al. [76] uses adaptive iteration threshold method twice to implement initial segmentation of the pulmonary parenchyma. Zhou et al. [99], Wang et al. [85] and Retico et al. [73] implemented a histogram-based thresholding to segregate the lung region from the adjacent structure.

Another distinct type of lung CT segmentation technique is region-based segmentation methods. These methods focus generally on the homogeneity of the image for determining object boundaries. Region growing is the most widely used technique. It examines adjacent pixels of initial seed points and determines whether the pixel neighbours should be added to the region and then the process is iterated on. Obviously, object of interest must have nearly constant or slowly varying intensity values to satisfy the homogeneity requirement, which is true for CT images.

Region growing was explored by Aggarwal et al. [1], Lee et al. [59], and Taher et al. [80] for lung tissue segmentation. Combining the region growing with morphological closing, Lin and Yan [62] and Lin et al. [63] succeed to fill the large indentation caused by blood vessel that could not be extracted by thresholding.

A part from region growing method, many other methods involving textural features have been implemented last years [18][20][28]. In [28] and [79] local binary patterns were used as textural features and regions of interest (ROls) were characterized by combining the intensity histograms. Devaki et al. used the SURF and the LBP descriptors to generate the features that describe the texture of common lung nodules [20].

Stochastic methods exploit the difference between the existing structures in the lung images statistically. They propose many techniques that attempt to fit the distribution of intensity values in an image to a set of mathematical statistical functions. Each function defines a class and the output of the function defines the probability of an intensity value belonging to it. This approach was used by Guo et al., who developed a lung segmentation method using expectation-maximization (EM) analysis in combination with morphological operations [40]. After computing the image's histogram, the authors apply the (EM) algorithm to estimate the appropriate threshold value for lung segmentation.

Another segmentation technique was proposed by El-Baz et al. [22]. It aims to isolate the lungs from the surrounding structures by using Gibbs Markov Random Field (GMRF). In the next step, the abnormalities in the lungs are detected by using adaptive template matching and genetic algorithm.

Contour-based methods were used to identify the boundaries of the objects in the CT images. The contour-based methods can be categorized into two groups, Deformable models and Gradient Based methods. Deformable models were implemented in [47] and in [49] to segment nodules images. In fact, Kim et al. [49] uses a set of segmentation methods, such as thresholding, mathematic morphology, and deformable model to detect the lung region. Bellotti et al. [9] employed region growing with contour following to isolate juxta-pleural nodules. Zhao et al. [95] improved the shape-based segmentation using nodule gradient and sphere occupancy measurements.

In [77], the segmentation algorithm is applied based Sobel edge detection method, in order to detect the cancer nodules from the extracted lung image, whereas, in [44], the snake algorithm was used to extract the nodules' boundaries. Later, Tariq et al. used gradient mean and variance based method for the extraction of lung background since gradient operator has high values for pixels belonging to the boundary between foreground and background [82].

Learning-based methods, known also as knowledge-based methods, use pattern recognition techniques to statistically estimate dependencies in the image. They aim to represent the knowledge about lung cancer in a form that the computer can deal with [58][90][4][45]. Leader et al. [58] developed a heuristic threshold-based scheme for initial lung segmentation and then they applied a rule-based process to correct the initial lung segmentation's result. In [4], the authors propose an anatomical model through a semantic network whose nodes are the anatomical structures in the lungs. Each node of this network contains information about a specific anatomical part, position relative to other structures, and gray level. Then, the authors describe these features by fuzzy sets.

In the same context, rule based technique is applied in [77] and a set of diagnosis rules are generated from the extracted features. In [19] Dehmeshki et al. proposed to use a fuzzy map to improve the contrast between nodules and surrounding structures, such as blood vessels.

In [45], Jaafar et al. implemented a genetic algorithm procedure to segment the lung part from the original image, then they used morphology and Susan thinning algorithm to detect lung's edges. In [90], the authors present an intelligent medical system for lung cancer cell identification based on a two-layer rule-based fuzzy knowledge model.

2) 3D-based approaches

Several approaches exist in literature regarding the volumetric lung nodule segmentation. They can be classified into five categories: thresholding [96], mathematical morphology, region growing, deformable model, and dynamic programming, as shown in Figure 4.

Thresholding approach was adopted by Zhao et al. [96] and Yankelevitz et al. [91][92], where the appropriate threshold values can be deduced either after applying the Kmean clustering in [91][92] or applying the average gradient magnitudes algorithm [96].

Mathematical Morphology was also used for detection lung nodules in 3D CT images. Kostis et al. [52, 53] and Kuhnigk et al. [56, 57] have proposed effective iterative approaches for binary morphological filtering with various combinations of these basic operators. Okada et al. [69] presented a data-driven method to determine the ellipsoidal structuring element from anisotropic Gaussian fitting. Fetita et al. [30] proposed a new gray-level mathematical morphology operator, in order to discriminate the volumetric lung nodules from other dense structures. In [38], Goodman et al. segmented the existing lung nodules using the watershed algorithm followed by a model-based analysis.

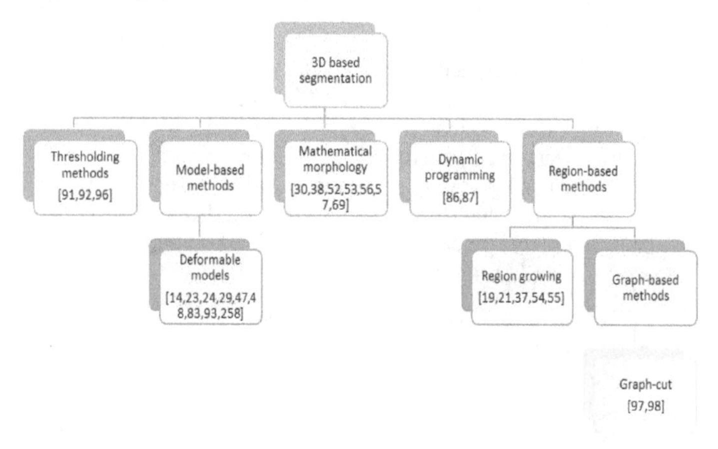

Fig. 4. An overview of the 3D segmentation methods

In other hands, more recent studies [19][21][54, 55] used the region growing approach as the main component of their overall segmentation algorithms. Dehmeshki et al. [19] proposed an adaptive region growing scheme on the fuzzy connectivity map computed from a prior segmented images. Diciotti et al. [21] proposed also a modified region growing algorithm designed with a geodesic distances. Kubota et al. [54, 55] used the same concept but with an Euclidean distance map. Later, Gong et al. segmented the lung lobes via a 3D region growing algorithm and then a number of regions of interest were extracted by using the Otsu threshold algorithm [37].Graph-Cuts is one of the well-known techniques of region-based segmentation. Zheng et al. [97, 98] applied graph-cuts to derive their initial 2D nodule segmentation in their coupled segmentation-registration method with B-spline registration.

Deformable models are widely applied methods for 3D segmentation purposes. They were implemented firstly by Kawata et al. [47, 48] who adopted the geodesic active contours approach introduced in [14]. El-Baz et al. [23, 24] adopted the energy minimization approach when designing an appearance model to segment the 3D lung nodules. Farag et al. [29] proposed a Level Sets solution with adaptive prior probability term for nodule segmentation. Yoo et al. [93] adopted the multiphase level sets framework introduced in [83] to present an asymmetric segmentation method for partially solid nodules. Active contours were also a widely used technique in image segmentation research community. In this context, Way et al. proposed in [88], an explicit active contour method which minimized energy that took into account 3D gradient, curvature, and penalized contours when growing against chest wall.

Dynamic Programming is another well-known technique for detecting optimal contours in images. Several methods extend this approach to a 3D surface detection process. In Wang et al. [87], a set of 2D dynamic programming iterations are applied to successive slices along the third dimension. In [86], the authors proposed to transform the 3D spherical lung volume to the 2D polar coordinate system before applying the standard 2D dynamic programming algorithm and this was in order to detect 3D lesion boundary.

According to Diciotti et al. [21], segmentation algorithms should be evaluated on large public databases with a well-defined ground truth for verification. Several of the existing studies utilized private databases. Therefore, a performance comparison between various methods is thus limited [59]. Usually, a nodule will appear in several slices of image in a CT scan. In 2D method, the slice with the greatest sized nodule is selected for analysis to differentiate between benign and malignancy. Compared with 2D method, the addition of extra dimension dramatically increases the operational complexity and computational cost for processing the entire 3D nodule volume. Thus, to reduce both the computational cost and radiation dose, the study in this paper tries to distinguish between benign and malignant nodules by using a 2Dapproach for a single post-contrast CT scan [64].

C. *Nodule extraction and classification*

Lung nodule detection aims to identify the location of the nodules if they exist. The most widely proposed approach is detection by classification and clustering. This approach comprises four categories: Fuzzy and neural network, K-nearest neighbour, Support vector machines and linear discriminant analysis, as shown in Figure 5.

Fuzzy rules were first designed by Brown et al. [13] who developed a knowledge-based, fully automated method for segmenting volumetric chest CT images. The method utilizes a modular architecture consisting of an anatomical model, image processing routines, and an inference engine. Later, Li et al. [61] and Dehmeshki et al. [19] implemented an automated rule-based classifier to classify nodules and non-nodules. The same approach was also adopted by Kostiset al. [52], Bong et al. [12] and Hosseini et al. [43]. In [12] Bong et al. propose and apply state-of-the-art fuzzy hybrid scatter search for segmentation of lung Computed Tomography (CT) image to identify the lung nodules detection. It utilized fuzzy clustering method with evolutionary optimization of a population size. Later in [43], the authors employed two fuzzy methods for the lung nodule CAD application: The Mamdani model and the Sugeno model of the fuzzy logic system. These methods were implemented and the classification results were compared and evaluated through ROC curve analysis and root mean squared error methods.

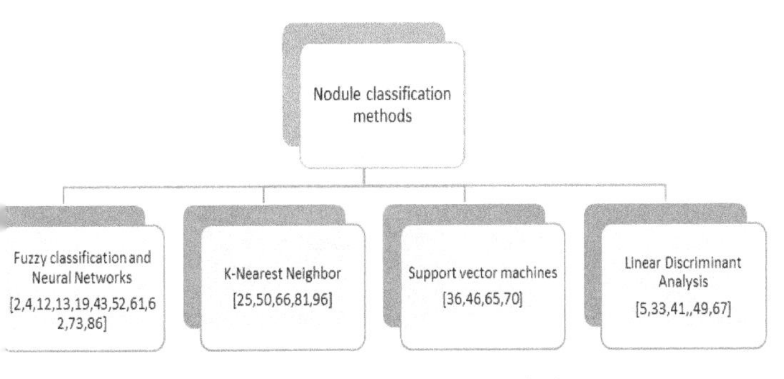

Fig. 5. An overview of the Nodule classification methods

Artificial neural networks were employed by Arimura et al. [6] for lung nodule detection. Reticoet al. introduces the identification of the pleural region by Directional-gradient concentration (DGC) and morphological opening, then, the features are extracted and candidate nodules are classified using Feed-forward Neural Network [73]. A two-level convolution neural network was proposed in Lin et al. [86]. Lin and Yan [62] and Lin et al. [63] combined fuzzy logic and neural networks for lung nodule detection and reported that the combination was superior to rule-base, convolution neural network, and genetic algorithm template matching approaches. Also, Antonelli et al. [4] adopted a decision fusion technique to develop a computer-aided detection (CAD) system for automatic detection of pulmonary nodules in low-dose CT images. In the classification stage, they built multi-classifier systems, aggregating the decisions of a feed forward four-layer neural network and a decision tree.

Recently, Akram et al. implemented an automated pulmonary nodule detection system a novel pulmonary nodule detection system using Artificial Neural Networks based on hybrid features consist of 2D and 3D Geometric and Intensity based statistical features [2].

A nearest cluster method was used by Ezoe et al. [25] and Tanino et al. [81] to classify the detected nodules candidate. Zhao et al. [96] applied boosting of the KNN classifier to estimate the probability density function of the intensity value of the trained ground glass opacity nodules. In [50], Kockelkorn et al. designed a user-interactive framework for lung segmentation with a k-nearest-neighbour (KNN) classifier. After that, Mabrouk et al. selected, in [66], a total of 22image features from the enhanced CT image, then, a fisher score ranking method was used as a feature selection method to select the best ten features and a K-Nearest Neighbourhood classifier was used to perform classification.

Support vector machines (SVM) were performed by Ginneken [36] to classify the nodule feature vector. It was also used by Lu et al. to classify the volumetric lung cancer from based on the concept of machine learning [65]. In 2013, Orozco et al. [70] presented a computational alternative to classify long nodules in frequency domain using Support Vector Machines. In the same year, Javed et al. proposed a new weighted SVM classifier in order to increase the accuracy of a lung tumour classification system [46].

The LDA classifier was employed by Gurcan et al. [41] and Armato et al. [5] to reduce the false positives produced by a rule-based classifier. A new feature with 3D gradient field was added to the LDA classifier by Ge et al. [33] to improve the false positives of Gurcan et al. [41]. In addition, Matsumoto et al. [67] implemented the same classifier using eight features to identify the candidate nodules. Kim et al. [49] classified the ground glass opacity nodule using LDA based on the Mahalanobis distance distribution.

III. Conclusion

This review gives an overview of the current detection techniques for CT images that may help researchers when choosing a given method. Certainly, lung analysis techniques have been improved over the last decade. However, there still are issues to be solved such as developing new and better techniques of contrast enhancement and selecting better criteria for performance evaluation is also needed.

References

[1] P. Aggarwal, R. Vig and H.-K. Sardana, Semantic and Content-Based Medical Image Retrieval for Lung Cancer Diagnosis with the Inclusion of Expert Knowledge and Proven Pathology, In proc. of the IEEE second international conference on Image Information Processing ICIIP'2013, pp. 346-351, 2013.

[2] S. Akram, M.-Y. Javed, U. Qamar, A. Khanum and A. Hassan, Arti_cial Neural Network based Classi_cation of Lungs Nodule using Hybrid Features from Computerized Tomographic Images, Appl. Math. Inf. Sci., Vol. 9, No. 1, pp. 183-195, 2015.

[3] V. Ambrosini,S. Nicolini, P. Carolia, C. Nannia, A. Massarob, M.-C. Marzolab, D. Rubellob and S. Fantia, PET/CT imaging in di_erent types of lung cancer: An overview, European Journal of Radiology, Vol. 81, pp. 988-1001, 2013.

[4] M. Antonelli, M. Cococcioni, B. Lazzerini and F. Marcelloni, Computer-aided detection of lung nodules based on decision fusion techniques, Pattern. Anal. Applic., Vo. 14, pp. 295310, 2011.

[5] H. Arimura, S. Katsuragawa and K. Suzuki, Computerized scheme for automated detection of lung nodules in low-dose computed

tomography images for lung cancer screening, Acad. Radiol., Vol. 11, pp. 617629, 2004.

[6] S.-G. Armato, F. Li and M.-L. Giger, Lung cancer: performance of automated lung nodule detection applied to cancers missed in a CT screening program, Radiology, Vol. 225, pp.685692, 2002.

[7] S. Ashwin, S.-A. Kumar, J. Ramesh and K. Gunavathi, E_cient and Reliable Lung Nodule Detection using a Neural Network Based Computer Aided Diagnosis System, In Proc. of the International Conference on Emerging Trends in Electrical Engineering and Energy Management (ICETEEEM'2012), pp. 135-142, Chennai, 13-15 Dec. 2012.

[8] K.-T. Bae, J.-S. Kim,, Y.-H. Na, Pulmonary nodules: automated detection on CT images with morphologic matching algorithm, preliminary results, Radiology, Vol. 236, pp. 286294, 2005.

[9] R. Bellotti, F. De Carlo and G. Gargano, A CAD system for nodule detection in low-dose lung cts based region growing and active contour models. Med. Phys., Vol. 34, pp. 49014910, 2007.

[10] N. Birkbeck, M. Sofka, T. Kohlberger, J. Zhang, J. Wetzl, J. Kaftan and S. Kevin Zhou, Robust Segmentation of Challenging Lungs in CT Using Multi-stage Learning and Level Set Optimization, Computational Intelligence in Biomedical Imaging, pp. 185-208, 2014.

[11] S.-C. Blo, M.-T. Freedman, J.-S. Lin and S.-K. Mun, Journal of Digital lmaging, Vol. 6, No. 1, pp. 48-54, 1993.

[12] C.-W. Bong, H.-Y. Lam and H. Kamarulzaman, A Novel Image Segmentation Technique for Lung Computed Tomography Images, Communications in Computer and Information Science, Vol. 295, pp. 103-112, 2012.

[13] M.-S. Brown, M.-F. McNitt-Gray and N.-J. Mankovich, Method for segmenting chest CT image data using an anatomical model: preliminary results, IEEE Trans. on Med. Imaging, vol. 16, No. 6, pp. 828839, 1997.

[14] V. Caselles, R. Kimmel, G. Sapiro and C. Sbert, Minimal surfaces based object segmentation, IEEE Trans. on Pattern Analysis and Machine Intelligence, Vol. 19, No. 4, pp. 394398, 1997.

[15] T. Chhabra, G.Dua, T. Malhotra, Comparative Analysis of Methods to Denoise CT Scan Images, International Journal of Advanced Research in Electrical, Electronics and Instrumentation Engineering, Vol. 2, No. 7, pp. 3363-3369, 2013.

[16] Y.-S. Chiou, Y.-M. FlemingLure, M.-T. Freedman and S. Fritz, Application of Neural Network Based Hybrid System for Lung Nodule Detection, In. proc of the Sixth Annual IEEE Symposium on Computer-Based Medical Systems, pp. 216-216, Ann Arbor, 13-16 Jun 1993.

[17] W.-J. Choi, A. Majid and T.-S. Choi, Computerized Detection of Pulmonary Nodule Based on Two-Dimensional PCA, Computational Science and Its Applications ICCSA 2009, Part II, Vol. 5593, pp. 693-702, 2009.

[18] M. Ceylan, Y. Ozbay, O.-N. Ucan and E. Yildirim, novel method for lung segmentation on chest CT images: complex-valued arti_cial neural network with complex wavelet transform, Turk J. Elec. Eng. and Comp. Sci., Vol.18, No.4, pp. 613-623, 2010.

[19] J. Dehmeshki, H. Amin, M.-V. Casique and X. Ye, Segmentation of pulmonary nodules inthoracic CT scans: A region growing approach, IEEE Trans. on Med. Imaging, Vol. 27, pp. 467480, 2008.

[20] K. Devaki, V. MuraliBhaskaran and M. Mohan, Segment Segmentation in Lung CT Images- Preliminary Results, Special Issue of International Journal on Advanced Computer Theory and Engineering (IJACTE), Vol. 2, No. 1, pp. 84-89, 2013.

[21] S. Diciotti, G. Picozzi, M. Falchini, M. Mascalchi, N. Villari and G. Valli, 3-D segmentation algorithm of small lung nodules in spiral CT images, IEEE Transactions on Information Technology in Biomedicine, Vol. 12, No. 1, pp. 719, 2008.

[22] A. El-Bazl, A. Farag, R. Falk and R. LaRocca, Automatic identification of lung abnormalities in chest spiral CT scans, In proc. of the international conference on Acoustics, Speech, and Signal Processing (ICASSP '03), Vol.2, pp. 261-264, 2003.

[23] A. El-Baz, A. Farag, G. Gimelfarb, R. Falk, M.-A. El-Ghar and T. Eldiasty, A framework for automatic segmentation of lung nodules from low dose chest CT scans, in Proc. of the 18th International Conference on Pattern Recognition (ICPR 06), Vol. 3, pp. 611614, 2006.

[24] A. El-Baz, G. Gimelfarb, R. Falk and M. Abo El-Ghar, 3D MGRF-based appearance modelling for robust segmentation of pulmonary nodules in 3D LDCT chest images, in Lung Imaging and Computer Aided Diagnosis, chapter 3, pp. 5163, Taylor and Francis edition, 2011.

[25] T. Ezoe, H. Takizawa and S. Yamamoto, An automatic detection method of lung cancers including ground glass opacities from chest X-ray CT images, In Proc. of SPIE, vol. 4684, pp. 16721680, 2002.

[26] K.-Z. Faizal and V. Kavitha, An E_ective Segmentation Approach for Lung CT Images Using Histogram Thresholding with EMD Refinement, Proceedings of International Conference on Internet Computing and Information Communications, Advances in Intelligent Systems and Computing, Vol. 216, pp. 483-489, 2014.

[27] A. Farag, J Graham, A. Farag and R. Falk, Lung Nodule Modelling A Data-Driven approach, Advances in Visual Computing, Vo. 5875, pp 347-356, 2009.

[28] A. Farag, A. Ali, J. Graham, S. Elhabian, A. Farag and R. Falk, Feature-Based Lung Nodule Classi_cation, ISVC 2010, Part III, Vol. 6455, pp. 7988, 2010.

[29] A. Farag, H. Abdelmunim, J. Graham, Variational approach for segmentation of lung nodules, in Proc. of the IEEE International Conference on Image Processing (ICIP 11), pp. 21572160, 2011.

[30] C.-I. Fetita, F. Prteux, C. Beigelman-Aubry and P. Grenier, 3D automated lung nodule segmentation in HRCT, In Proc. of the International Conference Medical Imaging Computing and Computer-Assisted Intervention (MICCAI 03), Vol. 2878, pp. 626634, 2003.

[31] Frangi, W. Niessen, K. Vincken, and M. Viergever, Multiscale vessel enhancement filtering, Med. Image Computing Computer Assisted Intervention, vol. 1496, pp. 130137, 1998.

[32] T. Gao, X. Sun, Y. Wang and S. Nie, A Pulmonary Nodules Detection Method Using 3D Template Matching, Foundations of Intelligent Systems, AISC, Vol. 122, pp. 625633, 2012.

[33] Z. Ge, B. Sahiner and H.-P. Chan, Computer aided detection of lung nodules: False positive reduction using a 3d gradient _eld method, In Proc. of SPIE, Vol. 5370, pp. 1076-1082, 2004.

[34] M.-L. Giger,K. Doi and H. MacMahon, Pulmonary Nodules: Computer-Aided Detection in Digital Chest Images, RadioGraphics, Vol. 10, pp. 41-51, 1990.

[35] M.-L. Giger, N. Ahn, K. Doi, H. MacMahon and C.-E. Metz, Computerized Detection of Pulmonary Nodules in Digital Chest Images: Use of Morphological Filters in Reducing False-Positive Detections, Med. Phys., Vol. 17, pp. 861-865, 1990.

[36] B.-V. Ginneken, Supervised probabilistic segmentation of pulmonary nodules in CT scans, In proc. of the 9th Medical Image Computing and Computer-assisted Intervention MICCAI Conference, Berlin, 2006.

[37] J. Gong, T. Gao, R.-R. Bu, X.-F. Wang and S.-D. Nie, An Automatic Pulmonary Nodules Detection Method Using 3D Adaptive Template Matching, Communications in Computer and Information Science, Vol. 461, pp. 3949, 2014.

[38] L.-R. Goodman, M. Gulsun, L. Washington, P.-G. Nagy and K.-L. Piacsek, Inherent variability of CT lung nodule measurements in vivo using semi-automated volumetric measurements, American Journal of Roentgenology, Vol. 186, No. 4, pp. 989994, 2006.

[39] I. Gori, R. Bellotti, P. Cerello, S.-C. Cheran, G. De Nunzio, M.-E. Fantacci, P. Kasae, G.-L. Masala, A. Martinez and A. Retico, Lung nodule detection in screening computed tomography, in Proc. of the IEEE Nuclear Science Symposium, Vol. 6, pp. 3489-3491, 2006.

[40] Y. Guo, C. Zhou, H.-P. Chan, A. Chughtai, J. Wei, L.-M. Hadjiiski and E.-A. Kazerooni, Automated iterative neutrosophic lung segmentation for image analysis in thoracic computed tomography, Med. Phys., Vol.40, No. 8, pp. 081912/1-081912/11, 2013.

[41] M.-N. Gurcan,B. Sahiner and N. Petrick, Lung nodule detection on thoracic computed tomography images: preliminary evaluation of a computer-aided diagnosis system. Med. Phys. 29, pp. 2552-2558, 2002.

[42] H. Haussecker and B. Jahne, A tensor approach for local structure analysis in multidimensional images in 3-D, Image Anal. Synthesis, pp. 171178, 1996.

[43] R. Hosseini, J. Dehmeshki, S. Barman and M. Mazinani, A Fuzzy Logic System for Classification of the Lung Nodule in Digital Images in

Computer Aided Detection, In proc. of the Fourth International Conference on Digital Society, pp. 255-259, 2010.

[44] Y. Itai, K. Hyoungseop, T. Ishida, A segmentation method of lung areas by using snakes and automatic detection of abnormal shadow on the areas, Int. J. Innov. Comput. Inf. Control, Vol. 3, 277-284, 2007.

[45] M.-A. Ja_ar, A. Hussain, F. Jabeen, M. Nazir and A.-M. Mirza, GA-SVM Based Lungs Nodule Detection and Classi_cation, Communications in Computer and Information Science, Vol. 61, pp 133-140, 2009.

[46] U. Javed, M.-M. Riaz, T.-A. Cheema and H.-F. Zafar, Detection of Lung Tumor in CE CT Images by using Weighted Support Vector Machines, In. proc. of the 10th International Bhurban Conference on Applied Sciences and Technology (IBCAST), pp. 113-116, 2013.

[47] Y. Kawata, N. Niki, H. Ohmatsu, Quantitative surface characterization of pulmonary nodules based on thin-section CT images. IEEE Trans. Nuclear Sci., Vol. 45, pp. 2132-2138, 1998.

[48] Y. Kawata, N. Niki, H. Ohmatsu and N. Moriyama, A deformable surface model based on boundary and region information for pulmonary nodule segmentation from 3-D thoracic CT images, IEICE Transactions on Information and Systems, Vol. 86, No. 9, pp. 1921-1930, 2003.

[49] H. Kim, T. Nakashima and Y. Itai, Automatic detection of ground glass opacity from the thoracic MDCT images by using density features. In proc. of the International Conference on Control, Automation and Systems, pp. 1274-1277. Seoul, 2007.

[50] J.-P. Kockelkorn, E.-M. Van Rikxoort, J.-C. Grutters and B. Van Ginneken, Interactive lung segmentation in CT scans with severe abnormalities, In Proc. of the 7th IEEE International Symposium on Biomedical Imaging: From Nano to Macro (ISBI '10), pp. 564567, 2010.

[51] T. Koller, G. Gerig, G. Szekely, and D. Dettwiler, Multiscale detection of curvilinear structures in 2-D and 3-D image data, in Int. Conf. Computer Vision, pp. 864869, 1995.

[52] W.-J. Kostis, A.-P. Reeves, D.-F. Yankelevitz and C.-I. Henschke, Three-dimensional segmentation and growth-rate estimation of small pulmonary nodules in helical CT images, IEEE Transaction on Medical Imaging, Vol. 22, No. 10, pp. 12591274, 2003.

[53] [53] W.-J. Kostis, D.-F. Yankelevitz, A.-P. Reeves, S.-C. Fluture and C.-I. Henschke, Small pulmonary nodules, reproducibility of three-dimensional volumetric measurement and estimation of time to follow-up CT, Radiology, Vol. 231, No. 2, pp. 446452, 2004.

[54] [54] T. Kubota, A.-K. Jerebko, M. Dewan, M. Salganico_ and A. Krishnan, Segmentation of pulmonary nodules of various densities with morphological approaches and convexity models, Medical Image Analysis, Vol. 15, No. 1, pp. 133154, 2011.

[55] [55] T. Kubota, A. Jerebko, M. Salganico_, M. Dewan and A. Krishnan, Robust segmentation of pulmonary nodules of various densities: from ground-glass opacities to solid nodules, in Proc. of the International Workshop on Pulmonary Image Processing, pp. 253262, 2008.

[56] J.-M. Kuhnigk, V. Dicken, L. Bornemann, D. Wormanns, S. Krass and H.-O. Peitgen, Fast automated segmentation and reproducible volumetry of pulmonary metastases in CT-scans for therapy monitoring, in Proceedings of the 7th International Conference on Medical Image Computing and Computer-Assisted Intervention (MICCAI 04), Vol. 3217, pp. 933941, 2004.

[57] J.-M. Kuhnigk, V. Dicken, L. Bornemann, Morphological segmentation and partial volume analysis for volumetry of solid pulmonary lesions in thoracic CT scans, IEEE Transaction on Medical Imaging, Vol. 25, No. 4, pp. 417434, 2006.

[58] J.-K. Leader, B. Zheng, R.-M. Rogers, F.-C. Sciurba, A. Perez, B.-E. Chapman, S. Patel, C.-R. Fuhrman and D. Gur, Automated lung segmentation in X-ray computed tomography: Development and evaluation of a heuristic threshold-based scheme, Acad. Radiol., Vol. 10, pp. 12241236, 2003.

[59] S.-L.-A. Lee, A.-Z. Kouzani and E.-J. Hu, Automated detection of lung nodules in computed tomography images: a review, Machine Vision and Applications, Vol. 23, pp. 151163, 2012.

[60] Q. Li, S. Sone, and K. Doi, Selective enhancement filters for nodules vessels, and airway walls in two- and three-dimensional CT scans, Med. Phys., Vol. 30, pp. 20402051, 2003.

[61] Q. Li, F. Li and K. Doi, Computerized detection of lung nodules in thin-section CT images by use of selective enhancement filters and an automated rule-based classifier, Acad. Radiol., Vol. 15, pp. 165175, 2008.

[62] D.-T. Lin and C.-R. Yan, Lung nodules identification rules extraction with neural fuzzy network, In Proc. of the 9th IEEE International Conference of Information Processing (ICONIP), Vol. 4, pp. 2049-2053, Singapore, 18-22 Nov.2002.

[63] D.-T. Lin, C.-R. Yan and, W.-T. Chen, Autonomous detection of pulmonary nodules on CT images with a neural network-based fuzzy system, Comput. Med. Imaging Graph., Vol. 29, pp. 447-458, 2005.

[64] P.-L. Lin, P.-W. Huang, C.-H. Lee and M.-T. Wu, Automatic classification for solitary pulmonary nodule in CT image by fractal analysis based on fractional Brownian motion model, Pattern Recognition, Vol. 46, pp. 32793287, 2013.

[65] X. Lu, G.-Q. Wei, J. Qian and A.-K. Jain, Learning-based Pulmonary Nodule Detection from Multislice CT Data, In proc. of the 18th International Congress and Exhibition, Chicago, 2004.

[66] M. Mabrouk, A. Karrar and A. Sharawy, Computer Aided Detection of Large Lung Nodules using Chest Computer Tomography Images, International Journal of Applied Information Systems (IJAIS), Vol. 3, No. 9, pp. 12-18, 2012.

[67] S. Matsumoto, H.-L. Kundel and J.-C. Gee, Pulmonary nodule detection in CT images with quantized convergence index filter, Medi. Image Anal., Vol. 10, pp. 343352, 2006.

[68] R.-A. Ochs, J.-G. Goldin and A. Fereidoun, Automated classification of lung bronchovascular anatomy in CT using Adaboost, Med. Image Anal., Vol. 11, pp. 315324, 2007.

[69] K. Okada, V. Ramesh, A. Krishnan, M. Singh and U. Akdemir, Robust pulmonary nodule segmentation in CT: improving performance for juxtapleural cases, in Proc. of the International Conference on Medical Imaging Computing and Computer-Assisted Intervention (MICCAI 05), Vol. 8, pp. 781789, 2005.

[70] H.-M. Orozco and O.-O. Villegas, Lung Nodule Classification in CT Thorax Images using Support Vector Machines, In proc. of the 12th Mexican International Conference on Artificial Intelligence, pp. 277-283, 2013.

[71] D.-S. Paik, C.-F. Beaulieu and G.-D. Rubin, Surface normal overlap: a computer-aided detection algorithm with application to colonic polyps and lung nodules in helical CT, IEEE Trans. Med. Imaging, Vol. 23, pp. 661675, 2004.

[72] J. Pu,, J. Roos and C. Yi, Adaptive border marching algorithm: automatic lung segmentation on chest CT images, Comput. Med. Imaging Graph., Vol. 32, pp. 452462, 2008.

[73] Retico, P. Delogu, M.-E. Fantacci, Lung nodule detection in low-dose and thin-slice computed tomography, Comput. Biol. Med., Vol. 38, pp. 525534, 2008.

[74] E.-M. vanRikxoort, B. vanGinneken, M. Klik, and M. Prokop, Supervised Enhancement Filters: Application to Fissure Detection in Chest CT Scans, IEEE trans. on Medical imaging, Vol. 27, No. 1, pp. 1-10, 2008.

[75] Y. Sato, C.Westin, A. Bhalerao, S. Nakajima, N. Shiraga, S. Tamura and R. Kikinis, Tissue classi_cation based on 3-D local intensity structure for volume rendering, IEEE Trans. Vis. Comput. Graphics, Vol. 6, No. 2, pp. 160180, 2000.

[76] H. Shao, L. Cao and Y. Liu, A Detection Approach for Solitary Pulmonary Nodules Based on CT Images, in Proc. of the 2nd International Conference on Computer Science and Network Technology, pp. 1253-1257, 2012.

[77] D. Sharma and G. Jindal, Identifying Lung Cancer Using Image Processing Techniques, in Proc. of the International Conference on Computational Techniques and Artificial Intelligence (ICCTAI'2011), pp. 115-120, 2011.

[78] Y. Song, W. Cai, Y. Wang, and D.-D. Feng, Location classification of lung nodules with optimized graph construction, in Proc. ISBI, pp. 1439-1442, May 2012.

[79] L. Sorensen, S.-B. Shaker, M. deBruijne, Quantitative Analysis of Pulmonary Emphysema Using Local Binary Patterns, IEEE Trans. on Medical Imaging, Vol. 29, No. 2, pp. 559-569, .2010.

[80] F. Taher, R. Sammouda, Identi_cation of Lung Cancer Based on Shape and Color, In proc. of the 4th International Conference, ICISP'2010, June 30-July 2, 2010.

[81] M. Tanino, H. Takizawa and S. Yamamoto, A detection method of ground glass opacities in chest X-ray CT images using automatic clustering techniques. In Proc. of SPIE, vol. 5032, pp. 17281737, 2003.

[82] Tariq, M.-U. Akram and M.-Y. Javed, Lung Nodule Detection in CT Images using Neuro Fuzzy Classi_er, in Proc. of the Fourth International IEEE Workshop on Computational Intelligence in Medical Imaging (CIMI), pp. 49-53, 2013.

[83] L.-A. Vese and T.-F. Chan, A multiphase level set framework for image segmentation using the Mumford and Shah model, International Journal of Computer Vision, Vol. 50, No. 3, pp. 271293, 2002.

[84] G. Vijaya and A. Suhasini, An Adaptive Pre-processing of Lung CT Images with Various Filters for Better Enhancement, Academic Journal of Cancer Research, Vol. 7, No. 3, pp. 179-184, 2014.

[85] P. Wang, A. DeNuzio, P. Okunie_, Lung metastases detection in CT images using 3D template matching, Med. Phys., Vol. 34, pp. 915922, 2007.

[86] J.Wang, R. Engelmann and Q. Li, Segmentation of pulmonary nodules in three-dimensional CT images by use of a spiral scanning technique, Medical Physics, Vol. 34, No. 12, pp. 46784689, 2007.

[87] Q.Wang, E. Song and R. Jin, Segmentation of lung nodules in computed tomography images using dynamic programming and multidirection fusion techniques, Academic Radiology, Vol. 16, No. 6, pp. 678688, 2009.

[88] T.-W. Way, L.-M. Hadjiiski and B. Sahiner, Computer-aided diagnosis of pulmonary nodules on CT scans: segmentation and classification using 3D active contours, Medical Physics, Vol. 33, No. 7, pp. 23232337, 2006.

[89] G.-Q. Wei, L. Fan and J. Qian, Automatic detection of nodules attached to vessels in lung CT by volume projection analysis, Medical Image Computing and Computer-assisted Intervention, Vol. 2488, pp. 746752, 2002.

[90] Y. Yang, S. Chen, H. Lin and Y. Ye, A Chromatic Image Understanding System for Lung Cancer Cell Identification Based on Fuzzy Knowledge, Innovations in Applied Artificial Intelligence, Vol. 3029, pp 392-401, 2004.

[91] D.-F. Yankelevitz, R. Gupta, B. Zhao, and C.-I. Henschke, Small pulmonary nodules: evaluation with repeat CT preliminary experience, Radiology, Vol. 212, No. 2, pp. 561566, 1999.

[92] D.-F. Yankelevitz, A.-P. Reeves, W.-J. Kostis, B. Zhao and C.-I. Henschke, Small pulmonary nodules: volumetrically determined growth rates based on CT evaluation, Radiology, Vol. 217, No. 1, pp. 251256, 2000.

[93] Y. Yoo, H. Shim, I.-D. Yun, K.-W. Lee and S.-U. Lee, Segmentation of ground glass opacities by asymmetric multi-phase deformable model, Medical Imaging: Image Processing, Vol. 6144, 2006.

[94] F. Zhang, W. Cai, Y. Song, M.-Z. Lee, S. Shan and D.-D. Feng, Overlapping Node Discovery for Improving Classi_cation of Lung Nodules, The 35th Annual International Conference of the IEEE EMBS Osaka, Japan, 3 - 7 July, 2013.

[95] B. Zhao, D. Yankelevitz and A. Reeves, Two dimensional multi-criterion segmentation of pulmonary nodules on helical CT images, Med. Phys., Vol. 26, pp. 889895, 1999.

[96] B. Zhao, A.-P. Reeves, D.-F. Yankelevitz and C.-I. Henschke, Three-dimensional multicriterion automatic segmentation of pulmonary nodules of helical computed tomography images, Optical Engineering, Vol. 38, No. 8, pp. 13401347, 1999.

[97] Y. Zheng, K. Steiner, T. Bauer, J. Yu, D. Shen and C. Kambhamettu, Lung nodule growth analysis from 3D CT data with a coupled segmentation and registration framework, in Proc. of the IEEE 11th International Conference on Computer Vision (ICCV 07), 2007.

[98] Y. Zheng, C. Kambhamettu, T. Bauer and K. Steiner, Accurate estimation of pulmonary nodules growth rate in ct images with nonrigid registration and precise nodule detection and segmentation, in Proc. of the IEEE Conference on Computer Vision and Pattern Recognition (CVPR 09), pp. 101108, 2009.

[99] X. Zhou, T. Hayashi, T. Hara, H. Fujita, R. Yokoyama, T. Kiryu and H. Hoshi, Automatic segmentation and recognition of anatomical lung structures from high-resolution chest CT images, Computerized Medical Imaging and Graphics, Vol. 30, pp. 299313, 2006.

[100] J. Zhou, S. Chang and D.-N. Metaxas, An automatic method for ground glass opacity nodule detection and segmentation from CT studies. In proc. of 28th IEEE EMBS Conference, pp. 30623065, USA, 2006.

A Directional Audible Sound System using Ultrasonic Transducers

Wen-Kung Tseng
Graduate Institute of Vehicle Engineering
National Changhua University of Education
Changhua City, Taiwan

***Abstract*—In general the audible sound has the characteristics of spreading, however the ultrasound is directional. This study used amplitude-modulating technique for an array of 8 ultrasonic transducers to produce directional audible sound beam. In this study sound field distribution for the directional audible sound beam has been investigated. The effect of different weightings varied with different frequency for the transducers on the directivity of the sound beam has also been evaluated. An H_∞ optimization method was used to calculate the optimal weightings of the transducers for better directivity of the sound beam. Different optimal weightings also added to the carrier and sideband frequencies to control the difference frequency's beam width and side lobe amplitude. The results showed that the beam width can be controlled and good directivity of the sound beam can be obtained by using the H_∞ optimization method.**

Keywords—ultrasound; amplitude-modulating; directional audible sound beam; weightings; H_∞ optimization method

I. INTRODUCTION

Generally, the audible sound frequency is the range from 20 Hz to 20 kHz; however, ultrasound is sound pressure with a frequency greater than the upper limit of human hearing i.e. 20 kHz as shown in Fig. 1. Recently the highly directional audible sound has been investigated for a few years. The audible sound has the characteristics of spreading, however the ultrasound is directional.

In the pass decades, the performance in the ultrasound systems and their beamforming has been studied. Beamforming is the concept of forming directional beams. Previous study [1] has discussed that two plane waves of different ultrasonic frequencies could generate the directional audible sound due to acoustical nonlinearity. There are mainly two new waves, one of which has a frequency equal to the decrease of the original two frequencies and the other equal to the difference frequency.

New ultrasonic waves whose frequencies correspond to the decrease and difference of the two ultrasonic signals will be produced. The nonlinear interaction of ultrasonic sound waves in air will produce this phenomenon. For instance, if there are two ultrasonic signals in the air at closely frequencies f_0 and f_1 as shown in Fig. 2, they will be transformed into $2f_0$, $2f_1$, $f_1 \pm f_0$ and other higher order harmonics in the signal [2]. The new components of $2f_0$, $2f_1$ and $f_1 + f_0$ in the air will be strongly attenuated rapidly with increasing distance from the transducer. However, the remaining frequency $f_1 - f_0$ is decay slowly with increasing distance from the transducer because of the relatively low absorption.

For simplicity, the ultrasonic transducer is fed with two ultrasonic signals f_0 at 40 kHz and f_1 at 41k Hz, then the new modulation frequency is 1 kHz ($f_1 - f_0$). The new frequency component of 1 kHz has lower absorption than other high-frequency terms. For this reason, this study will not discuss the new high-frequency terms. Instead, it will focus on and discuss the frequency subtraction component $f_1 - f_0$ of new frequency components.

Traditional in-car communication gadgets, such as cell phones or radios, generally adopt traditional loudspeakers to broadcast. This has several weaknesses. For instance, people in the car can all hear the sound. This makes it impossible to maintain privacy. Besides, people feel bothered and thus the ride quality would not be so pleasant. If cars are equipped with directional loudspeakers, each passenger can hear the different music or information. Thus, people would not feel bothered. Before directional loudspeakers are applied to vehicles, it is necessary to do through research on audible sound beams. Most of efforts were devoted to the preprocessing methods [3] and speaker design [4] for improving the system performance of the directional sound beam. On the other hand, several model equations have been presented to describe the propagation of finite-amplitude sound beams from the parametric array, which has been reviewed in Ref. [5]. More works were refer to the directional audible sound by using the parametric array [6, 7]. Yoneyama used an array of transducers to demodulate the broadband sound signals with reasonable loudness, and called the audio spotlight [5]. However, the approaches of beamforming require complex mathematics equations. As an alternative approach, the ultrasonic field is presented to a newly numerical computation [8, 9]. Also J. Yang. Et. [2] presented an algorithm to discuss the weighted parametric array.

Fig. 1. The range of ultrasound

There has been little work done on the beam width control in directional loudspeakers. Without the beam width control, the directivity of the sound beams is fixed, and it cannot meet the various needs while applying the directional loudspeakers to vehicles. In order to find a way to better control the directivity of the sound beams, papers concentrated on this issue have been closely reviewed so far, and the Chebyshev method has proven to be the only effective method on this issue [2, 8, 9].

However, this paper discusses the theoretical simulation controlling the directivity of the sound beams using the uniform linear array (ULA) composed of eight transducers. An H_∞ optimization method is used to investigate the weighting distribution of uniform linear array, and the influence on the spreading angle of the sound beam. The method proposed in the study can improve the performance of the directional audible sound for all frequencies. The optimal weighting values of the transducers corresponding to specific sound beam spreading angle can be used to control the beam width and the amplitude of the side lobe.

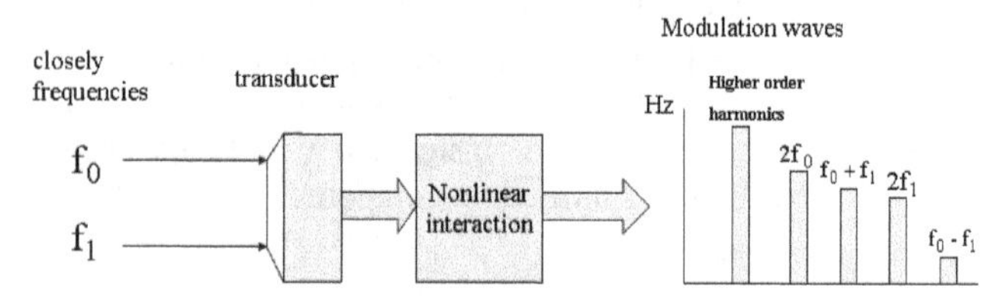

Fig. 2. Nonlinear phenomenon in the air

II. Design of Directional Audible Sound Systmes

In this section the formulation for designing directional audible sound systems using an H_∞ control method is presented. The directional audible sound can be expressed by the Khokhlov–Zabolotskaya–Kuznetsov (KZK) equation which accurately describes the diffraction, absorption, and nonlinearity of ultrasound in a parametric array as follows [10].

$$\frac{\partial^2 p}{\partial z \partial \tau}=\frac{c_0}{2}\nabla_\perp^2 p+\frac{\delta}{2c_0^3}\frac{\partial^3 p}{\partial \tau^3}+\frac{\beta}{2\rho_0 c_0^3}\frac{\partial^2 p^2}{\partial z^2}, \tag{1}$$

where

p = acoustic pressure
z = coordinate along the axis of the beam propagation direction
τ = retarded time
c_0 = small signal sound speed
ρ_0 = ambient density
δ = sound diffusivity
β = coefficient of nonlinearity
$\nabla_\perp^2$ = transverse Laplacian operator

Because of this successive approximations, a quasi-linear solution of the form $p = p_1 + p_2$ is assumed. Therefore p_1 is the linear solution of (1) for the primary pressure at frequency ω and p_2 is a small correction to p_1 at the second-harmonic frequency 2ω.

The definition becomes:

$$p_1(r,z,\tau)=\frac{1}{2j}[q_{1a}(r,z)e^{j\omega_a\tau}+q_{1b}(r,z)e^{j\omega_b\tau}]+c.c., \tag{2}$$

$$p_2(r,z,\tau)=\frac{1}{2j}[q_{2a}(r,z)e^{j2\omega_a\tau}+q_{2b}(r,z)e^{j2\omega_b\tau}+q_+(r,z)e^{j\omega_+\tau}+q_-(r,z)e^{j\omega_-\tau}]+c.c., \tag{3}$$

Then q_1 and q_2 are the complex pressure amplitudes, *c.c.* denotes the complex conjugate of preceding terms.

q_1 can be obtained from a homogeneous equation as:

$$\frac{\partial q_1}{\partial z}+\frac{i}{2k}\nabla_\perp^2 q_1+\alpha_1 q_1=0 \tag{4}$$

Then an inhomogeneous equation for q_2 is presented as:

$$\frac{\partial q_2}{\partial z}+\frac{i}{4k}\nabla_\perp^2 q_2+\alpha_2 q_2=\left(\frac{\beta k}{2\rho_0 c_0^2}\right)q_1^2 \tag{5}$$

Where $k=\omega/c$ becomes the wave number, whereas $\alpha_n=\delta n^2\omega/2c^3$ is the thermos viscous attenuation coefficient at frequency $n\omega$. For a parametric speaker, we are only interested in the audible sound beam at difference frequency. Assume that the sound beam is caused to be at difference frequency $\omega_- = \omega_a - \omega_b$ ($\omega_a > \omega_b$), then ω_a and ω_b become the two primary frequencies. The result of the complex pressure amplitude q_- is:

$$q_-(r',z')=-\frac{\pi\beta k_-}{\rho_0 c_0^2}\int_0^z\int_0^\infty q_{1a}(r',z')\,q_{1b}^*(r',z')\,G_-(r,z|r',z')\,r'dr'dz' \tag{6}$$

Thus $k_- = \omega_-/c$, q_{1a} q_{1b} are the complex pressure amplitudes for frequencies ω_a and ω_b, relatively, G-(r, z|r

---

z

)is the Green's function presented as:

$$G_-(r,z|r',z')=\frac{ik_-}{2\pi(z-z')}J_0\left(\frac{k_- rr'}{z-z'}\right)\exp\left[-\alpha_-(z-z')-\frac{ik_-(r^2+r'^2)}{2(z-z')}\right] \tag{7}$$

Then $\alpha_-=\delta\omega_-^2/2c^3$, and J_0 become the zeroth-order Bessel function. Being simple, we consider the sound beams produced by a primary source with Gaussian amplitude shading. While the complex pressure amplitude $q_1(r, 0)$ at primary frequency can be determined as:

$$q_1(r,0)=p_0\exp\left[-\left(\frac{r}{a}\right)^2\right] \tag{8}$$

On the other hand p_0 is the peak source pressure and a is the effective source radius.

Then, the linear solution can be deduced as:

$$q_1(r,z)=\frac{p_0 e^{-\alpha_1 z}}{1-iz/z_0}\exp\left[-\frac{(r/a)^2}{1-iz/z_0}\right] \tag{9}$$

Where $Z_0=\frac{1}{2}ka^2$. The far-field solution of the directivity is given by

$$D_1(k,\theta)=\exp\left[-\frac{1}{4}(ka)^2\tan^2\theta\right] \quad (10)$$

Assume that we only considered one far-field, according to a bi-frequency Gaussian source, the far-field directivity of different frequency is described by the product of the directivity function, i.e.:

$$D_-(\theta)=D_{1a}(\theta)D_{1b}(\theta) \quad (11)$$

Thus, for a bi-frequency Gaussian source the far-field directivity of the difference frequency is described by the product of the directivity functions of the primary waves.

Assume that a group of M ultrasonic transducers is arranged in a uniform linear array (ULA) with an inter element spacing of d as shown in Fig. 3. An observation point is set in the far-field of the array at an angle, θ with respect to the normal of the transducer array aperture. If each transducer is weighted with a weighting, ω_m for m=0, 1, 2, . . ., M−1, the array response function [11] can be derived as:

$$H(\omega\tau)=\frac{1}{M}\sum_{m=0}^{M-1}w_m e^{im\omega\tau} \quad (12)$$

Where,

$\tau = d/c\sin\theta_0$ is the time delay.

ω_m , m = 0,1,2,3.....M-1 is the weighting for each transducer

θ is the angel with respect to the axis of the beam.

In selecting proper weightings, the side lobes level can be made suitable at the expense of the beam width of the main lobe. From (12), it showed that the maximum of the main lobe exists on the broadside of the ULA ($\theta = 0$). However, the maximum of the main lobe can be changed by adding a phase shift or delay to each transducer. If the ULA is to be steered in the direction θ_0, time delay ($m\tau_0$) has to be added to mth transducer before transmitting the signal into the air. The time delay τ_0 can be calculated as $\tau = (d/c)\times\sin\theta_0$, and the array response of the delay-and-sum beamforming becomes:

$$H(\omega\tau)=\frac{1}{M}\sum_{m=0}^{M-1}w_m e^{-im\omega(\sin\theta_0-\sin\theta)d/c_0} \quad (13)$$

Then, the far-field directivity of the weighted primary sources array for frequency ω_a, $D_{1a}(\theta)$ can be appeared:

$$D_{1a}(\theta)=D_1(k_a,\theta)H(k_a,\theta) \quad (14)$$

When $D_1(k_a, \theta)$ is the aperture directivity shown in (10) for frequency ω_a, and the far-field array response $H(k_a, \theta)$ is indicated in (12) with w_{am} and ω_a instead of w_m and ω. Similarly, the far-field directivity for primary frequency ω_b, $D_{1b}(\theta)$ can be shown as:

$$D_{1b}(\theta)=D_1(k_b,\theta)H(k_b,\theta) \quad (15)$$

$D_1(k_b, \theta)$ and $H(k_b, \theta)$ are written as (10) and (12) for frequency ω_b with w_{bm} instead of w_{am}. Therefore, the beam pattern of the audible frequency is produced by substituting (12) and (13) into (14):

$$D_-(\theta)=D_1(k_a,\theta)H(k_a,\theta)D_1(k_b,\theta)H(k_b,\theta) \quad (16)$$

Thus, by substituting (10),(13) into (16) ,the directivity of far-field difference frequency equation can be written as:

$$\mathrm{D}\_(\theta)=\exp[-\frac{1}{4}(k_a*a)^2\tan^2\theta]*\frac{1}{M}\sum_{m=0}^{M-1}\omega_m e^{-jm\omega(\sin\theta_0-\sin\theta)d/c}*$$
$$\exp[-\frac{1}{4}(k_b*a)^2\tan^2\theta]*\frac{1}{M}\sum_{m=0}^{M-1}\omega_m e^{-jm\omega(\sin\theta_0-\sin\theta)d/c} \quad (17)$$

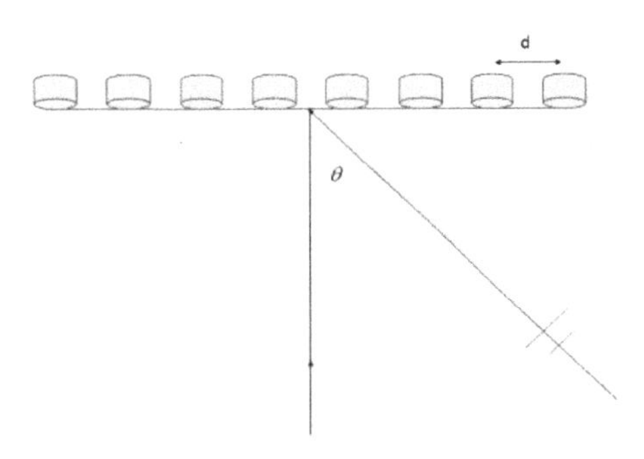

Fig. 3. Single array of M ultrasonic transducers

In this work the H_∞optimization method is used to control the directivity of ultrasonic transducers. The design formulation can be written as:

***Minimize* σ**
Subject to

$\|D\_(\theta_{sidelobe1})\|_\infty < \sigma$ -40<θ<-1/2Δ

$\|D\_(\theta_{sidelobe2})\|_\infty < \sigma$ 1/2Δ<θ<40

$\|D\_(\theta_{mainlobe})\|_\infty - \|D\_(\theta_{sidelobe1})\|_\infty > \varepsilon$ -1/2Δ<θ<1/2Δ

$\|D\_(\theta_{mainlobe})\|_\infty - \|D\_(\theta_{sidelobe2})\|_\infty > \varepsilon$ -1/2Δ<θ<1/2Δ (18)

where
σ is the real number.

Δ is the beam width, and ε is the amplitude difference between the main lobe and side lobe.

Substituting (17) into (18) gets:

$$\exp[-\frac{1}{4}(k_a * a)^2 \tan^2\theta_{sidelobe1}] * \frac{1}{M}\sum_{m=0}^{M-1} w_a(f_a)e^{-jm\omega(\sin\theta_0 - \sin\theta_{sidelobe1})d/c} *$$
$$\exp[-\frac{1}{4}(k_b * a)^2 \tan^2\theta_{sidelobe1}] * \frac{1}{M}\sum_{m=0}^{M-1} w_b(f_b)e^{-jm\omega(\sin\theta_0 - \sin\theta_{sidelobe1})d/c} < \sigma$$

-40<θ<-1/2Δ

$$\exp[-\frac{1}{4}(k_a * a)^2 \tan^2\theta_{sidelobe2}] * \frac{1}{M}\sum_{m=0}^{M-1} w_a(f_a)e^{-jm\omega(\sin\theta_0 - \sin\theta_{sidelobe2})d/c} *$$
$$\exp[-\frac{1}{4}(k_b * a)^2 \tan^2{}_{sidelobe2}] * \frac{1}{M}\sum_{m=0}^{M-1} w_b(f_b)e^{-jm\omega(\sin\theta_0 - \sin\theta_{sidelobe2})d/c} < \sigma \qquad 1/2\Delta<\theta<40$$

$$\exp[-\frac{1}{4}(k_a * a)^2 \tan^2\theta_{mainlobe}] * \frac{1}{M}\sum_{m=0}^{M-1} w_a(f_a)e^{-jm\omega(\sin\theta_0 - \sin\theta_{mainlobe})d/c} *$$
$$\exp[-\frac{1}{4}(k_b * a)^2 \tan^2\theta_{mainlobe}] * \frac{1}{M}\sum_{m=0}^{M-1} w_b(f_b)e^{-jm\omega(\sin\theta_0 - \sin\theta_{mainlobe})d/c} -$$
$$\exp[-\frac{1}{4}(k_a * a)^2 \tan^2\theta_{sidelobe1}] * \frac{1}{M}\sum_{m=0}^{M-1} w_a(f_a)e^{-jm\omega(\sin\theta_0 - \sin\theta_{sidelobe1})d/c} *$$
$$\exp[-\frac{1}{4}(k_b * a)^2 \tan^2\theta_{sidelobe1}] * \frac{1}{M}\sum_{m=0}^{M-1} w_b(f_b)e^{-jm\omega(\sin\theta_0 - \sin\theta_{sidelobe1})d/c} > \varepsilon$$

-1/2Δ<θ<1/2Δ

$$\exp[-\frac{1}{4}(k_a * a)^2 \tan^2\theta_{mainlobe}] * \frac{1}{M}\sum_{m=0}^{M-1} w_a(f_a)e^{-jm\omega(\sin\theta_0 - \sin\theta_{mainlobe})d/c} *$$
$$\exp[-\frac{1}{4}(k_b * a)^2 \tan^2\theta_{mainlobe}] * \frac{1}{M}\sum_{m=0}^{M-1} w_b(f_b)e^{-jm\omega(\sin\theta_0 - \sin\theta_{mainlobe})d/c} -$$
$$\exp[-\frac{1}{4}(k_a * a)^2 \tan^2\theta_{sidelobe2}] * \frac{1}{M}\sum_{m=0}^{M-1} w_a(f_a)e^{-jm\omega(\sin\theta_0 - \sin\theta_{sidelobe2})d/c} *$$
$$\exp[-\frac{1}{4}(k_b * a)^2 \tan^2\theta_{sidelobe2}] * \frac{1}{M}\sum_{m=0}^{M-1} w_b(f_b)e^{-jm\omega(\sin\theta_0 - \sin\theta_{sidelobe2})d/c} > \varepsilon$$

-1/2Δ<θ<1/2Δ (19)

The performance of optimization is according to the different concerns of the constraints. During the simulation, different constrains would lead to different levels of performances. We used optimization method to find the optimal weightings which varied with frequency. The constraints in (19) are not the only possible types of constrains. Commonly, the different types of constrains can results in different performance, i.e. the beam width and amplitude of the side lobe in this work, for example.

III. SIMULATION RESULTS

In this section the simulation results for the directivity of the audible sound beam created by using the H_∞ optimization method are presented, and then compared to those obtained by using Chebyshev weighting method [2]. In this study, the number of weightings is 16 for frequencies f_a and f_b as shown in Fig. 4. Therefore there are 16 design variables, i.e. $\{h_{a0}, h_{a1}, \ldots, h_{a7}\}\{h_{b0}, h_{b1}, \ldots, h_{b7}\}$,

The carrier frequency of the ultrasonic transducer array is set as 40 kHz, i.e. f_a = 40 kHz. The demodulated signal is at the frequency from 500 to 20,000 Hz with 500 Hz interval. A total of M = 8 ultrasonic transducer array is used with inter-element spacing, d = 9.7 mm. The effective source radius is set at a = 3.85 mm and the speed of sound c is 344 ms^{-1}. The weightings, h_{an} and h_{bn}, are calculated for difference frequency's beam width, $\theta_{-} = 20^0$, 40^0 and 60^0 using the proposed method.

Figs. 5, 6 and 7 show the difference frequency's directivity for $\theta_{-} = 20^0$, 40^0 and 60^0 respectively. Figs. 5(a), 6(a) and 7(a) are the difference frequency's directivity using Chebyshev weighting method [8], and Figs. 5(b), 6(b) and 7(b) are the difference frequency's directivity using the optimization method for weightings varied with frequency proposed in the paper. From the figures it can be seen that the amplitude in the side lobe using the proposed method is lower than that using Chebyshev weighting method. This is because the optimization method tried to find the optimal weightings which minimize the amplitude of the side lobe and subject to the amplitude difference between the main lobe and the side lobe. Therefore the optimization method proposed in this paper performs better than Chebyshev weighting method. As can be seen from the figures the beam width can also be controlled using the optimization method. This is because the difference frequency's directivity is the product of two primary frequency's directivities, its beam width always takes on the narrowest beam width of the two primary waves. It can obviously be seen that a constant beam width is achieved for all frequencies using the proposed method.

The highest side lobe amplitude with the proposed method gets attenuated more compared to the method with Chebyshev weighting. This is since the amplitude of the side lobe region is minimized through all the frequencies. Therefore the lower side lobe amplitude is obtained using the propose method.

IV. CONCLUSIONS

In this paper the H_∞ optimization method has been proposed for designing directional audible sound beam. The directional audible sound beam was generated by using a uniform linear array composed of eight ultrasonic transducers with different weightings varied with frequency. The performance of the beam width control using the proposed method has been evaluated. It can be seen that the proposed method could effectively control the beam width and the level of the side lobes for the audible sound beam. It is verified by the simulation results that the lower side lobes level could be obtained by using the proposed method. Therefore the proposed method could improve the system performance compared to that with the Chebyshev weighting method.

ACKNOWLEDGMENT

The study was supported by the National Science Council of Taiwan, the Republic of China, under project number MOST 103-2221-E-018 -005 -

REFERENCES

[1] P. J. Westervelt, ''Parametric acoustic array'', Journal of Acoustical Society of America, Vol. 35, no. 4, 1963, pp.535-537.

[2] Yang Jun, W. S. Gan, K. S. Tan, Er M. H., ''Acoustic beamforming of a parametric speaker comprising ultrasonic transducers'', Sensors and Actuators, A 125: 2005, pp.91-99.

[3] H. O. Berktay, ''Possible exploitation of nonlinear acoustics in underwater transmitting applications'', Journal of Sound & Vibration, Vol. 2, no. 4, 1965, pp. 435-461.

[4] Y. Roh and C. Moon,. "Design and fabrication of an ultrasonic speaker with thickness mode piezoceramic transducers", Sens. Actuators A 99, 2002, pp. 321–326.

[5] M. Yoneyama, and J. Fujimoto, ''The audio spotlight: An application of nonliear interaction of sound waves to a new type of loudspeaker design'', Journal of Acoustical Society of America, Vol. 73, no. 5, 1983, pp. 1532-1536.

[6] F. J. Pompei, Parametric audio system, WO 01/52437 A1, 2001.

[7] M. E. Spencer, and J. J. Croft, Modulator Processing for a Parametric Speaker System, WO 01/15491 A1, 2001.

[8] W. S. Gan, Yang Jun, K. S. Tan, Er M. H., ''A digital beamsteerer for difference frequency in a parametric array'', IEEE transactions on audio, speech, and language processing, 14(3), 2006, pp. 1018-1025.

[9] K. S. Tan, W. Gan, S. J. Yang, and Er, M. H.. "Constant beamwidth beam-former for difference frequency in parametric array", in Proc. ICASSP , 2003, pp.361–364.

[10] M. F. Hamilton and D.T. Blackstock, Nonlinear Acoustics, Academic press, San Diego, 1998. D. H. Johnson and D. E. Dudgeon, Array signal processing, Prentice Hall, New Jersey, 1993.

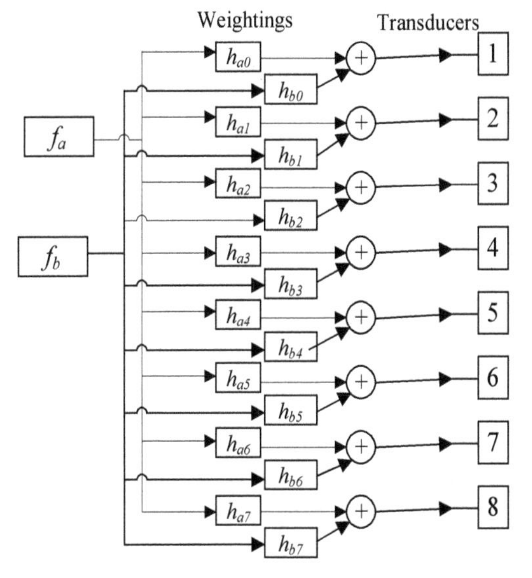

Fig. 4. Block diagram for beam width control systems with different weightings

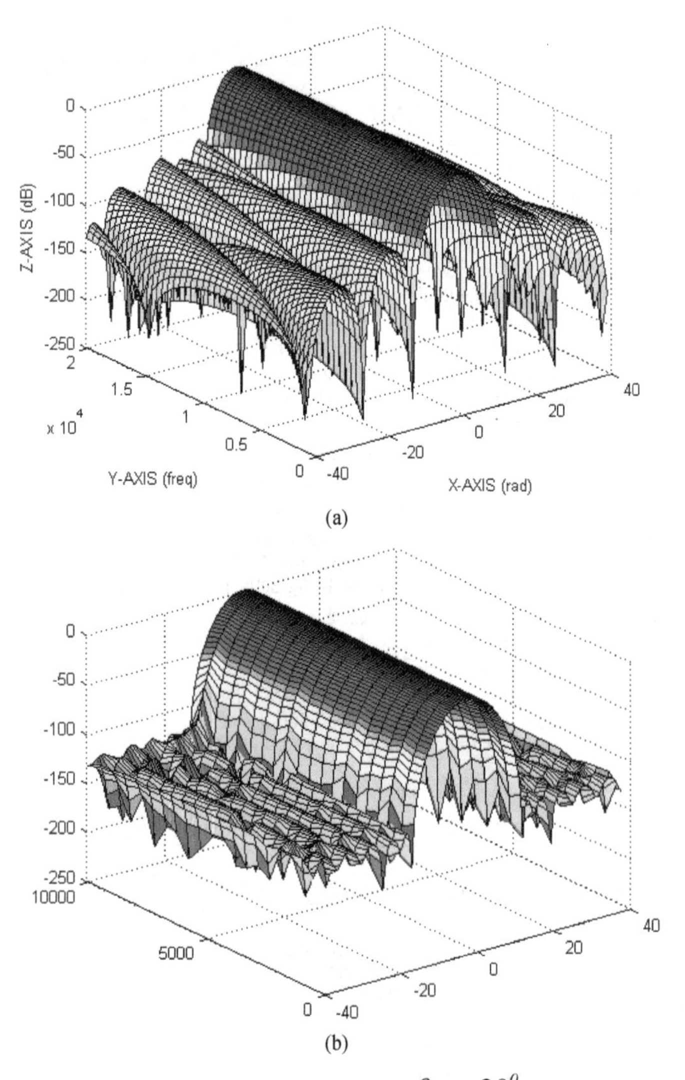

(a)

(b)

Fig. 5. Difference frequency's directivity for $\theta_{-} = 20^{0}$. (a) Chebyshev weighting method. (b) Optimization method for weightings varied with frequency

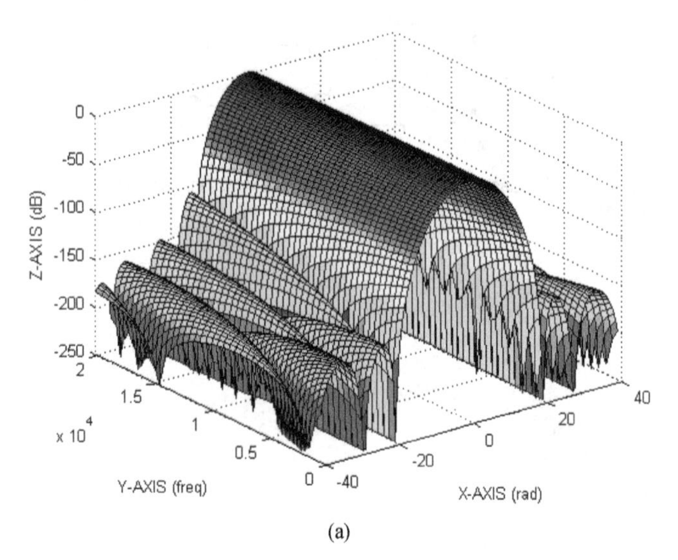

(a)

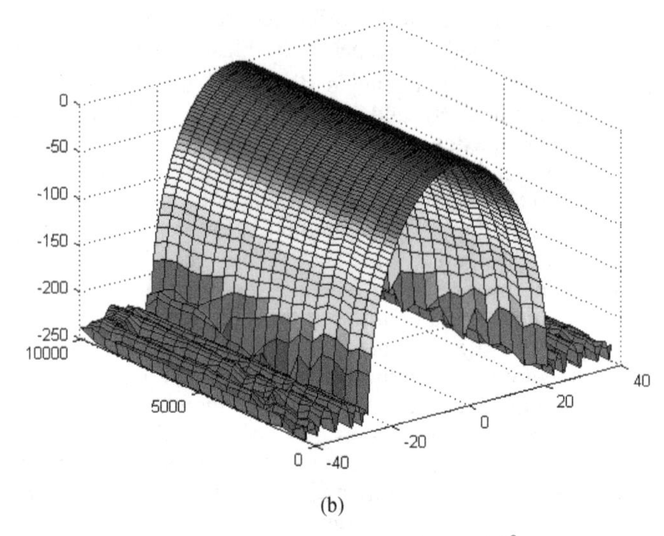

(b)

Fig. 6. Difference frequency's directivity for $\theta_{-} = 40^{0}$. (a) Chebyshev weighting method. (b) Optimization method for weightings varied with frequency

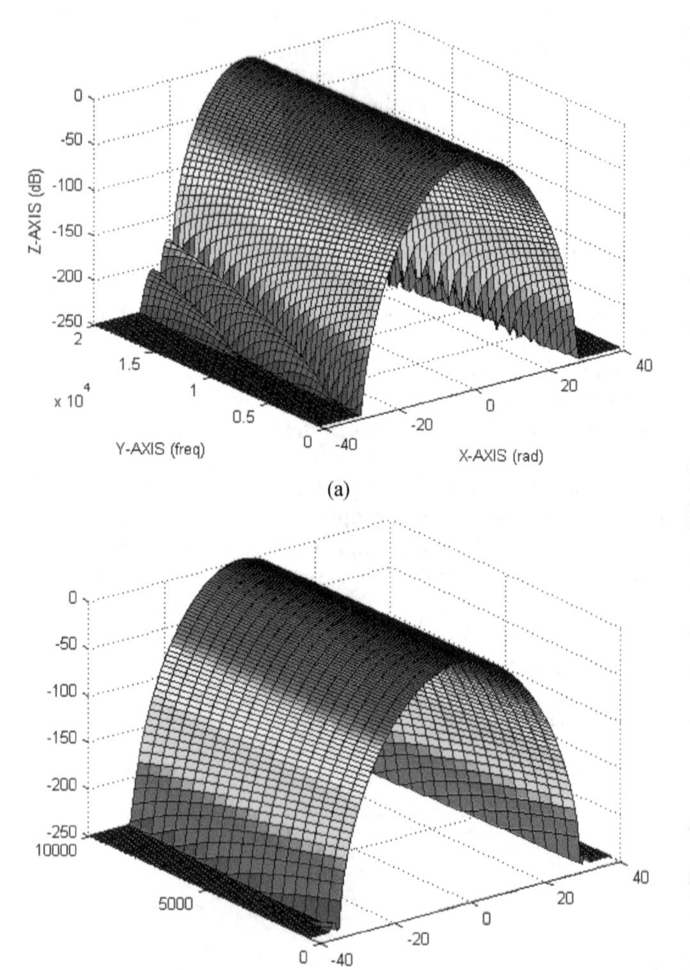

(a)

(b)

Fig. 7. Difference frequency's directivity for $\theta_{-} = 60^{0}$. (a) Chebyshev weighting method. (b) Optimization method for weightings varied with frequency

System for EKG Monitoring

Solution based on Arduino microcontroller

Jakub Ševčík
Dept. of Computers and Informatics
FEI TU of Kosice
Kosice, Slovak Republic

Peter Feciľak
Dept. of Computers and Informatics
FEI TU of Kosice
Kosice, Slovak Republic

Ondrej Kainz
Dept. of Computers and Informatics
FEI TU of Kosice
Kosice, Slovak Republic

František Jakab
Dept. of Computers and Informatics
FEI TU of Kosice
Kosice, Slovak Republic

Abstract—**In this paper the system for the electrocardiogram (EKG) monitoring based on the of Arduino microcontroller is presented. Detailed description of the electrocardiogram itself serves as a ground for building the proposed hardware and software solution. The software implementation is in a form of both, Mat-lab environment, and own application. Final output enables retrieval of the actual data in real time and further and provide the rudimentary diagnosis. Utilization of such device is for self-home diagnosis of arrhythmia.**

Keywords—Arduino; arrhythmia; C sharp; cardiovascular diseases; diagnosis; electrocardiogram; heart; Matlab

I. Introduction

Electrocardiogram machine got its popularity in the first half of the twentieth century. Examination at a cardiologist is considered these days to be nothing out of the ordinary. Despite the fact that EKG is only hundred years old, this technique is still the most reliable at determining an acute myocardial infarction (AMI). No other examination for diagnosis of such frequent and difficult problem as arrhythmia is being preferred. To make EKG machine work correctly, its proper construction and further configuration are necessary. Such device is used by people of different ages, for this reason, comfort, quick and exact usage should be provided. It is vital for the electrocardiogram to determine possible heart diseases in the shortest time possible.

The goal of this project is a construction of a machine analogous to the EKG machine and deployment of a software solution while still being able to achieve sufficient results. The brief introduction of EKG and hardware is introduced in the next sections with the subsequent rudimental implementation of both hardware and software.

II. Electrocardiogram in Detail

People have many things in common, and the electrocardiogram is one of those, providing no physiologic or pathologic factor is present. Labaš [1] describes it is a visual record representing the changes of electric potentials in the operation of cardiac muscle. The heart is a muscle in the body with special morphologic and functional role. Bada [2] refers to the myocardium as a syncytium of muscle cells is acting as a singular muscle fiber. During irritation – depolarization, which is gradually spread from atria to ventricles the potential differences originate, so-called action flows that create the electric field of the heart being spread by the surrounding structures of lungs and muscles. Due to the sensible registration machine, these flows are recordable also from body skin. Hampton presents in his book [3] dependency of the clinical diagnosis in particular on anamnesis and a lesser extent on a physical examination of the patient. EKG, as a tool, may support a diagnosis and in some cases is crucial for a concrete treatment of the patient. However, it is essential to perceive EKG as a tool and not as a goal itself. It helps with diagnosing the pain in the chest and also the proper utilization of thrombolysis is dependent on EKG. Further, it helps with cure of a heart attack or could contribute to a diagnosis of cardiac dyspnea. As a basis for the interpretation of EKG serves a recognition of the patterns based on several rules.

According to Pytliak [4] the electrical activity of the heart is visible by changes of electrical intensity also on the surface of the body. By sufficiently sensible galvanometer the changes on the surface of the body could be measured and registered, it is a principle of the creation of the EKG record. Summation of electric displays of all the cells in the given time interval results in a changes in the intensity.

III. Physiological Curve of EKG

Following Bada [2], the EKG curve is made by a group of positive and negative amplitudes so-called Wavelets. These are marked as given by international convention with letters [P,Q,R,S,T,U] just as were originally named by Einthoven. Horizontal passages of the curve are called intervals. Places between intervals are called segments. Physiologically, these passages of EKG are on the level of an isoelectric line. If the horizontal passages are above the level of the isoelectric line, it is the elevation, if the horizontal passages are under the level of the isoelectric line, it is the depression. Wavelets located above the level of the isoelectric line are called positive, and those under the level of the isoelectric line are called negative. Wavelets of which one part is positive and the second one negative are diphasic. Next, the relevant waves, intervals, and segments are described.

A. P wave

This wave is a sign of depolarization (contraction) of atrial muscle and also is the main criterion of the sinusoidal rhythm – when the initial impulse originated in the sinoatrial node and from there was carried to the atrial muscles.

B. Interval P-Q

From the beginning of the P wavelet to the beginning of the Q wavelet (if the wavelet Q is missing, then to the beginning of the wavelet R). This interval represents the time it takes for depolarization to spread over the muscle of the atria and the ventricles to the level of Purkinje fibers, this is referred to as a conversion time.

C. Q wave

This is the first negative wave after P wavelet or the first negative wave before the R wave. It is start of the ventricular complex and reflects the course of the depolarization of the interventricular barrier, which is of the whole ventricular muscle depolarized as the first one. Physiological Q wave tend to be just schematic, its absence is not pathological.

D. Q,R,S waves

Together they form the so-called initial ventricular complex of the same name, which reflects the depolarization of the ventricular muscle. It is bordered by the beginning of Q wave and ends with S wave.

E. S-T segment

This segment is the section from the end of the S-wave, where ends the ventricular QRS complex (J point) and the beginning of the T wave. Under physiological conditions, the S-T segment is on the level of the isoelectric line and reflects the early phase of repolarization of the ventricular muscle.

F. T wave

T wave has the same course as the ventricular complex, it is positive where the ventricular complex is positive. The T wave is a reflection of repolarization of the atrial muscles, thus it is always shown in the EKG. If the T wave is not present in the EKG, we say that it is on the level of the isoelectric line.

G. U wave

This wave is a non-constant part of the ECG curve, it tends to act as a small positive deflection behind wave. It is a reflection of so-called delayed repolarization of ventricular muscle. Lacking of U waves is never pathological. This wave is present in the EKG record from athletes.

H. Q-T interval

Interval, also known as electrical systole of heart, represents the time from the beginning of the ventricular QRS complex, i.e. from the beginning of the Q wave (in its absence from the beginning of the R wave) till the end of the T wave. It is a conclusion of depolarization and repolarization of the ventricular muscle. QT interval particularly depends on the heart rate.

IV. Arduino Microcontroller

Arduino is an open electronic platform [6], the principal reason of its success is user friendly hardware and software. Arduino is able to perceive the environment by utilizing various sensors and shields that may be connected to it. Arduino is a massive development kit based on the microprocessor ATMEGA328 and contains 13 digital input-output pins, 6 of them with support of pulse-width modulation (PWM) and 6 analog inputs [6]. Among all the free computer platforms for electronic projects such as Galileo or Raspberry was for this project chosen Arduino due to specific aspects, i.e. in conjunction with ECG it overruns other projects. Other aspects are: clearly and easily usable API with a simple programming language, then so-called Arduino shields that actually represent enhancements of the Arduino board [7]. Note that some of these shields are directly created for the purpose of human sensing signals. This very idea is the essential for our purpose, since it is possible to make own shield or to acquire already existing one. In our case, the shield provided by the Olimex is being utilized, bearing in mind the need to extract the highest possible signal quality. EKG shield can be connected in various ways, among the most basic are 12 - lead ECG.

V. Hardware Implementation

Based on the reviews of the available literature dealing with the EKG, and after comparative analysis of the options available to create an EKG device similar to a medical one, it was decided to utilize: Arduino UNO, shield Olimex EKG, electrodes designed for ECG shield. For signal processing is utilized Matlab environment and own software solution was also created.

As already stated, the project consists of two principal parts, i.e. hardware and software. In the hardware part, Arduino UNO with ECG Shield was combined. These two devices fit together very well. After their engagement in individual pins, Arduino was upgraded by the opportunity to capture the signals of the heart. Arduino was constantly connected to computer device and by this the need to use external battery was omitted, note that the energy was supplied through USB. It was proposed that that the EKG data is to be plotted directly on the screen using Matlab environment and our application. This condition removed necessity to use SD card for data recording, also use of mini displays or oscilloscope displays.

A. Electrodes

The original EKG devices used in medical practice in the art of cardiology use a combination of different electrodes. They combine electrodes connected to the legs, specifically to the ankles and wrists and electrodes that are attached directly to the chest near the heart. This project utilized electrodes from Olimex, which are also suitable for operation with Olimex shield. These electrodes are created exactly for connection in the form of Einthoven triangle. This type of connection includes three electrodes - one on the right hand, the second one on the left hand and the third one on the left leg. Three electrodes cross the notional heart. The right foot is taken as connection to the ground. Electrodes have also the option of renewable terminals. Terminals are affordable, thus a new set of terminals may be used for each tested person. The electrodes are connected directly to the ECG/EMG shield, specifically to the electrode jack.

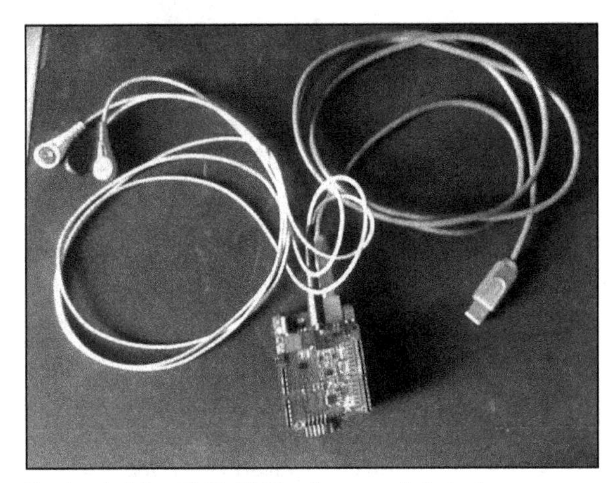

Fig. 1. Arduino, EKG shield and connected electrodes

Note that we can connect just one type of electrodes for a shield. Hardware component is depicted in the Fig. 1.

VI. Software Implementation

Software part utilized Matlab environment to output the direct sinusoid, specifically Matlab Data Acquisition Toolbox was used. Another approach included development of own secondary application, where the signals from the Arduino microcontroller were filtered and analyzed, note that application was developed in in C# language.

A. Matlab environemnt

In this case was created a simulation in which was simulated the RAW Arduino output via acquisition toolbox, which directly cooperates with the Arduino boards. In the simulation, specific COM port was selected to enable receiving of data and further is was opened. The entire simulation was portrayed in the cycle and a result was a graph. Graph itself was rather of chaotic nature, since we received input data from the RAW input along with noise.

B. Secondary application

The application can render down real-time ECG signal that can be filtered, also performs the calculation of the heart rate and enables detection of arrhythmia.

In the course of project, two programming languages for application development were considered, i.e. Java or C sharp. Both, Java and C Sharp, utilize RXTX libraries that can cooperate directly with Arduino boards. It was found that C sharp provides more advanced pre-build parts and principles that fit out needs. C sharp and Java in combination with Arduino facilitate us to connect Arduino to the USB port of the computer device and further its COM port can be found in the application. Once the COM port exists and is defined correctly, the application can send/receive messages from Arduino microcontroller. Communication takes place via the so-called serial interface. Main reason for utilization of C sharp was due to possibility of utilizing the advanced mathematical filters. These filters were in C sharp very well analyzed, and for this reason assisted in the implementation of the application in a significant way. The package OLIMEXINO-328 was used as well and helped in better synchronization with the shield.

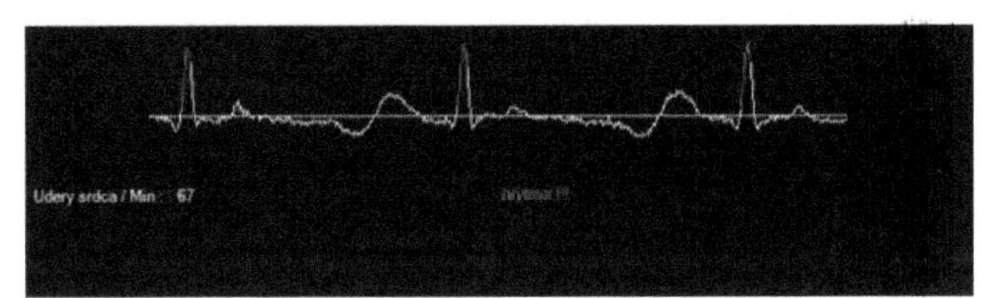

Fig. 2. Screen of application: Successful diagnosis

Screenshot of the secondary application is depicted in the Fig. 2.

Two types of filters were utilized:

1) Butterworth filter

Butterworth filter is a filter type for signal processing designed to provide the biggest possible flat frequency response. This is known as the maximum flat dimension filter. Originally it was described in 1930 by British engineer and physicist Stephen Butterworth.

2) Haar wavelet

Haar wave is a sequence recounting "square" functions, which form together a wavelet family or basis. Haar analysis is similar to Fourier analysis in allowing the target function in the interval to be expressed as an orthonormal function. The sequence of Haar was designed in 1909 by Alfred Haar.

C. Structure of secondary application

Application consists of eight main classes with basic Properties, References and app.configuration already included. Two of the principal classes are the classes of filters, one is for transformation, and the next one limits the size of the curves. Furthermore, there are classes for rendering down curves in real-time diagnostics, a class where the basic settings are and as the last one there is the main class.

Haar class creates a Haar wavelet filter. In this class there are three methods. First method is the one, which triggers a recursive method for calculating the Haar transform. The result of Haar transformation is a single integral number and a vector of coefficients. These coefficients are calculated from the highest to the lowest frequency however the Haar functions other way around, and therefore, we have a method which makes the inversion. The third method triggers the first two mentioned methods. Butter Worth is a class representing Butterworth filter. This class contains just one method, which takes information from the Fourier transformation of the class Fourier.cs. Consequently, a calculation to the original is in the progress. This method contains the conditions for the upper and lower limit of the filter. Fourier class contains an algorithm for calculation of fast Fourier transformation. Strop is class that sets scaling of the curve. The Class contains two methods. One method involves rendering down size of the curves. It also contains a condition for turning on and off scaling option. The second method runs in a cycle, and calls for the first one. Real Time is a class that contains a rendering of a curve in the real time. This class works with the package OLIMEXINO-328th. The main cycle can be found here, it receives information from the Arduino. Furthermore, there are cycles for applying Haar and Butterworth filters. In this class there are also saved settings for our filters along with scaling of the curve. There are also defined portrayals e.g. colors of lines and background, list of the heartbeat and the arrhythmia. Class called Diag deals directly with heartbeats and arrhythmia detection. Every single heartbeat can be detected and average heart rate calculated by the number of strokes per chosen time, which we calculate by the time between heartbeats and then convert it to a period of one minute. Arrhythmia is detected according to the time between heartbeats. If the time period is different between certain strokes, then it is evaluated as arrhythmia. Another class used was for communication with a serial port. This class concerns a method, which opens the COM port, which is from the beginning defined on COM.

D. Conclusion

The principal goal of this project was the construction and configuration the EKG-based device on Arduino microcontroller with an emphasis on its ease in controlling and applicability in routine practice, whether to professionals or the laymen. People are usually pretty familiar with EKG machine, which is used by cardiologists worldwide to determine different heart defects. Cardiovascular diseases take the top places in the list of the civilization diseases, one of the most common being the myocardial infarction (AMI). Such serious illnesses can be avoided by early diagnosis e.g. thanks to the devices such as EKG machine based on the Arduino.

The aforementioned sections introduced the advantage of this device being a quick diagnosing of cardiac arrhythmia. Generally, this examination is painless, and the patient has sufficient comfort and convenience. The greatest benefit would be for patients who need an immediate record of their health from the home. This benefit is the very consideration taken into account when developing such a device.

The disadvantages and related possible future improvements are the limited ability of the diagnosis since the exact type of arrhythmia cannot be determined. Next being the lack of storage of our measurement to a file or even lack of uploading it somewhere on the internet. However probably the biggest disadvantage of the current device is a large inaccuracy compared to the original EKG medical devices used in hospitals. These are the main concerns to be the principal focus in the future research.

ACKNOWLEDGMENT

We support research activities in Slovakia/This project is being co-financed by the European Union. Paper is the result of the Project implementation: University Science Park TECHNICOM for Innovation Applications Supported by Knowledge Technology, ITMS: 26220220182, supported by the Research & Development Operational Programme funded by the ERDF.

REFERENCES

[1] P. Labaš, "Podklady na prijímacie skúšky z chémie", Bratislava: ELÁN, 2013, 343 p.

[2] V. Bada, "Základy klinickej elektrokardiografie", Bratislava: UK, 2010, 124 p.

[3] J.R. Hampton, "EKG stručně, jasně, přehledně", Praha 7: Grada Publishing, a.s., 2013, 192 p.

[4] M. Pytliak, "Základy kardiografie pre štúdium ošetrovateľstva a ostatné zdravotnícke študijné programy", Košice: Univerzita Pavla Jozefa Šafárika v Košiciach, 2009, 134 p.

[5] Arduino Uno [online] available at: http://arduino.cc/en/Main/arduinoBoardUno

[6] Co je to Arduino [online] available at: http://www.czechduino.cz/?co-je-to-arduino,29

[7] Vývojové prostredie a programovanie Arduina [online] available at: http://uart.cz/90/ide-a-programovani-arduina/

A novel hybrid genetic differential evolution algorithm for constrained optimization problems

Ahmed Fouad Ali
Faculty of Computers and Information
Dept. of Computer Science, Suez Canal University
Ismailia, Egypt
Email: ahmed_fouad@ci.suez.edu.eg

***Abstract*—Most of the real-life applications have many constraints and they are considered as constrained optimization problems (COPs). In this paper, we present a new hybrid genetic differential evolution algorithm to solve constrained optimization problems. The proposed algorithm is called hybrid genetic differential evolution algorithm for solving constrained optimization problems (HGDESCOP). The main purpose of the proposed algorithm is to improve the global search ability of the DE algorithm by combining the genetic linear crossover with a DE algorithm to explore more solutions in the search space and to avoid trapping in local minima. In order to verify the general performance of the HGDESCOP algorithm, it has been compared with 4 evolutionary based algorithms on 13 benchmark functions. The experimental results show that the HGDESCOP algorithm is a promising algorithm and it outperforms other algorithms.**

Keywords*—*Constrained optimization problems, Genetic algorithms, Differential evolution algorithm, Linear crossover.

I. Introduction

Evolutionary algorithms (EAs) have been widely used to solve many unconstrained optimization problems [1], [3], [10], [15]. EAs are unconstrained search algorithms and lake a technique to handel the constraints in the constrained optimization problems (COPs). There are different techniques to handle constraints in EAs, these techniques have been classified by Michalewicz [13] as follows. Methods based on penalty functions, methods based on the rejection of infeasible solutions, methods based on repair algorithms, methods based on specialized operators and methods based on behavioral memory.

Differential evolutionary algorithm (DE) is one of the most widely used evolutionary algorithms (EAs) introduced by Stron and Price [17]. Because of the success of DE in solving unconstrained optimization problems, it attracts many researchers to apply it with their works to solve constrained optimization problems (COPs) [2], [18], [19]. In this paper, we proposed a new hybrid algorithm in order to solve constrained optimization problems. The proposed algorithm is called hybrid genetic differential evolution algorithm for solving constrained optimization problems (HGDESCOP). The HGDESCOP algorithm starts with an initial population consists of *NP* individuals, the initial population is evaluated using the objective function. At each generations, the new offspring is created by applying the DE mutation. In order to increase the global search behavior of the proposed algorithm and explore wide area of the search space, a genetic algorithm linear crossover operator is applied. In the last stage of the algorithm, the greedy selection is applied in order to accept or reject the trail solutions. These operations are repeated until the termination criteria satisfied.

The main objective of this paper is to construct an efficient algorithm which seeks optimal or near-optimal solutions of a given objective function for constrained problems by combining the genetic linear crossover with a DE algorithm to explore more solutions in the search space and to avoid trapping in local minima.

The reminder of this paper is organized as fellow. The problem definition and an overview of genetic algorithm and differential evolution are given in Section II. In Section III, we explain the proposed algorithm in detail. The numerical experimental results are presented in Section IV. Finally, The conclusion of the paper is presented in Section V.

II. Problem Definition and Overview of Genetic Algorithm and Differential Evolution Algorithm

In the following section and subsections, we give an overview of the constraint optimization problem and we highlight the penalty function technique, which are used to convert the constrained optimization problems to unconstrained optimization problems. Finally, we present the standard genetic algorithm and deferential evolutionary algorithm.

A. Constrained optimization problems

A general form of a constrained optimization is defined as follows:

$$\text{Minimize} \quad f(x), \quad x = (x_1, x_2, \cdots, x_n)^T, \tag{1}$$

Subject to

$$g_i(x) \leq 0, \quad i = 1, \cdots, m$$
$$h_j(x) = 0, \quad j = 1, \cdots, l$$
$$x_l \leq x_i \leq x_u$$

Where $f(x)$ is the objective function, x is the vector of n variables, $g_i(x) \leq 0$ are inequality constraints, $h_j(x) = 0$ are equality constraints, x_l, x_u are variables bounds. In this paper, we used the penalty function technique to solve constrained optimization problems [11]. The following subsection gives more details about the penalty function technique.

1) The Penalty function technique: The penalty function technique is used to transform the constrained optimization problems to unconstrained optimization problem by penalizing the constraints and forming a new objective function as follow:

$$f(x) = \begin{cases} f(x) & \text{if } x \in \text{feasible region} \\ f(x) + \text{penalty}(x) & x \notin \text{feasible region.} \end{cases} \quad (2)$$

Where,

$$\text{penalty}(x) = \begin{cases} 0 & \text{if no constraint is violated} \\ 1 & \text{otherwise.} \end{cases}$$

There are two kind of points in the search space of the constrained optimization problems (COP), feasible points which satisfy all constraints and unfeasible points which violate at least one of the constraints. At the feasible points, the penalty function value is equal the value of objective function, but at the infeasible points the penalty function value is equal to a high value as shown in Equation 2. In this paper, a non stationary penalty function has been used, which the values of the penalty function are dynamically changed during the search process. A general form of the penalty function as defined in [21] as follows:

$$F(x) = f(x) + h(k)H(x), \quad x \in S \subset \mathbb{R}^n, \quad (3)$$

Where $f(x)$ is the objective function, $h(k)$ is a non stationary (dynamically modified) penalty function, k is the current iteration number and $H(x)$ is a penalty factor, which is calculated as follows:

$$H(x) = \sum_{i=1}^{m} \theta(q_i(x))q_i(x)^{\gamma(q_i(x))} \quad (4)$$

Where $q_i(x) = \max(0, g_i(x)), i = 1, \ldots, m$, g_i are the constrains of the problem, q_i is a relative violated function of the constraints, $\theta(q_i(x))$ is the power of the penalty function, the values of the functions $h(.), \theta(.)$and$\gamma(.)$ are problem dependant. We applied the same values, which are reported in [21].

The following values are used for the penalty function:

$$\gamma(q_i(x)) = \begin{cases} 1 & \text{if } q_i(x) < 1, \\ 2 & \text{otherwise.} \end{cases}$$

Where the assignment function was

$$\theta(q_i(x))) = \begin{cases} 10 & \text{if } q_i(x) < 0.001, \\ 20 & \text{if } 0.001 \le q_i(x) < 0.1, \\ 100 & \text{if } 0.1 \le q_i(x) < 1, \\ 300 & \text{otherwise.} \end{cases}$$

and the penalty value $h(t) = t * \sqrt{t}$.

B. An overview of genetic algorithm

Genetic algorithm (GA) was introduced by Holland [8]. The basic principles of GA are inspired from the principles of life which were first described by Darwin [4]. GA starts with a number of individuals (chromosomes) which form a population. After randomly creating of the population, the initial population is evaluated using fitness function. The selection operator is start to select highly fit individuals with high fitness function score to create new generation. Many type of selection have been developed like roulette wheel selection, tournament selection and rank selection [12]. The selected individuals are going to matting pool to generate offspring by applying crossover and mutation. Crossover operator is applied to the individuals in the mating pool to produces two new offspring from two parents by exchanging substrings. The most common crossover operators are one point crossover [8], two point crossover [12], uniform crossover [12]. The parents are selected randomly in crossover operators by assign a random number to each parent, the parent with random number lower than or equal the probability of crossover ration P_c is always selected. Mutation operators are important for local search and to avoid premature convergence. The probability of mutation p_m must be selected to be at a low level otherwise mutation would randomly change too many alleles and the new individual would have nothing in common with its parents. The new offspring is evaluated using fitness function, these operations are repeated until termination criteria stratified, for example number of iterations. The main structure of genetic algorithm is presented in Algorithm 1

Algorithm 1 The structure of genetic algorithm

1: Set the generation counter $t := 0$.
2: Generate an initial population P^0 randomly.
3: Evaluate the fitness function of all individuals in P^0.
4: **repeat**
5: Set $t = t + 1$. { **Generation counter increasing**}.
6: Select an intermediate population P^t from P^{t-1}. {**Selection operator**}.
7: Associate a random number r from $(0, 1)$ with each row in P^t.
8: **if** $r < p_c$ **then**
9: Apply crossover operator to all selected pairs of P^t.
10: Update P^t.
11: **end if**{**Crossover operator**}.
12: Associate a random number r_1 from $(0, 1)$ with each gene in each individual in P^t.
13: **if** $r_1 < p_m$ **then**
14: Mutate the gene by generating a new random value for the selected gene with its domain.
15: Update P^t.
16: **end if**
{**Mutation operator**}.
17: Evaluate the fitness function of all individuals in P^t.
18: **until** Termination criteria satisfied.

1) Liner crossover operator: HGDESCOP uses a linear crossover [20] in order to generate a new offspring to substitute their parents in the population. The main steps of the linear crossover is shown in Procedure 1.

Procedure 1: Linear Crossover(p^1, p^2)

1. Generate three offspring $c^1 = (c_1^1, \ldots, c_D^1)$, $c^2 = (c_1^2, \ldots, c_D^2)$ and $c^3 = (c_1^3, \ldots, c_D^3)$ from parents

$p^1 = (p_1^1, \ldots, p_D^1)$ and $p^2 = (p_1^2, \ldots, p_D^2)$, where

$$c_i^1 = \frac{1}{2}p_i^1 + \frac{1}{2}p_i^2,$$
$$c_i^2 = \frac{3}{2}p_i^1 - \frac{1}{2}p_i^2,$$
$$c_i^3 = -\frac{1}{2}p_i^1 + \frac{3}{2}p_i^2,$$

$i = 1, \ldots, D$.

2. Choose the two most promising offspring of the three to substitute their parents in the population.
3. Return.

C. An overview of differential evolution algorithm

Differential evolution algorithm (DE) proposed by Stron and Price in 1997 [17]. In DE, the initial population consists of number of individuals, which is called a population size NP. Each individual in the population size is a vector consists of D dimensional variables and can be defined as follows:

$$\mathbf{x}_i^{(G)} = \{x_{i,1}^{(G)}, x_{i,2}^{(G)}, \ldots, x_{i,D}^{(G)}\}, \quad i = 1, 2, \ldots, NP. \quad (5)$$

Where G is a generation number, D is a problem dimensional number and NP is a population size. DE employs mutation and crossover operators in order to generate a trail vectors, then the selection operator starts to select the individuals in new generation $G+1$. The overall process is presented in details as follows:

1) Mutation operator: Each vector $\mathbf{x}_i$ in the population size create a trail mutant vector $\mathbf{v}_i$ as follows.

$$\mathbf{v}_i^{(G)} = \{v_{i,1}^{(G)}, v_{i,2}^{(G)}, \ldots, x_{i,D}^{(G)}\}, \quad i = 1, 2, \ldots, NP. \quad (6)$$

DE applied different strategies to generate a mutant vector as fellows:

$$DE/rand/1: \quad \mathbf{v}_i^{(G)} = \mathbf{x}_{r_1}^{(G)} + F \cdot (\mathbf{x}_{r_2} + \mathbf{x}_{r_3}) \quad (7)$$
$$DE/best/1: \quad \mathbf{v}_i^{(G)} = \mathbf{x}_{best}^{(G)} + F \cdot (\mathbf{x}_{r_1} + \mathbf{x}_{r_2}) \quad (8)$$
$$DE/currenttobest/1: \quad \mathbf{v}_i^{(G)} = \mathbf{x}_i^{(G)} + F \cdot (\mathbf{x}_{best} - \mathbf{x}_i) + F \cdot (\mathbf{x}_{r_1} - \mathbf{x}_{r_2}) \quad (9)$$
$$DE/best/2: \quad \mathbf{v}_i^{(G)} = \mathbf{x}_{best}^{(G)} + F \cdot (\mathbf{x}_{r_1} - \mathbf{x}_{r_2}) + F \cdot (\mathbf{x}_{r_3} - \mathbf{x}_{r_4}) \quad (10)$$
$$DE/rand/2: \quad \mathbf{v}_i^{(G)} = \mathbf{x}_{r_1}^{(G)} + F \cdot (\mathbf{x}_{r_2} - \mathbf{x}_{r_3}) + F \cdot (\mathbf{x}_{r_4} - \mathbf{x}_{r_5}) \quad (11)$$

where r_d, $d = 1, 2, \ldots, 5$ represent random integer indexes, $r_d \in [1, NP]$ and they are different from i. F is a mutation scale factor, $F \in [0, 2]$. $\mathbf{x}_{best}^{(G)}$ is the best vector in the population in the current generation G.

2) Crossover operator: A crossover operator starts after mutation in order to generate a trail vector according to target vector $\mathbf{x}_i$ and mutant vector $\mathbf{v}_i$ as follows:

$$u_{i,j} = \begin{cases} v_{i,j}, & \text{if } rand(0,1) \leq CR \text{ or } j = j_{rand} \\ x_{i,j}, & \text{otherwise} \end{cases} \quad (12)$$

Where *CR* is a crossover factor, $CR \in [0, 1]$, j_{rand} is a random integer and $j_{rand} \in [0, 1]$

3) Selection operator: The DE algorithm applied greedy selection, selects between the trails and targets vectors. The selected individual (solution) is the best vector with the better fitness value. The description of the selection operator is presented as fellows:

$$\mathbf{x}_i^{(G+1)} = \begin{cases} \mathbf{u}_i^{(G)}, & \text{if } f(\mathbf{u}_i^{(G)}) \leq f(\mathbf{x}_i^{(G)}), \\ \mathbf{x}_i, & \text{otherwise} \end{cases} \quad (13)$$

The main steps of DE algorithm are presented in Algorithm 2

Algorithm 2 The structure of differential evolution algorithm

1: Set the generation counter $G := 0$.
2: Set the initial value of F and CR.
3: Generate an initial population P^0 randomly.
4: Evaluate the fitness function of all individuals in P^0.
5: **repeat**
6: Set $G = G + 1$. {**Generation counter increasing**}.
7: **for** $i = 0; i < NP; i++$ **do**
8: Select random indexes r_1, r_2, r_3, where $r_1 \neq r_2 \neq r_3 \neq i$.
9: $\mathbf{v}_i^{(G)} = \mathbf{x}_{r_1}^{(G)} + F \times (\mathbf{x}_{r_2}^{(G)} - \mathbf{x}_{r_3}^{(G)})$. {**Mutation operator**}.
10: $j = rand(1, D)$
11: **for** $(k = 0; k < D; k++)$ **do**
12: **if** $(rand(0,1) \leq CR$ or $k = j$ **then**
13: $u_{ik}^{(G)} = v_{ik}^{(G)}$ {**Crossover operator**}
14: **else**
15: $u_{ik}^{(G)} = x_{ik}^{(G)}$
16: **end if**
17: **end for**
18: **if** $(f(\mathbf{u}_i^{(G)}) \leq f(\mathbf{x}_i^{(G)}))$ **then**
19: $\mathbf{x}_i^{(G+1)} = \mathbf{u}_i^{(G)}$ {**Greedy selection**}.
20: **else**
21: $\mathbf{x}_i^{(G+1)} = \mathbf{x}_i^{(G)}$
22: **end if**
23: **end for**
24: **until** Termination criteria satisfied.

III. THE PROPOSED HGDESCOP ALGORITHM

HGDESCOP algorithm starts by setting the parameter values. In HGDESCOP, the initial population is generated randomly, which consists of *NP* individuals as shown in Equation 5. Each individual in the population is evaluated by using the objective function. At each generation (G), each individual in the population is updated by applying the DE mutation operator by selecting a random three indexes r_1, r_2, r_3, where $r_1 \neq r_2 \neq r_3 \neq i$ as shown in Equations 6, 7. After updating the individual in the population, a random number r from $(0, 1)$ is associated with each individual in the population by applying the genetic algorithm linear crossover operator as shown in Procedure 1. The greedy selection operator is starting to select the new individuals to form the new population in next generation as shown in Equation 13. These operations are repeated until termination criterion satisfied, which is the number of iterations in our algorithm.

Algorithm 3 The proposed HGDESCOP algorithm

1: Set the generation counter $G := 0$.
2: Set the initial value of F, p_c and NP.
3: Generate an initial population P^0 randomly.
4: Evaluate the fitness function of all individuals in P^0.
5: **repeat**
6: Set $G = G + 1$. {**Generation counter increasing**}.
7: **for** ($i = 0; i < NP; i{+}{+}$) **do**
8: Select random indexes r_1, r_2, r_3, where $r_1 \neq r_2 \neq r_3 \neq i$
9: $\mathbf{v}_i^{(G)} = \mathbf{x}_{r_1}^{(G)} + F \times (\mathbf{x}_{r_2}^{(G)} - \mathbf{x}_{r_3}^{(G)})$ {**DE mutation operator**}.
10: **end for**
11: **for** ($j = 0; j < NP; j{+}{+}$) **do**
12: Associate a random number r from $(0, 1)$ with each $\mathbf{v}_j^{(G)}$ in $P^{(G)}$.
13: **if** $r < P_c$ **then**
14: Apply Procedure 1 to all selected pairs of $\mathbf{v}_i^{(G)}$ in $P^{(G)}$. {**GA linear crossover operator**}.
15: Update $\mathbf{u}_i^{(G)}$.
16: **end if**
17: **end for**
18: **for** ($k = 0; k < NP; k{+}{+}$) **do**
19: **if** ($f(\mathbf{u}_k^{(G)}) \leq f(\mathbf{x}_k^{(G)})$) **then**
20: $\mathbf{x}_k^{(G+1)} = \mathbf{u}_k^{(G)}$ {**Greedy selection**}.
21: **else**
22: $\mathbf{x}_k^{(G+1)} = \mathbf{x}_k^{(G)}$
23: **end if**
24: **end for**
25: Update $P^{(G)}$
26: **until** $Itr_{no} \leq Maxitr$ {**Termination criteria satisfied**}.

IV. NUMERICAL EXPERIMENTS

The general performance of the proposed HGDESCOP algorithm is tested using 13 benchmark function $G_1 - G_{13}$, which are reported in details in [5], [7], [13]. These functions are listed in Table I as follows.

TABLE I: Constrained benchmark functions.

Function	*D*	Type of function	Optimal
G_1	13	quadratic	-15.000
G_2	20	nonlinear	-0.803619
G_3	10	polynomial	-1.000
G_4	5	quadratic	-30665.539
G_5	4	cubic	5126.498
G_6	2	cubic	-6961.814
G_7	10	quadratic	24.306
G_8	2	nonlinear	-0.095825
G_9	7	polynomial	680.630
G_{10}	8	linear	7049.248
G_{11}	2	quadratic	0.75
G_{12}	3	quadratic	-1.000
G_{13}	5	nonlinear	0.053950

TABLE II: HGDESCOP parameter settings.

Parameters	Definitions	Values
NP	Population size	30
p_c	Crossover probability	0.8
F	Mutation scale factor	0.7
$Maxitr$	Maximum number of iterations	1000

A. Parameter settings

The parameters used by HGDESCOP and their values are summarized in Table II. These values are either based on the common setting in the literature or determined through our preliminary numerical experiments.

B. Performance analysis

In order to test the general performance of the proposed HGDESCOP algorithm, we applied it with 13 benchmark functions $G_1 - G_{13}$ and the results are reported in Table III. Also, six functions have been plotted as shown in Figure 1.

1) The general performance of the HGDESCOP algorithm: The best, mean, worst and standard deviation values are averaged over 30 runs and reported in Table III. We can observe from the results in Table III, that HGDESCOP could obtain the optimal solution or very near to optimal solution for all functions $G_1 - G_{12}$ for all 30 runs, However HGDESCOP could obtain the optimal solution with function G_{13} for 9 out of 30 runs. Also in Figure 1, we can observe that the function values are rapidly decrease as the number of function generations increases.

We can conclude from Table III and Figure 1, that HGDESCOP is an efficient algorithm and it can obtain the optimal or near optimal solution with only few number of iterations.

C. HGDESCOP and other algorithms

In order to evaluate the performance of HGDESCOP algorithm, we compare it with four evolutionary based algorithms, All results are reported in Table IV, and the results of the other algorithms are taken from their original papers. The four algorithms are listed as follows.

- Homomorphous Mappings (HM) [9]
  This algorithm, incorporates a homomorphous mapping between n-dimensional cube and a feasible search space.
- Stochastic Ranking (SR) [16]
  This algorithm introduces a new method to balance objective and penalty functions stochastically, (stochastic ranking), and presents a new view on penalty function methods in terms of the dominance of penalty and objective functions.
- Adaptive Segregational Constraint Handling EA (ASCHEA) [6]
  This algorithm is called ASCHEA and it is used after extending the penalty function and introducing a niching techniques with adaptive radius to handel multimodel functions. The main idea of the algorithm is to start for each equality with a large feasible

TABLE III: Experimental results of HGDESCOP for $G_1 - G_{13}$

$Function$	optimal	best	mean	worst	std
G_1	-15.000	-15.000	-15.000	-15.000	0.0e+00
G_2	-0.803619	-0.8036187	-0.7993549	-0.7861574	0.0062361
G_3	-1.000	-1.0005001	-1.0005000	-1.0004992	$2.7368237e^{-07}$
G_4	-30665.539	-30665.538	-30665.538	-30665.538	0.0e+00
G_5	5126.498	5126.496858	5126.496728	5126.49671	$4.5552e^{-05}$
G_6	-6961.814	-6961.813875	-6961.813875	-6961.813875	$1.9173e^{-12}$
G_7	24.306	24.306209	24.306209	24.306209	$4.706924e^{-13}$
G_8	-0.095825	-0.095825	-0.095825	-0.095825	$1.223905e^{-17}$
G_9	680.630	680.630057	680.630057	680.630057	$3.789561e^{-14}$
G_{10}	7049.248	7049.248020	7049.248020	7049.248020	$6.264592e^{-12}$
G_{11}	0.75	0.749900	0.749900	0.749900	$1.170277e^{-16}$
G_{12}	-1.000	-1.000	-1.000	-1.000	0.0e+00
G_{13}	0.053950	0.084356	0.372933	0.438802	0.139366

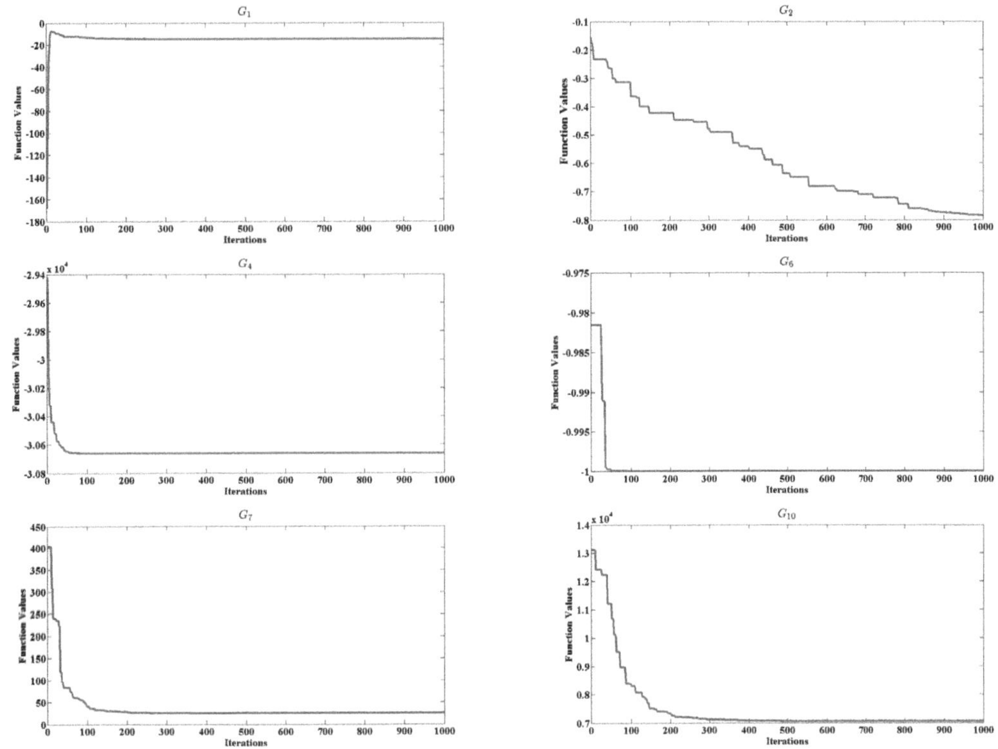

Fig. 1: The general performance of HGDESCOP algorithm.

domain and to reduce it progressively in order to bring it as close as possible to null measure domain.

- Simple Multimembered Evolution Strategy (SMES) [14].
  This algorithm is based on a multimembered ES with a feasibility comparison mechanism.

1) Comparison between HM, SR, ASCHEA, SMES and HGDESCOP: The best, mean, worst results of the five comparative algorithms are averaged over 30 runs and reported in Table IV. The evaluation function values for HM, SR, ASCHEA and SMES algorithms are 1,400,000, 350,000, 1,500,000 and 250,000 respectively. However the maximum evaluation function value for HGDESCOP algorithm is 120,000. We can observe from Table IV, that HGDESCOP results are better than the other algorithms for all functions $G_1 - G_{12}$ except the last function G_{13}. In term of evaluation function values, it is clear that HGDESCOP is faster than the other algorithms.

V. Conclusion

In this paper, a new hybrid genetic differential evolution algorithm to solve constrained optimization problems is presented. The proposed algorithm is called hybrid genetic differential evolution algorithm for solving constrained optimization problems (HGDESCOP). The proposed algorithm combines the differential evolution algorithm and the genetic linear crossover operator in order to improve the exploration ability of the DE algorithm and to avoid trapping in local minima. To verify the efficiency of the proposed algorithm, it has been compared with 4 Evolutionary based algorithm on 13 benchmark functions. The experimental results show that the HGDESCOP algorithm is a robust and efficient algorithm

TABLE IV: Experimental results of HGDESCOP and other EA-based algorithms for problems $G_1 - G_{13}$

$Function$	optimal		HM	SR	ASCHEA	SMES	HGDESCOP
G_1	-15.000	Best	-14.7864	-15	-15	-15	-15
	-15.000	Mean	-14.7082	-15	-14.84	-15	-15
	-15.000	Worst	-14.6154	-15	N.A.	-15	-15
G_2	-0.803619	Best	0.79953	0.803515	0.785	0.803601	-0.8036187
	-0.803619	Mean	0.79671	0.781975	0.59	0.751322	-0.7993549
	-0.803619	Worst	0.79119	0.726288	N.A.	0.751322	-0.7861574
G_3	-1.000	Best	0.9997	1.000	1.000	1.001038	1.000500
	-1.000	Mean	0.9989	1.000	0.99989	1.000989	1.0005000
	-1.000	Worst	0.9978	1.000	N.A.	1.000579	1.0004992
G_4	-30665.539	Best	-30664.5	-30665.539	-30665.5	-30665.539062	-30665.538
	-30665.539	Mean	-30655.3	-30665.539	-30665.5	-30665.539062	-30665.538
	-30665.539	Worst	-30645.9	-30665.539	N.A.	-30665.539062	-30665.538
G_5	5126.498	Best	-	5126.497	5126.5	5126.599609	5126.496728
	5126.498	Mean	-	5128.881	5141.65	5174.492301	5126.496728
	5126.498	Worst	-	5142.472	N.A.	5304.166992	5126.49671
G_6	-6961.814	Best	-6952.1	-6961.814	-6961.81	-6961.813965	-6961.813875
	-6961.814	Mean	-6342.6	-6875.940	-6961.81	-6961.283984	-6961.813875
	-6961.814	Worst	-5473.9	-6350.262	N.A.	-6961.481934	-6961.813875
G_7	24.306	Best	24.620	24.307	24.3323	24.326715	24.306209
	24.306	Mean	24.826	24.374	24.6636	24.474926	24.306209
	24.306	Worst	25.069	24.642	N.A.	24.842829	24.306209
G_8	0.095825	Best	0.0958250	0.095825	0.09582	0.095826	0.095825
	0.095825	Mean	0.0891568	0.095825	0.09582	0.095826	0.095825
	0.095825	Worst	0.0291438	0.095825	N.A.	0.095826	0.095825
G_9	680.630	Best	680.91	680.630	680.630	680.631592	680.630057
	680.630	Mean	681.16	680.656	680.641	680.643410	680.630057
	680.630	Worst	683.18	680.763	N.A.	680.719299	680.630057
G_{10}	7049.248	Best	7147.9	7054.316	7061.13	7051.902832	7049.248020
	7049.248	Mean	8163.6	7559.192	7497.434	7253.047005	7049.248020
	7049.248	Worst	9659.3	8835.655	N.A.	7638.366211	7049.248020
G_{11}	0.75	Best	0.75	0.750	0.75	0.749090	0.749900
	0.75	Mean	0.75	0.750	0.75	0.749358	0.749900
	0.75	Worst	0.75	0.75	N.A.	0.749830	0.749900
G_{12}	-1.000	Best	-0.999999875	-1.00000	N.A	-1.000000	-1.000000
	-1.000	Mean	-0.999134613	-1.00000	N.A.	-1.00000	-1.00000
	-1.000	Worst	-0.991950498	-1.00000	N.A.	-1.00000	-1.00000
G_{13}	0.053950	Best	N.A.	0.053957	N.A	0.053986	0.084356
	0.053950	Mean	N.A.	0.057006	N.A	0.166385	0.372933
	0.053950	Worst	N.A	0.216915	N.A	0.468294	0.438802

and can obtain the global minima or near global minima faster than other algorithms. AS part of our future work, in this paper we are using linear crossover to improve the performance of DE, whether the HGDESCOP algorithm could be improved by using more advanced GA crossover operators. Also we can apply the proposed algorithm with many real-life applications such as engineering design, finance, economics.

REFERENCES

[1] A.F. Ali, A. E. Hassanien, Minimizing molecular potential energy function using genetic Nelder-Mead algorithm, 8th International Conference on Computer Engineering & Systems (ICCES), Cairo, pp. 177-183, 2013.

[2] M.M. Ali and W.X. Zhu. A penalty function-based differential evolution algorithm for constrained global optimization, Computational Optimization and Applications, Vol. 54, No. 3, pp. 707–739, April 2013.

[3] P. Caamano, F. Bellas, J. A. Becerra, and R. J. Duro, Evolutionary algorithm characterization in real parameter optimization problems, Applied Soft Computing, vol. 13, no. 4, pp. 1902-1921, 2013.

[4] C. Darwin, On the Origin of Species. London: John Murray, 1859.

[5] C.A. Floudas and P.M. Pardalos, A collection of test problems for constrained global optimization algorithms. In P. M. Floudas (Ed.), Lecture notes in computer science, Vol. 455. Berlin: Springer, 1987.

[6] S. B. Hamida and M. Schoenauer, ASCHEA: New rsults using adaptive segregational constraint handling, in Proceedings of the Congress on Evolutionary Computation (CEC2002), Piscataway, New Jersey, IEEE Service Center, pp. 884-889, 2002.

[7] W. Hock, K.Schittkowski, Test examples for nonlinear programming codes. In Lecture notes in economics and mathematical systems (Vol. 187). Berlin: Springer, 1981.

[8] J. H. Holland, Adaptation in Natural and Artficial Systems, University of Michigan Press, Ann Arbor, MI, 1975.

[9] S. Koziel and Z. Michalewicz, Evolutionary algorithms, homomorphous mappings, and constrained parameter optimization, Evolutionary Computation 7(1), 19-44, 1999.

[10] T.S. Metcalfe, P. Charbonneau, Stellar structure modeling using a parallel genetic algorithm for objective global optimization, Journal of Computational Physics 185, 176-193, 2003.

[11] Z. Michalewicz, A Survey of Constraint Handling Techniques in Evolutionary Computation Methods, Evolutionary Programming, Vol.4, pp. 135, 1995.

[12] Z. Michalewicz, Genetic algorithms + data structures = evolution

programs. Berlin: Springer, 1996.

[13] Z. Michalewicz and M. Schoenauer, Evolutionary algorithms for constrained parameter optimization problems, Evolutionary Computation 4(1), 132, 1996.

[14] E. M. Montes and C. A. Coello Coello, A simple multi-membered evolution strategy to solve constrained optimization problems, IEEE Transactions on Evolutionary Computation, vol. 9, no. 1, pp. 117, 2005.

[15] Y. F. Ren and Y. Wu, An efficient algorithm for high-dimensional function optimization, Soft Computing, vol. 17, no. 6, pp. 995-1004, 2013.

[16] T.P. Runarsson and X. Yao, Stochastic Ranking for Constrained Evolutionary Optimization, IEEE Transactions on Evolutionary Computation 4(3), 284-294, 2000.

[17] R. Storn, K. Price, Differential evolutiona simple and efficient heuristic for global optimization over continuous spaces. J Glob Optim 11:341-359, 1997.

[18] Y. Wang and C. Zixing, A hybrid multi-swarm particle swarm optimization to solve constrained optimization problems, Frontiers of Computer Science in China, Vol. 3, No. 1, pp. 38-52, March, 2009.

[19] Y. Wang, Zixing Cai and Yuren Zhou. Accelerating adaptive trade-off model using shrinking space technique for constrained evolutionary optimization, International Journal for Numerical Methods in Engineering, Vol. 77, No. 11, pp. 1501-1534, March 2009.

[20] A. Wright, Genetic Algorithms for Real Parameter Optimization, Foundations of Genetic Algorithms 1, G.J.E Rawlin (Ed.) (Morgan Kaufmann, San Mateo), 205-218, 1991.

[21] J.M. Yang, Y.P. Chen, J.T. Horng and C.Y. Kao. Applying family competition to evolution strategies for constrained optimization. In Lecture Notes in Mathematics Vol. 1213, pp. 201-211, New York, Springer, 1997.

Comparison of Classifiers and Statistical Analysis for EEG Signals Used in Brain Computer Interface Motor Task Paradigm

Oana Diana Eva
Faculty of Electronics, Telecommunications and Information Technology
"Gheorghe Asachi" Technical University of Iasi
Iasi, Romania

Anca Mihaela Lazar
Faculty of Medical Bioengineering
"Grigore T. Popa" University of Medicine and Pharmacy
Iasi, Romania

Abstract—**Using the EEG Motor Movement/Imagery database there is proposed an off-line analysis for a brain computer interface (BCI) paradigm. The purpose of the quantitative research is to compare classifier in order to determinate which of them has highest rates of classification. The power spectral density method is used to evaluate the (de)synchronizations that appear on Mu rhythm. The features extracted from EEG signals are classified using linear discriminant classifier (LDA), quadratic classifier (QDA) and classifier based on Mahalanobis distance (MD). The differences between LDA, QDA and MD are small, but the superiority of QDA was sustained by analysis of variance (ANOVA).**

Keywords—brain computer interface; electroencephalogram; event related (de)synchronization

I. Introduction

Brain computer interface (BCI) facilitates a direct communication between brain and an external device. The system - hardware and software - enables humans to interact with their surroundings without involvement of peripheral nerves and muscles, by using control signals generated by brain activity [1]. The interface enhances the possibility of communication for people with severe neuromuscular and motor disabilities. The variety of BCI applications includes: environmental control, locomotion, entertainment and multimedia.

The artificial intelligence system recognize a certain set of patterns in brain signals following the stages: signal acquisition, preprocessing, feature extraction, classification and the control interface. Different methods such as electroencephalogram (EEG), magnetoencephalogram (MEG), positron emission tomography (PET), single photon emission computed tomography (SPECT) are used in measuring and studying the brain activity. The EEG is the most convenient method used in BCI systems: because it is non-invasive, it has relative low costs, the real-time analysis may be performed and can be used in a portable device. EEG based BCIs use a set of sensors that pick up the EEG signals from different brain areas.

EEG signals contain a wide range of frequency spectrum. The oscillatory activity in the EEG is classified according to frequency bands or rhythms: Delta (1-4 Hz), Theta (4-8 Hz), Alpha and Mu (8-12 Hz), Beta (13-25 Hz), Gamma (25-40 Hz) [2]. Mu rhythm (8-12 Hz) is affected by movements or movement imagery.

Preparing a movement or imagining movement can cause changes in the sensorimotor rhythms (SMR). The SMR refer to oscillations recorded in brain activity concentrated in certain frequency bands [3].

The event-related desynchronizations (ERD) are changes that appear while executing or imagining the movement. ERD starts when the subject begins to imagine a movement and manifests itself as a power decrease in Mu rhythm band. After that, a different phenomenon occurs, event-related synchronization (ERS) - an increase in power when the subject stops executing or imagining a movement.

The phenomenon of ERD/ERS related to motor imagery is stronger for the contralateral hemisphere and weaker in the ipsilateral hemisphere.

In the section II a presentation of the database is completed, how the features are extracted and how the statistical methods are applied. The paper ends with a conclusion and some recommendations based on our results (section III).

II. Methodology

A. Database description

The EEG Movement/Imagery Database (eegmmidb) was downloaded from www.physionet.org [4]. It contains recordings from 109 subjects, who executed real or imagined tasks. The EEG signals were recorded using International System 10-20 with 64 electrodes placed on the scalp. Subjects 43, 84, 88, 89, 92 and 100 were excluded because the contained recordings are not reliable for further processing. We have considered only FC1, FC2, FC3, FC4, C1, C2, C3, C4, CP1, CP2, CP3, CP4 electrodes, reported in the literature for enhancing Mu desynchronization. Every subject performed 14 experimental tasks: 2 runs of 1 minute for relaxation (one with eyes closed and one with eyes open) and 4 runs of 2 minutes for each of the following tasks: opening and closing left/right fist when a target appears on the screen followed by relaxation, imagining opening and closing left/right fist, opening and closing both fists, imagining opening and closing

SC CEMENTLAB SRL, www.c-labortech.com

both fists. In order to implement the proposed methods, there were used the first two sets described above.

Each signal is coded as follows: T0 corresponds to the resting period, T1 corresponds to movement/movement imagery left wrist, T2 corresponds to onset of motion (real or imagined) of the right wrist. The EEG signals are sampled at 160 Hz. There are three trials for wrist movement (named 3, 7, 11) and other three for wrist movement imagery (named 4, 8, 12).

B. Data Processing

Signals loaded from database are filtered with a 8-12 Hz band pass filter corresponding to the Mu rhythm frequency range. No artifact rejection or corrections were performed. We selected segments from the EEG signals (2 s after the stimulus appearance) according to annotation for each mental task (T2, T1) extracting the information corresponding to right/left wrist movement. For the relaxation period (T0) sequences of 2 s following right/left wrist movement are extracted.

The most widely used methods for EEG signal feature extractions are based on frequency analysis, for example discrete Fourier transform (DFT) or power spectral density (PSD). We use a method based on PSD to find the desynchronization during movement. Power spectral densities were calculated for all the useful mentioned channels and for all trials 3, 7, 11 which correspond to right/left wrist movement. The average of these trials was calculated using pwelch function from Matlab with a Hanning window [5]. The same procedure was applied both for computing the PSD during the movement period, denoted by $PSD_{MOVEMENT}$ and for the relaxation period which comes after right wrist movement and left wrist movement respectively, denoted by PSD_{REST}. The resulted value, denoted by ERD, is used to assess the desynchronization/synchronization which appears on the pair of electrodes during right or left wrist movement.

$$ERD = \frac{PSD_{REST} - PSD_{MOVEMENT}}{PSD_{REST}} \quad (1)$$

The feature vector was formed from each pair of electrodes on the left/right hemisphere in the following way: ERD calculated for right wrist movement for the signal recorded from left hemisphere (FC1, FC3, C1, C3, CP1 or CP3), ERD calculated for left wrist movement for the paired electrode from left hemisphere (FC2, FC4, C2, C4, CP2 or CP4), ERD calculated for left wrist movement for the electrode from left hemisphere, ERD calculated for left wrist movement for the electrode from right hemisphere.

C. Classifiers

Linear discriminant analysis (LDA) is one of the most popular classification algorithms for BCI application, and has been successfully used in a large number of BCI systems such as motor imagery based BCI, P300 speller and steady state visual evoked potentials based BCI [6]. In essence, LDA linearly transforms data from a high dimensional space to a low dimensional space, and finally the decision is made in the low dimensional space, thus the definition of the decision boundary plays an important role on the recognition performance [7].

Linear classifier is suitable for offline and online BCIs. Moreover, LDA, is simple to use and provides satisfactory results whether we are referring to a large or small databases.

Quadratic discriminant analysis (QDA) is closely related to LDA, where it is assumed that the measurements from each class are normally distributed. QDA makes no assumption that the covariance of each of the classes are identical [8]. Although it is not reported and used as much as linear classifier in BCI systems, the quadratic classifier reported satisfactory and encouraging results.

Mahalanobis distance (MD) is a statistical distance function. In mathematical terms, the Mahalanobis distance is equal to the Euclidean distance when the covariance matrix is the unit matrix. The use of the Mahalanobis distance removes several of the limitation of linear classifiers based on Euclidean metric, since it automatically account for the scaling of the coordinate axes, as well as for the correlation between the different considered features [9]. Mahalanobis classifier is simple but at the same time robust and leads to good results, as shown in [10]. Despite its good performance, it is still rarely used in the literature on brain computer interfaces.

Classifiers LDA, QDA and MD were used for all six pairs of electrodes. The steps described above were also accomplished for the trials corresponding to right/left imagery of wrist movement. The classification error obtained for the test set was surveyed for all the subjects, for movement/ imagery of movement, pair of electrodes and classifier.

D. Statistical Analysis

A two-way Analysis of Variance (ANOVA) was performed using Statistical Package for the Social Sciences (SPSS) [10] on the error values obtained on movement/movement imagery. The first main factor was CLASSIFIERS with levels LDA, QDA, and MD, while the second main factor was ELECTRODES with levels FC1-FC2, FC3-FC4, C1-C2, C3-C4, CP1-CP2, CP3-CP4. The Levene test was used for testing homogeneity of variances and the Tukey's test was used as post-hoc test at the 2% level of significance.

III. RESULTS

Table 1 shows the means of error rates for all subjects, classifiers and pair of electrodes.

TABLE I. MEAN OF TEST ERROR RATE PERFORMANCE FOR CLASSIFIERS AND PAIR OF ELECTRODES

Task	Classifier	Pair of electrodes					
		FC3-FC4	FC1-FC2	C3-C4	C1-C2	CP3-CP4	CP1-CP2
Movement	LDA	14,96	15,53	15,59	16,18	17,12	17,44
	QDA	11,31	12,62	11,92	12,26	11,81	13,32
	MD	15,05	15,91	15,12	17,26	16,38	17,62
Movement Imagery	LDA	13,63	15,19	19,85	17,17	18,02	17,46
	QDA	12,82	12,80	15,12	14,29	14,15	15,05
	MD	15,16	15,19	18,48	16,49	17,75	17,76

For movement, the smallest error of 11,31% was obtained with the quadratic classifier for FC3-FC4, while the largest error 17,62% was obtained with the classifier based on Mahalanobis distance for C3-C4.

For the imaginary of movement there were attained the following errors: the lowest, 12,82%, with the quadratic classifier for FC1 - FC2 and 19,85%, the highest value with classifier LDA for C3-C4. High errors could be explained due to the imperfect contact of the electrodes on the scalp or as the Mu rhythm could not be developed as specified in [11].

On 58% of 103 analyzed subjects the smallest classification errors were obtained for wrist movement, 27% of subjects were able to perform better the imposed task for imagining motor movement than movement. Small errors were achieved for movement as well as imagining the movement for 15% of subjects.

The errors obtained after applying the quadratic classifier were better than those obtained using linear classifier and classifier based on Mahalanobis distance, both for real and imagined motor task. Differences between results for LDA classifier and MD classifier are very small (Fig. 1).

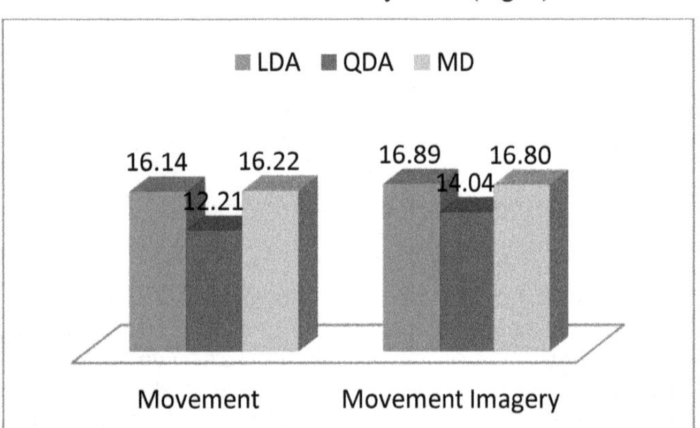

Fig. 1. The error rate for movement and movement imagery using LDA, QDA and MD

We have selected the subjects who attained, even with the quadratic classifier, low classification rates when the EEG signal was passed through a 8-12 Hz band-pass filter. We suppose that these subjects could elicit low or high Beta rhythm instead of Mu rhythm. So, for these subjects only the data were filtered with a 13-18 Hz eighth order Butterworth band pass filter. Filtering on 13-18 Hz was performed on subjects 2, 21, 36, 42, 54, 64, 74, 78, 82, 87, 102, 106. The errors achieved for subject 54 are shown in Table 2. For subject 54, as he attained the best results, the errors decrease significantly if the filter is on 13-18 Hz instead on 8-12 Hz. We can conclude that subject 54 achieved low Beta rhythm. Beta rhythm was also developed by subjects 21, 74, 82, 102. The results show that frequency band other than that of the Mu range may contain useful information. Notable changes on errors were not found for the other seven subjects on 13-18 Hz or 19-26 Hz. In conclusion these subjects could not attain low Beta rhythm or high Beta rhythm.

TABLE II. THE ERRORS FOR SUBJECT 54 WHEN THE SIGNALS ARE BAND PASS FILTERED ON 8-12HZ AND ON 13-18HZ

Subject 54								
Task	Trial	Classifier	Electrodes					
			FC3 - FC4		C3 - C4		CP3 - CP4	
			8-12 Hz	13-18 Hz	8-12 Hz	13-18 Hz	8-12 Hz	13-18 Hz
Movement	3	LDA	38,89	5,56	61,11	5,56	61,11	5,56
		QDA	30,00	5,56	50,00	5,56	55,00	5,56
		MD	38,89	5,56	44,44	5,56	44,44	5,56
	7	LDA	61,11	16,67	61,11	22,22	50,00	27,78
		QDA	27,78	16,67	22,22	22,22	50,00	27,78
		MD	44,44	22,22	44,44	22,22	44,44	33,33
	11	LDA	61,11	5,56	61,11	0,00	61,11	5,56
		QDA	61,11	0,00	61,11	11,11	61,11	5,56
		MD	61,11	11,11	61,11	16,67	61,11	16,67
Movement Imagery	4	LDA	50,00	5,56	61,11	5,56	61,11	22,22
		QDA	50,00	5,56	11,11	5,56	61,11	22,22
		MD	50,00	11,11	44,44	5,56	61,11	22,22
	8	LDA	50,00	50,00	61,11	16,67	61,11	44,44
		QDA	44,44	44,44	61,11	16,67	61,11	44,44
		MD	55,56	38,89	61,11	16,67	61,11	38,89
	12	LDA	50,00	5,56	44,44	27,78	61,11	33,33
		QDA	50,00	5,56	27,78	22,22	52,22	38,89
		MD	50,00	11,11	55,56	27,78	50,00	33,33

TABLE III. THE ERRORS FOR SUBJECTS 2, 74, 87 ON TEST AND TRAINING DATA

Subject	Data	Pair of electrodes					
		FC3-FC4	FC1-FC2	C3-C4	C1-C2	CP3-CP4	CP1-CP2
2	Test Data	16.67	16.98	21.61	19.41	19.44	27.16
	Training Data	16.63	16.05	17.90	24.07	22.84	25.78
74	Test Data	0.93	8.02	4.01	5.25	12.04	10.80
	Training Data	1.85	4.94	9.26	1.87	9.57	11.42
87	Test Data	22.84	23.15	21.91	28.70	19.44	29.94
	Training Data	19.14	21.60	17.90	23.77	19.16	21.30

Table 3 shows the results we had for training and test data for subjects 2, 74 and 87 on pairs of electrodes FC3-FC4, FC1-FC2, C3-C4, C1-C2, CP3-CP4, CP1-CP2. At subject 87 and 2 the errors are higher on test data, then those obtained on training data and there are consistent for all paired of electrodes. The errors obtained on training data for pairs FC3-FC4, C3-C4 at subject 74 are higher that errors achieved on the others pairs if electrodes.

The ANOVA demonstrated that the use of different classifiers improves the error values. The p-value obtained for CLASSIFIERS was 0.001 for movement and 0.002 for movement imagery. On ELECTRODES the p-value was 0.001 for both tasks.

No interactions between CLASSIFIERS x ELECTRODES was found (p=0.815 for movement and p=0.649 for movement imagery).

Since differences were found in between classifiers, we performed the post-hoc statistical analysis comparison (Turkey test) to determine which classifier is the best. The tests shown that the classification rates obtained for movement/movement imagery with quadratic classifier are higher than those obtained with LDA and MD. At the $\alpha = 0.02$ significance level, there is not enough evidence to conclude that the used electrodes have a significant interaction effect on errors obtained for both tasks.

In Table 4 are depicted the differences between p values obtained with different post hoc tests for classifiers LDA, QDA and DM. Differences between p values obtained with Tukey and Scheffe test are small. The ratio for movement is 0.836/0.850 and for movement imagery 0.965/0.968. Thus, we are confident 98% that with classifier QDA we attained small error that when using LDA and MD classifier. The use of LDA and MD will yield to higher errors since they are similar (Table 4). Although Scheffe procedure is most popular due its conservatism and flexibility, leads to type II errors. Tukey procedure is used mostly for means comparison and leads to type I erors.

TABLE IV. CLASSIFIER COMPARISON WITH POST HOC TESTS SCHEFFE AND TUKEY

Post Hoc Tests	Clasifiers Comparison		p value	
			Movement	Movement Imagery
Tukey	LDA	QDA	0,000	0,002
		MD	0,836	0,965
	QDA	LDA	0,000	0,002
		MD	0,000	0,004
	MD	LDA	0,836	0,965
		QDA	0,000	0,004
Scheffe	LDA	QDA	0,000	0,003
		MD	0,850	0,968
	QDA	LDA	0,000	0,003
		MD	0,000	0,006
	MD	LDA	0,850	0,968
		QDA	0,000	0,006

In [12] and [11], using the same dataset relevant results were reported, but also some drawbacks (there are unknown the timing between runs, the age of the subjects or if there are right or left handed subjects).

In another work [13] the classification results were reported only for 30 subjects and were not applied statistical tests.

In most papers regarding BCI research, the classification is performed using a single classifier. A recent trend involves using several classifiers. The combination of multiple classifiers has the advantage of obtaining lower classification errors [6].

IV. CONCLUSIONS

Using power spectral density on the EEG signals contained in EEG Motor/ Movement Imagery Dataset we have studied if desynchronizations appear in the frequency band 8-12 Hz. The classifiers LDA, QDA and MD applied on feature vector were used to determine the classification errors for all six pairs of electrodes.

The results of classification errors vary from subject to subject. The differences among classifiers as LDA and MD are small and reasonable results were attained considering the large database (103 subjects were tested). The used method showed the best performance for the QDA classifier. The performances could be altered because some subjects cannot concentrate well in performing each task. Sometimes they can be absent-minded, ocular or muscles artifacts can occur or they may not have the capacity to imagine movement. Movement and imagining involves sustained mental effort. Also it is important to notice that the recorders contained in database were made on healthy subjects. As some studies revealed that the people who suffer different disabilities can develop Mu rhythm better than the healthy ones, it is possible to get higher classification rates for these persons.

Future work will be focused on a combination of classifiers used in this paper in order to reduce the classification error.

REFERENCES

[1] L. Nicolas-Alonso and L. Gomez-Gil, “Brain computer interfaces, a review”, Sensors, vol. 12, no. 2, pp. 1211–79, Jan. 2012.

[2] A. Kachenoura, L. Albera, L. Senhadji, P. Como., “ICA: A Potential Tool for BCI Systems”, IEEE Signal Processing Magazine, Institute of Electrical and Electronics Engineers (IEEE), 25 (1), pp.57-68, 2008.

[3] R. Aldea and O.D. Eva, “Detecting sensorimotor rhythms from the EEG signals using the independent component analysis and the coefficient of determination”, Proceedings of the 11th IEEE International Symposium on Signals, Circuits and Systems (ISSCS’13), 11, pp. 13–16, 2013.

[4] G. B. Moody, R. G. Mark and A. L. Goldberger, “PhysioNet: Physiologic Signals, Time Series and Related Open Source Software for Basic, Clinical, and Applied Research”, Proceedings of the 33rd Annual International Conference of the IEEE EMBS Boston, Massachusetts USA, August 30 - September 3, 2011.

[5] H. Schmid, “How to use the FFT and Matlab’s pwelch function for signal and noise simulations and measurements,” Institute of Microelectronics, University of Applied Sciences NW Switzerland, no. August 2012.

[6] F. Lotte, M. Congedo, A. Lécuyer, F. Lamarche and B. Arnaldi, “A review of classification algorithms for EEG-based brain-computer interfaces”, Journal of Neural Engineering, vol. 4, no. 2, pp. R1–R13, June 2007.

[7] R. Zhang, P. Xu, L. Guo, Y. Zhang, P. Li and D. Yao, "Z-score linear discriminant analysis for EEG based brain-computer interfaces," PLoS One, vol. 8, no. 9, p. e74433, Jan. 2013.

[8] Hastie Trevor, Tibshirani Robert, and J. Friedman, "The Elements of Statistical Learning - Data Mining, Inference, and Prediction", Springer Series in Statistics, 2009.

[9] F. Babiloni, L. Bianchi, F. Semeraro, J. R. Millán, J. Mouriño, A. Cattini, S. Salinari, M. G. Marciani, and F. Cincotti, "Mahalanobis Distance-Based Classifiers Are Able to Recognize EEG Patterns by Using Few EEG Electrodes", Engineering in Medicine and Biology Society, Proceedings of the 23rd Annual International Conference of the IEEE, vol. 1, pp. 651–654, 2001.

[10] R. W. Walters, "Database Management, Graphing and Statistical Analysis Using IBM-SPSS Statistics", Creighton University, 2012.

[11] J. Sleight, P. Pillai, and S. Mohan, "Classification of Executed and Imagined Motor Movement EEG Signals," Ann Arbor: University of Michigan, pp. 1–10, 2009, retrived from http://www.scribd.com/doc/82045737/ICA.

[12] A. Loboda, A. Margineanu, G. Rotariu and A. M. Lazar, "Discrimination of EEG-Based Motor Imagery Tasks by Means of a Simple Phase Information Method", International Journal of Advanced Research in Artificial Intelligence(IJARAI), 3(10), 2014.

[13] O.D. Eva, R. Aldea, A.M. Lazar., "Detection and classification of Mu rhythm for motor movement/imagery dataset", Bulletin Of The Polytechnic Institute Of Jassy, vol. LX(LXIV), no. 2, pp. 36-44, 2014.

System for Human Detection in Image Based on Intel Galileo

Rastislav Eštók
Dept. of Computers and Informatics
FEI TU of Kosice
Kosice, Slovak Republic

Miroslav Michalko
Dept. of Computers and Informatics
FEI TU of Kosice
Kosice, Slovak Republic

Ondrej Kainz
Dept. of Computers and Informatics
FEI TU of Kosice
Kosice, Slovak Republic

František Jakab
Dept. of Computers and Informatics
FEI TU of Kosice
Kosice, Slovak Republic

Abstract—**The aim of this paper is a comparative analysis of methods for motion detection and human recognition in the image. Authors propose the own solution following the comparative analysis of current approaches. Then authors design and implement hardware and software solution for motion detection in the video with human recognition in the picture. The development board Intel Galileo serves as the basis for hardware implementation. Authors implement own software solution for motion detection and human recognition in the image, resulting in the evaluation of proposed implementation.**

Keywords—image processing; Intel Galileo; motion detection; object recognition

I. Introduction

Motion detection and objects recognition in the image is an active research being done as a part of computer science. The principal goal of intelligent systems for objects recognition and tracking is the use of these systems in various sectors of computer engineering, e.g. image processing, security, and robotics. People can focus on a specific object, and the aim of the research is to implement this ability in intelligent systems.

The object is in motion when it is compelled to change that state by force impressed on it. Detecting the motion is the object is done by [1]: sound, optical, vibration, geomagnetism, refection of transmitted energy and electromagnetic induction. Detection of motion by sound follows the principle that every object being in movement causes sound and detector can localize the source of sound in the area.

The most used methods for motion detection are optical methods compare frames from continuous video stream and detect moving objects. Video sequence source is in MJPEG and there are techniques to convert this stream into JPEG images. In this paper authors compare two frames images in bitmap image file which support RGB color model. Project consisted of both, hardware and software implementation. The first, utilize the Intel Galileo board and IP camera and the latter introduces and approach for detection of person directly by the proposed device. In the following sections, the well-known approaches in the area of detection are presented.

II. Motion Detection Methods in Video Sequence

There is a great number of the algorithms for tracking objects in video sequence, most of them uses comparison of one frame from video to the previous one. Three common methods are used for motion detection in video sequence [1]. These methods are background subtraction, optical flow and temporal differences. Systems for motion detection needs to handle a number of critical problems. The most common described in [2] are variations of the lighting conditions, small non-static movements such as tree branches in the wind, low image quality source making noise picture; movements in ghost regions in the image or sudden changes in the light conditions inside by presence of a light switch and outside by clouding. There is no system that would provide exact output and work without errors. The goal of every technique is to reach the best results without mistakes caused by described critical problems. It is recommended to combine two or more techniques to acquire better results.

A. Background Subtraction

There are a lot of algorithms for background subtraction methods and majority of them is based on principles described in [3] and shown on flow diagram in Fig. 1.

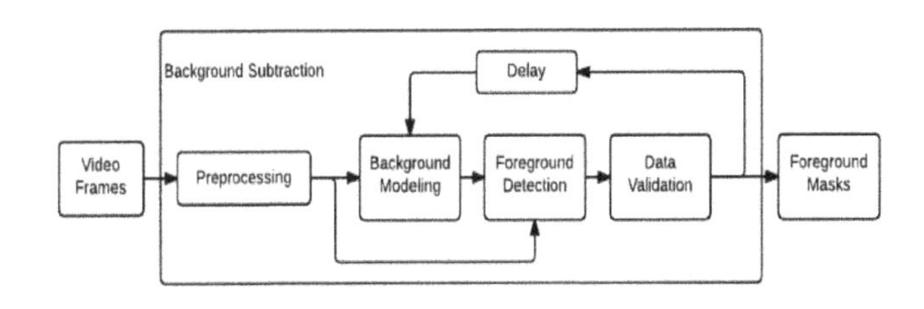

Fig. 1. Flow diagram of a generic background subtraction algorithm [3]

Fig. 1 represents a simple flow diagram that describes generic algorithm for the background subtraction. It is composed of four main parts: preprocessing, background modeling, foreground detection and data validation. Preprocessing is a process for setting videos properties. Video from camera is captured in real-time in required format with image processing. The video is divided to frames in this part of background subtraction process. The next part is background modeling. This is the main part of all process. There has been many researched towards the finding of robust background modeling techniques. The aim of these researches is described in [3] and it is to seek a balanced technique which is robust against environmental changes in background, but enough sensitive to identify all the moving object of interest. Foreground detection is part where are identified all pixels from image which are candidates for foreground pixels. The outputs of this part are objects in foreground. Subsequently there is a data validation part where candidates for foreground are identified and verified.

There are two different types of background modeling in background subtraction techniques: non-recursive and recursive techniques.

Non-recursive techniques use a buffer in a sliding-window which stores previous frame and estimates the background image based on temporal variations of each pixel in a buffer. These methods are adaptive, because they do not depend on history beyond those frames stored in the buffer. There is a large storage requirement depending on buffer size. In [3] the authors describe common non-recursive techniques for background modeling, i.e.: frame differencing, median filter, linear predictive filter and non-parametric model.

Recursive techniques do not use buffer for background estimation. They recursively updates background model in each input frame. Input frames from past could have an effect on new inputs. Recursive techniques do not need such storage. If an error occurs in recursive technique, it last more time to eliminate this error. Authors in [3] describe these recursive techniques for background modeling: approximated median filter, Kalman filter and mixture of Gaussians.

B. *Optical flow*

Method was described by Lucas and Kanade in [4] and based on this approaches there are three assumptions described in [5], these are: brightness constancy, temporal persistence and spatial coherence.

C. *Temporal differences*

This method is also called method of frame differences. Principles of this method are described by author in [6]. Method compares two frames and uses special techniques for electing referral object for motion detection. This technique is based templates and it is called template matching. Author in [1] describes two template matching techniques: static, used in background subtraction and dynamic template matching (DTM). This study uses frame differences method by comparing each of RGB components: red, green and blue. Method in this study compares all pixels from two frames and based on this detects the motion.

III. Methods for Face and Human Recognition in Static Image

The aim of methods for face detection is to find out if there is a face in a static picture. In positive case the face is localized and system returns face location and parameters. Also in face recognition there are many factors that influence the results. These factors are described in [2] and [7]. Face location may vary by camera position and angle of coverage. This can cause that faces are covered by other objects or are damaged. For example a nose may vary from any angle of coverage. Author in [8] described next factor. Presence or absence of structural components such as beards, mustaches or glasses which may or may not be present on face and there is a high variability of these components including shapes, color and size. Faces may vary by a person's facial expressions. Faces in picture may be occluded by other objects. There is also big possibility to damaging facing because of low image quality. Bed effect to face recognition is caused by lightning conditions that can vary very fast.

Author in [2] describes four common methods for face recognition in image: knowledge-based methods, feature invariant approaches, template matching methods and appearance-based methods.

A. *Knowledge-based Methods*

Knowledge-based methods use rules to define human faces properties. The rules describe relationships between parts of a face. These methods are used mainly for face localization in image. In general, it is simple to describe human face and relationships between each part of face. It is known that face is composed of one pair of eyes that are symmetric to each other, one nose, one mouth and other good known parts of face. Based on this information are created rules of relationship between each part. It is known that faces can be described with common rules. It is very important to set rules very effective. If relationships are described very exactly, there are a lot of mistakes in face recognition. On the other side, if rules are more benevolent, there are a lot of false positive. It is not difficult to find faces in image based on this technique and it is required to use verification process to find false results. The most important thing in this process is to set rules because of lot of variations of faces and poses.

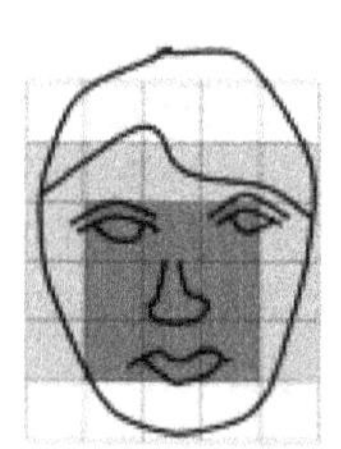

Fig. 2. Typical face in knowledge-based methods [7]

B. *Feature Invariant Approaches*

Feature invariant approaches are looking for special morphological parts on face which do not change their position while movements, angles of view or lighting conditions. These methods are used mainly for locating faces in image. Feature invariant approaches are described in [7] and they are easier to implementation.

C. *Template Matching Methods*

Methods based on comparing templates – template matching methods need own database in which are described faces and parts of faces. Method is based on correlation between new frame and templates from local database. There is a need to update database with new faces to gain better results and databases require a lot of storage. These methods are easier to implement and they are used mainly for face location but authors in [7] describe these methods also for face detection in image.

D. Appearance-based Methods

Template based methods have own database with faces, on the other side appearance-based methods gain and build folder with faces by learning detected faces from image. There is a training set of faces and these methods learn based on this training set. Appearance-base methods are used mainly for face detection in image. Author described these methods for face detection in [7].

E. Human Detection with History of Oriented Gradients

Method History of Oriented Gradients (HOG) is used by authors in [9][10] for human detection in image. This method is not recommended to use in real-time because of high need of system resources. There are many approaches that describe HOG in combination with other method to detect human in real-time. Author in [10] describes combination of HOG witch optical flow method. Authors in [9] propose using HOG with Kalman filter to use it in real-time.

IV. Proposed Method for Motion Detection and Human Recognition

This section of paper describes method used for software motion detection and human recognition in image sequence. Motion detection is based on analyzing each of pixels from reference image and comparing it with new image. In this paper authors use microcontroller Intel Galileo like a hardware solution.

A. Hardware Part of System

Hardware implementation is based on a development board. Authors proposed to use development board from Intel named Intel Galileo Gen 1. The device communicates with camera via IP protocol. IP camera is powered with POE 802.3af which makes it more able to move. In this implementation is used IP camera Vivotek FD7131 with resolution 640x480.

Development board Intel Galileo supports operating system Linux. Operating system is located on MicroSD card which is inserted in slot in development board. MicroSD is formatted on FAT or FAT32 and with maximum capacity of 32 GB.

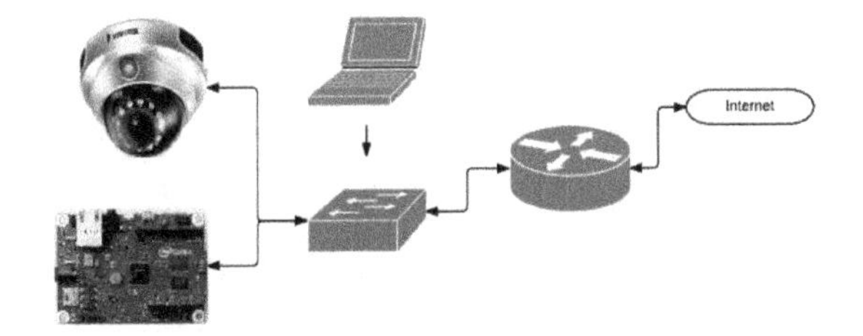

Fig. 3. Hardware part of system for motion detection and human recognition

Fig. 3 depicts the hardware implementation of the proposed system. A computer device is used to control software and to start the program. This solution is designed to be accessible from the Internet.

B. Overall Software Implementation of Proposed System

The implementation part is based on analysis of methods for motion detection and human recognition in picture. Authors based on analysis carry out the implementation, which is composed of two parts: software motion detection and software human recognition. The main program has to be compiled directly on the device Intel Galileo and the program is written in programming language C. The program does not use any other libraries, only standard libraries from C. Software implementation is compiled on device using GNU compiler. The software part is divided to smaller following parts:

1) *Installation of operating system on development board Intel Galileo using MicroSD card.*
2) *Configuration of IP camera and test connection between IP camera and development board.*
3) *Daemonize program to run in system like a process.*
4) *Software motion detection based on comparing two following frames.*
5) *Software human recognition based on analyzing frames.*
6) *Create web interface to show results from program and gallery with pictures with positive detection.*
7) *Automation of solution.*

C. Implementation of Main Program

The main program has multiple parts and together they create complete solution. The main part is program to motion detection and human recognition implemented in C and compiled on development board. The main program has to serve to different parts which are:

1) *Download first frame from camera which become reference image.*
2) *Program has cycle in which are downloaded images from camera. Program has to have access to two images in one moment. All images are saved on development board.*
3) *Program compare new image with reference image and all pixels are being compared.*
4) *If there is detected motion then system tries to recognize persons in image. If there is a person in image then image is saved and shown in gallery with positives detections.*
5) *Program is running as a daemon in operating system and results from motion detection and human recognition are logged in text file.*

The program is running in cycle. There are two cycles. The first one is for downloading actual image from camera. The second one is used for motion detection and human recognition if needed. The main program has two parts: motion detection and human recognition.

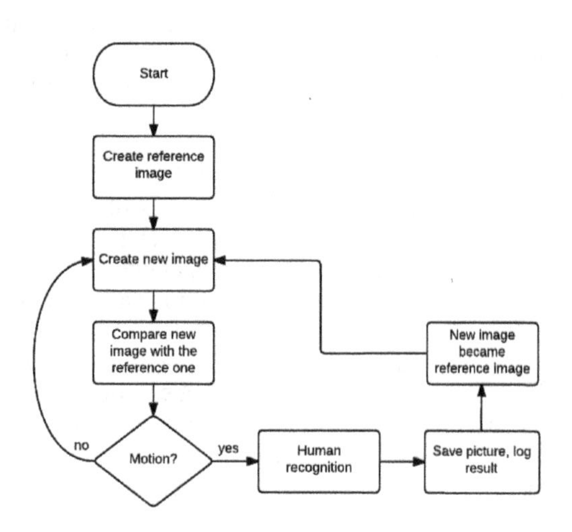

Fig. 4. Lifecycle of proposed program

D. Proposed Function for Motion Detection in the Two Following Video Frames

One of two major functions of the main program is motion detection. Reference image is created after program starts and it is converted from JPEG in to BMP format in which program can works witch pixels color scales. New image from camera is downloaded in cycle. Motion detection is implemented like comparison of color scale from each pixel in following image frames. Pixel has changes if value of one of color part – red, blue or green has been changed. Value of sensitivity can be set on the start of the program. Right settings for sensitivity are very important for true motion detection in the picture. If value for sensitivity is too low, there are a lot of fake pixels detected like in a motion. On the other side, if the value is too high, pixels for detection must be changed more. One pixel has been changed if every color part has been changed in comparison with pixel from reference image in the same location. Motion is detected if there is a number of changed pixels in interval for motion detection. And based on previous function, the next function tries if there is a number of changed pixels which indicates motion detection in two following frames.

E. Proposed Function for Human Recognition in Image

The second function in the main program is for human recognition in picture. Human is detecting in image only if the result of motion detection is positive, on the other side, program continues with motion detection on the next image. Human recognition works on principles that the program has location of changed pictures and their count. Changed pixels must be next to themselves. The human silhouette is described in invariant parameters and program tests if the square with pixels is in shape of person. Based on testing, program knows how many pixels on image create one silhouette. Program tries to eliminate object which are not persons with different restrictions for pixels location. Silhouette is described like a square m x n. The size of square was determined by testing in different locations and in different poses. Results of this method are logged into text file with timestamp and number of counted persons. The modified output from program is shown in the Fig. 5 with marked persons who were identified by method to human recognition.

Fig. 5. Modified output with marked persons

V. Results and Conclusion

The aim of this paper was the implementation of the system for motion detection and human recognition in video frames using development board. The system proposed by authors of this paper satisfies requirements. Development board Intel Galileo and video source from IP camera are used as a basis for the hardware implementation. Operating system Linux on the Intel Galileo development board is a good solution. However, there are not enough system resources to implement better methods for motion detection and human recognition. Authors in this implementation show the way of realization system for motion detection and human recognition on the development board. Implemented solution was tested in two different scenes to gain better results. In each scene was program tested with three different values for sensitivity to change color value of a pixel. The results show that implemented method for human recognition works. The results from two different scenes summarize the following tables:

TABLE I. Human Recognition Test Results: Scene 1

Scene 1, total images: 1200, detections: 186	Low sensitivity	Medium sensitivity	High sensitivity
True positive	57 / 79%	67 / 85%	26 / 72%
False positive	15 / 21%	11 / 15%	10 / 18%

TABLE II. Human Recognition Test Results: Scene 2

Scene 2, total images: 850, detections: 141	Low sensitivity	Medium sensitivity	High sensitivity
True positive	27 / 60%	46 / 82%	36 / 90%
False positive	18 / 40%	10 / 18%	4 / 10%

Two types of scenes were considered, both indoors, the first scene was a computer networks laboratory. Camera in this case was attached to the ceiling of the room. Sun light had a great impact on the number of false positives. The second room was storeroom, same as for the first scene; camera was attached to the ceiling. Number of true positives was in this case better due to better lightning conditions and lesser complexity of the room. For both cases, presence of artificial light enhanced the overall results.

In the future, we plan to eliminate just stated issue with nature light and enable utilization of system in real conditions. Implementation of a system for tracking of people in the laboratories of Technical university of Kosice is also considered, the solution itself has already been implemented as a part of the surveillance system at the university's dormitory.

ACKNOWLEDGMENT

We support research activities in Slovakia/This project is being co-financed by the European Union. Paper is the result of the Project implementation: University Science Park TECHNICOM for Innovation Applications Supported by Knowledge Technology, ITMS: 26220220182, supported by the Research & Development Operational Programme funded by the ERDF.

REFERENCES

[1] P. Labaš, "Podklady na prijímacie skúšky z chémie", Bratislava: ELÁN, 2013, 343 p.

[2] V. Bada, "Základy klinickej elektrokardiografie", Bratislava: UK, 2010, 124 p.

[3] J.R. Hampton, "EKG stručně, jasně, přehledně", Praha 7: Grada Publishing, a.s., 2013, 192 p.

[4] M. Pytliak, "Základy kardiografie pre štúdium ošetrovateľstva a ostatné zdravotnícke študijné programy", Košice: Univerzita Pavla Jozefa Šafárika v Košiciach, 2009, 134 p.

[5] Arduino Uno [online] available at: http://arduino.cc/en/Main/arduinoBoardUno

[6] Co je to Arduino [online] available at: http://www.czechduino.cz/?co-je-to-arduino,29

[7] Vývojové prostredie a programovanie Arduina [online] available at: http://uart.cz/90/ide-a-programovani-arduina/

Using Mining Predict Relationships on the Social Media Network: Facebook (FB)

Dr. Mamta Madan
Professor
Vivekananda Institute of Professional Studies
GGSIP University,India

Meenu Chopra
Assistant Professor
Vivekananda Institute of Professional Studies
GGSIP University,India

***Abstract*—The objective of this paper is to study on the most famous social networking site Facebook and other online social media networks (OSMNs) based on the notion of relationship or friendship. This paper discussed the methodology which can used to conduct the analysis of the social network Facebook (FB) and also define the framework of the Web Mining platform. Lastly, various technological challenges were explored which were lying under the task of extracting information from FB and discuss in detail the about *crawling agent* functionality.**

Keywords—Online Social Media Networks (OSMNs); Facebook (FB); Data Mining; Crawling Process; Protocol

I. INTRODUCTION

The web mining architecture called as crawler agent, that allow us to pull out the various different specimens of the popularly known, SNS (social networking site) Facebook and to study the network topology anatomy of the above social network graph. To be more concise, the two main techniques of OSMN (online social media network) are, the first one based on the idea of visual extraction (called as uniform sampling based on rejection policy without bias) and the second one based on sampling procedure (called as Breadth-first Search or Traversal having bias).

II. BACKGROUND AND RELATED WORK

The process of mining and analyzing data from OSMNs has attracted many researchers from the world wide [1] [2] [3]. Our focus is to discuss the techniques which are used to crawl huge and complex social networks and extract the data from them. Then this collected data is mapped with the graph data structures with the aim of understanding their structural traits. Kleinberg [4], laid the foundation for all efforts, by indicating that the geographical properties of social graphs may be the trustworthy indicators of user's behaviors. The spectrum of targeted research queries arising from the analysis of OSMNs is unlimited. But for our research paper is focusing on the three important themes which are as follows:

A. OSMNs Dataset

The task of extracting relevant data from Web mining Platforms by means of OSMNs web extraction techniques. Since OSMNs Datasets resides in back-end servers and are not available publicly, so they are accessible only through Web interface. The research done on the friendship graph of the FB by Gjoka et al. [5] using many visiting algorithm for example (Random Walk or BFS) with the aim to produce a uniform sample of the FB graph. Our focus in this paper is to creep the little part of the social network graph like FB and to figure out the structural characteristics of the crawled data. In [6], researchers crawled data from the complex SMNs like Live Journal, Flickr and Orkut.

B. Uniform Node Detection (UND)

The task of acquiring the extent of uniformity of two nodes or users in SM graphs. Finding users of common properties and also to calculate their uniformity is by means of Jaccard coefficient similarity metrics on the sets of their neighbors [7].But the disadvantage of this coefficient is firstly, not taking global information into consideration, secondly, it showed the similarity between nodes even if nothing real similarity exists between them because of the fact that nodes having high number of acquaintances would have high probability of sharing. In [8], authors suggested uniformity between two users increases, if one user exchanges acquaintances with another who have less number of acquaintances. Many other methods have explored in this like *Regular Equivalence* (two nodes are uniform or similar if they have uniform acquaintances too), in [9] authors, used the approaches Katz coefficient, Simrank [10], provides a method on iterative fix point, where in [11], researchers, have given the nodes uniformity as optimization problem and in [12], they worked upon directed graphs and exploited an iterative approach.

The other approaches for the node similarity in social media network analysis are *Formal Concept Analysis* (it depends upon the formal relationship between nodes and then calculate the nodes similarity which is hard to compute because it rely on the concept of number of common friend between the nodes) and *Singular value Decomposition (SVD)*[13] which used a technique from Linear Algebra and able to compute the uniformity degree of two nodes even if number of friendship relationship they share is less or close to zero.

C. Effective User Detection (EUD)

The process of discovering users having potential of charging others users to participate discussions/events/activities in their network. Few algorithms being designed for blog analysis such as HITS algorithm[14], Random Walk technique to search for initiators, HP Labs researchers [15], used Twitter to analyze behavior of the users in a network, in [16] authors found the concept of initiator i.e. user who starts the conversation in the network and last but not least in [17], authors recommended a model which represent blogosphere as a graph and consist of nodes and edges where former represent the bloggers and later represents the blogger cites.

III. Exploring the Graph Structure of FB

As of March, 2014 (the data is collected) Facebook1 has 802 million (Daily), 1.28 billion (monthly) active users, 609 million (daily) and 1.01 billion (monthly) mobile active users. Approximately 81.2% of our daily active users are outside the U.S. and Canada. Our interest in exploiting the characteristics and the properties of this social network on a wide-scale. To achieve this goal, first is to collect the data from this online platform and then perform the analysis on it.

A. The Structure of the Online Social Network

The network layout of FB is simple. Every node is connected to each other by a relation called friendship. The social network graph is called as unimodal because it doesn't follow any hierarchy whereas friendship is called as bilateral reason being the relationship confirms among them. This FB graph is represented by G= (V, E): where V->End Users: E-> Edges (relationship). The graph is having two features, firstly, unweighted (Because within the network all the relationships have same value) and secondly undirected graph. In [18] adopted this kind of model for FB social network which has no loops simple unweighted undirected graph. In contrast to FB, the configuration or structure of other online social networks is more complex. For e.g. Nobii [19], YouTube and Flickr [20]. Twitter represents a multiplex directed network reason being it represent different types of relationships among users like "mention", "reply to" ,"following" etc.

This paper tries to explore the two things Firstly, Network Structural Information Retrieval Process of the FB network, secondly, FB data extraction process.

B. How to Retrieve the Structural Information of the FB

Various options are available to extract the information about the structure of FB, like one of option is acquire the data directly from the social networking company, which is not viable solution. Another option is acquire the data, directly from the platform itself, which is needed to reconstruct the model of the network; actually, we could take the representative sample of the social network, which further predicts its structure. Using various web mining techniques, this solution is viable, but the drawback of this option is that, large computational overhead of a large and complex Web Mining task. Moreover, network is not static; it is evolving, so its structure keeps on changing every time, because of this dynamism property of the network, the resultant sample would be a snapshot of the structure of the graph only at the time of data collection process.

There are many different data sampling algorithms that can be used for above mention task, but for our paper we zero down to only two approaches discuss in Table 1, firstly, "Breadth-First-Sreach (BFS) (Biased Approach)" and secondly, "Uniform (Un-Biased Approach)". Following are the characteristics of the above mention sampling algorithms.

TABLE I. Types of Approaches for Fetching Structural Information Extraction

Attributes	BFS Algorithm	Uniform Sampling Algorithm
1. Definition	Uninformed Traversal	Rejection-based Sampling
2. Advantages	• Easy to implement • Efficient • Optimal solution for un-weighted graphs [25,26,63,28,277, 27]	• Easily estimate the probability of a user by statistically[1,6] • To fetch the desired dimension of a sample, we randomly generate no. of User-Ids.
3. Hypothesis	Produces Biased Data towards high degree nodes [24]	Unbiased and Comparable Sample
4. Description	• User-Id's maintained in FIFO queue. • Time constraint is Adopted	• Parallelize the process of extraction. • User-Ids were stored in different queues.

C. How to Extract the Facebook Data

Once data collected could be used for comparing and analyzing their properties, behavior and quality. The quality parameters on which the collected data samples can be evaluated are: i) Significance with respect to statistical or mathematical models, ii) The quality of agreeing with results with other similar research studies. Because of the privacy and protection of data in FB, Twitter, etc., companies running these social networking services do not shared their data about users [21, 22]. We can access the information through graphical user interface with some technical glitches for example, using an asynchronous script; the friend-list can be crawled. Some of other online services like "Graph API (Application Programming Interface) [2]" etc., provided by FB developers team in 2010 and in by the end of 2011, using the Web data Mining techniques, we can able to access the structure of FB.

IV. The Sampling Framework of FB

Figure 1 depicts the architecture of Web data mining process, which is composed of the following components.

1) A web-server executing Agents for Mining,

2) A Java based platform independent application, which executes the code of the agent,

3) An Apache interface, which controls and manages the flow of information through online network.

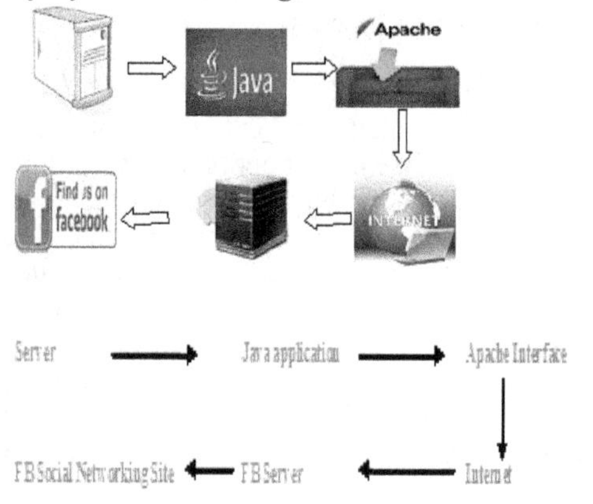

Fig. 1. Topology of the data Mining Platform

While executing, the mining agents inspect the FB server(s) to fetch the list of web pages of the friends connected to the specific requested users, reforming the structure of friendship among them. Finally, the data which has been collected would be stored on the web server and thereafter, goes post-processing task and delivered in an XML-format [23] for further processing.

A. Facebook Crawling Process

Figure 2 shows the architecture of FB Crawler, it is a cross-platform java based agent which actually crawl the GUI of the Facebook (front platform) and also the crucial part of the web data mining process. The given figure 2 below depicts the logic of the java agent, irrespective of the sampling algorithm executed. For the crawling agent execution, which is first preparative step in data mining process, includes two things, firstly chosen sampling algorithm and secondly, setting up some of technical parameters like maximum execution time, existing criteria etc. Therefore, the crawling process can initiate or start from the previous back-step. During the process of execution the java based crawling agent visits the friend-list web page of the requested user, obeying the rules of the selected sampling algorithm directives, for searching the social network or graph. To save I/O operations, all the data about newly discovered nodes and relationships among them are saved in a compact format. Termination of the process of crawling takes place when termination condition met.

Figure 2 shows the flowchart, which depicts the process of HTTP requests flow of the crawler with proper authentication and mining steps. First step, in the data mining process, is the front-end platform uses the Apache HTTP Request Library[3] to have a communication with the FB server(s). Second step, after establishing a secure connection (i.e. an authentication phase) and obtaining "cookies" for logging into the FB platform, finally getting the HTML web pages of the friend-list of the user through HTTP requests. This process is describes in Table 2.

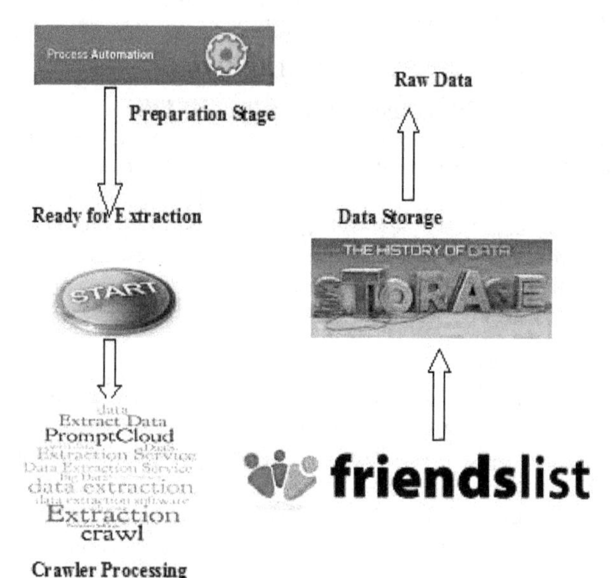

Fig. 2. The Flow diagram of the Data Mining task

The web-crawler has two executing modes:

a) HTTP Request-Based Execution: This mode is faster on large-scale of extraction.

b) Extraction Based on Visual Component: In this crawler embeds a Firefox Browser via XPCOM[4] and XULRunner[5]. The advantage of using this mode is its ability to perform asynchronous requests, for e.g. AJAX scripts but disadvantage of slower execution, time-consuming in rendering the web page.

At last, the paper discuss about the technical constraint imposed by the FB, which has been noticed during data mining task, is the limit of the generated friend-list web pages (which is not above 400 friends), through or via HTTP requests. To decrease network traffic, this limitation is put on, and if friends exceeds by 400, then asynchronous scripts fills the web page, this will led to a non-reproducible crawler or agent based on HTTP requests. This problem can be rectify by using different mining approach, for example use of visual crawler which is less cost effective and not viable for large-scale data mining tasks.

TABLE II. THE MINING AND AUTHENTICATION STEP OF THE CRAWLER VIA HTTP REQUESTS

Action taken		Protocol/ Method	URI
1.	Access the FB page	HTTP/GET	www.facebook.com/
2.	Authentication/Login	HTTPS/POST	Login.facebook.com/login.php
		HTTP/GET	/home.php
3.	Visit Friend-List	HTTP/GET	/friend-list/ajax/friends.php?id =#&filter=afp

V. CONCLUSION

The analysis as well as extraction of data from OSMN is a challenging task. This paper had discussed and explored, the different sampling algorithms that have been implemented to search or examine the social network graph Facebook that consist of countless friend-friend relationships. Out of the two sampling techniques, the visiting technique, BFS is known to deliver biasness in the scenario of incomplete traversal. Lastly, this paper described the random FB crawler agent, which could be used to generate samples of anonymous types. Analysis of these samples, SNA (social network analysis) using *graph theory (nodes and relations), diameter metrics, degree distribution and coefficient of clustering distribution* is the part of future discussion.

VI. WEB REFERENCES

1) *http://www.facebook.com/press/info.php?statistics*
2) *http://developers.facebook.com/docs/api*
3) *http://httpd.apache.org/apreq*
4) *https://developer.mozilla.org/en/xpcom*
5) *https://developer.mozilla.org/en/XULRunner*
6) *http://www.google.com/adplanner/static/top1000/*

REFERENCES

[1] Albert, R., Barabasi, A.: Statistical mechanics of complex networks. Reviews of Modern Physics 74(1), 47-97 (2002).

[2] Garton, L., Haythornthwaite, C., Wellman, and B.: Studying online social networks. Journal of Computer-Mediated Communication 3(1) (1997).

[3] Ye, S., Lang, J., Wu, and F.: Crawling Online Social Graphs. In: Proc. of the 12th International Asia-Pacific Web Conference, pp. 236-242. IEEE (2010).

[4] Kleinberg, J.: The small-world phenomenon: an algorithm perspective. In: Proc. of the 32nd annual symposium on Theory of computing, pp. 163-170. ACM (2000).

[5] Gjoka, M., Kurant, M., Butts, C., Markopoulou, and A.: Walking in Facebook: a case study of unbiased sampling of OSNs. In: Proc. of the 29th conference on Information communications, pp. 2498-2506. IEEE (2010)

[6] Mislove, A., Marcon, M., Gummadi, K., Druschel, P., Bhattacharjee, and B.: Measurement and analysis of online social networks. In: Proc. of the 7th SIGCOMM conference on Internet measurement, pp. 29-42. ACM (2007)

[7] Han, J., Kamber, M., Pei, J.: Data mining: concepts and techniques. Morgan Kaufman Pub (2011)

[8] Adamic, L., Adar, E.: Friends and neighbors on the web. Social networks 25(3), 211-230 (2003)

[9] Blondel, V., Gajardo, A., Heymans, M., Senellart, P., Van Dooren, P.: A measure of similarity between graph vertices: Applications to synonym extraction and web searching.Siam Review pp. 647-666 (2004)

[10] Jeh, G., Widom, and J.: Simrank: a measure of structural-context similarity. In: Proc. Of the 8th SIGKDD international conference on Knowledge discovery and data mining, pp. 538-543. ACM (2002)

[11] Batagelj, V., Doreian, P., Ferligoj, A.: An optimization approach to regular equivalence .Social Networks 14(1-2), 121-135 (1992)

[12] Blondel, V., Gajardo, A., Heymans, M., Senellart, P., Van Dooren, P.: A measure of similarity between graph vertices: Applications to synonym extraction and web searching.Siam Review pp. 647-666 (2004)

[13] Golub, G., Van Loan, C.: Matrix computations, vol. 3. Johns Hopkins University Press (1996)

[14] Kleinberg, J.: Authoritative sources in a hyperlinked environment. Journal of the ACM 46(5), 604-632 (1999)

[15] Romero, D., Galuba, W., Asur, S., Huberman, B.: Influence and passivity in social media. In: Proc. of the 20th International Conference Companion on World Wide Web, pp. 113-114. ACM (2011)

[16] Mathioudakis, M., Koudas, N.: Efficient identification of starters and followers in social media. In: Proc. of the International Conference on Extending Database Technology, pp. 708-719. ACM (2009)

[17] Song, X., Chi, Y., Hino, K., Tseng, B.: Identifying opinion leaders in the blogosphere. In: Proc. of the 16th Conference on Information and Knowledge Management, pp. 971-974. ACM (2007).

[18] Goldenberg, A., Zheng, A., Fienberg, S., Airoldi, E.: A survey of statistical network models. Foundations and Trends in Machine Learning 2(2), 129-233 (2010)

[19] Aiello, L.M., Barrat, A., Cattuto, C., Ruffo, G., Schifanella, R.: Link creation and profile alignment in the aNobii social network. In: Proc. of the 2nd International Conference on Social Computing, pp. 249-256 (2010)

[20] Mislove, A., Marcon, M., Gummadi, K., Druschel, P., Bhattacharjee, B.: Measurement and analysis of online social networks. In: Proc. of the 7th SIGCOMM conference on Internet measurement, pp. 29-42. ACM (2007)

[21] Gross, R., Acquisti, A.: Information revelation and privacy in online social networks. In: Proc. of the Workshop on Privacy in the Electronic Society, pp. 71-80. ACM (2005).

[22] McCown, F., Nelson, M.: What happens when Facebook is gone? In: Proc. of the 9th Joint Conference on Digital Libraries, pp. 251-254. ACM (2009).

[23] Brandes, U., Eiglsperger, M., Herman, I., Himsolt, M., Marshall, M.: GraphML Progress report structural layer proposal. In: Graph Drawing, pp. 109-112. Springer (2002).

[24] Kurant, M., Markopoulou, A., Thiran, P.: On the bias of breadth first search (bfs) and of other graph sampling techniques. In: Proc. of the 22nd International Teletrafic Congress, pp. 1-8 (2010).

[25] Catanese, S., De Meo, P., Ferrara, E., Fiumara, G.: Analyzing the Facebook friendship graph. In: Proc. of the 1st International Workshop on Mining the Future Internet, vol. 685, pp. 14-19 (2010) 4, 52

[26] Catanese, S., De Meo, P., Ferrara, E., Fiumara, G., Provetti, A.: Crawling Facebook for social network analysis purposes. In: Proc. of the International Conference on Web Intelligence, Mining and Semantics, pp. 52:1-52:8. ACM (2011).

[27] D'haeseleer, P.: How does gene expression clustering work? Nature Biotechnology 23(12),1499-1502 (2005).

[28] Mislove, A., Marcon, M., Gummadi, K., Druschel, P., Bhattacharjee, B.: Measurement and analysis of online social networks. In: Proc. of the 7th SIGCOMM conference on Internet measurement, pp. 29-42. ACM (2007).

[29] Wilson, C., Boe, B., Sala, A., Puttaswamy, K., Zhao, B.: User interactions in social networks and their implications. In: Proc. of the 4th European Conference on Computer Systems, pp. 205-218. ACM (2009)

Analysis of Security Protocols using Finite-State Machines

Dania Aljeaid
School of Science and Technology
Nottingham Trent University
Nottingham, United Kingdom

Xiaoqi Ma
School of Science and Technology
Nottingham Trent University
Nottingham, United Kingdom

Caroline Langensiepen
School of Science and Technology
Nottingham Trent University
Nottingham, United Kingdom

***Abstract*—This paper demonstrates a comprehensive analysis method using formal methods such as finite-state machine. First, we describe the modified version of our new protocol and briefly explain the encrypt-then-authenticate mechanism, which is regarded as more a secure mechanism than the one used in our protocol. Then, we use a finite-state verification to study the behaviour of each machine created for each phase of the protocol and examine their behaviours together. Modelling with finite-state machines shows that the modified protocol can function correctly and behave properly even with invalid input or time delay.**

Keywords—identity-based cryptosystem; cryptographic protocols; finite-state machine

I. Introduction

Security protocols are becoming the core subject in communication systems and verifying them has gained significant attention by researchers and developers. Security analysis aims to formally guarantee these protocols to satisfy their specifications and they can function soundly. Security evaluation is a fundamental step in the development of security protocols. The methods used to analyse security protocols can be categorised into two groups: methods based on analytical approach and methods based on simulation. The analytical approach offers accurate results and provides a clear perception of the system characteristics. However, this approach becomes unreliable when dealing with high complex system. Therefore, the latter approach, simulation approach, has become more popular in system analysis. Simulation tools, such as finite-state machines and Petri nets, expose progress in two directions: one related to the development of faster methods during execution of mathematical algorithms [1], and the other associated with the effectiveness simulation presentations and results [2].

Protocol modelling is a crucial step in designing security protocols. It contributes to diminishing ambiguity and misinterpretation of protocol specifications. For example, modelling a protocol using finite-state machine can help to understand how it will interact with the changes and how it will behave with invalid inputs. A Finite-State Machine (FSM) is a powerful tool to simulate software architecture and communication protocols. FSM can only model the control part of a system and consists of a finite number of states, a finite number of events, and a finite number of transitions.

Modelling with finite-state machine helps to understand the behaviour of complex protocol. Also, it offers accurate results and provides a clear perception of the system characteristics. The analysis presented in this paper covers the process of the three-way handshake used to negate the session key and examine the behaviours of the protocol and enumerates all possible states it can reach.

The structure of this paper is organised as follows. In Section 2, we briefly review previous works on extended finite-state machine and briefly discuss the weakness in our new protocol and present modified version of it. In Section 3, we model the modified protocol using EFSM. We then provide a brief discussion on security analysis in Section 4. Finally, the conclusions are given in Section 5.

II. Review of Related Work

A. Extended Finite-State Machines

In order to model the complex behaviour of the proposed protocol, an extended model of FSM is considered. According to [3], EFSM helps to comprehend the *state space* complexity of a system when the number of states and transitions increases Also, they emphasise the importance of introducing *state variables* in FSM models. State variable play a key role in modelling because they can "define a range of arithmetic and logical operators to manipulate state variables and trigger transitions based on logical primitives" [3]. Moreover, EFSM with variables can transfer variable values from one model to another. Consequently, the produced output value from one machine can be consumed by other machines. With the introduction of variables, EFSM allows one to model a system with conditions. Transitions may have guards and predicates, which consist of operations or Boolean-valued expressions that can depend on input variables [3].

A formal definition of an EFSM is as follows [3, 4]:

An Extended Finite State Machine (EFSM) M is a tuple (S, T, E, V) where,

S is a set of states,
T is a set of transitions,
E is a set of events, and
V is a store represented by a set of variables.

Transitions have a source state $source(t) \in S$, a target state $target(t) \in S$ and a label $lbl(t)$. Transition labels are of the form $e1[c]/a$ where $e1 \in E$, c is a condition and a is a sequence of actions.

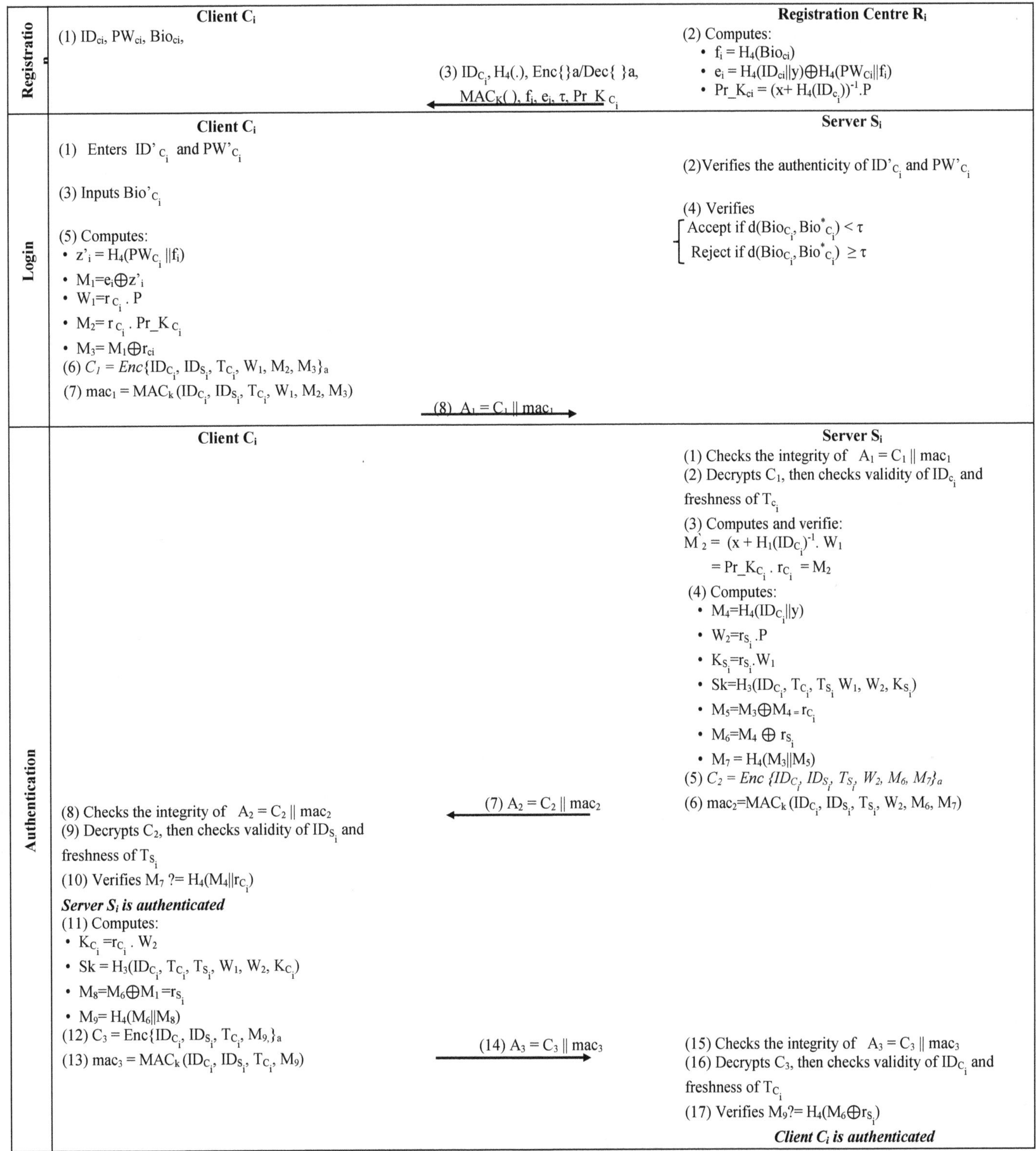

Fig. 1. The modified proposed protocol

B. Review of Proposed Protocol

In our previous work [5], we have developed a new authentication protocol that allows remote mutual authentication with key agreement. Our new protocol is based on biometric verification and ID-based Cryptograph. However, it is not secure against chosen-ciphertext attacks. The new protocol needs modifications to initiate secure authentication between the client and server.

The modified version of the proposed protocol should improve security and provide users with better authentication and data confidentiality. To address and correct the perceived security weakness in the proposed protocol, authenticating the ciphertext by applying encrypt-then-authenticate mechanism is considered to be one of the secure methods for security protocols. The previous message exchange in the proposed protocol was constructed like this:

Encrypt (Message || MAC)

The new modification for the message exchange will be constructed as this [7,8]:

Encrypt (Message) || MAC

This way the MAC is covering the entire ciphertext to preserve the integrity of the cipher message. The MAC value is then appended to the encrypted message. When the recipient receives the authenticated encrypted message, the MAC should be evaluated before attempting to decrypt the ciphertext. If the MAC verification fails, the recipient will terminate the session immediately. This process will be efficient by eliminating the time spent to going through the manipulated data. The enhancements for the proposed protocol will only affect part of the registration phase and the authentication and key agreement phase. Additionally, enclosing the identity of the server along with the client's identity can mitigate the impact of masquerading attack. The ID's of entities must be unique in the network. Thus, the entities that wish to communicate are aware of each other. The modified protocol is summarised in Fig. 1. Based on the investigation above, we need to modify the state machine described in [5,6] according to the new enhancements.

IV. Protocol Model and State Machine

The EFSM is used to model the communication channel of the proposed protocol between the Client C_i and the Server S_i. Since the exchange of packets follows a pattern defined by a finite set of rules, each principal in the protocol has a corresponding state machine: $EFSM_{server}$, $EFSM_{register}$ and $EFSM_{client}$.

A. Verifier EFSM

The EFSMverifier is an embedded machine within $EFSM_{client}$ and $EFSM_{server}$ where states themselves can have other machines. To be precise, it is a set of sub-states that are integrated as a nested finite state machine which are inside the states S5 and S6 in $EFSM_{server}$ and state C6 in the $EFSM_{client}$ .It is only activated when the authentication and key agreement have started. The $FSM_{verifier}$ is triggered when it obtains authentication information from FSM_{client} or FSM_{server}. It represents various transitions during the authentication and validation process. This machine is modelled using 5 states and 8 transitions. Table 1 describes the transitions specifications and Fig.2 illustrates the verifier modelled by EFSM.

- State V0: this state accepts the authentication information that needs to be verified and sends an authenticity-checking request to V1.
- State V1: the EFSMverifier verifies the integrity of the received cipher message by recalculating the MAC value of the received message and comparing it with the attached MAC value. If the MAC values appeared to be identical, the machine triggers itself to the next state, V2, since the condition is fulfilled. However, if the comparison shows a different result, this would trigger to invalid state that then leads to termination.
- State V2: while in this state, EFSMverifier decrypts the ciphertext since MAC integrity check has been successful. After decryption is successful, the EFSMverifier transitions to the state V3.
- State V3: the $EFSM_{verifier}$ checks the freshness of T via $T^{`} - T_{C_i} \leq \Delta T$. If the freshness is valid, the $EFSM_{verifier}$ triggers itself to the next state. Otherwise, it produces invalid input if the freshness of $T^{`} - T_{C_i} \geq \Delta T$ and changes to state V0.
- State V4: while in state V4, the $EFSM_{verifier}$ checks the validity of ID and based on the result it changes to state V0 either with event of valid ID or invalid ID.

TABLE I. The Transitions Specification of the Verifier EFSM

Transition	Transition Direction	Guards/Condition
Validate	C5 → V0 S5 → V0 S6 → V0	
Authenticity check	V0 → V1	
Invalid	V1 → V0	Client_MAC != Server_MAC
Integrity checked	V1 → V2	Client_MAC== Server_MAC
Decrypted the ciphertext	V2 → V3	
Freshness checked	V3 → V4	$T^{`} - T_{C_i} \leq \Delta T$
Invalid	V3 → V0	$T^{`} - T_{C_i} > \Delta T$
ID valid	V4 → V0	
ID invalid	V4 → V0	Invalid ID

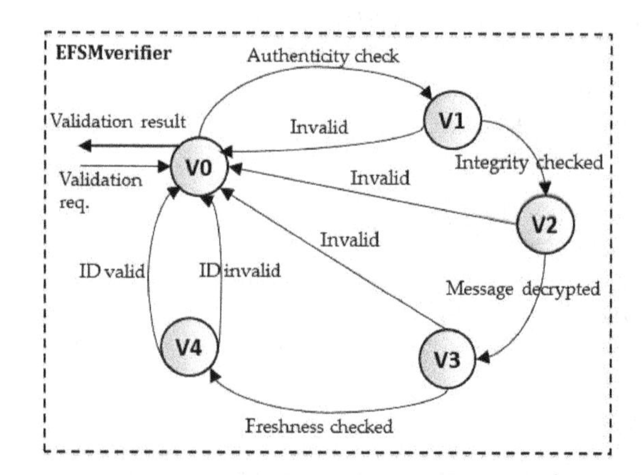

Fig. 2. The verifier machine modelled by EFSM

B. Server EFSM

The FSM at the server side represents the various on-going communications with the client at any point in time. It is modelled using 10 states and 24 transitions and one nested EFSM as detailed below. Table 2 describes the transitions specifications Fig. 3 shows the transitions diagram for the $EFSM_{server}$.

1) The $EFSM_{server}$ will loop continuously while the server is waiting for clients. The machine advances to the next state once it is triggered by a login/enrol transition.

2) When the $EFSM_{server}$ is in the state S1, it checks the validity of the received ID. If ID is proved to be incorrect, S_i will request C_i to enter the valid ID for three times and $EFSM_{server}$ will loop until C_i enters the valid ID up to three times. In the latter case, the C_i's account will be blocked and $EFSM_{server}$ will change to state S4 from state S1. Generally, three attempts are made through our protocol steps to allow common errors.

3) When the $EFSM_{server}$ is in the state S2, it is triggered by a valid ID and it is now waiting for a valid PW. Once S_i receives PW, it verifies the validity of PW. If PW is proved to be invalid, S_i will request C_i to enter the valid PW for three times and $EFSM_{server}$ will loop until C_i enters the valid PW or if the attempts exceed three times. In the latter case, the C_i's account will be blocked and $EFSM_{server}$ changes state to S4 from state S2.

TABLE II. THE TRANSITIONS SPECIFICATION OF THE SERVER-SIDE EFSM

Transition	Transition Direction	Guards/Condition
Waiting for clients	S0 → S0	
Request to enrol	S0 → R0	ClientEnrol == True
Client is registered	S0 → S1 R0→S0	ClientReg == True
Enter ID	S0 → S1	ID Valid
Enter Password	S1 → S2	Password Valid
Submit Biometric	S2 → S3	Biometric Valid
Request client login (SYN received)	S3 → S5	
Re-enter ID/Password/ Biometric	S2→S2 S3→S3 S4→S4	ID_attempt < 3, ID_attempt = ID_attempt +1 PW_attempt < 3, PW_attempt = PW_attempt +1 Bio_Attempt == < 3, Bio_attempt = Bio_attempt +1
Invalid ID/Password/Biometric	S2→S4 S3→S4 S4→S4 S5→S4 S6→S4	ID_attempt == 3 PW_attempt == 3 Bio_Attempt == 3 Invalid ID
Send SYN/ACK and C2	S5→S6	Validation check is valid
Client ACK and C3 received	S6→S7	Validation check is valid
Terminate	S5→S8 S6→S8	
Timeout	S1→ S0 S2→S0 S3→S0	

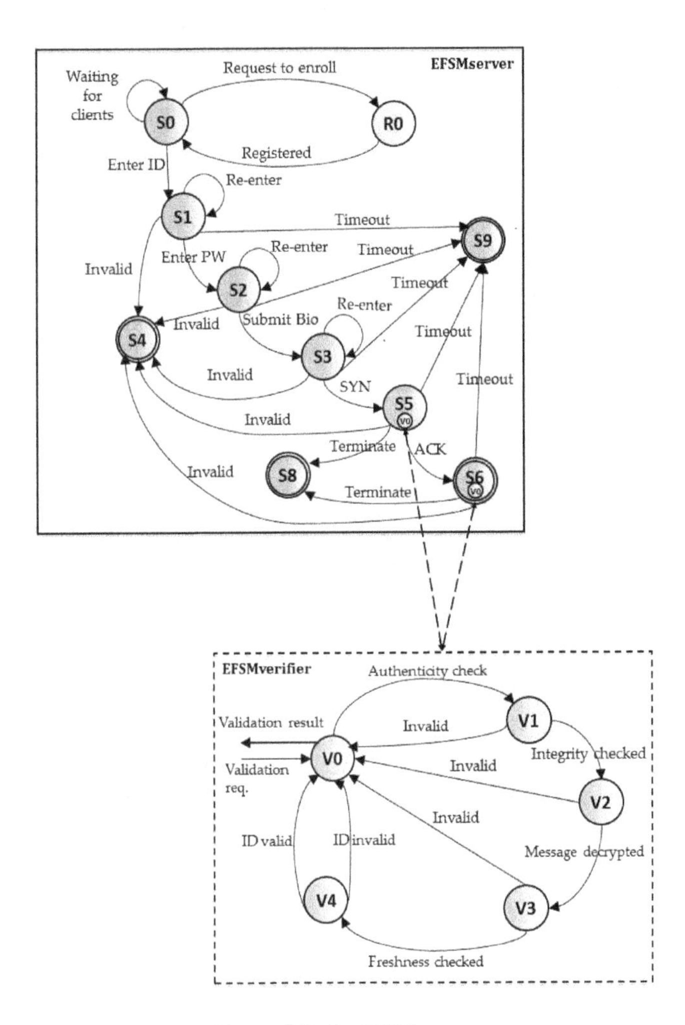

Fig. 3. The server machine modelled by EFSM

4) When the $EFSM_{server}$ is in the state S3, it is triggered by a valid PW and it is now waiting for a valid Bio. Once S_i receives Bio, it verifies the validity of Bio by comparing the imprinted Bio with the template stored. If Bio does not match the stored template, S_i will request C_i to enter the valid Bio up to three times and the $EFSM_{server}$ will loop until C_i enters the valid Bio or if the attempts exceed three times. In the latter case, the C_i's account will be blocked and the $EFSM_{server}$ changes state to S4 from state S3.

5) In state S5, the $EFSM_{server}$ waits until it receives the login request $SYN = A_1 = C_1 \| mac_1$ from the FSM_{client} to establish a connection by performing three-way handshake.

6) While in state S5, the $EFSM_{server}$ activates the nested $EFSM_{verifier}$ and waits for the validation check result.

7) Once the validation has proved to be true. S_i generates a random number and timestamp, then S_i replies with authenticated $SYN/ACK = A_2 = C_2 \| mac_2$ to the $EFSM_{client}$, which is a combination of $C_2 = Enc\ \{ID_{C_i}, ID_{S_i}, T_{S_i}, W_2, M_6, M_7\}_a$ and $Mac_2 = MAC_k(ID_{C_i}, ID_{S_i}, T_{S_i}, W_2, M_6, M_7)$.

8) In state S6, $EFSM_{server}$ waits until it receives ACK from the $EFSM_{client}$. Once the authenticated $ACK = A_3 = C_3 \| mac_3$

is received, the $EFSM_{server}$ activates the nested $EFSM_{verifier}$ and waits for the validation check result.

9) Once the validation check is proved to be true, the $EFSM_{server}$ verifies $M_9 \stackrel{?}{=} H_4$ $(M_6 \,||\, r_{S_i})$. At this point, S_i authenticates C_i as a legitimate user.

10) At state S5 and state S6, $EFSM_{server}$ terminates the current session if any of the following situations occurs:

- The client ID is invalid
- The freshness of $T' - T_{C_i} \geq \Delta T$
- A negative result when checking the integrity of mac_1 and mac_3
- $M2 \mathrel{!}= (x + H_1(ID_{C_i})^{-1}.\ W_1$
- $M9 \mathrel{!}= H_4\ (M_6 \,||\, r_{S_i})$

At any stage of $EFSM_{server}$ activity, $EFSM_{server}$ aborts the current session and changes to state S9 if the timeout exceeds the defined TIME_WAIT while waiting for packets. This feature helps to prevent an infinite wait when the $EFSM_{client}$ fails to respond.

C. Client EFSM

The EFSM at the client side represents the various on-going transmissions with the server at any point in time. It is modelled using 9 states, 22 transitions, and one nested EFSM as detailed below. Fig. 4 shows the transition diagram for the $EFSM_{client}$ and Table 3 describes the transitions specifications.

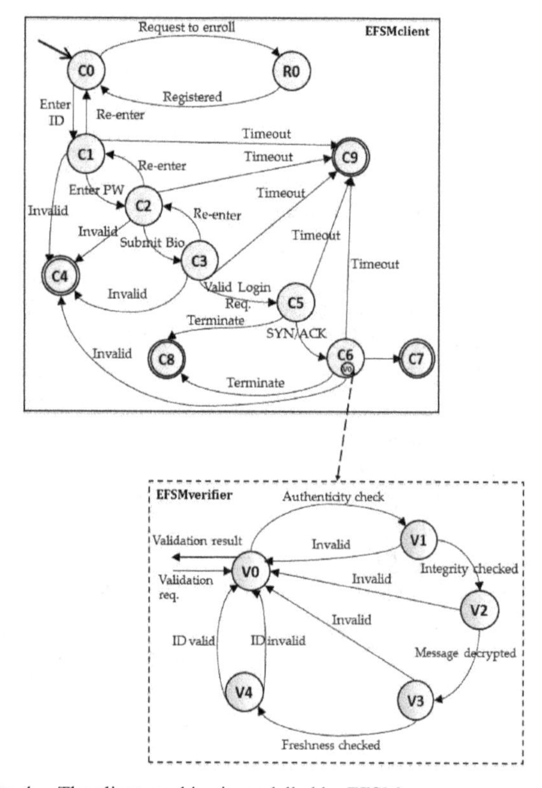

Fig. 4. The client machine is modelled by EFSM

TABLE III. THE TRANSITIONS SPECIFICATION OF THE CLIENT-SIDE EFSM

Transition	Transition Direction	Guards/Condition
Request to enrol	C0 → R0	ClientEnrol == True
Client is registered / Enter ID	C0 → C1	ClientReg == True
Enter Password	C1 → C2	ID valid
Submit Biometric	C2 → C3	Password valid
Send login request SYN (C_1)	C3 → C5	Biometric valid
Re-enter ID/Password/ Biometric	C1→C1	ID_attempt < 6, ID_attempt = ID_attempt +1
	C2→C2	PW_attempt < 3, PW_attempt = PW_attempt +1
	C3→C3	Bio_Attempt < 3, Bio_attempt = Bio_attempt +1
Invalid ID/Password/Biometric	C1→C4 C2→C4 C3→C4	ID_attempt == 6 PW_attempt == 3 Bio_Attempt == 3
Client receives SYN/ACK (C2)	C5→C6	
Client sends ACK (C3)	C6→C7	Validation check is valid
Authenticated by server	C7→C8	
Terminate	C5→C8 C6→C8	
Timeout	C1→C0 C2→C0 C3→C0	

1) First, the $EFSM_{client}$ is in the initial state C0. That is when the request for register/login is initiated by itself. While in state C0, the $EFSM_{server}$ checks whether C_i is enrolled or not. The next state will be determined according to the condition ClientReg == True.

2) In state C1, C2, C3, the FSM_{client} is waiting for validating ID, PW, and Bio. Once the client credentials are validated, the $EFSM_{client}$ triggers itself and changes to state C5.

3) In states C1, C2, C3, the client may be required to re-enter ID, PW, Bio in cases where they were incorrect. However, the client's account will be blocked if the number of attempts exceeds three, which changes the above states to state C4.

- ID_attempt < 3, ID_attempt = ID_attempt +1
- PW_attempt < 3, PW_attempt = PW_attempt +1
- Bio_Attempt < 3, Bio_attempt = Bio_attempt +1

4) In state C5, The $EFSM_{client}$ generates a random number and a timestamp to calculate the encrypted login request $\{ID_{C_i}, ID_{S_i}, T_{C_i}, W_1, M_2, M_3\}_a$ and then computes $mac_1 = MAC_k\,(ID_{C_i}, ID_{S_i}, T_{C_i}, W_1, M_2, M_3)$. It sends $A_1 = C_1 \,||\, mac_1$ to the $EFSM_{server}$. This request represents the SYN part in the three-way handshake procedure.

5) While in state C5, the FSM_{client} is waiting for the $EFSM_{server}$ to respond after sending the login request to establish the connection. Once the authenticated SYN/ACK = $A_2 = C_2 \,||\, mac_2$ is received, the FSM_{client} changes to state C6.

6) In state C6, The $EFSM_{client}$ activates the nested $EFSM_{verifier}$ and waits for the validation check result. Once the validation check is proved to be true, the $EFSM_{client}$ is validating the $EFSM_{server}$ response $M_7 \stackrel{?}{=} H_4 (M_4 \,||\, r_{C_i})$. If S_i is proved to be honest, C_i authenticates S_i at this stage.

7) While in state C6, the $EFSM_{client}$ computes the shared session key $sk = H_3(ID_{C_i}, T_{C_i}, T_{S_i}, W_1, W_2, K_{C_i})$ and finalises the handshake procedure by sending authenticated encrypted ACK = $A_3 = C_3 \,||\, mac_3$ to S_i, which is a combination of $C_3 = Enc\{ID_{C_i}, ID_{S_i}, T_{C_i}, M_9\}_a$ and $Mac_3 = MAC_k (ID_{C_i}, ID_{S_i}, T_{C_i}, M_9)$.

8) In state C7, the $EFSM_{client}$ is waiting to be authenticated by S_i.

9) In state C8, the client terminates the current session if one of the following occurs:

- Negative result when checking the integrity of mac_2
- $T' - T_{S_i} \geq \Delta T$
- The server ID is invalid
- $M_7 \neq H_4 (M_4 \,||\, r_{C_i})$

D. Register EFSM

The EFSM at the registration side represents the various on-going transmissions with the server and client at any point in time. It is modelled using EFSM with 4 states and 7 transitions. Fig. 5 shows the states and transitions diagram for the $EFSM_{register}$.

1) First, the $EFSM_{register}$ is triggered if the client is not enrolled at state R0. That is when the request for registration is initiated by $EFSM_{client}$. While in state C0, the $EFSM_{server}$ checks whether C_i is enrolled or not.

2) Once C_i enters ID, $EFSM_{register}$ changes to state R1 and validates the format of ID. Then $EFSM_{register}$ triggers itself asking C_i to enter PW and changes to state R2.

3) In state R2, on receiving PW for the first time, $EFSM_{register}$ requires C_i to re-enter PW for confirmation. Then it triggers itself and changes to the state R3.

4) In state R3, C_i is required to submit multiple scans of the biometric data to increase accuracy. Once the acquisition process is complete, $EFSM_{register}$ triggers itself and sends a message to $EFSM_{register}$, which indicates that the enrolment is successful.

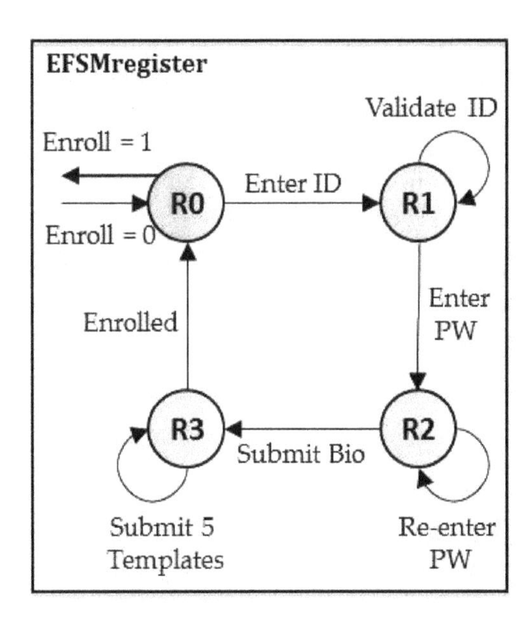

Fig. 5. The client machine is modelled by EFSM

III. Security Analysis

The capability to detect errors and vulnerability is substantial in protocol design implementation. Since communication protocols are partially specified, the finite state approach provides a flexible way to handle invalid inputs and ambiguous specifications, which are usually unspecified or vague in protocol design. Testing the proposed protocol with FSM helps to verify whether the protocol complies with its specification or not. Modelling with FSM shows that the proposed protocol can function correctly and behave properly even with invalid input or time delay.

The state machine in Fig. 6 represents the result of combing the three machines together. The composite model executes efficiently and handles errors in a safe way and it performs certain actions in case of unreliable state. Each valid and reachable state generates a valid protocol state and the transitions can be triggered by either events or guards. Based on the equivalent behaviour, each machine may follow nondeterministic behaviour and produce different outputs according to the original input. For example, if $EFSM_{client}$ generates an illogical input for the authentication process then $EFSM_{client}$ rejects the session and goes to the *terminate state.* Predicating and considering all possible combinations of both desirable and undesirable states are one mean to fully understand the complexity of the proposed protocol.

Note that the states S9 and C9 are defined in terms of a timeout being reached with an inability to complete the mutual authentication.

The states S4 and C4 are defined in terms of an invalid input being injected due to invalid ID, wrong password, or unmatched biometric. The states S8 and C8 are defined in the case of unreliable actions being performed for example, if the integrity or validity check failed. Furthermore, a state machine hierarchy or hierarchical FSM is used to provide a more concrete level of refinement; $FSM_{register}$ can be refined by introducing an "Enrol" feature. This state determines if the client is pre-enrolled or not. The state R0 becomes a new EFSM with three states R1, R2, R3 as described previously.

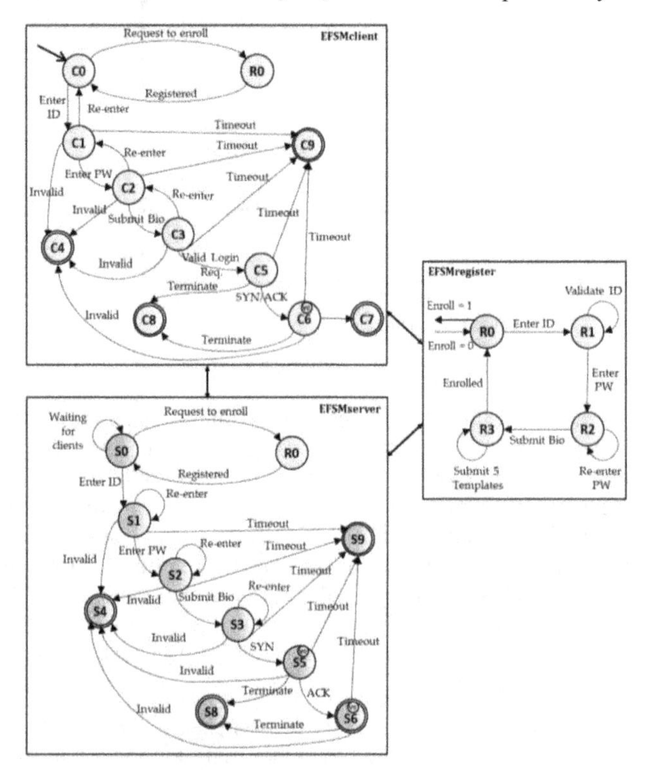

Fig. 6. The modified protocol modelled by EFSM

Based on the parallel behaviour, each machine goes through stages until it reaches the final state. For example, after successful authorisation, the $EFSM_{client}$ switches to the authorised state and proceeds to reach the next state, which is authentication. This comprehensive analysis distinguishes three types of errors that can be detected the protocol run:

- Type I: Timeout errors

This error occurs when the waiting time exceeds the predefined time interval or it occurs when the freshness check exceeds ΔT.

- Type II: Invalid errors

This error is generated in case of invalid inputs, for example, invalid ID, invalid password, or invalid biometric.

- Type III: Terminate error

This error detects if something suspicious occurs in cases where the values did not match. A typical example of this error can be found in the integrity check, when the recomputed MAC value does not match the received MAC value. Another example is when there is a discrepancy in the results of the following equations:

- $M_2 \neq (x + H_1(ID_{C_i})^{-1}.\ W_1$
- $M_7 \neq H_4(M_3 \parallel r_{C_i})$
- $M_9 \neq H_4(M_6 \parallel r_{S_i})$

This error can pose serious threat because it would occur if the data has been modified or injected.

IV. CONCLUSIONS

This paper started by giving a brief definition of extended finite state machines (EFSM). Then it elaborates the details of the finite-state verification of the modified protocol and identifies the functionality of each phase. Also, it studies the behaviour of each machine created for each phase and how they interrelate.

The composite model executes efficiently and handles error in a safe way according to their types. The modified protocol connection progresses from one state to another based on the data pertained from the message exchanged. EFSM helps to understand the behaviour of the protocol and logs any unwanted behaviours. This mechanism is very useful for determining the types of errors the protocol experiences during running and it can be useful to later on investigate what causes these errors and learn from them.

In future, an in-depth security analysis and evaluation will be conducted via **P**etri **N**et (PN). PN will be used to simulate the communication patterns between the server and the client as well as to validate the protocol functionality. First, we will model the protocol without an intruder. Then, we will add the intruder to the model and implement a token-passing scheme. At this stage, we will test different attacks, such as impersonation attack, man-in-the-middle attack, and replay attack against the modified protocol and verify the security requirements. After analysis with PN, we will do a comparison between the previous protocol [5] and the modified version of it.

ACKNOWLEDGMENT

This research has been funded by Saudi Arabian Cultural Bureau in London and King Abdul Aziz University in Saudi Arabia.

REFERENCES

[1] Chiola, G. and Ferscha, A., 1993. Distributed simulation of Petri nets. *IEEE Concurrency,* **1**(3), pp. 33-50.

[2] Genter, G., Bogdan, S., Kovacic, Z. and Grubisic, I., 2007. Software tool for modeling, simulation and real-time implementation of Petri net-based supervisors, *Control Applications, 2007. CCA 2007. IEEE International Conference on* 2007, IEEE, pp. 664-669.

[3] Androutsopoulos, K., Clark, D., Harman, M., Li, Z. and Tratt, L., 2009. Control dependence for extended finite state machines. Fundamental Approaches to Software Engineering. Springer, pp. 216-230.

[4] Alagar, V.S., 2011. Specification of software systems. 2nd edn. England: Springer.

[5] Aljeaid, D., Ma, X. and Langensiepen, C., 2014. Biometric identity-based cryptography for e-Government environment, *Science and Information Conference (SAI), 2014* 2014, IEEE, pp. 581-588.

[6] Aljeaid, D., Ma, X. and Langensiepen, C., Modelling and Simulation of a Biometric Identity-Based Cryptography. *International Journal of Advanced Research in Artificial Intelligence (IJARAI),* **3**(10),.

[7] KRAWCZYK, H., 2001. The order of encryption and authentication for protecting communications (or: How secure is SSL?), *Advances in Cryptology—CRYPTO 2001* 2001, Springer, pp. 310-331.

[8] KATZ, J. and LINDELL, Y., Introduction to Modern Cryptography 2007.

Silent Speech Recognition with Arabic and English Words for Vocally Disabled Persons

Sami Nassimi
Electrical and Electronic Engineering
University of Bahrain
Isa Town, Bahrain

Walaa AbuMoghli
Electrical and Electronic Engineering
University of Bahrain
Isa Town, Bahrain

Noora Mohamed
Electrical and Electronic Engineering
University of Bahrain
Isa Town, Bahrain

Mohamed WaleedFakhr
Electrical and Electronic Engineering
University of Bahrain
Isa Town, Bahrain

***Abstract*—This paper presents the results of our research in silent speech recognition (SSR) using Surface Electromyography (sEMG); which is the technology of recording the electric activation potentials of the human articulatory muscles by surface electrodes in order to recognize speech. Though SSR is still in the experimental stage, a number of potential applications seem evident. Persons who have undergone a laryngectomy, or older people for whom speaking requires a substantial effort, would be able to mouth (vocalize) words rather than actually pronouncing them. Our system has been trained with 30 utterances from each of the three subjects we had on a testing vocabulary of 4 phrases, and then tested for 15 new utterances that were not part of the training list. The system achieved an average of 91.11% word accuracy when using Support Vector Machine (SVM) classifier while the base language is English, and an average of 89.44% word accuracy using the Standard Arabic language.**

Keywords*—Surface Electromyography; Support Vector Machine; Hidden Markov Models; Silent Speech Recognition

I. INTRODUCTION

Automatic speech recognition (ASR) is a computer-based speech-to-text process, in which speech is recorded with acoustical microphones by capturing air pressure changes. ASR has now matured to a point where it is successfully deployed in a wide variety of every-day life applications, including telephone based services and speech-driven applications on all sorts of mobile personal digital devices [1]-[2].

Despite this success, speech-driven technologies still face two major challenges: first, recognition performance degrades significantly in the presence of noise. Second, confidential and private communication in public places is difficult due to the clearly audible speech. But most importantly, the performance is poor if the system having any form of speech disabilities [1]-[2].

Coming from a relative experience, elder people suffer a lot while speaking, and talking becomes a very challenging task that they have to face on a daily basis. Also, people who have undergone s laryngectomy which is surgical removal of the larynx due to cancer suffer a lot to communicate with others. These facts have motivated us to investigate the possibility of developing a Silent-Speech Recognition system (SSR) which will be able to recognize phrases that describe the basic needs of a person especially if he's spending most of his time in a care/nursing home.

The proposed approach for our project is by using the surface ElectroMyoGraphy (EMG); which stands for the technique concerned with the recording and analysis of electric signals taken from articulatory muscles using surface electrodes [2-4]. In contrast to many other technologies, EMG is a low cost, non-invasive, and portable technology.

The remainder of this paper is organized as follows: In section 2, we give an overview of previous related works. Section 3 provides a brief introduction about our methodology (sEMG) and presents our data acquisition, and section 4 presents our experiments and results. In section 5, we conclude the paper and propose possible future work.

II. RELATED WORKS

Research in the area of sEMG-based speech recognition has only a short history. Jorgensen et al. [2] investigated the recognition of non audible speech. Their idea is to intercept nervous signal control signals sent to speech muscles using surface EMG electrodes placed on the larynx and sublingual areas below the jaw. Initially, they demonstrated the potential of non-audible speaker dependent isolated word recognition based on the MES with a Neural Network classifier. They reported recognition rates of 92% for six control words and of 73% on an extended vocabulary which additionally contains the ten English digits.

More recently, there have been some serious efforts to enhance EMG based speech recognition and to make it user-independent as well as open vocabulary [3-5].

III. SEMG BASICS

The abbreviation EMG stands for Electro (electric), Myo (muscle), Graphy (writing). ElectroMyoGraphy is a technology that allows to measure and record the electrical activity of the muscles and the nerve cells that control them (motor neurons). The last transmit electrical signals at the movement of the muscle and an EMG translates it into graphs, or numerical data [6].

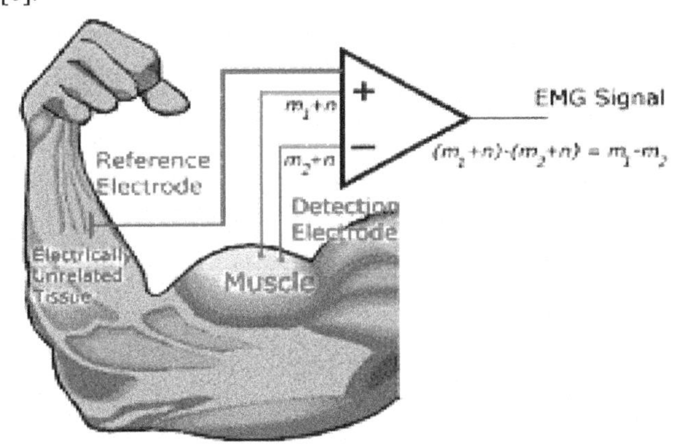

Fig. 1. Measuring an EMG signal

Surface EMG refers to the same process but using surface (non-implemented) electrodes. These electrodes work as a camera that transmits to us the electrical activity of the muscle. Since the amplitude of the signal can range from 0 to 1.5 mV (rms), an amplifier is needed. Amplified electrical signals are then fed into electronic devices for further processing [7]. However, because the EMG signal is based upon action potentials at the muscle membrane; a differential amplifier subtracts the signals from two detection sites and amplifies the difference voltage [8]. Consequently, any signal that originates far away from the detection site will appear as common signal - will have zero output- and thus removed, while the useful information carried out by the EMG signal will be different from both sites and thus amplified.

IV. SEMG SETUP & DATA CORPUS

sEMG Setup:

For sEMG recording, we used the MP system from BIOPAC Systems, Inc. the MP150 system serves as a data acquisition unit that converts analog signals (speech) into digital signals for further processing [9]. The Universal Interface Module UIM100C was used as the main interface between the MP150 and the external devices which for the purpose of our research has been the Electromyogram amplifier. At the early stage of our research, four EMG100C amplifiers have been used to amplify the electrical activity at four different detection sites. The EMG's have been connected to Ag-AgCl lead electrodes that can be directly attached to the skin of the user.

The electrodes positions have been adopted from (Maier-Hein et al., [4,5]). The channels captures signals from the levatorangulisoris (EMG2 & 3), the zygomaticus major (EMG2 & 3), the platysma (EMG4), and the orbicularis (EMG5). Later on, our research focused on investigating the performance of the system using only one channel which has been positioned with classical bipolar configuration with a 2cm center-to-center electrode spacing was used as shown in EMG2. The common ground reference has been connected to the wrist.

For the purpose of impedance reduction at the electrode-skin junction a small amount of electrode gel was applied to each electrode.

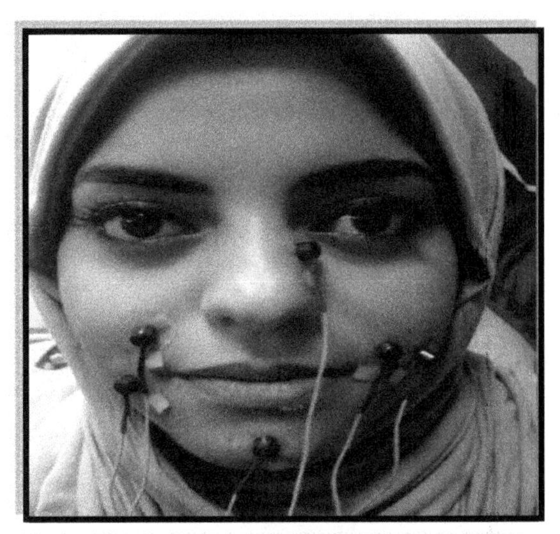

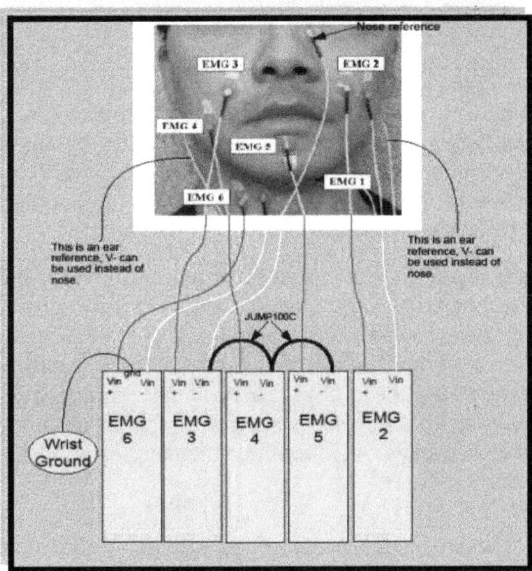

Fig. 2. Electrodes Positions (top: one of our volunteers, below: original connection adopted from (Maier-Hein et al., 2005 [5]).

The gain of the amplifiers has been set to 2000. The usable energy of the signal is limited to 0 to 500 Hz frequency range, thus, 1000Hz has been chosen as our sampling frequency, and a low pass filter with a cut-off frequency at 500Hz was used. To remove motion artifacts, a high-pass filter with cut-off frequency of 10Hz was used. Finally, all signals have been filtered with a notch filter at line frequency of 50 Hz because it is to be considered a dominant source of electrical noise.

A. Data Corpus:

All signal data used for our experiments was collected in so-called *recording sessions*. A recording session is defined as a set of utterances collected in series by one particular speaker. All settings (no. of channels, sampling rate, speech mode) remain constant during all sessions. For our research, three subjects that varied in age, nationality, and thus mother-tongue with no known speech disorders participated to construct our database. In all sessions, the subject has been asked to pronounce the phrases non-audibly, i.e. without producing any sound. In this research, isolated-word recognition was performed. Thus, a word list was selected containing all phrases a speaker need to record during each session. The list can optionally be randomized. The phrases in this list have been chosen carefully to serve the focus of our project which has been elderly people who spend most of their time in a nursing home. The list consisted of the following four phrases:

TABLE I. THE FOUR PHRASES IN ENGLISH AND ARABIC LANGUAGES

The four phrases in English and Arabic Languages	
English language	**Arabic language**
I feel dizzy	أشعرُ بالدّوار
Take me outside	خذني إلى الخارج
I want to go to the toilet	أريد الذهاب إلى الحمام
I need water	أريدُ ماءً

From each subject, a total of forty five utterances have been collected making sure that each is not bounded by any silence at the beginning or end of it. The subject would start recording each phrase at a sign from our team, and he has been asked to repeat the phrase at each repetition of the sign.

Since any slight changes in electrodes position, temperature or tissue properties may alter the signal significantly. So, in order to make comparisons of amplitudes possible, we needed to apply a normalization procedure at each recording that compensates for these changes. A simple approach that we followed was to find the maximum of the absolute value of each utterance and divide the whole utterance by it.

V. DATA TRAINING

To ensure comparability of results from different experiments the same number of samples was used for each classifier for training stage, namely thirty exemplars of each phrase. Throughout our research, we used two classifiers:

A. *Hidden Markov Model (HMM) Modeling technique*

First order HMMs with Gaussian mixture models are used in most conventional speech recognition systems as classifiers because they are able to cope with both, variance in the time-scale and variance in the shape of the observed data. Each phrase in the list has been trained using a seven state left-to-right Hidden Markov Model with 3 Gaussians per state using the Expectation Maximization (EM) algorithm [10]. The number of iterations was chosen to be N=20.

To recognize an unknown signal the corresponding sequence of feature vectors was computed. Next, the Viterbi alignment for each vocabulary word was determined and the word corresponding to the best Viterbi score was output as the hypothesis. Feature extraction, HMM training, and signal recognition were performed using the Hidden Markov Model Toolkit (HTK) [11]. For this reason, a conversion of the cropped, cleaned, and normalized utterances to wave files was needed.

B. *Support Vector Machine Classifier*

SVMs are widely used as soft margin classifiers that find separating hyperplanes with maximal margins between classes in high dimensional space [12]. A kernel function is used to describe the distance between two data points. In order to find some statistically relevant information from incoming data, it is important to have mechanisms for reducing the information of each audio signal into a relatively small number of parameters, or segments. The classifier has been trained and tested with pre-segmented data with number of segments chosen to be N=20. Since each input signal was of different length, the segmentation procedure was as follows; first finding the length of each input signal, divide it by 20 to know the exact number that will produce 20 segments out of each input signal, and lastly dividing the signal by that number. After segmenting the utterances, we transformed each segment into frequency spectrum with FFT to extract some features. Feature extraction is the process of transforming raw signals into more informative signatures or fingerprints. We extracted the mean and the variance out of each segment, and then concatenated all features as one vector of attributes. All this segmentation and feature extraction part has been done using MATLAB [13-14]

The following block diagram represents the whole process of our work flow with its main steps.

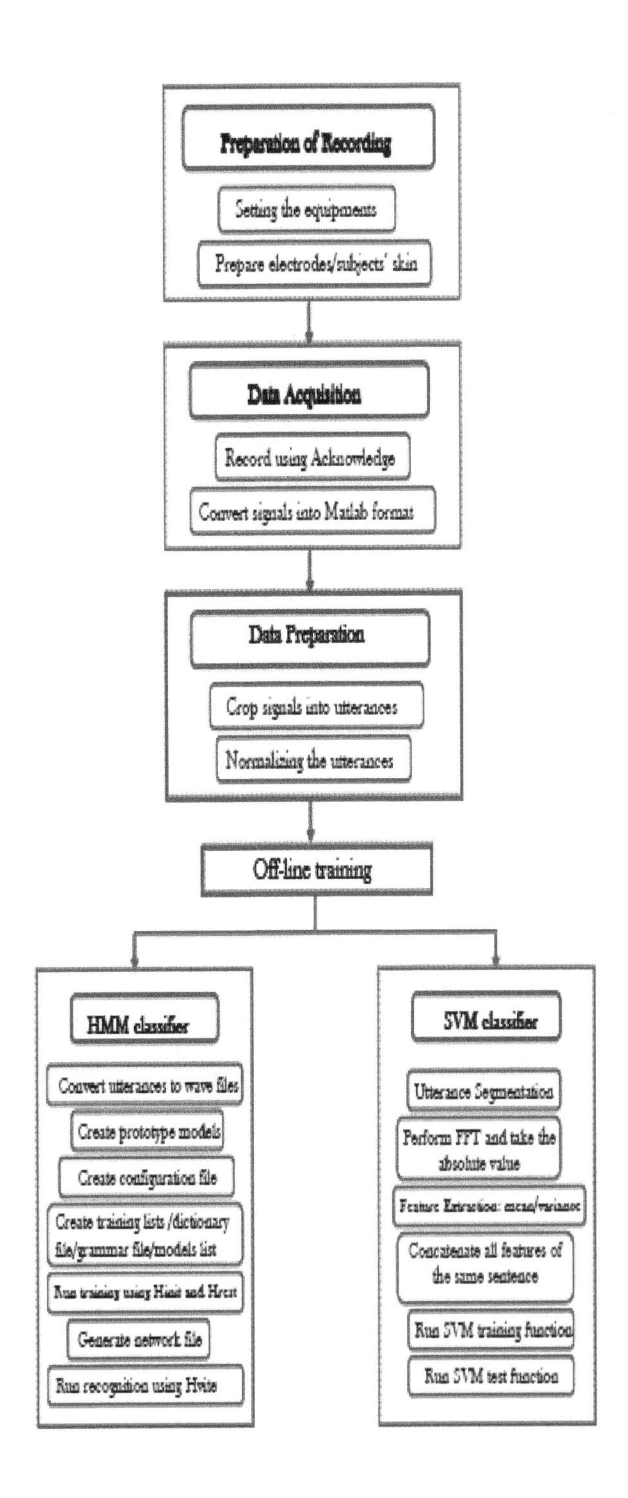

Fig. 3. The overall Block Diagram

VI. Experimental Results

Since we used thirty utterances from each subject for the training, we have been left with fifteen utterances for the test. The results varied between each phrase and among speakers. The results have been as follows:

A. Base Language: English

Our initial experiments have been conducted using HMM, and the results for each subject and each phrase were as shown in Figure 4 below. Although for few phrases the recognition was poor, the system achieved an average performance for all three subjects of 76.663%.

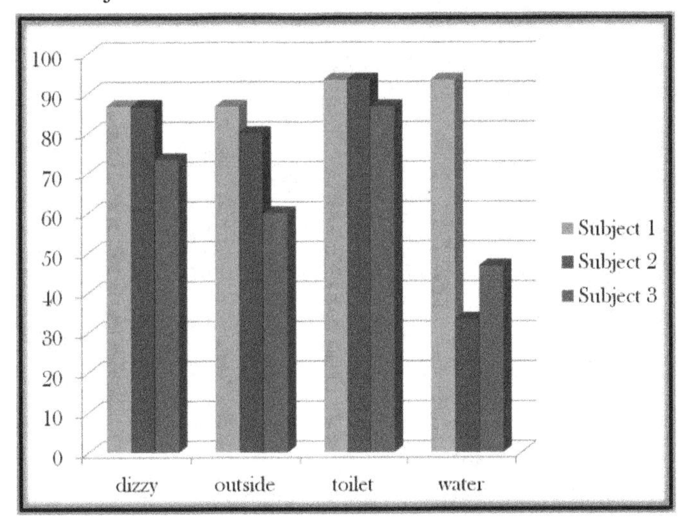

Fig. 4. Results using HMM for English system

Secondly, we investigated the performance of the system using SVM, and a clear improvement has been seen for all subjects. The system achieved an average of 91.11% word accuracy. The results in detail are shown in Figure 5.

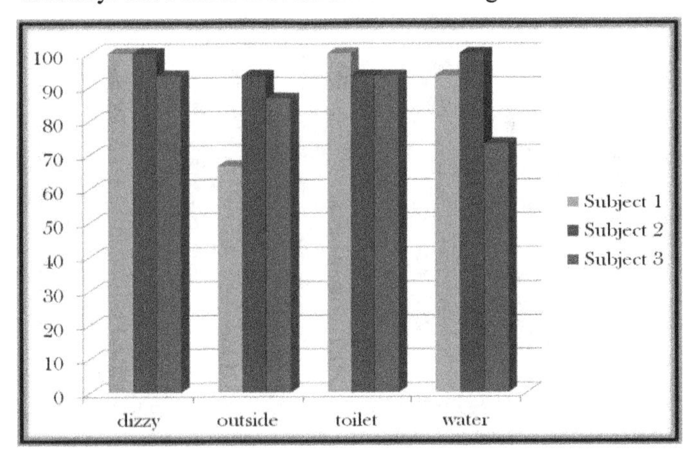

Fig. 5. Results using SVM for English system

B. Base Language: Arabic

Since studies and researches of Arabic-based SSR systems are poor compared to other similar languages, we have been motivated and curious to investigate and develop an Arabic-based system.

For the same number of test utterances used to examine the English system, a similar one was used for the Arabic. We have also checked the performance of the system using the same two classifiers and the results using HMM were as shown in Figure 6.

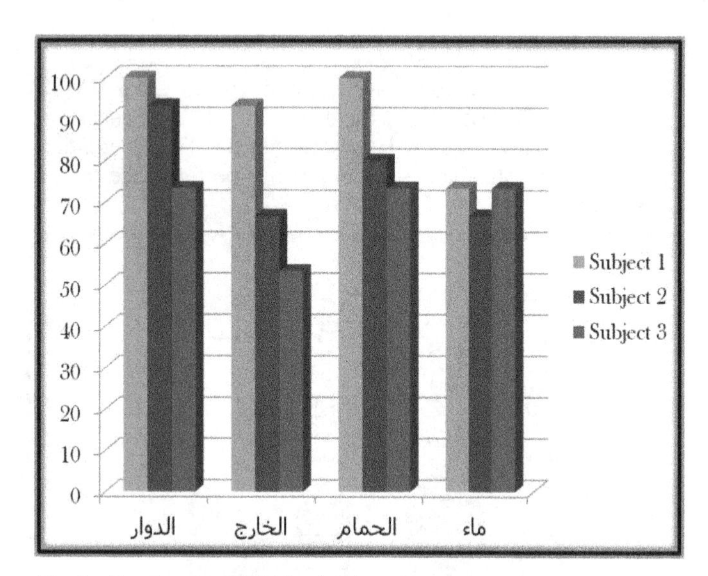

Fig. 6. Results using HMM for Arabic system

Averaging the results of all three subjects, the system achieved an average of 78.89% word accuracy.

Similarly, we experienced an improvement in the results when we trained and tested the system using SVM classifier, where the system has achieved 89.44% word accuracy. The results of each subject are shown in Figure 7.

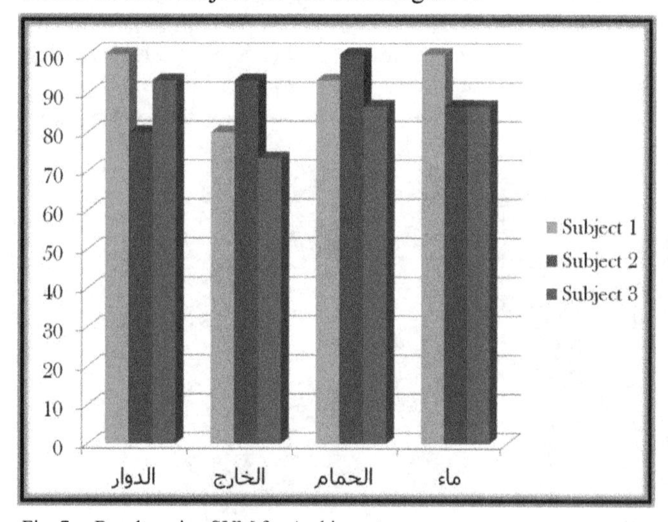

Fig. 7. Results using SVM for Arabic system

VII. Conclusion and Future Work

We have presented in this paper the results of our work for developing an isolated word Silent Speech Recognition System for both Arabic and English words. The technology is based on Surface Electromyography; which is capturing and recording of electrical potentials that arise from the muscle activity using surface electrodes attached to the skin. The concept of our work is still in the research area, so this work can be seen as a feasibility study. Moreover, we have investigated several state-of-the-art tools to check the performance of our system; such as: Hidden Markov Model and Support Vector Machine classifiers. Our experimental results indicate the effectiveness and efficiency of our proposed whole-sentence recognition system mainly using SVM algorithm in contrast to HMM classifier.

For the English system, an average of 91.11% word accuracy has been obtained when using SVM compared to the 76.663% obtained using HMM. Likewise, the Arabic system has achieved an average performance of 89.44% when using SVM compared to the 78.89% obtained while using HMM.

Though the obtained results are encouraging, this research does not claim completeness and it has lots of room for improvements. For example, Comparative experiments indicate that applying more than one electrode is crucial in order to construct a more robust system. Also, to demonstrate the potentials of this technology, EMG based speech recognition should move beyond isolated-word speech recognition and approach continuously spoken large vocabulary tasks.

Acknowlegement

The authors would like to acknowledge the research grant (2012/07)by the University of Bahrain Research Deanship which was used to purchase all the equipment.

References

[1] B. Denby, T. Schultz, K. Honda, T. Hueber, and J. Gilbert, *Silent Speech Interfaces,* Speech Communication, vol. 52, 2010.

[2] C. Jorgensen, D. Lee, and S. Agabon, *Sub Auditory Speech Recognition Based on EMG/EPG Signals*, in Proc. of the International Joint Conference on Neural Networks, 2003.

[3] Denby, B., Schultz, T., Honda, K., Hueber, T., Gilbert, J.M., Brumberg, J.S., *Silent Speech Interfaces*, Speech Communication (2009), doi: 10.1016/j.specom.2009.08.002.

[4] Maier-Hein, L. (July, 2005). Diplomarbeit (Thesis). *Speech Recognition Using Surface Electromyography.* Retrieved October 5th, 2013 from http://citeseerx.ist.psu.edu/viewdoc/download?doi=10.1.1.63.2803&rep=rep1&type=pdf

[5] Maier-Hein, L., Metze, F., Schultz, T., and Waibel, A. (November 2005). *Session Independent Non-audible Speech Recognition Using Surface Electromyography.* In Proc. ASRU, San Juan, Puerto Rico.

[6] Al-Mulla, M. R., Sepulveda, F. and Colley, M. (2012b). *sEMG Techniques to Detect and Predict Localised Muscle Fatigue*, In: EMG Methods for Evaluating Muscle and NerveFunction, Mark Schwartz, pp. 157-186, InTech, ISBN 978-953 307-793-2, Rijeka, Croatia.

[7] Day, S. (2010). Important *Factors in Surface EMG Measurement.* Retrieved October, 2nd, 2013 from http://www.andrewsterian.com/courses/214/EMG_measurement_and_recording.pdf

[8] De Luca, C. (2002). *Surface Electromyography: Detection and Recording.* DelSys Incorporated. Retrieved February 28th, 2013 from http://www.delsys.com/Attachments_pdf/WP_SEMGintro.pdf

[9] http://www.biopac.com/emg-electromyography.

[10] Gales, M., and Young, S. (2008). *The Application of Hidden Markov Models in Speech Recognition.*Foundations and Trends® in Signal Processing, Vol. 1, No. 3 (2007) 195 – 304.

[11] Young, S., Evermann, G., Gales, M., Hain, T., Kershaw, D., Moore, G., Odell, J., Ollason, D., Povey, D., Valtchev, V., and Woodland. P. (April 2005). *The HTK Book.*Cambridge University Engineering Department.

[12] Scholkopf, B., and Smola, A.J., *Learning with Kernels*, MIT Press, Cambridge, MA. 2002.

[13] Rakotomamonjy, A. (2006).*SVM and Kernel Methods Matlab Toolbox.* http://asi.insa-rouen.fr/enseignants/~arakoto/toolbox/index.html.

[14] Mathworks "Train support vector machine classifier". http://www.mathworks.com/help/toolbox/bioinfo/ref/svmtrain.html (4/6/2013).

A New Technique to Manage Big Bioinformatics Data Using Genetic Algorithms

Huda Jalil Dikhil
Dept. of Comoputer Sciecne
Applied Science Private University
Amman, Jordan

Mohammad Shkoukani
Dept. of Comoputer Sciecne
Applied Science Private University
Amman, Jordan

Suhail Sami Owais
Dept. of Comoputer Sciecne
Applied Science Private University
Amman, Jordan

***Abstract*—The continuous growth of data, mainly the medical data at laboratories becomes very complex to use and to manage by using traditional ways. So, the researchers started studying genetic information field in bioinformatics domain (the computer science field, genetic biology field, and DNA) which has increased in past thirty years. This growth of data is known as big bioinformatics data. Thus, efficient algorithms such as Genetic Algorithms are needed to deal with this big and vast amount of bioinformatics data in genetic laboratories. So the researchers proposed two models to manage the big bioinformatics data in addition to the traditional model. The first model by applying Genetic Algorithms before MapReduce, the second model by applying Genetic Algorithms after the MapReduce, and the original or the traditional model by applying only MapReduce without using Genetic Algorithms. The three models were implemented and evaluated using big bioinformatics data collected from the Duchenne Muscular Dystrophy (DMD) disorder. The researchers conclude that the second model is the best one among the three models in reducing the size of the data, in execution time, and in addition to the ability to manage and summarize big bioinformatics data. Finally by comparing the percentage errors of the second model with the first model and the traditional model, the researchers obtained the following results 1.136%, 10.227%, and 11.363%, respectively. So the second model is the most accurate model with less percentage error.**

Keywords—Bioinformatics; Big Data; Genetic Algorithms; Hadoop MapReduce

I. INTRODUCTION

The important evaluation of the Bioinformatics and genetics field in the recent years has helped scientists and doctors to understand illnesses and diagnose it the better way and discover the reasons behind many diseases and genetic mutations, including muscular degeneration, which causes disability of many children around the world. To diagnose genetic diseases at medical laboratories, it requires a comparison procedure between the defective genes with the natural ones by alignment and matching sequence of Nucleated (nitrogenous bases) in the genes through National Center for Biotechnology Information (NCBI), which consider as the largest database and repository of genes.

Processing medical data due to the large size of bioinformatics data is hard to manage and it is not easy to reduce the size of needed data. For this and other reasons, it becomes important to develop such models and algorithms that can manage big bioinformatics data that are produced by genetic laboratories, and have the ability to find the defective gene in less time with less error because medical application requires high accuracy.

So, for managing big bioinformatics data, the authors proposed two new models. The original model used only the Hadoop MapReduce. Since Genetic Algorithms GAs have many benefits especially in optimization problems, the authors tried to propose two new models by applying Genetic Algorithms before and after MapReduce. So, the first model was by applying Genetic Algorithms before MapReduce, and the second model was by implementing Genetic Algorithms after the MapReduce.

The paper consists of eight sections. The first section is an introduction. The second Section discusses the Big Data, its characteristics, and the architectures. The third section demonstrates the Bioinformatics. The fourth section explains the Genetic Algorithms and its features. The fifth section presents the problem statement of the research. The sixth part discussed the two proposed models. Section seven explains the data description. Part eight explored the results of the proposed models.

II. BIG DATA

Big Data is a term used to describe the enormous and massive amounts of data that could not be handled and processed using traditional methods. Big Data size has increased conspicuously in various fields over the past twenty years, where the volume of the generated and duplicated data has grown more than ten times in the over years, which cannot be predicted because data continuously increased to be double every two years. Data in Big Data are structured and unstructured. Thus, it needs more complicated tools rather than the traditional ones to be analyzed and managed. Managing this data brings more challenges and requires more efficient methods [1, 2].

Managing Big Data is one of the main challenges that faces large corporations, and has attracted the interest of researchers in the past years [1, 2]. Big Data has several characteristics known by *nV's* characteristics, and it has several type of architectures for Big Data analysis.

A. Big Data Characteristics

There are several characteristics (5Vs) that distinguish Big Data from standard set of data: Volume, Variety, and Velocity,

Value, and Veracity [14, 15]. The 5Vs characteristics of the Big Data were illustrated in Fig. 1.

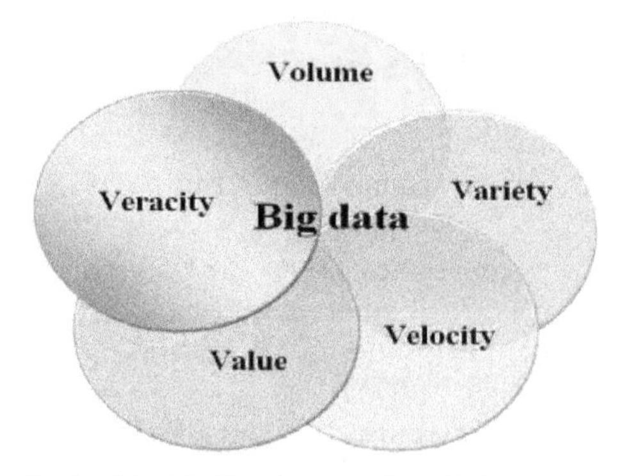

Fig. 1. The 5Vs Characteristics of the Big Data

Some other researchers have different views and consider only 3Vs (Volume, Velocity, and Variety) as fundamental features. And others add Veracity as 4Vs [1, 2, 3, 4].

B. *Architectures of Big Data Analytics*

On the different type of sources and different structures. There are three main types of architectures for Big Data analysis, MapReduce architectures, fault-tolerant graph architectures and streaming graph architectures. Some of the characteristics of these styles as shown in Table 1 in terms of the used memory type: if the used memory is local memory or shared global memory, in addition to the fault tolerance [2].

III. Bioinformatics

Bioinformatics is relatively an old field, it started before more than a century and introduced by the Austrian scientist Gregor Mendel, who known as the "father of genetics." Since then, the understanding of genetic information has increased, especially in past thirty years. The researches and studies in the domain of bioinformatics led to the creation of the largest international organization (HUGO), the first international organization that published the first complete map of the genome of sustainable in bacteria. Bioinformatics is the relationship between computer science and biology [5, 6, 7].

TABLE I. Characteristics of Different Types of Architectures for Big Data Analytics

Characteristics	Architectures		
	MapReduce architecture	**Fault tolerant graph architecture**	**Streaming graph architecture**
Memory	Local memory Global memory	Global Memory	Data not need to be stored into disks
Fault Tolerance Allow	Allow	Allow	Not
Operations Synchronization	Synchronization	Synchronization	Asynchronous

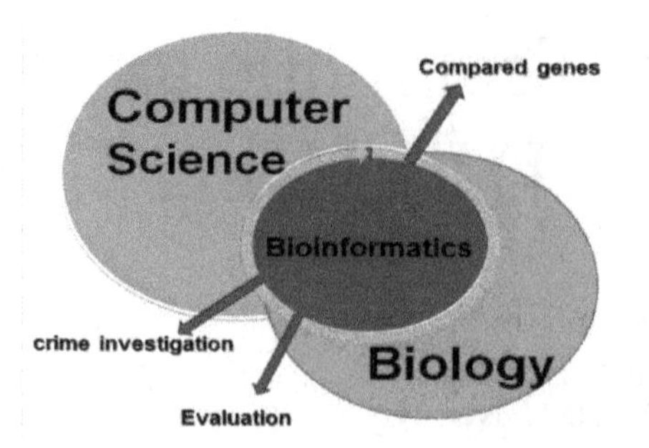

Fig. 2. Bioinformatics in general [6, 7, 8]

Fig. 2 shows that the bioinformatics in general which is an intersection between the biology and the computer science. It can be used in different fields such as crime scene investigation, comparing genes, and evaluation.

Bioinformatics characteristic debate the collaborative resources that work together for such task [8].

The theory "structure prediction" as a technique to recruit computer tools and algorithms is the most important objective of bioinformatics (Molecular Bioinformatics) in addition to being an alternative method and attractive [8]. Molecular bioinformatics logic and dealing with the concepts of biology regarding molecules and the application of the "information" to understand the technology and organization associated with these biomolecules in cells and organisms information. New genes discovered by searching for systematic data available to genome sequence, so through the sequence identity algorithms are appointed the supposed new genes function [8].

IV. Genetic Algorithms

Genetic algorithm GAs is one of the most powerful computer algorithms that based on natural living genes combining, producing and inheritance acts; it has vital importance in Computer Sciences branches like Artificial Intelligent and Computer Vision techniques. Genetic algorithms are useful tools for search and optimization problems. [9, 10].

The most important characteristic of genetic algorithms is solving hard problems with an optimal solution. Fig. 3 presents the simple Genetic Algorithm flowchart which it briefly describes the four basic operators to resolve a problem as follows: fitness function, selection operator, crossover operator, and mutation operator [9, 10, 11, 12, 13, 14].

V. Problem Statement

There is a tight relationship between big data and bioinformatics since there is a vast data in bioinformatics especially the DNA, which each human genome sequence approximately 200 gigabytes [2].

The development of Computer Science (CS) helped other scientific fields and became a key and essential part in most biological and medical experimentations. With the continuous growth of data, especially the medical data at laboratories (lab), it becomes very hard to use this data and manage it using the traditional ways, so efficient algorithms are needed to deal with this large and vast amount of bioinformatics data in genetic laboratories, which includes a gene and protein sequence. Thus, the researchers used one of the evolutionary algorithms which are genetic algorithms.

VI. PROPOSED MODELS

Data management is a very arduous task, especially when you have an enormous amount of data such as DNA. The proposed model based on genetic algorithm and Hadoop MapReduce. The researchers presented two models, the first one (GAHMap) by applying Genetic Algorithm before Hadoop MapReduce. The second one (HMapGA) by executing Genetic Algorithm after Hadoop MapReduce.

C. Model 1 (GAHMap): GAs before Hadoop MapReduce

Fig. 4 demonstrates the stages of the first proposed model (GAHMap) as follows:

1) The input of prototype 1 is the big bioinformatics data which is denoted by (M).

2) Applying Genetic Algorithms on (M), which will produce an optimized data which is indicated by (M').

3) Carrying out Hadoop MapReduce on (M'), the result will be the reduced data which is denoted by (M'').

D. Model 2 (HMapGA): GAs after Hadoop MapReduce.

The second paradigm (HMapGA) presented in Fig. 5 and it has the following stages:

1) The input of prototype II is the big bioinformatics data which is denoted by (M).

2) Applying Hadoop MapReduce on (M), the result will be the reduced data which is denoted by (M').

3) Executing Genetic Algorithms (M'), which will produce an optimized data which is indicated by (M'').

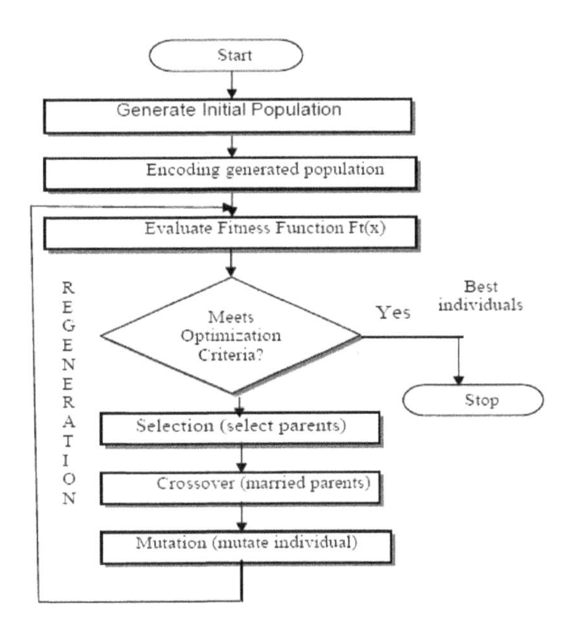

Fig. 3. Flowchart of Genetic Algorithms [9]

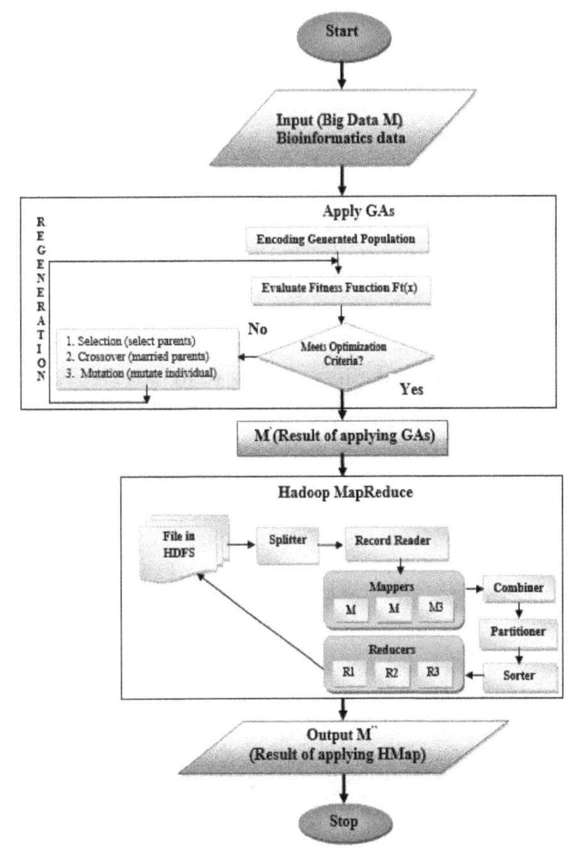

Fig. 4. Proposed Model I GAHMap stages

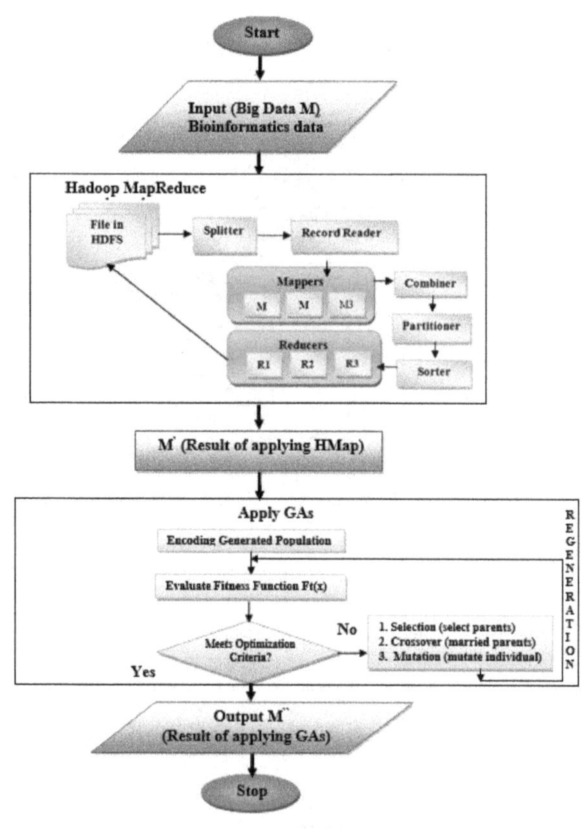

Fig. 5. Proposed Model II HMapGA Stages

The two models were fully implemented using MATLAB and applied as graphical user interface (GUI) system as shown in Fig. 6. The system mainly consists of two parts: the first part is the Create Data Server that reads the big bioinformatics data as input dataset (M). The second part consists of three options to manage the Big Data: the first alternative to execute the first model GAHMap, the second option to perform the second model HMapGA, and the third option to proceed the original model which uses Hadoop MapReduce without genetic algorithms (HMap).

The results from the first and second models will be compared to identify the model that can give the best result which reduces the size of the data with better accuracy. After that, the outcome of the chosen model will be compared with the result from the original Hadoop MapReduce. Finally, the outcomes will determine the best model among them.

VII. Data Description

The dataset of genes used in this research acquired from the Genetics Center at Specialty Hospital–Amman, Jordan. The dataset is related to Duchene Muscular Dystrophy (DMD), which is a popular and widespread genetic disease in the country as well as all over the world. It was an 88 sample from 88 individuals (genes). Each gene in the dataset represents 108 gigabytes of DNA tape; gene number 19 was obtained for this research and saved in text file format (txt file). Each gene and file contain 2,220,388 nucleotides; the nucleotides consist of a base (one of four chemicals and amino acids: cytosine, guanine, adenine, and thymine). The dataset with the 88 genes has been already diagnosed and alignment using the global location of the genes NCBI website by the genetic center to find the defective and normal genes. The results showed that 48 of the genes were defective and suffer from DMD disease, and the other 40 genes were normal. A sample of one of the genes from the DNA sequence was saved in a text file as shown in Fig. 7.

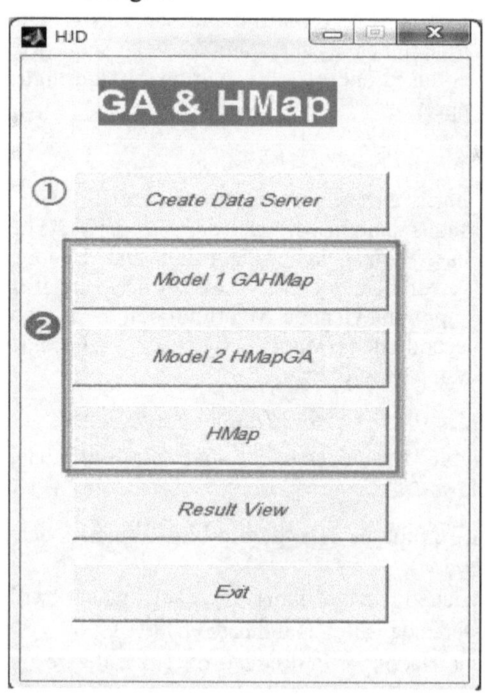

Fig. 6. Graphical User Interface of the Proposed Method

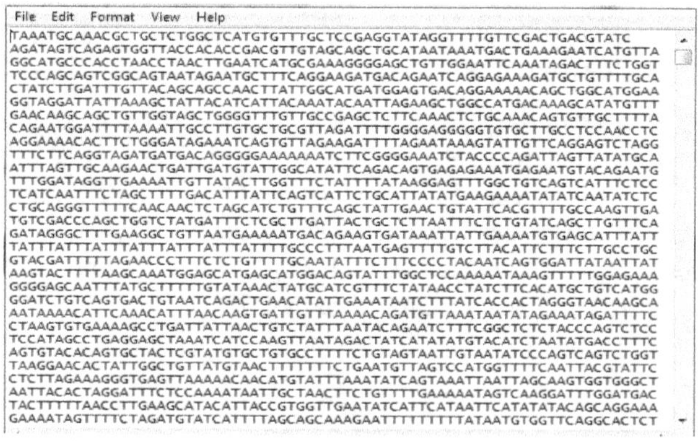

File Edit Format View Help

Fig. 7. Gene text file with nucleotides sequence

VIII. System Test Result

The system was tested using dataset within 88 genes (40 normal genes and 48 defective genes) as an input of Bioinformatics' Big Data (M) to find which model will be better. So, after data server reads the dataset, it was tested by applying the three models on the same dataset. During execution, each model will display the result with the following information:

- TP: True Positive, which means the number of standard genes.
- TN: True Negative, which means the number of genes which is defective.

- FP: False Positive, which means the number of genes which is regular and detected as defective.
- FN: False Negative, which implies the number of genes which is defective and detected as deranged.

Fig. 8 shows the outcome of GAHMap implementation, which concludes that there are 40 TP and zero FP which implies that there are no normal genes revealed as defective genes, and there are 39 TN and 9 FN which means that there are 9 defective genes detected as standard genes which listed in FN sequence.

Fig. 9 displays the result of second model HMapGA execution, which concludes that there are 40 TP and zero FP which means that there are no natural genes revealed as defective genes, and there are 47 TN and 1 FN which signifies that there is only one deficient gene exposed as standard genes.

Fig. 10 presents the consequence of the original model HMap enforcement, which determines that there are 40 TP and zero FP which implies that there is no normal genes appeared as defective genes, and there are 38 TN and 10 FN which means that there are 10 deficient genes detected as normal genes.

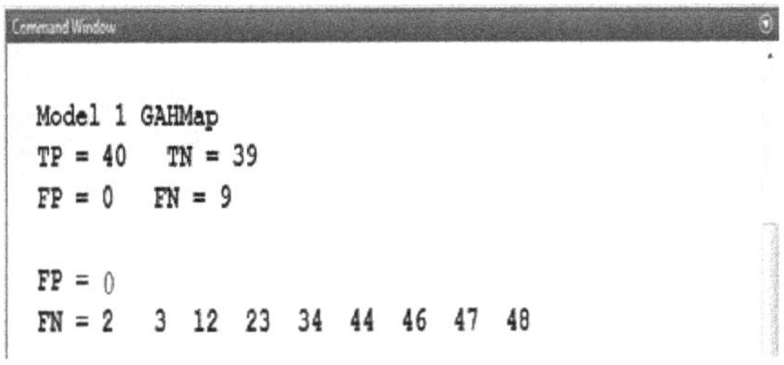

```
Command Window

Model 1 GAHMap
TP = 40   TN = 39
FP = 0   FN = 9

FP = 0
FN = 2   3  12  23  34  44  46  47  48
```

Fig. 8. GAHMap Model 1 Result

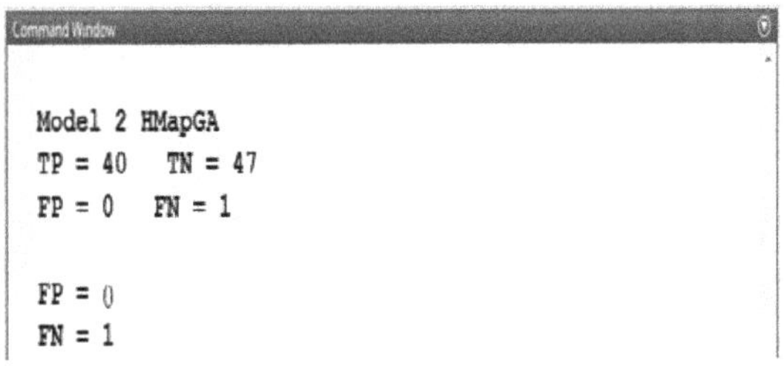

```
Command Window

Model 2 HMapGA
TP = 40   TN = 47
FP = 0   FN = 1

FP = 0
FN = 1
```

Fig. 9. HMapGA Model Results

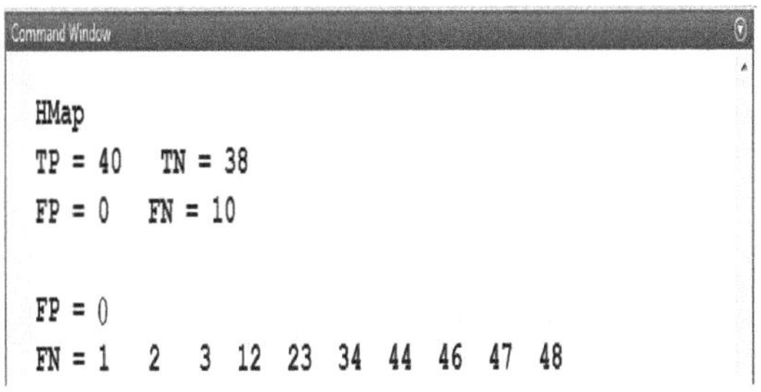

```
Command Window

HMap
TP = 40   TN = 38
FP = 0   FN = 10

FP = 0
FN = 1   2   3  12  23  34  44  46  47  48
```

Fig. 10. HMap Model Results

Table 2 summarizes the results of the three models GAHMap, HMapGA, and HMap.

TABLE II. THREE MODELS SUMMARY RESULTS

Model	TP	TN	FP	FN
GAHMap	40	39	0	9
HMapGA	40	47	0	1
HMap	40	38	0	10

For more accuracy of the results, the percentage error (%Error) was calculated for each model by using the mathematic formula which is the difference between the experimental value and theoretical value divided by theoretical value as shown in equation (1) [15]:

$$\%Error = \frac{|\text{TheoreticalValue} - ExperimentalValue|}{|TheoreticalValue|} \times 100\% \quad (1)$$

Where, the *TheoreticalValue* means the total number of genes used in the research (natural + defective), and the *ExperimentalValue* means the total number of correctly detected genes (natural + defective) by the system.

As shown in Table 3 by applying the percentage error equation, it was found that the percentage error of the first model GAHMap =|(Total number of genes used in the research (normal + defective)) - (total number of correctly detected genes (normal + defective)|/ (total number of genes used in the thesis (normal + defective))*100

=| (40+48) – (40+39)| / (40+48) * 100%
= |88-79| / 88 * 100
= 10.227 %
For the second model HMapGA % Error =| (40+48) – (40+47)| / (40+48) * 100
= |88-87| / 88 * 100
= 1.136%
For the original model HMap % Error =| (40+48) – (40+38)| / (40+48) * 100
= |88-77| / 88 * 100
= 11.363%.

TABLE III. THREE MODELS RESULTS OF PERCENTAGE ERROR

Model	GAHMap	HMapGA	HMap
% Error	10.227%	1.136%	11.363%

According to the results in Table 3 the researchers conclude that the HMapGA is better than the GAHMap, and if the HMapGA compared with the original model HMap it found that the HMapGA is also better than the original one HMap. So the HMapGA proved to be the most accurate model with less percentage error and succeed in achieving the objectives of this research which includes organizing big bioinformatics data by matching and finding normal and defective genes with less time and less percentage error.

IX. CONCLUSION

This paper proposed two models to manage big bioinformatics data of DMD disorder. In the first model, GAHMap was implemented genetic algorithms before Hadoop MapReduce. In the second model, HMapGA has executed genetic algorithms after Hadoop MapReduce. The proposed models in addition to the original model were tested using the real dataset of 88 genes related to DMD disorder.

By comparing the results of the three paradigms, the researchers found that the number of genes which is natural and revealed as defective (FP) was zero for all models, but the number of genes which is faulty and detected as normal (FN) were 9, 1, and 10 defective genes for the first, second, and original models respectively. The researchers conclude that the HMapGA detected less number of defective genes as natural ones.

Also, when comparing the percentage error for the three models, the second model has 1.136 % which is the most accurate model with the less percentage error.

Finally, the researchers conclude that the second model HMapGA is the best model since it succeeds in matching and finding normal and defective genes in less time and less percentage error. So HMapGA provides an efficient technique to manage and reduce the size of big bioinformatics data in the laboratory.

Reference

[1] M. Chen, S. Mao, Y. Zhang, and V. C. Leung, Big data: related technologies, challenges and future prospects. Springer, 2014.

[2] H. Kashyap, H. A. Ahmed, N. Hoque, S. Roy, and D. K. Bhattacharyya, Big data analytics in bioinformatics: a machine learning perspective. Arxiv preprint arxiv:1506.05101, 2015.

[3] M. Moorthy, R. Baby, and S. Senthamaraiselvi, "An analysis for big data and its technologies", International Journal of Computer Science Engineering and Technology (IJCSET), Vol. 4, no 12, pp. 412-418, December 2014.

[4] J. Hurwitz, A. Nugent, F. Halper , M. Kaufman, Big Data For Dummies, 2013.

[5] S. M. Thampi, Introduction to Bioinformatics, LBS College of Engineering, 2009.

[6] Ralf Hofestädt, Bioinformatics: german conference on bioinformatics, GCB'96, Leipzig, Germany, Springer Science & Business Media, vol. 1278, September 30-October 2, 1996.

[7] D. C. Rubinsztein, Annual review of genomics and human genetics, 2001.

[8] P. Narayanan, Bioinformatics: A Primer. New Age International, 2006.

[9] S. S. Owais, P. Krömer, and V. Snásel, Query optimization by Genetic Algorithms. In Proceedings of the Dateso 2005 Annual International Workshop on databases, pp. 125-137, April 2005.

[10] R. Kaur, and S. Kinger, Enhanced genetic algorithm based task scheduling in cloud computing. International Journal of Computer Applications, vol. 101, no 14, pp. 1-6, 2014.

[11] S. N. Sivanandam, and S. N. Deepa, Introduction to genetic algorithms, Springer Science & Business Media, 2007.

[12] A. E. Eiben, and J. E. Smith, Introduction to evolutionary computing. Springer Science & Business Media, 2003.

[13] C. Y. Jiao, and D. G. Li, Microarray image converted database-genetic algorithm application in bioinformatics. In biomedical Engineering and Informatics, International Conference, vol. 1, pp. 302-305, May 2008.

[14] M. Chen, S. Mao, Y. Zhang V. C. Leung. Big data related technologies, challenges and future prospects. SpringerBriefs in Computer Science, New York Dordrecht London, 2014.

[15] Suhail Sami Owais, And Nada Sael Hussein. Extract Five Categories CPIVW from the 9V's Characteristics of the Big Data. International Journal of Advanced Computer Science and Applications (IJACSA), vol. 7, no. 3, pp. 254-258, 2016.

A Semantic-Aware Data Management System for Seismic Engineering Research Projects and Experiments

Md. Rashedul Hasan
Department of Civil, Environment and Mechanical Engineering
University of Trento, Italy

Feroz Farazi
Department of Information Engineering and Computer Science
University of Trento, Italy

Oreste Bursi
Department of Civil, Environment and Mechanical Engineering
University of Trento, Italy

Md. Shahin Reza
Department of Civil, Environment and Mechanical Engineering
University of Trento, Italy

Ernesto D'Avanzo
Department of Political, Social and Communication Sciences
University of Salerno,Italy

Abstract—**The invention of the Semantic Web and related technologies is fostering a computing paradigm that entails a shift from databases to Knowledge Bases (KBs). There the core is the ontology that plays a main role in enabling reasoning power that can make implicit facts explicit; in order to produce better results for users. In addition, KB-based systems provide mechanisms to manage information and semantics thereof, that can make systems semantically interoperable and as such can exchange and share data between them. In order to overcome the interoperability issues and to exploit the benefits offered by state of the art technologies, we moved to KB-based system. This paper presents the development of an earthquake engineering ontology with a focus on research project management and experiments. The developed ontology was validated by domain experts, published in RDF and integrated into WordNet. Data originating from scientific experiments such as cyclic and pseudo dynamic tests were also published in RDF. We exploited the power of Semantic Web technologies, namely Jena, Virtuoso and VirtGraph tools in order to publish, storage and manage RDF data, respectively. Finally, a system was developed with the full integration of ontology, experimental data and tools, to evaluate the effectiveness of the KB-based approach; it yielded favorable outcomes.**

Keywords—Ontology; Knowledge Base; Earthquake Engineering; Semantic Web; Virtuoso

I. INTRODUCTION

This is an extended version of the following paper: Hasan et al. 2013. The inventor of the Web, Tim Berners-Lee, envisioned a more organized, well connected and well integrated form of the Web data that are suitable for humans to read and for machines to understand. This new form of the Web is called the Semantic Web (T. Berners-Lee, 1999; T. Berners-Lee et al., 2001). On the Semantic Web data can be published using Resource Description Framework (RDF) and Web Ontology Language (OWL). Traditional databases are a persistent storage mechanism that enables large scale of data; however, they were not originally designed for managing RDF and OWL data or ontologies. KBs can do this job effectively. Ontologies are intended to be stored in the KBs, which can offer better user experience by supporting reasoning over ontological data and semantics. Moreover, KB-based systems provide mechanism to manage information and semantics thereof that can make systems semantically interoperable and as such can exchange and share data between them. To overcome the interoperability issues and to exploit the benefits offered by the state of the art technologies, we moved to the KB-based system.

In fact, we have developed an ontology named as Earthquake Engineering Research Projects and Experiments using a faceted approach that gives emphasis on research project management and experiments. Following the validation of the ontology by domain experts, it was published in the knowledge representation language RDF and integrated into the generic ontology WordNet[1]. The experimental data coming from, inter alia, the cyclic and pseudo-dynamic tests were also published in RDF. We used Jena[2] , Virtuoso[3] and VirtGraph[4] tools for ontology and data publishing, storage and management, respectively. Finally, a system was developed to verify the effectiveness of the approach through the integration of the aforementioned tools, ontologies and data.

The rest of the paper is organized as follows. Section II depicts an ontology based information management system development approach. Section III describes the ontology development steps and the created ontology (partially). In Section IV, we provide a brief description of the ontology representation languages RDF and OWL. In Section V, we present existing ontology/thesaurus that are relevant for this work and as such worth discussing them. Section VI provides

[1] http://wordnet.princeton.edu
[2] https://jena.apache.org
[3] http://virtuoso.openlinksw.com
[4] http://docs.openlinksw.com/jena/virtuoso/jena/driver/VirtGraph.html

the ontology integration approach and Section VII describes experimental data collection procedure. While Section VIII demonstrates the architecture of the final system that was built on top of the integrated ontology, Section IX reports evaluation results that show the effectiveness of the ontology. In section X we briefly describe related work and in Section XI we conclude the paper.

II. APPROACH

Figure 1 describes an ontology based information management system development approach that involves standard three-tier architecture. KB works as a backend of the system hosting ontologies represented in RDF, while query processing, inference mechanism and reasoning are incorporated in the business logic layer. Issuing queries and showing the corresponding results are supported by the User Interface (presentation) layer. However, for ontology development (see Section III) we follow the DERA (Domain, Entity, Relation, Attribute) methodology (Giunchiglia and Dutta, 2011), for ontology representation (see Section IV) in RDF we use Jena and for ontology integration (see Section VI) we implemented a facet based algorithm.

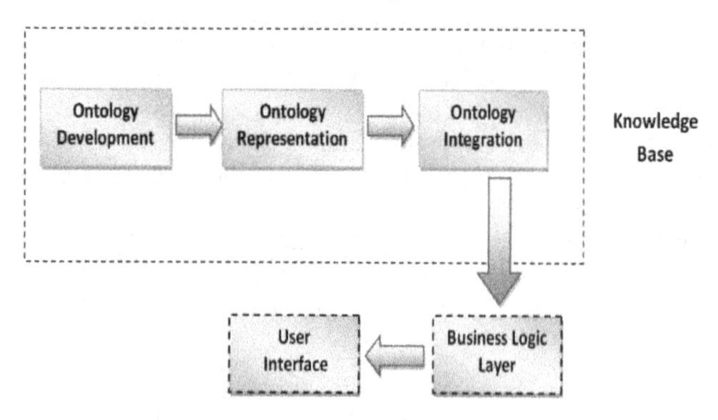

Fig. 1. Ontology based development Approach

III. ONTOLOGY DEVELOPMENT

We use the DERA methodology (Giunchiglia and Dutta, 2011) for ontology development in fact it is known to be extendable and scalable and some ontologies including GeoWordNet were developed following this approach (Giunchiglia, 2010).

DERA methodology allows for building domain specific ontologies. Domain is an area of knowledge in which users are interested in. For example, earthquake engineering, oceanography, mathematics and computer science can be considered as domains. In DERA, a domain is represented as a 3-tuple $\boldsymbol{D} = <\boldsymbol{E}, \boldsymbol{R}, \boldsymbol{A}>$, where $\boldsymbol{E}$ is a set of entity-classes that consists of concepts (e.g., device and experiment) and entities (e.g., an instance of device and an instance of experiment); $\boldsymbol{R}$ is a set of relations that can be held between entity-classes (e.g., *IS_A* and *PART_OF*) and $\boldsymbol{A}$ is a set of attributes of the entities (e.g., *number of devices and name of the experiment*).

In this three basic components concepts, relations and attributes are organized into facets; hence, the ontology is based on faceted methodology. Facet is a hierarchy of homogeneous concepts describing an aspect of a domain. S. R. Ranganathan, who was an Indian mathematician-librarian, was the first to introduce faceted approach capable of categorizing books in the libraries (Ranganathan, 1967).

Note, however, that a domain can alternatively be called as domain ontology. Henceforth in this paper it will be referred to as domain ontology. Among the macro-steps to develop each component of a domain ontology, we used the following ones.

In the first step (*identification*) towards building an ontology, we identified the atomic concepts of terms collected from research papers, books, existing ontological resources and experts belonging to Earthquake Engineering domain giving emphasis on research projects and experiments aspects. We found terms such as device, shaker, experiment, dynamic test, etc., and identified the atomic concept for each of them. We bootstrapped our Knowledge Base with the concepts and relations of WordNet. The term device has 5 different concepts in it. In our case, we selected the one that has following description: *device -- (an instrumentality invented for a particular purpose*). We have found 193 atomic concepts. In the second step (analysis) we analyzed the concepts, i.e., we studied their characteristics to understand the similarity and differences between them. Once the analysis was completed, in the third step (*synthesis*) we organized them into some facets according to their characteristics. For example, shaker is more specific than device, actuator is more specific than device, motor is a part of electric actuator and we assigned the following relationships between them: shaker *IS_A* device, actuator *IS_A* device, motor *PART_OF* electric actuator. This is how we built device fact. In this way, we built 11 facets. A partial list of the facets is as follows: device, experiment, specimen, experimental computation facility, project, project person and organization. Device and experiment facets are shown in Fig. 2. In the fourth step (*standardization*), we marked concepts with a preferred name in the cases of availability of synonymous terms. For example, while experiment and test are used to refer to the same concept, we assigned the former term as the preferred one. Finally, the ontology was validated by domain experts. They suggested a number of changes, e.g., the inclusion of the concepts *shaker-based test* and *hammer-based test* in the experiment facet, the exclusion of the concept *simulation* from the same facet.

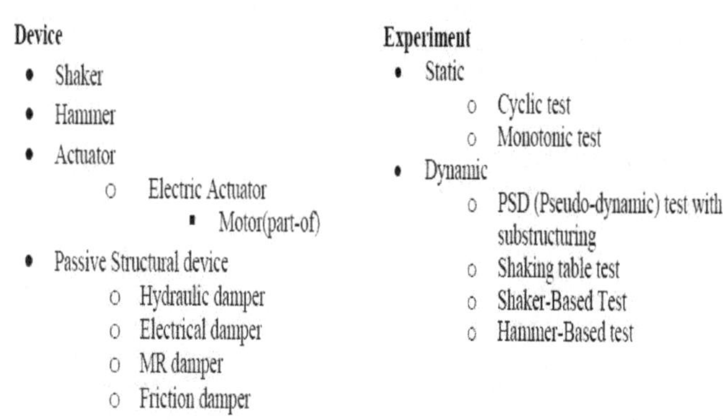

Fig. 2. The device and experiment facets

Note that in Fig. 2, concepts which are connected by PART_OF relation with the concepts one level above in the hierarchy are explicitly written, for example, motor is PART_OF electric actuator. In the other cases, IS_A relation holds between them, for example, electric actuator IS_A actuator.

IV. Ontology Presentation

In the following subsections, we describe the Knowledge Representation Languages RDF and OWL in terms of their capacity in representing ontologies of varied kinds.

A. RDF

The Resource Description Framework (RDF) is a data model used to represent information about resources in the World Wide Web (WWW) and can be used to describe the relationships between concepts and entities. It is a framework to describe metadata on the web. Three types of things are in RDF: resources (entities or concepts) that exist in the real world, global names for resources (i.e. URIs) that identify entire web sites as well as web pages, and RDF statements (triples, or rows in a table) (Klyne, 2004). Each triple includes a subject, an object and a predicate. RDF is designed to represent knowledge in a distributed way particularly concerned with meaning. The following RDF statements describe the resources *Hammer* and *Damper*.

```
<rdf:Description rdf:about="http://earthquake.linkeddata.it/resource/Hammer">
    <rdfs:subClassOf rdf:resource="http://earthquake.linkeddata.it/resource/Device"/>
    <ontology:desciption rdf:datatype="http://www.w3.org/2001/XMLSchema#string">A hand
    tool with a heavy rigid head and a handle; used to deliver an impulsive force by strik-
    ing</ontology:desciption>
    <rdf:type rdf:resource="http://www.w3.org/2002/07/owl#Class"/>
</rdf:Description>

<rdf:Description rdf:about="http://earthquake.linkeddata.it/resource/Damper">
    <rdfs:subClassOf rdf:resource="http://earthquake.linkeddata.it/resource/Device"/>
    <ontology:desciption rdf:datatype="http://www.w3.org/2001/XMLSchema#string">A de-
    vice that decreases the amplitude of electronic, mechanical, acoustical or aerodynamic oscil-
    lations</ontology:desciption>
    <rdf:type rdf:resource="http://www.w3.org/2002/07/owl#Class"/>
</rdf:Description>
```

Fig. 3. RDF statements describe the resources Hammer and Damper

The above example represented relationship between Hammer and Device concepts; and the rdfs: sub Class Of property is used to relate the former concept to its more generic later concept.

B. OWL

Web Ontology Language is designed to represent comparatively complex ontological relationships and to overcome some of the limitations of RDF such as representation of specific cardinality values and disjointness relationship between classes (Giunchiglia et al. 2010). The language is characterized by formal semantics and RDF/XML based serializations for the web. As an ontology representation language, OWL is essentially concerned with defining terms that can be used in RDF documents, i.e., classes, properties and instances (Antoniou et al. 2004). It serves two purposes: first, it helps identifying current document as an ontology and second it serves as a container of metadata about the ontology. This language focuses on reasoning techniques, formal foundations and language extensions. OWL uses URI references as names and constructs these URI references in the same manner as that used by RDF. The W3C allows OWL specification to include the definition of three variants of OWL, with different levels of expressiveness. These are OWL Lite, OWL DL and OWL Full ordered by increasing expressiveness.

V. Existing Ontology/ Thesaurus

Ontologies and thesaurus, which are germane to our Earthquake Engineering ontology, are described in terms of the amount of concepts they have and the types of relations that exist between concepts.

A. WordNet

WordNet (Miller et al. 1990) is an ontology that consists of more than 100 thousand concepts and 26 different kinds of relations, e.g., hyponym, synonym, antonym, hypernyms and meronyms. It was created and is being maintained at the Cognitive Science Laboratory of Princeton University. The most obvious difference between WordNet and a standard dictionary is that its concepts are organized into hierarchies, like professor *IS_A* kind of person and person *IS_A* kind of living thing. It can be used for knowledge-based applications. It is a generic knowledge base and as such does not have good coverage for domain specific applications. It has been widely used for a number of different purposes in information systems including word sense disambiguation, information retrieval and automatic text summarization.

B. NEES Thesaurus

The Network for Earthquake Engineering Simulation NEES is one of the leading organizations for Earthquake Engineering in the USA.

TABLE I. NEES Earthquake Engineering Thesaurus

Term	Broader Term	Narrower Term
AASHTO_LRFD_Bridge_Design_Specifications	AASHTO_2001	
Peak_Base_Acceleration	Acceleration	
Dynamic_Actuator	Actuator	
Static_Actuator	Actuator	
Cyclic_Axial_Load	Axial_Load	
Preformed_Fabric_Pads	Bearings	Cotton_Duck_Bearing_Pads

(https://nees.org/)

They developed an earthquake engineering thesaurus, which is based on Narrower and Broader terms. It contains around 300 concepts and we have integrated in our ontology 75 of them. Table I reports a small portion of NEES thesaurus.

VI. Ontology Integration

Developed facets include concepts that were selected from NEES thesaurus to be incorporated into our ontology. This integration was accomplished in fact when we built the facets. In this Section, we describe how we integrated our developed ontology with Wordnet. Basically, we applied the semi-automatic ontology integration algorithm proposed in Farazi et al. (2011). In particular, we implemented the following macro steps:

*1) **Facet concept identification**:* For each facet, the concept of its root node is manually mapped to WordNet, in the case of availability.

*2) **Concept Identification**:* For each atomic concept C of the faceted ontology, it checks if the concept label is available in WordNet. In the case of availability, it retrieves all the concepts connected to it and maps with the one residing in the sub-tree rooted at the concept that corresponds to the facet root concept.

*3) **Parent Identification:*** In the case of unavailability of a concept it tries to identify the parent. For each multiword concept label it checks the presence of the header, and if it is found within the given facet, it identifies it as a parent. For instance, in WordNet it does not find hydraulic damper for which damper is the header and that is available there in the hierarchy of device facet. Therefore, it recognizes the damper with the description damper, muffler -- (a device that decreases the amplitude of electronic, mechanical, acoustical, or aerodynamic oscillations), as the parent of the hydraulic damper.

VII. EXPERIMENTAL DATA COLLECTION

In this section, an experimental test on a piping system under earthquake loading carried out by Reza et al. (2013) is briefly discussed to provide the reader with an overview of experimental Data Acquisition (DAQ) procedure.

Fig. 4. Experimental set-up of a piping system tested under earthquake loading (Reza et al., 2013)

Fig. 4 illustrates the relevant set-up of the experiment. As can be seen in this figure, the test specimen, i.e. the piping system, is excited with earthquake loading by means of two actuators which are controlled via an MTS controller. The test specimen is mounted with several sensors, such as strain gauges and displacement transducers, in order to observe its responses under applied seismic loading. In this particular experiment, four Spider8 DAQ systems were used to collect data from the sensors. Generally, output from a sensor, e.g. displacement transducer, is found in voltage, which is then transformed in another unit, such as mm, through a predefined calibration made in the DAQ measurement software. This data are then stored in a computer in an easily manageable format, such as Matlab (.mat) excel or ASCII, which are published in the ontology.

VIII. EXPERIMENTAL SET-UP

In Fig. 5, we describe the process of creating the KB. The domain specific ontology that we developed was published into RDF by means of Jena (a Semantic Web tool for publishing and managing ontologies) and integrated with WordNet RDF using the approach described in Section VI. In order to increase the coverage of the background knowledge in the KB, we performed the integration of the two ontologies. The outcome of the ontology integration was put in Virtuoso triple store.

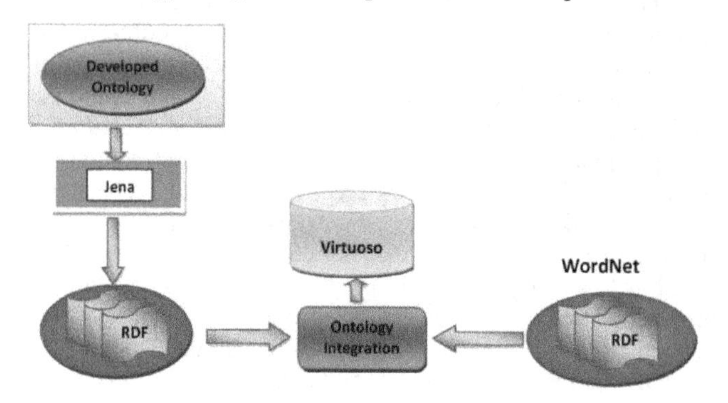

Fig. 5. Ontology Integration and Population to KB

Fig.6 illustrates the architecture of our KB-based information management system that uses Semantic Web tools and technologies. As presented in the figure, the system is organized into three layers, which are User Interface (UI), Middleware and KB.

To execute any user request, for example, visualizing the whole ontology or part of it, the corresponding service is called from the middleware. Each service communicates with the KB using SPARQL query. SPARQL is a query language especially designed to query RDF representations. It allows add, update and delete of RDF data.

User Interface: Developed user interface allows people to perform the following operations on the ontological TBoxes: edit, search, integration and visualization, which are shown in the upper-most layer of Fig. 6 alongside the following operations defined to be performed on the ABoxes: edit entity, entity navigation and experimental result visualization. With the edit ontology operation, concepts and relations can be created, deleted and updated. With the search ontology operation, concepts can be queried with their natural language labels. For the aggregation of an external ontology with the ones already present in the KB we perform the integration operation. In order to view and surf any of the ontologies, we employ (ontology) visualization operation. Note that in the KB until now we have two ontologies, WordNet and EERPE.

Edit entity operation is designed to help perform create, delete and update entities. Existing entities can be viewed and browsed with the entity navigation operation and experimental results can be shown with the corresponding visualization operation.

Middleware: All the functionalities germane to the operations that can be requested and eventually be performed from the user interface are implemented as services and deployed on a web server.

Each service is basically communicating with the KB to execute one or more of the CRUD (create, read, update and delete) operations on its knowledge objects.

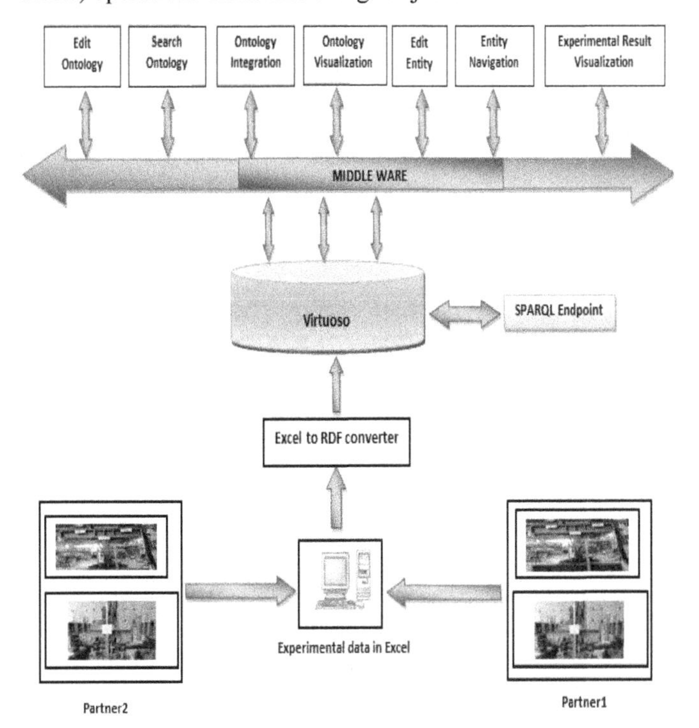

Fig. 6. KB-based System Architecture

KB: This is our Knowledge Base hosting the ontologies consists of concepts and relations thereof, entities and their attributes and relations, and exogenous data from our own experimental setup and the one of our partner university, the University of Napoli.

IX. RESULTS

In Table II, we report the detailed statistics about EERPE ontology. This ontology consists of 11 facets, 193 entity classes, 6 relations and 13 attributes. Note that each of the entity classes, relations and attributes represents an atomic concept. Hence, in total we found 212 atomic concepts in the ontology and out of them 100 concepts are available in WordNet.

TABLE II. STATISTICS ABOUT EERPE ONTOLOGY

Object	Quantity
Facet	11
Entity class	193
Relation	6
Attribute	13
Concept	212
Concepts found in WordNet	100

Moreover, we describe basically what sort of advantages users can get with KB-based systems over traditional DB systems. In particular, we performed synonym search and more specific concept search.

Synonym Search: when a concept is represented with two or more terms, they are essentially synonymous and can be represented in RDF with ***owl:equivalentClass***. For example, test and experiment represent the same concept and in the ontology they are encoded accordingly with equivalent relation. Therefore, as can be seen in Fig. 7, user query for test can also return experiment because they are semantically equivalent.

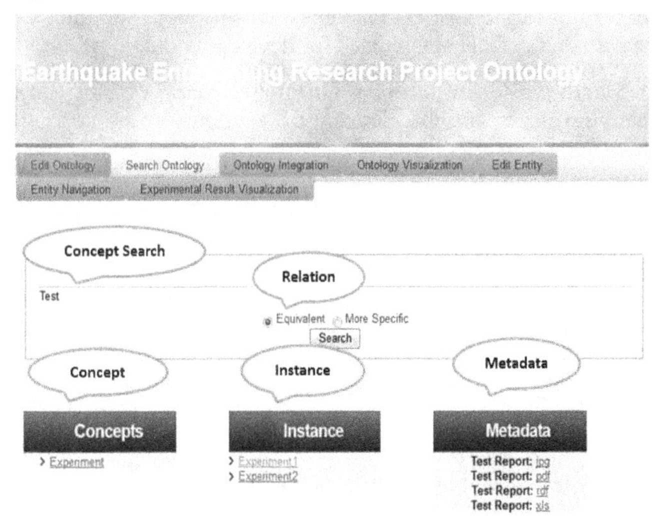

Fig. 7. Synonymus relationship of Test

More specific concept search: In our ontology concept hierarchies are represented using ***rdfs:subClassOf***. For example, hammer and damper are more specific concepts of device, hence, they are represented as follows: hammer ***rdfs:subClassOf*** device; and damper ***rdfs:subClassOf*** device.

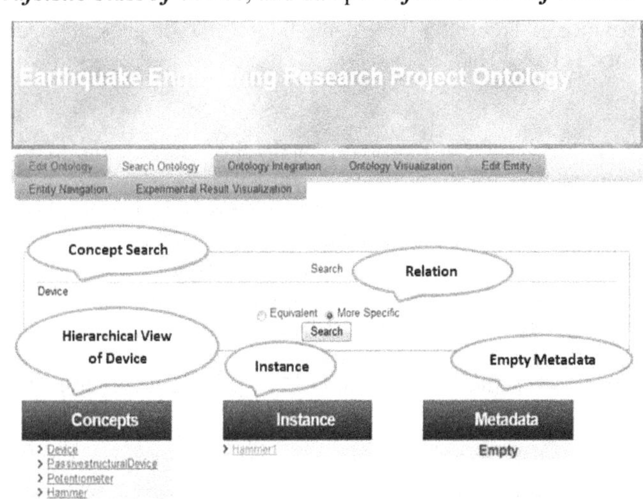

Fig. 8. Transitive Relationship of Device

Moreover, hydraulic damper is more specific than damper and it is encoded as hydraulic damper ***rdfs:subClassOf*** damper. Note that ***rdfs:subClassOf*** is a transitive relation. Using OWL inference engine, we can utilize the power of transitivity and for a given concept we can retrieve all the more specific concepts that are directly or indirectly connected by ***rdfs:subClassOf***. Therefore, a search for device retrieved all of its more specific concepts as shown in Fig. 8.

In addition to the search facility, we have implemented ontology editing, integration and visualization, entity editing and navigation and experimental result visualization functionalities. We have tested them with the help of a number of users. Their feedbacks were satisfactory.

X. Related Work

We have classified the related works into two kinds. One covers the earthquake engineering ontology topic and another focuses on the faceted approach for developing ontologies.

Earthquake Engineering Ontology: NEES ontology has been developed in the domain of earthquake engineering. However, it is mainly a thesaurus encoding broader and narrower relations that cannot capture ontological details. For instance, it cannot be clarified in thesaurus whether a relation between two concepts is *IS_A* or *PART_OF*. As a result ontologies represents as thesaurus might lead to some unexpected results. DB Pedia is an example that uses broader/narrower relations and ended up establishing connection between Telecommunication, and Flora and Fauna. In contrast, the ontology developed in this paper does not suffer from this issue; rather it provides better clarification because it exploits ontological relations.

Faceted ontology development: This approach was followed in developing Geo WordNet, a faceted ontology aimed at building geospatial Semantic Web and enhancing interoperability among numerous information systems developed in isolations dealing with data of the geographic domain (Giunchiglia et al. 2010). By taking into account the advantages offered by this approach, such as easy to follow and linear time requirement, it was employed in the creation of some other ontologies including the one for the Autonomous Province of Trento for developing their semantic geo-catalogue (Farazi et al. 2011).

XI. Conclusion

In this paper, we provided a detailed description of the development of Earthquake Engineering Projects and Experiments ontology. We followed DERA methodology for building this domain specific ontology. We exploited an ontology integration algorithm that was employed to incorporate our ontology into WordNet. It helped to increase the coverage of the Knowledge Base. On top of the integrated ontology that is kept in an instance of Vrituoso, we experimented the semantic and ontological capabilities of the developed system and interesting results were found.

The need for ontologies in Earthquake Engineering is demonstrated, and it has been shown that ontology can be a useful tool for knowledge codification, management, sharing and reuse. We have planned the following future works. We will improve the query performing capabilities using Natural Language Processing (NLP) techniques. We will also include automatic ontology updating feature employing supervised machine learning approach.

Acknowledgment

The work presented in this paper is part of ReLUIS research project, funded by the Italian Department of Civil Protection.

References

[1] Grigoris, A. and Harmelen, F. V. Semantic Web Primer. the MIT Press, 2004.

[2] Farazi, F., Maltese, V. , Giunchiglia, F. , Ivanyukovich, A. A faceted ontology for a semantic geo-catalogue, Extended Semantic Web Conference (ESWC), 2011.K. Elissa, "Title of paper if known," unpublished.

[3] Giunchiglia, F., Dutta, B. DERA: A Faceted Knowledge Organization Framework, March 2011

[4] Giunchiglia, F., Dutta, B., Maltese,V., Farazi, F. A Facet-Based Methodology for the Construction of a large-scale GEO-SPATIAL Ontology, 2012.

[5] Giunchiglia, F., Farazi, F., Tanca, L. , R. de Virgilio. The Semantic Web Languages. Semantic Web Information Management, Springer Verlag, Berlin, 2010.

[6] Giunchiglia, F., Maltese, V., Farazi, F., Dutta, B.. GeoWordNet: a resource for geo-spatial applications. In The Semantic Web: Research and Applications. Springer Berlin Heidelberg, 2010.

[7] Miller, G. A., Beckwith, R. , Fellbaum, C. D. , Gross, D., Miller, K. WordNet: An online lexical database. Int. J. Lexicograph. 3, 4, pp. 235–244, . 1990.

[8] Klyne, G., Carroll, J. (Editors). Resource Description Framework (RDF): Concepts and Abstract Syntax, W3C Recommendation, 10 February 2004.

[9] Ranganathan, S. R. Prolegomena to library classification. Asia Publishing House,1967.

[10] Berners-Lee, T. Weaving the Web. Orion Business Books, 1999.

[11] Berners-Lee, T., Hendler, J. A. and Lassila, O. The Semantic Web. In Scientific American Journal, 284(5), 34-43, (2001).

[12] Reza, Md Shahin, et al. "Pseudo-Dynamic Heterogeneous Testing With Dynamic Substructuring of a Piping System Under Earthquake Loading." ASME 2013 Pressure Vessels and Piping Conference. American Society of Mechanical Engineers, 2013.

[13] Hasan, M.R., et al. A Semantic Technology-Based Information Management System for Earthquake Engineering Projects and Experiments. In 5th International Conference on Advances in Experimental Structural Engineering, Taipei, Taiwan: National Center for Research on Earthquake Engineering, 2013.

[14] Sure, Y., Angele, J., Staab, S. OntoEdit: Multifaceted Inferencing for Ontology Engineering, Journal on Data Semantics, LNCS 2800, Springer, 2003, pp. 128-152.

[15] Maedche, S. Staab.: KAON: The Karlsruhe Ontology and Semantic Web Meta Project, Künstliche Intelligenz. Special Issue on Semantic Web 3/2003, pp. 27-30.

[16] Gennari, John H., Mark A. Musen, Ray W. Fergerson, William E. Grosso, Monica Crubézy, Henrik Eriksson, Natalya F. Noy, and Samson W. Tu. "The evolution of Protégé: an environment for knowledge-based systems development." International Journal of Human-computer studies 58, no. 1 (2003): 89-123.

[17] Giunchiglia, F. et al. "GeoWordNet: a resource for geo-spatial applications." The Semantic Web: Research and Applications. Springer Berlin Heidelberg, 2010. 121-136.

[18] Giunchiglia, F., Dutta, B., Maltese, V. "Faceted lightweight ontologies." Conceptual Modeling: Foundations and Applications. Springer Berlin Heidelberg, 2009. 36-51.

30

Mobile Subscription, Penetration and Coverage Trends in Kenya's Telecommunication Sector

Omae Malack Oteri
Department of Telecommunication and Information Engineering
Jomo Kenyatta University of Agriculture & Technology
Nairobi, Kenya

Langat Philip Kibet
Department of Telecommunication and Information Engineering
Jomo Kenyatta University of Agriculture & Technology
Nairobi, Kenya

Ndung'u Edward N.
Department of Telecommunication and Information Engineering
Jomo Kenyatta University of Agriculture & Technology
Nairobi, Kenya

Abstract—**Communication is the activity of conveying information through the exchange of thoughts, messages, or information, as by speech, visuals, signals, writing, or behavior. In Kenya the mobile subscription, penetration and coverage have been growing since the first mobile operators started operating in 1999. The current mobile operators in Kenya are given by Safaricom Ltd, Airtel Networks Kenya Ltd, Essar Telecom Kenya Ltd and Telkom Kenya Ltd (TKL-Orange).**

This paper discusses the present condition of the telecommunication sector in Kenya and the trends in subscription, penetration and coverage since 1999 when the first two mobile operators (Safaricom Ltd and Celtel now as Airtel Networks Kenya Ltd) started their operations and later the introduction of two other operators Essar Telecom Kenya Ltd and Telkom Kenya Ltd (Orange brand) in 2008. The paper also tries to find out what is likely to happen in a few years to come and provision of a set of recommendations based on the analysis. The study was based on extensive literature review and secondary data sources mainly from Communication Authority of Kenya (CAK). The data obtained was analyzed using Matlab and Microsoft excel to obtain the relevant graphical representations as given in results and discussions.

Keywords—Communication; Telecommunication; Cellular phone; Civilization; Trends

I. Introduction

Communication has been part and parcel of the human race since time immemorial. Telecommunication is the science of transferring information over long distances by electrical and electromagnetic waves which started with the telegraph, telephone, radio, television, satellites, internet and now commonly mobile communication. It can also be said to represent any process or group of processes that allow the transmission of audio, video information or data over long distance by means of electromagnetic or electrical signals. The invention of a practical telegraph in the late 1830s and the telephone in 1876 brought humanity into the era of electrical telecommunications. After over 100 years of triumph, public telegraphy left the communications market. But telex took over and evolved into massively popular forms such as SMS and email, while telephony remains the most widely used means of communication in the world today. Over the past century or so, communications technologies have advanced from manual to automatic switching and from analog to digital communications, alongside many other major technological revolutions.

However, the greatest changes in consumer experience did not occur until mobile phone and the Internet came out and became common elements of life, and which continue to generate even larger changes [1].

In our study we shall look at it as the transfer of information from one location to the other using mobile phone or cellular phone.

In earlier times a "Telephone" was a symbol of status. It was quite a difficult and lengthy process for one to have a telephone connection at his/her home. This time provided by only telecom Kenya up to 1999 when we had the entrance of mobile operators Safaricom and Celtel. In case of an emergency when someone needed to call abroad there was so much harassment in getting a line or if you were lucky to get it, was not clear, distance call rate was so high that sometimes poor people could not afford. Even locally for those who were not lucky enough to get a line had to make long queues behind a telephone booth (located at long distances from one's home) to make a call. But now the situation is totally changed in that technology is available which can avail the information within a very short time thanks to mobile phones. This has drastically changed our lives by making it easy for us as everyone can attest to this. Employment opportunities have increased whereby most graduates get placed in formal employment in the telecommunication sector which can take graduates from various fields.

For instance telecommunication, information technology, human resource and other graduates can easily fit into this sector. The Government has also greatly benefited from the revenues from this sector taking an example of companies like Safaricom contributing up to a tune of several billions in a year. Also information technology has got a big boost by mobile phone technology growth more so in the mobile application software development section. These rapid changes have been made possible with the mobile technology growth. From the information found in the yearly reports from Communications Authority of Kenya (CAK) reveals that mobile phone subscription in Kenya reached the 32.2 million subscriptions by June 2014 [2]. The mobile penetration has not been left behind in which case it has also been increasing from 1999 to march 2014 where it hit a high of 78%.

A. Statement of the problem

Information communication technology is an important sector in any country where it plays a very important role in a nation's economy. Researchers have not looked into the study of the trends of mobile subscription and penetration over the period from 1999 to 2013. This research is intended to have a detailed study on the trends of this industry in Kenya and maybe world over. This can help the telecommunication industry understand the trends and what is likely to happen in the near future so that they can give it a strategic response.

The telecommunications industry in Kenya has been undergoing rapid changes. The Communications Authority of Kenya has in the past 15 years licensed four mobile operators (Safaricom, Airtel and Essar (Yu); all of which are global operators) and several internet service providers like Wananchi and Jamii Telkom. Clearly, competition in this sector has greatly intensified in both the voice and data service provision.

B. Research objectives

The broad objective of the paper is to make an extensive study on the trends experienced in the telecommunication sector in Kenya. The specific objectives are:

- Determine the mobile subscription trend in Kenya.
- Determine the mobile penetration trend in Kenya.
- Analyze the above two.
- To find out the determinants that affects the expansion (growth) of the sector.

C. Importance of the study

The study would be valuable to several stakeholders for the following reasons:

1) Enable the telecommunication companies to know the trends in mobile subscription in Kenya.

2) Help companies to make informed future plans.

3) Enable the telecommunication companies to know the trends in mobile penetration in Kenya.

4) Telecommunication companies can use the findings to develop appropriate policies and strategic responses to the challenges as result of the rate of subscription and penetration.

5) Enable CAK to give proper guidance to the mobile phone operators.

The study would also provide a source of inspiration to a researcher for self-professional development and enrichment.

II. LITERATURE REVIEW

A. Telecommunication sector in Kenya

Up to date there are four mobile operators in our country namely Safaricom Ltd, Airtel Networks Kenya Ltd, Essar Telecom Kenya Ltd and Telkom Kenya Ltd (Orange). According to the mobile subscription and profitability Safaricom Ltd is in the top position among the four operators [2]. The other companies have lower market shares as shown in the study but their main companies are the world's famous and big organizations like Airtel has very high subscription in most Asian countries including India (highest with 183.61 million subscribers as of November 2012) and Bangladesh (third with 5.1 million subscribers as of June 2011) [3]. They have invested a lot and also have more plans for investment having in mind that their key objective is to earn profits. By the end of the third quarter of the 2013/14 financial year (June. 14), there were a total of 32.2 million subscriptions in the mobile telephony market segment up from 31.8 million posted in the previous quarter. This represents an increase of 5.6 percent during the period. Mobile penetration grew to reach 79 percent from 78 recorded at the close of the previous quarter.

B. Mobile subscription

Mobile cellular telephone subscriptions are subscriptions to a public mobile telephone service using cellular technology, which provide access to the public switched telephone network. Post-paid and prepaid subscriptions are included [4]. The number of mobile subscribers usually gives an indication of how vibrant the telecommunication sector of a country is. It also shows the rate of growth of the sector. It can help many companies determine their stage of growth and respond strategically to the different challenges that come with each stage. The market share for each player in this field can also be determined using this very important indicator.

C. Mobile penetration

Mobile phone penetration rate is a term generally used to describe the number of active mobile phone numbers (usually as a percentage) within a specific population [5]. This value can go beyond 100% due the fact that one person can have more than one SIM-card. This can be noted from countries like Qatar which has 170% [6] and most of Europe with 128% [7].

D. Mobile coverage

In telecommunications, the coverage of a radio station is the geographic area where the station can communicate. Broadcasters and telecommunications companies frequently produce coverage maps to indicate to users the station's intended service area. Coverage depends on several factors, such as orography (i.e. mountains) and buildings, technology, radio frequency, transmitted power and distance from the station. Some frequencies provide better regional coverage, while other frequencies penetrate better through obstacles, such as buildings in cities.

The ability of a mobile phone to connect to a base station depends on the strength of the signal. That may be boosted by higher power transmissions, better antennae and taller antenna masts. Signals will also need to be boosted to pass through buildings, which is a particular problem designing networks for large metropolitan areas with modern skyscrapers. Signals also do not travel deep underground, so specialized transmission solutions are used to deliver mobile phone coverage into areas such as underground parking garages and subway trains. This is the same case with indoor coverage [8]. This is also an important parameter in showing the number of people who are likely to receive a signal depending on their location. This parameter can go up to a maximum of 100% since it deals with the area occupied by a country.

III. RESEARCH METHODOLOGY

A. *Research design and data collection*

This study basically covers a period of 13 years starting from 1999 to 2013. An attempt has also been made to include the latest information whenever available.

Much of the information for this research was obtained from secondary sources i.e. the internet e-journals, e-books and websites. This was done by collecting data from the reports of the Communications Commission of Kenya (CAK), via the internet.

Annual reports of different telecom companies, articles published in newspapers, conference papers and seminars proceedings were also carefully studied to procure the needed information. The report only presents simple frequency and quantitative tables. Various statistical tools and techniques have been applied for the analysis and interpretation of data.

B. *Data analysis*

For this study, the content analysis technique was employed to analyze the data. Matlab and Microsoft Excel Spread Sheet, with the associated trend analysis and graphical representation techniques were used to analyze quantitative data. The full report on the key findings of this study by the researcher is presented in section below.

IV. FINDINGS AND DISCUSSIONS

A. *Introduction*

This section deals with analysis and discussion of the research findings. Data was collected from the CAK website in which case the mobile subscription and penetration are analyzed. Following is a report on the key findings of this study by the researcher.

B. *Market share analysis of telecom sector in Kenya:*

TABLE I. TOTAL MOBILE SUBSCRIPTIONS

Subscription Type	June-14	March-14	Quarterly Variation (%)
Prepaid Subscriptions	31,580,696	31,222,434	1.1
Post-Paid Subscriptions	665,697	607,569	9.5
Total Mobile Subscriptions	**32,246,393**	**31,830,003**	**1.3**

Mobile subscription by June 2014

By June 2014 the total number of mobile subscriptions was recorded as 32.2 million up from 31.8 million posted in the previous quarter (March the same year). This represents an increase of 5.6 percent during the period. The continued growth in mobile subscriptions indicates that there is still opportunity for growth in the mobile telephony services. However, the rate of growth in the subscriber base is flattening as the sector progressively tends towards maturity. Consistent, attractive promotions and special offers coupled with competitive retail tariffs could have contributed to the increase in subscriptions during this period. The prepaid subscriptions grew by 1.1 percent during the period compared to 9.5 percent growth recorded in post-paid subscriptions. Even though the growth rate for post-paid subscriptions was more than that for pre-paid, the ratio of prepaid subscriptions continued to dominate and represented 98.0 percent of the total mobile subscriptions. This is consistent with the trends in developing countries where prepaid service is preferred due to the ease and convenience of subscription compared to postpaid which has requirements that are not within the reach of the majority of the population. The reduction of the value of prepaid calling cards to as low as KES 5.00 has made prepaid services a choice for most low income subscribers. Besides the availability of low value cards one can now easily access credit through money transfer methods like M-pesa and advance credit like Okoa Jahazi.

The growth of mobile subscriptions is shown in Table I.

Among the operators at the end of June 2014, Safaricom had the highest subscriber with 21,928,450, Airtel is in the second positions with the total subscriber base of 5,068,765, Essar Telecom Kenya Limited with 2,563,810 and then Telkom Kenya Limited (Orange) with 2,685,368 subscribers which stands as the fourth largest mobile phone operator in Kenya.

During the period under review, three mobile operators recorded positive gains in subscriptions. Telkom Kenya Limited (Orange) recorded the highest gains in new subscriptions of 231,470, representing a growth of 9.4 percent compared to the previous quarter. Safaricom Limited gained 361,062 (1.7% growth) new subscribers while Airtel Networks

Kenya Limited lost 130,005 (-3.5 % growth). Essar Telecom Kenya Limited on the other hand gained 511,526 subscriptions, representing 0.2 percent increase from the previous quarter. The following table shows the number of mobile subscribers in Kenya as of June 2014 [2].

TABLE II. MOBILE SUBSCRIPTIONS PER OPERATOR

Name of operator	June-14		
	Pre-paid	**Post-paid**	**Total**
Safaricom Limited	21,405,667	522,783	21,928,450
Airtel Networks Kenya Limited	4,930,774	137,991	5,068,765
Essar Telecom Kenya Limited	2,562,339	1,471	2,563,810
Telkom Kenya Limited (Orange)	2,681,916	3,452	2,685,368
Total	**31,580,696**	**665,697**	**32,246,393**

March-14			Quarterly Variation (%)
Pre-paid	**Post-Paid**	**Total**	
21,094,414	472,974	21,567,388	1.7
5,121,082	130,005	5,251,087	-3.5
2,556,110	1,520	2,557,630	0.2
2,450,828	3,070	2,453,898	9.4
31,222,434	**607,569**	**31,830,003**	1.3

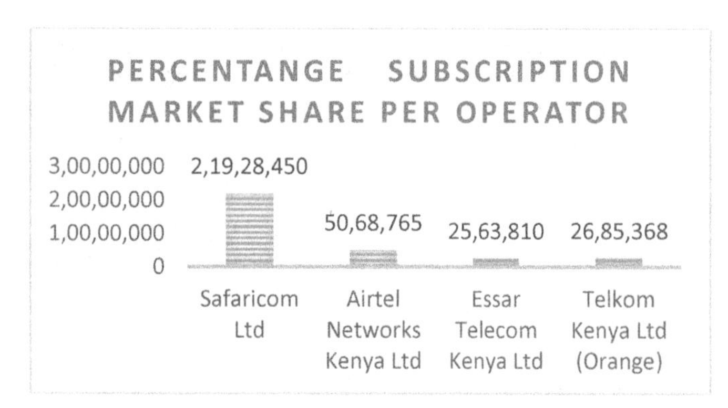

Fig. 1. Percentage market share per operator as of June 2014

The level of market share measured by subscription is still dominated by Safaricom. Safaricom Ltd market share by subscription was 68 percent during the period under review. Telkom's (Orange) market share percentage points stands at 8 percent at the end of the quarter.

Airtel Networks Kenya Limited's market share is given by 16 percent while that of Essar Telecom's is at 8 percent. The market share by subscription by operator is shown in Figure 1.

Mobile subscription from 1999 to 2013

The table below summarizes the key findings on the mobile subscription from the year 1999 to 2013. The data provided is on yearly basis taking subscription as of December each year.

Since the Communication Authority of Kenya (CAK) then CCK issued the two mobile licenses in 2000 (Safaricom and Celtel now Airtel), the mobile telecommunication continued to be popular with consumers. For the first five years the sub sector registered over 60% annual growth with over 16,233,833 subscribers at the end of December 2008. In 2005 the growth rate shot up to 106% but then started declining from 2006 to December 2011 where it was 12%. The sudden increase could be attributed by the fact that around this time there were so many promotions. The difference between the subscribers in 2008 and 2007 was close to 4million because of the introduction of the other two mobile operators TKL-Orange brand and Essar Telcom Kenya limited in 2008.

TABLE III. MOBILE SUBSCRIPTIONS, 1999 TO 2013

Year	**1999**	**2000**	**2001**	**2002**
Number of mobile subscribers	15,000	114,000	585,131	1,325,222
2003	**2004**	**2005**	**2006**	**2007**
1,590,785	2,546,157	5,263,676	7,340,317	11,440,077
2008	**2009**	**2010**	**2011**	**2012**
16,233,833	19,364,559	24,968,891	28,080,771	30,731,754
2013				
31,830,003				

The highest difference was between 2010 and 2009 attributed most probably by the fact that the Orange grew with a higher rate and also the competition of the new entrants with the already existing mobile operators. From 2012 to 2013 increase was only 3.6 percent mostly as a result of sim card registration in Kenya where deadline was 2013 December.

Another reason is due to Mobile tariffs reducing significantly over the period registering an average of KES 2.65 for on-net calls per minute from KES 4.78 per minute in the previous period and KES 2.5 for post-paid subscribers at the end of the period under review.

This represents 33.4 per cent reduction on pre-paid tariffs and 55.5 per cent on post-paid tariffs from the previous period.

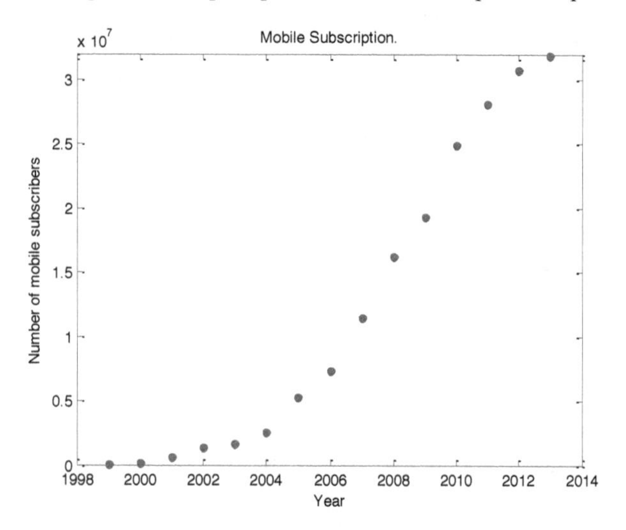

Fig. 2. Mobile subscription from 1999 to 2013

The tariff decline is attributed to an interconnection determination by the Commission during the period that saw mobile termination rates reduced to KES 2.21 from KES 4.42 coupled with increased competition among operators in the mobile market which followed the acquisition of Zain by Bharti Airtel whose business strategy seemed to target the mass market. The figure below is a graphical representation of the number of subscribers against the years.

TABLE IV. MOBILE SUBSCRIPTIONS GROWTH RATE

Year	**1999**	**2000**	**2001**	**2002**	**2003**	**2004**	**2005**
Mobile Subscriptions growth rate		660 %	413 %	126 %	20%	60%	106 %
2006	**2007**	**2008**	**2009**	**2010**	**2011**	**2012**	**2013**
39%	56%	42%	19%	29%	12%	8%	4%

Source: Research findings

From the graph we can be able to see that at the initial stages the growth was not as high as the later stages. Between 2004 and 2011 the growth rate was very high this is due to still there were several people in the country without mobile phone considering the total population of about 40 million. From 2011 onwards the subscription seems to start leveling. The leveling could be due to expected market maturity. Because of this mobile operators should introduce value added services to stay afloat. The services include additions to money transfer and different other mobile technologies (m-technologies). We can conclude that the graphical changes seem to follow a natural trend of exponential growth.

This is because at the initial stages there is a slow growth, the middle is characterized by high growth with steep gradient and towards the end the growth starts to flatten.

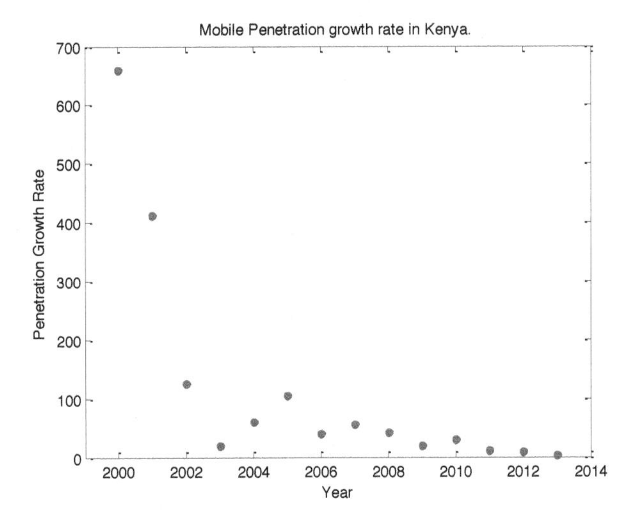

Fig. 3. Mobile subscriptions yearly growth rate

The figure above is a graphical representation of the penetration growth rate against the years. The graph shows that the period between 1999 and 2002 was high but decreasing by the year. This could be attributed by the fact that during this period the business was at its initial stages of growth. From 2003 to 2005 the growth rate increased. From 2005 to 2012 there were slight variations probably due market maturity in which most of the new subscribers are the youth.

Mobile penetration

Over the last thirteen years mobile penetration has registered an exponential growth from 0.053 in 1999 to 80 in December 2013. The increase in mobile penetration can be attributed to a number of factors. First, the reduction in the value of calling cards from the lowest of Kshs 250 in 1999 to Kshs 5 by September 2012 which has made calling cards affordable to low income earners thus stimulating this positive trend. Secondly, the average costs of making calls has declined from Kshs 27 to Kshs 1 to the same network. This has led to increased uptake of mobile phones as costs of calls become affordable thus increasing subscription rates and penetration. International call charges, on the other hand, have also changed over the period because of the use of VOIP technology. Third, even as the mobile operators adjust their tariffs, the mobile coverage has also increased with services now available to a higher population. The increase in mobile penetration can also be attributed to increase in the number of mobile operators from two in 1999 to four in 2008, increased mobile coverage.

The table above shows the mobile penetration in % in Kenya from 1999 to 2013. The mobile penetration per operator as of December 2013[2]

TABLE V. MOBILE PENETRATION

Year	**1999**	**2000**	**2001**	**2002**	**2003**	**2004**	**2005**
Mobile penetration %	0.053	0.38	1.90	4.20	4.95	7.77	15.74
2006	**2007**	**2008**	**2009**	**2010**	**2011**	**2012**	**2013**
21.62	33.65	43.64	49.7	63.2	71.3	78	79

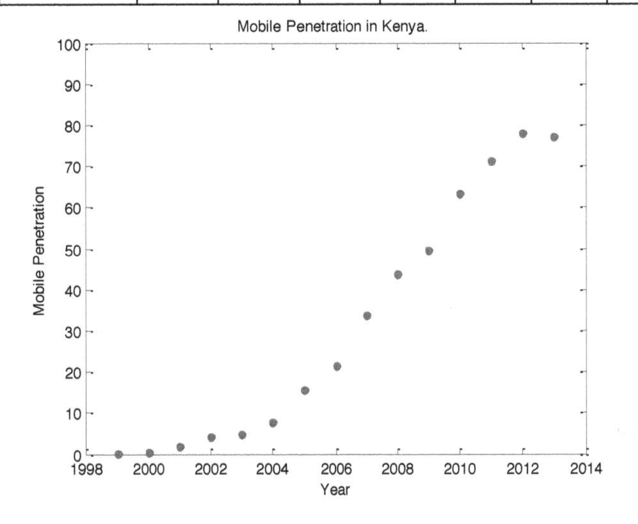

Fig. 4. Mobile penetration in Kenya

The graphical representation above is for mobile penetration against the years from 1999 to 2013. It can be noted that the rate is very low at initial stages being the time when the first two mobile operators were licensed in Kenya. This trend continued up to 2004 when the rate started growing. The growth saw high mobile penetration rate from this time to 2013. This could be attributed by the fact that there were high levels of competition between the first two operators which was even higher after the entrance of two other new operators Orange and Essar. In 2013 it declined probably because of the government policy that required that all mobile phone users and their sim cards be registered by December 2013 where the unregistered cases were to be locked out.

Mobile coverage

The current level of competition in the country has witnessed a network expansion by the four operators to levels that have surpassed their license conditions. This could be attributed to entry of the two mobile operators within the period of December 2008 which may have compelled the existing operators to expand their network coverage in order to solidify their market positions.

Establishment of new sites in areas hitherto considered uneconomical, has in effect increased the level of population coverage including some rural areas.

However, as is the case in most developing countries, this coverage is concentrated in areas with high population densities and promising economic potential especially in urban areas and along major highways. With a national coverage of about 77% of the population, the mobile industry invariably covers over 31 million people in the country. However, the 38% geographic coverage implies that many parts of the country are not covered especially the arid and semi-arid areas. The challenge is how to make these services affordable to the wider population and to encourage investment in high cost areas.

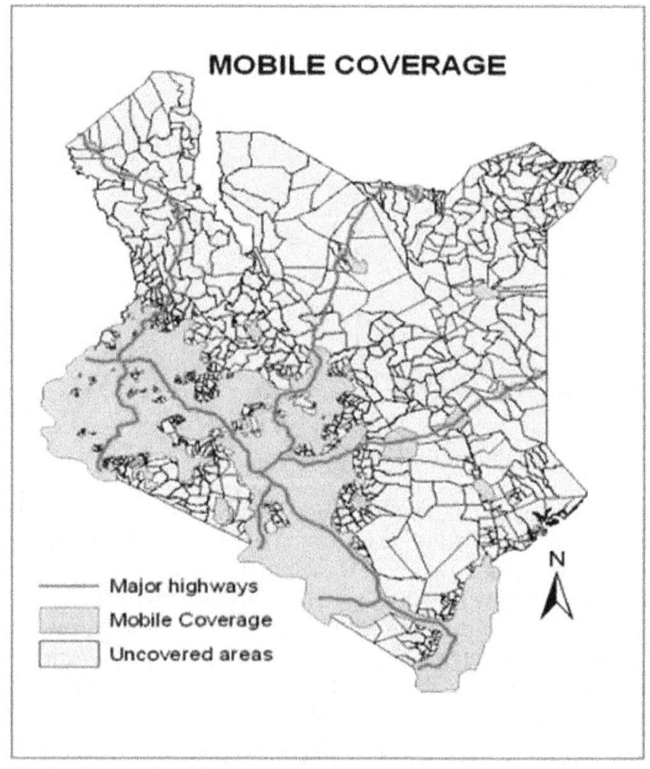

Fig. 5. A map of mobile coverage

As shown in the figure above the covered areas count for a mere 38% by June 2012 while the rest of the country remains uncovered.

V. Summary, Conclusions and Recommendations

A. Summary

The objectives of this study were to analyze the mobile subscription and penetration trends in the telecommunication industry in Kenya. The researcher found out that subscription and penetration rate have been growing since the first two companies Safaricom and Celtel (now Airtel) were licensed. By June 2014 the total number of mobile subscriptions was recorded as 32.2 million up from 15 thousand posted in 1999. Both mobile penetration and coverage have also been steadily growing.

B. Conclusion

There has been significant growth in telecommunication sector particularly in the use of mobile telephony. Competition among the operators, unification of the licenses and the application of new technologies in mobile market segment has witnessed diversification of services by the operators, reduced tariff rates and increased affordability of communication services by a large population. This is further seen as a movement towards closing the digital divide.

The telecommunication landscape in Kenya has continued to be shaped by technological developments arising out of convergence and the new market structure. The subsequent shifts in consumer needs and expectations has compelled aggressive network roll out and infrastructure upgrades using technologies that support high capacity services. Increased competition among the operators has also contributed to the high level of product and service innovations as a means of customer acquisition and customer retention. This is as witnessed in the increase in subscriber base by the fixed network operator.

The outlook of the communication sector in Kenya continues to be strong, and growth will be achieved through gaining new customers, offering new services, and in general capitalizing on the growing role of telecommunications in people's everyday lives. The mobile sector in the country continues to grow increasingly competitive. This is shown with the steady growth in subscriptions.

Increased competition in the mobile sector has resulted in steady growth of this market segment as the services become more affordable. Consequently, the number of mobile subscriptions as well as mobile coverage patterns has continued to demonstrate a positive growth over the period. This trend is likely to continue in future as operators continue employing innovative ways of creating market niche and retaining customers.

The mobile data and internet market experienced an exponential growth on the number of internet users. This is a step forward towards bridging the digital divide in the country with mobile phones being the instrumental platform to accessing internet. The development has greatly contributed in facilitating ease of doing business for small and medium enterprises as operators offer tailor made services for this market at increasingly affordable costs. In addition, internet connectivity has been enhanced by the operationalization of the submarine cables which has led to a reduction to the cost of bandwidth. This in effect has spurred growth in the data market as is witnessed with increased utilization of e-services. Some notable initiatives include the development in e-transactions such as the Mobipay that enables customers buy and pay for goods through the internet, the introduction of e-ticketing services by some bus companies. There is opportunity for further growth with the expected connectivity through an additional cable in the future.

With competitive pressure likely to remain intense among the four service providers, growth in subscriptions is expected to continue to rise further even though it is headed for maturity. This growth is expected to increase as more and more people especially the youth continue to join the social networks like face book, twitters and blogs.

The Information communication technology (ICT) sector continues to experience growth as witnessed in the increase in subscriptions. The sector is largely driven by the mobile telephony, which continues to dominate the sector. Considering this trend, the coming periods are likely to continue recording growth as operators devise innovative products and services expected to entice subscribers and thus propel this sector even further. The steady growth in the mobile industry sector as recorded by increased subscriptions and traffic is expected to continue with constant technological innovations and continued demand for cellular services. The influential factors in the evolution of mobile services are but not limited to the following; rapid growth in the internet, increasingly data hungry applications, data overtakes voice traffic, users want mobility, prices for voice decrease, huge potential for mobile internet, bleak future for voice revenues and access needed to desktop applications when mobile. Even as operators offer attractive promotions and special offers, this will undoubtedly bring about increased subscriptions and growth in traffic during the coming periods. The mobile money transfer service continues to gain popularity due to its convenience and effectiveness. Owing to its growth pattern, operators will be keen to extend the service to the un-banked population and this is expected to expand this service further.

The ease of subscription coupled by the ease of accessing the service through the mobile phone has enhanced growth in this market segment. Moreover, there is still unexploited capacity and potential in this market segment. Consideration for projects geared towards optimal utilization of this capacity could be valuable as this will ultimately stimulate further growth in this market segment.

C. Recommendations

In the light of the foregoing findings by the researcher, Telecommunication companies should keenly implement the following:

- Concentrate their energies in value added services in order lock in the already existing subscribers and also attract new subscription since the market is headed for saturation.
- Look for ways of improving mobile penetration and coverage while balancing on their returns.
- Research on the different ways of increasing quality of service to significantly new competitive levels.

D. Limitations of the study

Major limitation of the study is lack of available information and previous workings on the topic. There are not enough supportive articles to make an extensive literature review. Also, most of the secondary data obtained and used were scattered. Kenya being a dual-SIM market (where over half the subscribers own three or four lines), counting each SIM card bought could be misleading as well as the fact that the data obtained includes unregistered SIM-cards.

E. Areas of further study

Future research should include forecasting the future of telecommunications firms in Kenya using different methods. Another area of study would be to look at the impact of market maturity on the competitiveness of Kenyan telecommunications companies.

F. Future outlook

The future outlook of the telecommunication is positive as the Commission did the implementation of the Kenya Communications Amendment Act 2008. Telecommunication infrastructure will have a major boost with the completion of the National Optic Fibre Backbone Infrastructure. This key infrastructure is expected to increase bandwidth capacity, in most parts of the country which should have a positive impact on internet diffusion and mobile coverage in rural and remote areas. With the recent focus on local content by data service providers' internet usage in the country is expected to rise. Because of the expected market maturity the telecommunication companies should invest more in value added services and improvement of quality of service.

REFERENCES

[1] Hequan, W. (2009). Telecommunications: challenges & transformation [Technology Leaders Forum]. Communications Magazine, IEEE, 47(1), 10-13.

[2] CAK (2008-2014). Quarterly Sector Statistics Reports. (Online) Available: http://www.cck.go.ke/resc/statcs.html

[3] R. Jai Krishna and Prasanta Sahu (2010-01-12). "Bharti Airtel to Buy Warid Telecom for $300 Million - WSJ.com"

[4] The World Bank Group. (Online) Available: http://data.worldbank.org/indicator/IT.CEL.SETS.P2

[5] Pearson press financial times, (online) available: http://www.ftpress.com/articles/article.aspx? p=463943&seqNum=5

[6] Qatar, (online) available: http://www.ameinfo.com/qatar-mobile-penetration-reached-170-q3-330556

[7] (Online) available: http://www.parseco.com/worlds-mobile-penetration/

[8] (Online) available: http://www.askcomreg.ie/mobile/ coverage.44.LE.asp

Automatic Recognition of Human Parasite Cysts on Microscopic Stools Images using Principal Component Analysis and Probabilistic Neural Network

Beaudelaire Saha Tchinda
Laboratoire d'Automatique et d'Informatique Appliquée (LAIA), IUT-FV de Bandjoun, Université de Dschang-Cameroun, B.P. 134 Bandjoun

Daniel Tchiotsop
Laboratoire d'Automatique et d'Informatique Appliquée (LAIA), IUT-FV de Bandjoun, Université de Dschang-Cameroun, B.P. 134 Bandjoun

René Tchinda
Laboratoire d'Ingénierie des Systèmes Industriels et de l'Environnement (LISIE), IUT-FV de Bandjoun, Université de Dschang-Cameroun, B.P. 134 Bandjoun

Didier WOLF
CRAN UMR CNRS 7039, ENSEM Université de Lorraine, Nancy, France.

Michel NOUBOM
Département des sciences Biomédicales, Faculté des sciences, Université de Dschang

***Abstract*—Parasites live in a host and get its food from or at the expensive of that host. Cysts represent a form of resistance and spread of parasites. The manual diagnosis of microscopic stools images is time-consuming and depends on the human expert. In this paper, we propose an automatic recognition system that can be used to identify various intestinal parasite cysts from their microscopic digital images. We employ image pixel feature to train the probabilistic neural networks (PNN). Probabilistic neural networks are suitable for classification problems. The main novelty is the use of features vectors extracted directly from the image pixel. For this goal, microscopic images are previously segmented to separate the parasite image from the background. The extracted parasite is then resized to 12x12 image features vector. For dimensionality reduction, the principal component analysis basis projection has been used. 12x12 extracted features were orthogonalized into two principal components variables that consist the input vector of the PNN. The PNN is trained using 540 microscopic images of the parasite. The proposed approach was tested successfully on 540 samples of protozoan cysts obtained from 9 kinds of intestinal parasites.**

Keywords—Human Parasite Cysts; Microscopic image; Segmentation; Parasite extraction; feature extraction; Principal component analysis; probabilistic neural Network; Parasite Recognition

I. Introduction

The intestinal parasite is a form of human parasite. It is one cause of medical consultations in tropical countries, especially in underdeveloped countries. It was estimated about to 4 billion the number of people infected worldwide [1]. This pathology causes death or physical and mental disorders in children and immune-deficient individuals [1]. The diagnosis of parasitical diseases is performed in the laboratory by the visualization of stools samples through the optical microscopy. Intestinal parasites are classified taxonomically into protozoa and helminths. The protozoa can be seen in stools either on the vegetative form or as the resistant cyst. Helminths are found in the stool in the statement of eggs or larvae. The identification of a parasite is done by the comparison of the morphology observed with the known shapes. This practice is very tedious and is not without consequences for the eyes of laboratory technicians. Also, it is time-consuming and is subject to many diagnosis errors. The identification of amoebic cysts remains the most difficult. Indeed, the cysts are smaller than the helminth eggs, and their distinguishing criteria are more complex. Unlike helminths for which the size is a determinant parameter distinction, many amoeba cysts have almost the same dimensions, and we must use other types of parameters such as the number of the nucleus, for example, to identify them.

During the last decades, several studies employed microscopic image analysis to automatically diagnose the human parasites [2; 3; 4]. Since many parasitic organisms present developmental stages that have a well-defined and reasonably homogeneous morphology, they are amenable to pattern recognition techniques. Each study can be distinguishing from other by the species of parasite concern in the classification, the classification tools and the type of feature using by the classifier. Yang et al. [2] addressed the identification of human helminth eggs by artificial neural network (ANN). Avci et al. [3] and Dogantekin et al. [4] addressed the recognition of human helminth eggs using support vector machines (SVMs) and a fuzzy inference system based on adaptive network, respectively. While these methods are limited to helminths, Castanon et al. [5] used Bayesian classification for the identification of seven species of Eimeria (a protozoan of the domestic fowl); Ginoris et al. [6; 7] used ANN to recognize protozoa and metazoa that are typically found in sludge; and Widmer et al. [8] addressed the identification of Cryptosporidium oocysts and Giardia cysts in water using ANN and immunofluorescence microscopy. These works, however, do not address the identification of human intestinal protozoa in feces, and the segmentation of the parasites is manual. Recently, Suzuki et al. [9] proposed a first solution for automatic identification of the 15 most common species of protozoa and helminths in Brazil.

In this paper, we propose an automated method of human parasite diagnosis via image analysis and an artificial neural network system. Our approach relies on three main steps after image acquisition: edge detection, image segmentation and object recognition. In [10] and [11], we proposed a solution for the two first steps. The present work focuses on the parasite recognition using the results of the precedent works. The main difference of our method from the previously parasite recognition method is the type of feature descriptor used in this work. Our feature descriptor uses the image pixel directly and not need to process other parameters from this image. Our classification tool combines principal component analysis (PCA) for dimensionality reduction and Probabilistic neural network. Our method was applied successfully on nine amoebic cysts types.

The rest of the paper is organized as follow. In section 2, we describe the principles of the methodology. The algorithms of parasite recognition are described in section 3. Experimental results are presented and discussed in section 4. Finally, section 5 presents the conclusions of the work.

II. Methods

A. Edge detection, image segmentation and parasite extraction

Parasite extraction is a crucial step preceding the recognition. An image may contain multiple parasites and we need to identify them individually. Object extraction consists to separate the image region of interest (ROI) from it background. This is done after the segmentation process. The common methods of segmentation can be divided into two categories: the method based on the region and method based on contours. The region oriented segmentation is based on the intrinsic properties of objects to be extracted. This method is highly dependent on the characteristics of the image and shape to extract. The method using contours consists to seek the boundaries of the region to be extracted by exploiting the discontinuity of intensity levels. The usual techniques of edge detection are based on either the gradient or the Laplacian of the image intensity function. We distinguish the following detectors: Sobel, Robert, Prewitt, and Canny. It is shown in [10] that the edge detection based on multi-scale wavelet is better than the other classic detectors when applying on the microscopy image of intestinal parasite. The other advantage for applying the wavelet transform to the detection of edges in an image is the possibility of choosing the size of the details that will be detected [10].The Hough transform has the ability to extract the parametric forms in an image. The Hough Transform has been applied to a wide variety of problems in machine vision, including: line detection, circle detection, detection of general outlines, the detection of surfaces and the estimation of 2 and 3-D motion. The Hough transform (HT) has been used in [12] to segment ultrasound images of longitudinal and transverse sections of the carotid artery. Certain intestinal parasites are circular in its shape. The circular Hough transform can be easily used to locate and extract them. However the location of other random forms remains partial. Another segmentation method uses the active contours model. The active contour method is very effective for the detection of boundaries. An example of the active contours model implementation is presented in [13]. The strong dependence on initial contour has long been considered as its main drawback. An initial contour close to the target contour promotes greater convergence. It is possible to combine the technique of Hough transform to the active contour technique. The Hough transform allows to this new method to automatically locate the region of interest of the parasite. This first result will be the initial contour for active contours model. This approach had been used successfully in [11] to extract intestinal parasites on the microscopy images.

The first step is the edge detection obtained from the multi-scale wavelet transform. The second step process the circular Hough transform from the edge. In this step, each edge element votes for all the circles that it could lie on and the circle corresponding to the maximum vote is retained. This circle locates the region of interest around of which the parasite is situated. In the third step, the active contour based on gradient vector flow model is computed. The active contour uses the circle obtained from the Hough transform as its initial contour. This initial contour is deformed and attracted towards the target contour by various forces that control the shape and location of the snake within the image. The last step uses the final contour to get the mask corresponding to the interior of the area delimited by the contour. The parasite extracted corresponds to a logic operation of the mask with the original image. Besides the microscopic image, our scheme also use as input the length of the radius to find the parasite, the number of convolutions (analyzing scale) to find the edge map by the multi-scale wavelet transform and the threshold of the edge detection. The detailed version of the parasite extraction algorithm can be seen in [11].

B. Feature extraction and dimensionality reduction

Feature extraction aims to represent candidate objects with a simple and representative manner to discriminate an object from other. There is several type of feature descriptor [14]. The main difference of our method in relation to the previously parasite recognition method is the type of feature descriptor used in this work. Our feature descriptor uses the image pixel directly and not need to process other parameters from this image. The dimension of such vectors is very large. There is the need to map those vectors to the lower dimensional space; thereby reducing the computational complexity for further classification. The feature dimension reduction consists of mapping the input vectors of observations $I \in \mathbb{R}^m$ onto a new feature description $X \in \mathbb{R}^n$, with $n < m$. The dimensionality reduction based on the projection technique is considered on this paper. The projection technique is achieved by using transformation matrix M. The Principal Component Analysis (PCA) is a representative of the unsupervised learning method which yields the linear projection [15; 16].

Let consider an input vector $I \in \mathbb{R}^m$ which need to be mapping onto a new feature description $X \in \mathbb{R}^n$. The projection is given by the matrix $M[m \times n]$ and the bias vector $k[n \times 1]$.

Let $T_I = \{I_1, ..., I_l\}$ be a set of training vector from the m-dimensional input space $\mathbb{R}^m$.

$$T_I = \begin{cases} I_1 = [I_{11}, I_{12}, ... I_{1m}]^T \\ . \\ . \\ . \\ I_l = [I_{l1}, I_{l2}, ... I_{lm}]^T \end{cases} \quad (1)$$

The set of vectors $T_X = \{X_1, ..., X_l\}$ is a lower dimensional representation of the input training vectors T_I in the n-dimensional space $\mathbb{R}^n$.

$$T_X = \begin{cases} X_1 = [X_{11}, X_{12}, ... X_{1n}]^T \\ . \\ . \\ . \\ X_l = [X_{l1}, X_{l2}, ... X_{ln}]^T \end{cases} \quad (2)$$

The vectors of T_X are obtained by the linear orthonormal projection

$$X = M^T I + k, \quad (3)$$

where the matrix M and the vector k are parameters of the projection. The reconstruction vector $T_{\tilde{I}} = \{\tilde{I}_1, ..., \tilde{I}_l\}$ are computed by the linear back projection

$$\tilde{I} = M(X - k), \quad (4)$$

obtained by inverting (3). The mean square reconstruction error

$$e_{MS}(M, k) = \frac{1}{l} \sum_{i=1}^{l} \left\| I_i - \tilde{I}_i \right\|^2, \quad (5)$$

is a function of the parameters of the linear projections (3) and (4). The principal component analysis is the linear orthonormal projection (3) which allows for the minimal mean square reconstruction error (5) of the training data T_I. The parameters (M, k) of the linear projection are the solution of the optimization task

$$(M, k) = \underset{M', k'}{\text{argmin}} \left(e_{MS}(M', k') \right), \quad (6)$$

such as $\langle M_i M_j \rangle = \delta(i, j), \ \forall i, j \in \mathbb{N}$,

where M_i, $i = 1, ..., n$ are column vectors of the matrix $M = [M_1, ..., M_n]$ and $\delta(i, j)$ is the Kronecker delta function. The solution of (6) is the matrix $M = [M_1, ..., M_n]$ containing the n eigenvectors of the sample covariance matrix which have the largest eigen values. The vector k equals to $M^T \mu$, where μ is the sample mean of the training data. The PCA is treated throughout for instance in [17].

C. Object classification using the Artificial Neural Network

Artificial Neural Networks are analogous to their namesake, biological neural networks, in that both receive multiple inputs and respond with a single output. ANN classifies input vector into a specific class according to the maximum probability to be correct. These networks have diverse applications in machine vision [18]. One of their applications is in classification and decision making based on existing data [19]. For this goal, the input data is often divided into two parts for training and testing.

In an ANN, multiple neurons are interconnected to form a network and facilitate distributed computing. Each neuron partition constitutes a layer. Networks may contain an input layer, an output layer, and a hidden inner layer. Additional hidden layers may be added to increase the complexity of the network. Weights are assigned to each of the links between neurons, and they are updated as part of the learning process.

The configuration of the interconnections can be described efficiently with a directed graph. A directed graph consists of nodes and directed arcs. The topology of the graph can be categorized as either acyclic or cyclic [20; 21]. A neural network with acyclic topology consists of no feedback loops. Such an acyclic neural network is often used to approximate a nonlinear mapping between its inputs and outputs. A neural network with cyclic topology contains at least one cycle formed by directed arcs. Such a neural network is also known as a recurrent network. Due to the feedback loop, a recurrent network leads to a nonlinear dynamic system model that contains internal memory. Recurrent neural networks often exhibit complex behaviors and remain an active research topic in the field of artificial neural networks. Network topology architecture was systematically chosen in terms of variance of classification results and its complexity. The Probabilistic Neural Networks (PNN) is used in this paper to classify the type of intestinal parasite. This choice is adopted for its advantages [22]. The main advantage of PNN is that training is easy and instantaneous [23; 24; 25]. The architecture for this system is shown in Fig.1. We adopt the symbols and notations used in the book "neural network toolbox for use in Matlab" [26].

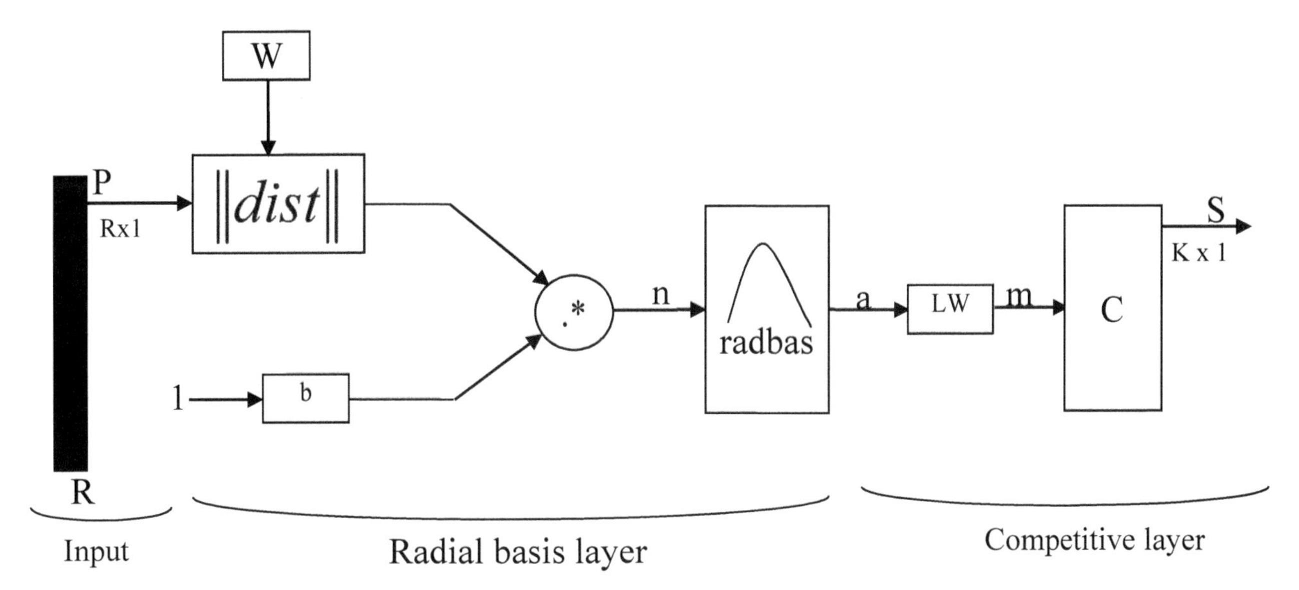

Fig. 1. Neural Network architecture

Our PNN has three layers: the Input layer, Radial Basis Layer and the Competitive Layer.

The input layer has two units, corresponding to the two features. The input vector, denoted as P, is presented as the black vertical bar in Fig. 1. Its number of element R corresponds to the number of neurons in the Radial Basis Layer.

In Radial Basis Layer, the vector distances between input vector P and the weight vector W is calculated. This operation is represented in the Fig. 1 by the $\|dist\|$ box and its output gives $\|W-P\|$. The bias vector b and the output $\|W-P\|$ are combined by an element-by-element multiplication, represented as ".*" in Fig.1. The result is denoted as $n=\|W-P\|.*b$. The transfer function in the radial basis network is define as

$$radbas(n)=\exp(-n^2) \quad (7)$$

Each element of n is substituted into (7) and produces corresponding element of a, the output vector of Radial Basis Layer. We can represent the i-th element of a as

$$a_i = radbas(\|W_i - p\|.*b_i) \quad (8)$$

where Wi is the vector made of the i-th row of W and b_i is the i-th element of bias vector b. An input vector close to a training vector is represented by a number close to 1 in the output vector a. If an input is close to several training vectors of a single class, it is represented by several elements of a that are close to 1.

The competition layer classifies each input in each of the K class of protozoan cysts use during the training phase. There is no bias in the competitive layer. In competitive layer, the vector a is firstly multiplied with layer weight matrix LW, producing an output vector m. The competitive function, denoted as C in Fig. 1, produces a 1 corresponding to the largest element of m, and O's elsewhere. The output vector of competitive function is denoted as S. The index of K in the output is the number of parasite that our system can classify. It can be use as the index to look for scientific name of the parasite. The dimension of output vector is K=9 in this paper.

III. System Architecture

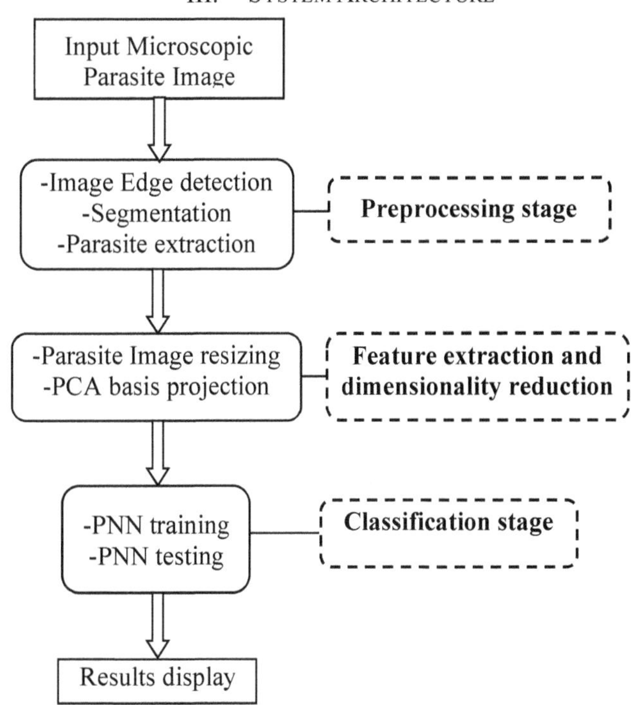

Fig. 2. The overall processing stages of proposed scheme for intestinal human parasite diagnosis

The main goal of this work is to recognize the types of intestinal parasites in the microscopic image of stools. To achieve this purpose, a block diagram is designed based on this type of dataset. Specially, we focus on the classification and recognition of nine types of protozoan cysts. The block diagram of the proposed scheme is illustrated in the Fig. 2. As indicated in the figure, this method has the following main steps:

Step1: image edge detection, segmentation and parasite extraction

Step 2: feature extraction and dimensionality reduction

Step 3: classification and recognition

These steps are explained in the following sections.

A. Step1: image edge detection, segmentation and parasite extraction

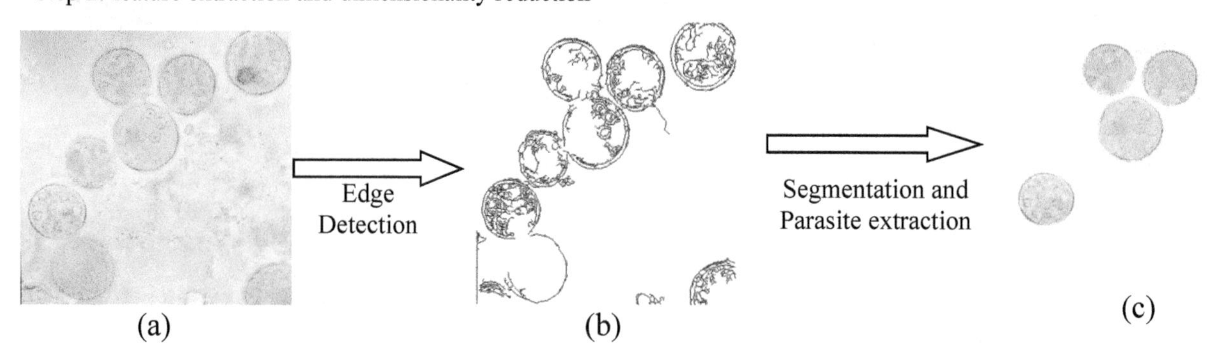

Fig. 3. Representative illustration of a microscopic image preprocessing. (a) Microscopic image of stools, (b) image of edge detected, (c) image of the extracted parasites

Since in an image, we can have several objects which cannot be necessary interest us. Also, image can contain different parasites. We need to extract individually a parasite before the recognition. The parasite extraction is doing via the image segmentation. The segmentation use the contour edge map of the image to separate the parasite from is background. A logic operation is used to suppress the parasite background. Our segmentation method is based on the active contour initialized by the Hough transform. The edge detection method is based on the multi-scale wavelet transform. The Fig. 3 presents the illustration of the edge detection, parasite segmentation and extraction on an image containing protozoan parasite cysts.

B. Step 2: image resizing, feature extraction and dimensionality reducing

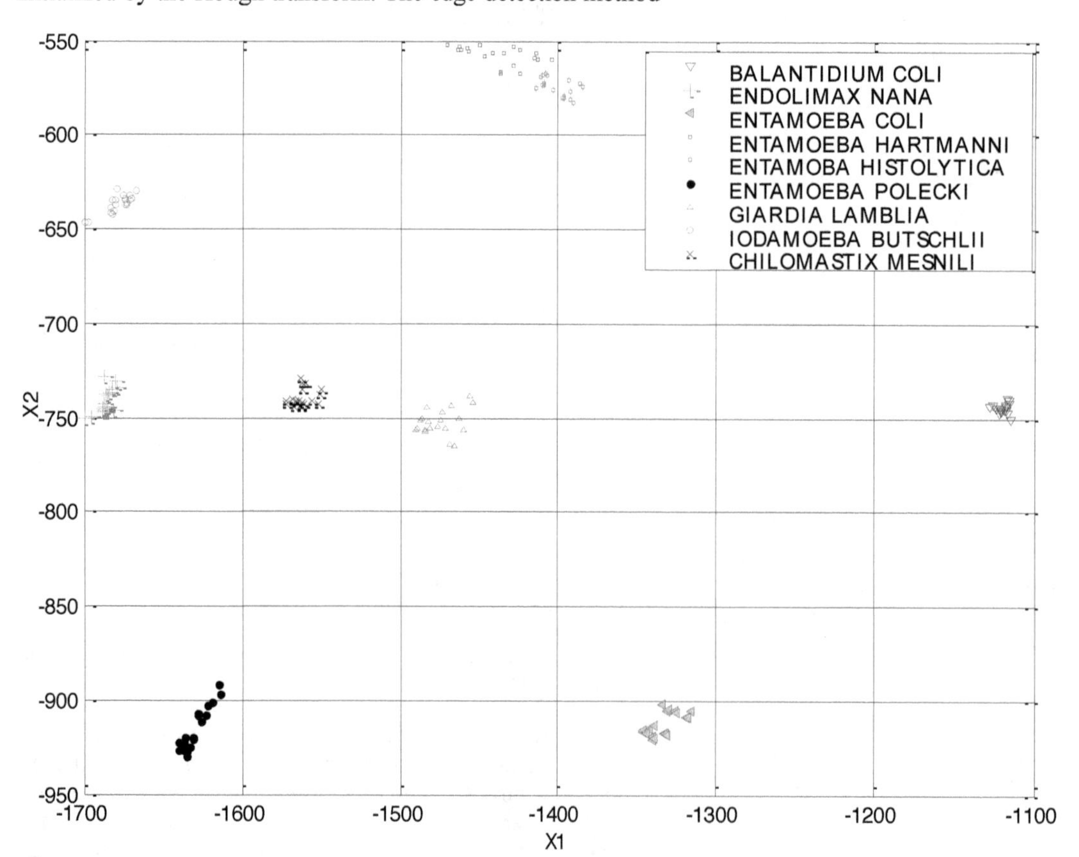

Fig. 4. Two dimensionality projection with two Eigen vectors corresponding to the largest two Eigen values in PCA basis

Our feature descriptor uses the image pixel and does not need to process other parameters from this image. Nevertheless, the image needs to be resized. All image dimension of the extracted parasite has been reduced to the size 12x12. The resizing uses the Bicubic interpolation method. With this method, the output pixel value is a weighted average of pixels in the nearest 4-by-4 neighborhood.

In order to reduce the complexity of the classification and recognition phase, the feature using as input need to be reduced. PCA is used to project the 12x12 features on the new feature space. The goal of PCA is to reduce the dimensionality of the data space (12x12) to smaller intrinsic dimensionality of feature space, which is needed to describe the data economically. For this purpose, the dimensionality of the new feature space is chosen according to the classification rate and the system complexity. In this paper, the first 2 principal components have been used. When using our algorithm, one can use the mapping $f:\mathbb{R}^{144} \rightarrow \mathbb{R}^{2}$ to obtain the values of components in the new coordinate system. The Fig. 4 shows the result obtained in the two dimensionality projection with two Eigen vectors corresponding to the largest two Eigen values in PCA basis. This transformation is applied on 20 trained images of each of the nine types of parasite. In this figure, we can distinct the nine types of parasite. Also, as shown in this figure, each type of parasite is grouped separately from other. The capability of the new features in separating of the 9 classes of parasites can be qualitatively evaluated.

C. Step 3: parasite classification and recognition

In this phase, the reduced features are applied to train and test the neural network. Totally 1080 microscopic images of protozoan cysts are used. These samples were divided in 540, with 60 of each type of 9 parasites for training, and 540 for the test.

For our neural network training, Radial Basis Layer Weights is set to the transpose of R x Q matrix of Q training vectors. Each row of W consists of 2 principal variables of one training samples. Since 540 samples are used for training. In this paper, Q = 540.

In radial basis layer, all biases are all set to $\sqrt{-\ln 0.5}/s$ resulting in radial basis functions that cross 0.5 at weighted inputs of $\pm$s. s is called the spread constant of PNN.

Competitive Layer Weights LW is set to K x Q matrix of Q target class vectors. K is the dimension of the output vector, equal to 9 in this paper. The target class vectors are converted from class indices corresponding to input vectors. This process generates a sparse matrix of vectors, with one 1 in each column, as indicated by indices.

IV. Experimental Results

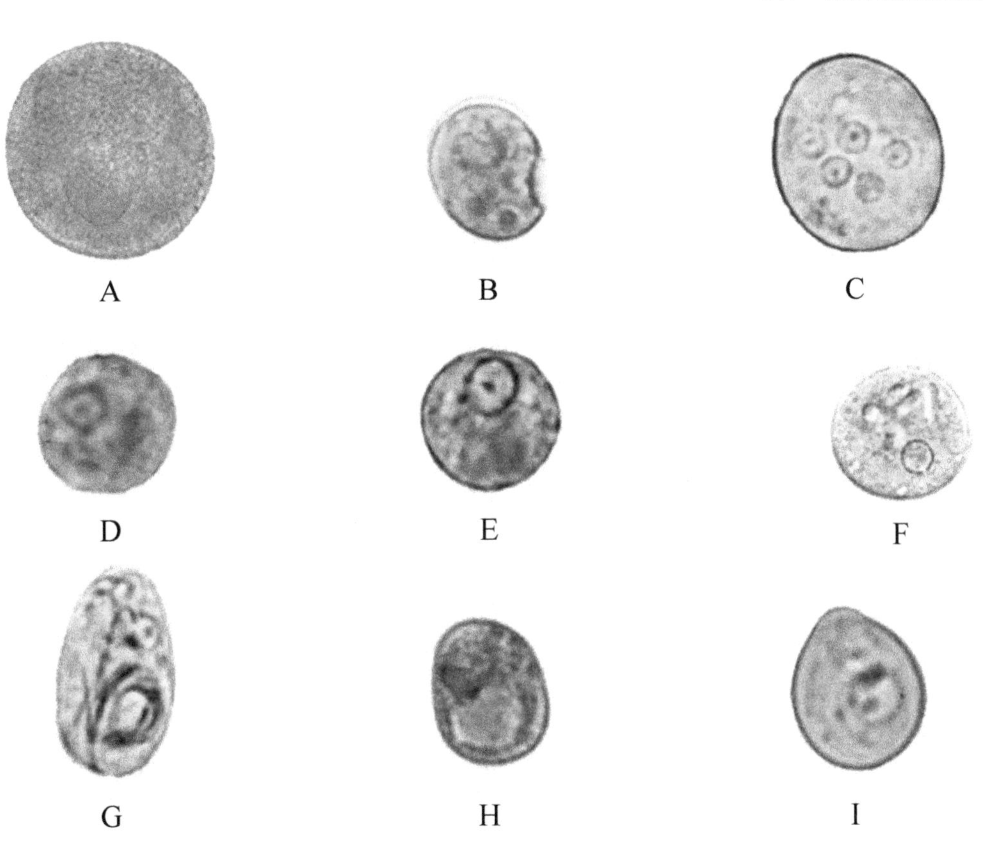

Fig. 5. Digital microscopic images of each of parasite cyst types. A: BALANTIDIUM COLI; B: ENDOLIMAX NANA; C: ENTAMOEBA COLI; D: ENTAMOEBA HARTMANNI; E: ENTAMOBA HISTOLYTICA; F: ENTAMOEBA POLECKI; G: GIARDIA LAMBLIA; H: IODAMOEBA BUTSCHLII; I: CHILOMASTIX MESNILI

In this experimental study, 9 different types of human parasite cyst are used. These types of parasite are: BALANTIDIUM COLI, ENDOLIMAX NANA, ENTAMOEBA COLI, ENTAMOEBA HARTMANNI, ENTAMOEBA HISTOLYTICA, ENTAMOEBA POLECKI, GIARDIA LAMBLIA, IODAMOEBA BUTSCHLII, and CHILOMASTIX MESNILI. Digital microscopic images of each of parasite cyst types are given in Fig.5.

All these parasite cysts are rotated in steps of 30° from 0°-150, five different scales were chosen. Also, these images were bruited with 4 types of image noise ('Gaussian', 'Poisson', 'salt & pepper', 'speckle'). In this way, 120 microscopic images were obtained for each parasite cysts types. For training the classifier, randomly chosen half of the database was used. The spread parameter was chosen to 0.4, for the best performance, after several different experiments.

The rest of the database was used in testing stage. For each kind of parasite, 60 pieces of cyst are used to test the accuracy of our algorithm. The confusion matrix of the proposed expert diagnosis system is shown in table 1. As shown in table 1, all the nine parasite cysts type images were classify with 100 % correct classification rate. This demonstrates the effectiveness of the proposed feature descriptor in the protozoan cyst recognition system based on the principal component analysis and the probabilistic neural network.

TABLE I. CONFUSION MATRIX OF THE CLASSIFICATION SCHEME

	Predicted class									
	Class	**A**	**B**	**C**	**D**	**E**	**F**	**G**	**H**	**I**
Actual class	**A**	60	0	0	0	0	0	0	0	0
	B	0	60	0	0	0	0	0	0	0
	C	0	0	60	0	0	0	0	0	0
	D	0	0	0	60	0	0	0	0	0
	E	0	0	0	0	60	0	0	0	0
	F	0	0	0	0	0	60	0	0	0
	G	0	0	0	0	0	0	60	0	0
	H	0	0	0	0	0	0	0	60	0
	I	0	0	0	0	0	0	0	0	60

V. CONCLUSION

This paper presented an image analysis system to recognize nine kinds of human parasites. Artificial Neural Network can classify automatically the intestinal parasite via the microscopic images of stools loaded from digital cameras or scanner. The probabilistic neural network is adopted for hit has fast speed on training and simple structure. Contrary to previous work, 12x12 features vectors were obtained directly from image pixel. This features that form the input vector of PNN have been projected in a Principal Component Analysis basis for dimensionality reduction.

Results indicate that the image pixel features of microscopic image of human parasites are achievable, and it offers remarkable accuracy when using in a probabilistic neural network classifier with dimensionality reduction of the features by projection on the PCA basis. Further works will be focused on the recognition of other types of human parasites.

REFERENCES

[1] World Health Organization (2010). Working to overcome the global impact of neglected tropical diseases. First WHO report on neglected tropical diseases.

[2] Yang, Y. S., Park, D. K., Kim, H. C., Choi, M., & Chai, J. (2001). Automatic identification of human helminthes eggs on microscopic fecal specimens using digital image processing and an artificial neural network. IEEE Transactions on Biomedical Engineering, 48(6), 718–730.

[3] Avci, D. Varol, A. (2009). An expert diagnosis system for classification of human parasite eggs based on multi- class SVM. Expert Systems with Applications, 36(1), 43–48.

[4] Dogantekin, E., Yilmaz, M., Dogantekin, A., Avci, E. and Sengur, A. (2008). Arobust technique based on invariant moments—ANFIS for recognition of human parasite eggs in microscopic images. Expert Syst. Appl., 35(3), 728–738.

[5] Castanon, C. A. B., Fraga, J. S., Fernandez, S., Gruber A. and Costa L. da F. (2007). Biological shape characterization for automatic image recognition and diagnosis of protozoan parasites of the genus Eimeria. Pattern Recognit., 40, 1899–1910.

[6] Ginoris, Y. P., Amaral, A. L., Nicolau, A., Coelho, M. A. Z. and Ferreira, E. C. (2007a) Development of an image analysis procedure for identifying protozoa and metazoa typical of activated sludge system. Water Res., 41(12), 2581–2589.

[7] Ginoris, Y. P., Amaral, A. L., Nicolau, A., Coelho, M. A. Z. and Ferreira, E. C. (2007b). Recognition of protozoa and metazoa using image analysis tools, discriminant analysis, neural networks and decision trees. Anal. Chim. Acta, 595(1/2), 160–9.

[8] Widmer, K., W. and Srikumar, D., Pillai, S. D. (2005). Use of artificial neural networks to accurately identify cryptosporidium oocyst and giardia cyst images. Appl. Environ. Microbiol., 71(1), 80–84.

[9] Suzuki, C. T. N., Gomes J. F., Falcao, A. X. Papa, J. P. and Hoshino-Shimizu S. (2013). Automatic Segmentation and Classification of Human Intestinal Parasites from Microscopy Images. IEEE Transactions on Biomedical Engineering, 60 (3), 803-812.

[10] Tchiotsop, D., Saha T. B., Tchinda, R. and Kenné G. (2014). Edge detection of intestinal parasites in stool microscopic images using multi-scale wavelet transform. SIViP, DOI 10.1007/s11760-014-0716-6.

[11] Saha T. B., Tchiotsop, D., Tchinda, R. and Kenné G. (2014). Automated Extraction of the Intestinal Parasite in the Microscopic Images using Active Contours and the Hough Transform. Current Medical Imaging Reviews. In press.

[12] Golemati S., Stoitsis J., Sifakis E., Balkizas T., Nikita K. (2007). Using the Hough transform to segment ultrasound images of longitudinal and transverse sections of the carotid artery. Ultrasound in Medicine & Biology; 33 (12):1918–32.

[13] Stoitsis J., Golemati S., Kendros S., Nikita K.S. (2008). Automated detection of the carotid artery wall in B-mode ultrasound images using active contours initialized by the Hough transform. In: Conf. Proc. IEEE Eng. Med. Biol. Soc. 3146-9.

[14] Nixon, S. M. and Aguado, A. S. (2012). Feature extraction and image & processing for computer vision, third edition. Academic Press, Elsevier.

[15] Theodoridis, S. & Koutroumbas, K. (2010). An introduction to pattern recognition: a MATLAB approach, 5th ed.,. Academic Press, Elsevier.

[16] Engelbrecht, A.p. (2007). Computational Intelligence an Introduction, (2nd ed.).. John Wiley & Sons Ltd, The Atrium, Southern Gate, Chichester.

[17] Jollife, I.T., (1986). Principal Component Analysis. Springer-Verlag, New York.

[18] Sumathi, S. and Paneerselvam, S. (2010). Computational intelligence paradigms theory and applications using matlab. CRC Press, Taylor & Francis Group.

[19] Adhikari, A.; Adhikari, J. ; Pedrycz, W. (2014). Data Analysis and Pattern Recognition in Multiple Databases. Intelligent Systems Reference Library, Vol. 61. Springer International Publishing, Switzerland.

[20] Kasabov, N.K. (1998). Foundations of Neural Networks, Fuzzy Systems, and Knowledge Engineering, Second printing. A Bradford Book, the

Massachusetts Institute of Technology Press Cambridge, Massachusetts London, England.

[21] Hu, Y. H. and Hwang, J.N. (2002). Handbook of Neural Network Signal Processing. CRC PRESS LLC. Boca Raton London New York Washington, D.C.

[22] Hagan M. T., Demut, H. B. and Beale M. H. (2002). Neural Network Design

[23] Kramer, C.; Mckay B. and Belina, J. (1995). Probabilistic neural network array architecture for ECG classification. in Proc. Annu. Int. Conf. IEEE Eng. Medicine Biol., 17, 807–808.

[24] Mao, K. Z.; Tan K.-C. and Ser W. (2000). Probabilistic Neural-Network Structure Determination for Pattern Classification. IEEE transactions on neural networks, 11 (4), 1009-1016.

[25] Wu S.G., Bao, F. S., Xu, E. Y., Wang, Y., Chang, Y. and Xiang, Q. (2007). A Leaf Recognition Algorithm for Plant Classification Using Probabilistic Neural Network. IEEE International Symposium on Signal Processing and Information Technology.

[26] Demuth, H. and Beale, M. (2000). Neural Network Toolbox for Use with MATLAB.

Accurate Topological Measures for Rough Sets

A. S. Salama
Department of Mathematics, Faculty of Science,
Tanta University,
Tanta, Egypt
Department of Mathematics, Faculty of Science and Humanities
Shaqra University
Al-Dawadmi, KSA

***Abstract*—Data granulation is considered a good tool of decision making in various types of real life applications. The basic ideas of data granulation have appeared in many fields, such as interval analysis, quantization, rough set theory, Dempster-Shafer theory of belief functions, divide and conquer, cluster analysis, machine learning, databases, information retrieval, and many others. Some new topological tools for data granulation using rough set approximations are initiated. Moreover, some topological measures of data granulation in topological information systems are defined. Topological generalizations using $\delta\beta$ -open sets and their applications of information granulation are developed.**

Keywords—component; Knowledge Granulation; Topological Spaces; Rough Sets; Rough Approximations; Data Mining; Decision Making

I. INTRODUCTION

Granulation of the universe involves the decomposition of the universe into parts. In other words, the grouping individual elements or objects into classes, based on offering information and knowledge [7, 14,15, 21, 36,37, 42-45]. Elements in a granule are pinched together by indiscernibility, similarity, proximity or functionality [43]. The starting point of the theory of rough sets is the indiscernibility of objects or elements in a universe of concern [14,15, 17-20, 51,52, 21-22].

The original rough set theory was based on an equivalent relation on a finite universe U. For practical use, there have been some extensions on it. One extension is to replace the equivalent relation by an arbitrary binary relation; the other direction is to study rough set via topological method [8, 14]. In this work, we construct topology for a family covering rough sets.

In [40] addressed four operators on a knowledge base, which are sufficient for generating new knowledge structures. Also, they addressed an axiomatic definition of knowledge granulation in knowledge bases.

Rough set theory, proposed by Pawlak in the early 1980s [18, 51-52], is an expansion of set theory for the study of intelligent systems characterized by inexact, uncertain or insufficient information. Moreover, this theory may serve as a new mathematical tool to soft computing besides fuzzy set theory [42-45] and has been successfully applied in machine learning, information sciences, expert systems, data reduction, and so on [28-33,34, 1-13]. In recent times, lots of researchers are interested to generalize this theory in many fields of applications [1-10].

In Pawlak's novel rough set theory, partition or equivalence (indiscernibility) relation is an important and primeval concept. But, partition or equivalence relation is still limiting for many applications. To study this matter, several interesting and having an important effect generalization to equivalence relation have been proposed in the past, such as tolerance relations, similarity relations [51], topological bases and subbases [52, 2,6] and others [4,5,11]. Particularly, some researchers have used coverings of the universe of discourse for establishing the generalized rough sets by coverings [11-14]. Others [24-26,27-33] combined fuzzy sets with rough sets in a successful way by defining rough fuzzy sets and fuzzy rough sets. Furthermore, another group has characterized a measure of the roughness of a fuzzy set making use of the concept of rough fuzzy sets [34-38]. They also suggested some possible real world applications of these measures in pattern recognition and image analysis problems [24,41-46].

Topological notions like semi-open, pre-open, $\beta-$open sets are as basic to mathematicians of today as sets and functions were to those of last century [48-52]. Then, we think the topological structure will be so important base for knowledge extraction and processing.

The topology induced by binary relations on the universes of information systems is used to generalize the basic rough set concepts. The suggested topological operations and structure open up the way for applying affluent more of topological facts and methods in the process of granular computing. In particular, the notion of topological membership function is introduced that integrates the concept of rough and fuzzy sets [17-20].

In this paper, we indicated some topological tools for data granulation by using new topological tools for rough set approximations. Moreover, we introduced using general binary relations a refinement data granulation instead of the classical equivalence relations. Section 1 gives a brief overview of data granulation structures in the universe using equivalence and general relations. Fundamentals of rough set theory under general binary relations are the main purpose of Section 2. Section 3 studies the topological data granulation properties of topological information systems. Explanation of topological data granulation in information systems appears in

Section 4. In Section 5 we are given some more accurate topological tools for data granulation using $\delta\beta-$ open sets approach. The conclusions of our work are presented in Section 6.

II. Essentials of Rough Set Approximations Under General Binary Relations

In rough set theory, it is usually assumed that the knowledge about objects is restricted by some indiscernibility relations. The Indiscernibility relation is an equivalence relation which is interpreted so that two objects are equivalent if we can't distinguish them using our information. This means that the objects of the given universe U indiscernible by R into three classes with respect to any subset $X \subseteq U$:

Class 1: the objects which surely belong to X,

Class 2: the objects which possibly belong to X ,

Class 3: the objects which surely not belong to X ,

The object in Class 1 form the lower approximation of X , and the objects of Class 1 and 3 form together its upper approximation. The boundary of X consists of objects in Class 3. Some subsets of U are identical to both of them approximations and they are called crisp or exact; otherwise, the set is called rough.

For any approximation space $A=(U,R)$, where R is an equivalence relation, lower and upper approximations of a subset $X \subseteq U$, namely $\underline{R}(X)$ and $\overline{R}(X)$ are defined as follows:

$$\underline{R}(X)=\{x \in U : [x]_R \subset X\},$$
$$\overline{R}(X)=\{x \in U : [x]_R \cap X \neq \phi\}.$$

The lower and upper approximations have the following properties:

For every $X,Y \subset U$ from the approximation space $A=(U,R)$ we have:

1. $\underline{R}(X) \subseteq X \subseteq \overline{R}(X)$,
2. $\underline{R}(U)=\overline{R}(U)=U$,
3. $\underline{R}(\phi)=\overline{R}(\phi)=\phi$,
4. $\overline{R}(X \cup Y)=\overline{R}(X) \cup \overline{R}(Y)$,
5. $\underline{R}(X \cup Y) \supseteq \underline{R}(X) \cup \underline{R}(Y)$,
6. $\overline{R}(X \cap Y) \subseteq \overline{R}(X) \cap \overline{R}(Y)$,
7. $\underline{R}(X \cap Y)= \underline{R}(X) \cap \underline{R}(Y)$,
8. $\overline{R}(-X)=-\ \underline{R}(X)$,
9. $\underline{R}(-X)=-\overline{R}(X)$,
10. $\overline{R}(\overline{R}(X))= \underline{R}(\overline{R}(X))=\overline{R}(X)$,
11. $\underline{R}(\underline{R}(X))=\overline{R}(\underline{R}(X))= \underline{R}(X)$,
12. $If\ X \subseteq Y, then\ \overline{R}(X) \subseteq \overline{R}(Y)\ and\ \underline{R}(X) \subseteq \underline{R}(Y)$.

The equality in all properties happens when $\underline{R}(X)=\overline{R}(X)=X$. The proof of all these properties can be found in [17-23,51].

Furthermore, for a subset $X \subseteq U$, a rough membership function is defined as follows: $\mu_X(x)=\frac{|[x]_R \cap X|}{|[x]_R|}$, where $|X|$ denotes the cardinality of the set X . The rough membership value $\mu_X(x)$ may be interpreted as the conditional probability that an arbitrary element belongs to X given that the element belongs to $[x]_R$.

Based on the lower and upper approximations, the universe U can be divided into three disjoint regions, the positive $POS(X)$, the negative $NEG(X)$ and the boundary $BND(X)$,where:

$$POS(X)=\underline{R}(X)$$
$$NEG(X)=U-\overline{R}(X)$$
$$BND(X)=\overline{R}(X)-\underline{R}(X)$$

Considering general binary relations in [18,52] is an extension to the classical lower and upper approximations of any subset X of U . $\beta=\{R_x : x \in X\}$ is the base generated by the general relation defined in [17,52]. The general forms based on β are defined as follows:

$$\underline{R}_\beta(X)=\cup\{B : B \in \beta_x, B \subset X\},$$
$$\overline{R}_\beta(X)=\cup\{B : B \in \beta_x, B \cap X \neq \phi\},\ \text{where}$$

$\beta_x=\{B \in \beta : x \in B\}$.

For data granulation by any binary relation, in [E. Lashein (2005)] a rough membership function is defined as follows:

$$\mu_X(x)=\frac{|X \cap (\cap \beta_x)|}{|\cap \beta_x|}.$$

III. Rough Sets of Equivalence and General Binary Relations

Indiscernibility as defined by equivalence relation represents a very restricted type of relationships between elements and universes. The procedure to granule the universe by general binary relations is introduced in [6].

A topological space [1,2] is a pair (X,τ) consisting of a set X and a family τ of subset of X satisfying the following conditions:

(1) $\phi, X \in \tau$,
(2) τ is closed under arbitrary union,
(3) τ is closed under finite intersection.

The pair (X,τ) is called a topological space. The elements of X are called points . The subsets of X belonging to τ are called open sets. The complement of the open subsets are called closed sets. The family τ of all open subsets of X is also called a topology for X . $cl(A)=\bigcap\{F\subseteq X: A\subseteq F \text{ and } F \text{ is closed}\}$ is called τ -closure of a subset $A\subset X$.

Obviously, $cl(A)$ is the smallest closed subset of X which contains A . Note that A is closed iff $A=cl(A)$. $int(A)=\bigcup\{G\subseteq X: G\subseteq A \text{ and } G \text{ is open}\}$ is called the τ -interior of a subset $A\subseteq X$. Manifestly, $int(A)$ is the union of all open subsets of X which contained in A . Make a note of that A is open iff $A=int(A)$. $b(A)=cl(A)-int(A)$ is called the τ -boundary of a subset $A\subseteq X$.

For any subset A of the topological space (X,τ) , $cl(A)$, $int(A)$ and $b(A)$ are closure, interior, and boundary of A respectively. The subset A is exact if $b(A)=\phi$, otherwise A is rough. It is clear that A is exact iff $cl(A)=int(A)$. In Pawlak space a subset $A\subseteq X$ has two possibilities either rough or exact.

In later years a number of generalizations of open sets have been considered [21-23]. We talk about some of these generalizations concepts in the following definitions.

Let U be a finite universe set and R is any binary relation defined on U , and $rR(x)$ be the set of all elements which are in relation to certain elements x in U from right for all $x\in U$, in symbols $rR(x)=\{xR, x\in U\}$ where $xR=\{y:(x,y)\in R; x,y\in U\}$.

Let β be the general knowledge base (topological base) using all possible intersections of the members of $rR(x)$. The component that will be equal to any union of some members of β must be misplaced.

IV. Topological Generalizations of Rough Sets

Let $A=(U,R)$ be an approximation space where R is any binary relation defined on U . Then we can define two new approximations as follows:

$$\underline{\tau}_\beta(X)=X\bigcap \underline{R}_\beta(\overline{R}_\beta(X)),$$
$$\overline{\tau}_\beta(X)=X\bigcup \overline{R}_\beta(\underline{R}_\beta(X)).$$

The topological lower and the topological upper approximations have the following properties:

For every $X,Y\subset U$ and every approximation space $A=(U,R)$ we have:

1. $\underline{\tau}_\beta(X)\subseteq X\subseteq \overline{\tau}_\beta(X)$,
2. $\underline{\tau}_\beta(U)=U=\overline{\tau}_\beta(U)$,
3. $\overline{\tau}_\beta(\phi)=\underline{\tau}_\beta(\phi)=\phi$,
4. $\overline{\tau}_\beta(X\cup Y)\supset \overline{\tau}_\beta(X)\cup\overline{\tau}_\beta(Y)$,
5. $\underline{\tau}_\beta(X\cup Y)\supset \underline{\tau}_\beta(X)\cup\underline{\tau}_\beta(Y)$,
6. $\overline{\tau}_\beta(X\cap Y)\subset \overline{\tau}_\beta(X)\cap\overline{\tau}_\beta(Y)$,
7. $\underline{\tau}_\beta(X\cap Y)\subseteq \underline{\tau}_\beta(X)\cap\underline{\tau}_\beta(Y)$,
8. $\overline{\tau}_\beta(-X)=-\overline{\tau}_\beta(X)$,
9. $\underline{\tau}_\beta(-X)=-\underline{\tau}_\beta(X)$,
10. $\overline{\tau}_\beta(\overline{\tau}_\beta(X))=\overline{\tau}_\beta(X)$,
11. $\underline{\tau}_\beta(\underline{\tau}_\beta(X))=\underline{\tau}_\beta(X)$,
12. *If* $X\subseteq Y$, *then* $\overline{\tau}_\beta(X)\subseteq\overline{\tau}_\beta(Y)$ *and* $\underline{\tau}_\beta(X)\subseteq\underline{\tau}_\beta(Y)$.

Given that topological lower and topological upper approximations satisfy that: $\underline{R}_\beta(X)\subseteq\underline{\tau}_\beta(X)\subseteq X\subseteq\overline{\tau}_\beta(X)\subseteq\overline{R}_\beta(X)\subseteq U$ this enables us to divide the universe U into five disjoint regions (granules) as follows: (See Figure 1)

1. $POS_\beta(X)=\underline{R}_\beta(X)$,
2. $\tau-POS(X)=\underline{\tau}_\beta(X)-\underline{R}_\beta(X)$,
3. $\tau-BND(X)=\overline{\tau}_\beta(X)-\underline{\tau}_\beta(X)$,
4. $\tau-NEG(X)=\overline{R}_\beta(X)-\overline{\tau}_\beta(X)$,
5. $NEG_\beta(X)=U-\overline{R}_\beta(X)$.

The following theorems study the properties and relationships among the above regions namely boundary, positive and negative regions.

Theorem 4.1 let $IS = (U, A, \tau_R)$ be a topological information system and for any subset $X \subset U$ we have:

(1) $\tau - BND(X) \cap \underline{\tau}_\beta(X) = \phi$,

(2) $\tau - BND(X) \cap \tau - NEG(X) = \phi$,

(3) $\overline{\tau}_\beta(X) = \underline{\tau}_\beta(X) \cup \tau - BND(X)$,

(4) $\underline{\tau}_\beta(X), \tau - NEG(X)$ and $\tau - BND(X)$ are disjoint granules of U.

Proof: You can make use of Figure 1.

Theorem 4.2 let $IS = (U, A, \tau_R)$ be a topological information system and for any subsets $X, Y \subset U$ we have:

(1) $\tau - BND(U) = \phi$,

(2) $\tau - BND(X) = \tau - BND(U - X)$,

(3) $\tau - BND(\tau - BND(X)) \subset \tau - BND(X)$,

$\tau - BND(X \cap Y) \subset \tau - BND(X) \cup \tau - BND(Y)$

Proof: (1) and (2) is obvious, by definitions.

(3)

$$\tau - BND(\tau - BND(X)) = \tau - BND(\overline{\tau}_\beta(X) \cap \overline{\tau}_\beta(U - X))$$

$$= \overline{\tau}_\beta(\overline{\tau}_\beta(X) \cap \overline{\tau}_\beta(U - X)) \cap \overline{\tau}_\beta(U - (\overline{\tau}_\beta(X) \cap \overline{\tau}_\beta(U - X))) \subset \overline{\tau}_\beta(X) \cap \overline{\tau}_\beta(U - X) = \tau - BND(X).$$

(4)

$$\tau - BND(X \cap Y) = \overline{\tau}_\beta(X \cap Y) \cap \overline{\tau}_\beta(U - X \cap Y)$$

Theorem 4.3 let $IS = (U, A, \tau_R)$ be a topological information system and for any subset $X, Y \subset U$ we have:

(1) $U = \tau - NEG(\phi)$,

(2) $\tau - NEG(X) = \underline{\tau}_\beta(U - X)$,

(3) $X \cap \tau - NEG(X) = \phi$,

(4) $\tau - NEG(U - \tau - NEG(X)) = \tau - NEG(X)$,

(5) $\tau - NEG(X \cup Y) \subset \tau - NEG(X) \cup \tau - NEG(Y)$,

(6) $\tau - NEG(X \cap Y) \supset \tau - NEG(X) \cap \tau - NEG(Y)$

Proof: (1), (2), (3) and (4) are obvious.

(5)

$$\tau - NEG(X \cup Y) = U - \overline{\tau}_\beta(X \cup Y) \subset U - (\overline{\tau}_\beta(X) \cup \overline{\tau}_\beta(Y))$$

$$= (U - \overline{\tau}_\beta(X)) \cap (U - \overline{\tau}_\beta(Y)) \subset \tau - NEG(X) \cup \tau - NEG(Y)$$

(6)

$$\tau - NEG(X) \cap \tau - NEG(Y) = (U - \overline{\tau}_\beta(X)) \cap (U - \overline{\tau}_\beta(Y))$$

$$= U - (\overline{\tau}_\beta(X) \cup \overline{\tau}_\beta(Y)) \subset U - \overline{\tau}_\beta(X \cap Y) = \tau - NEG(X \cap Y)$$

Example 4.1 let $U = \{u_1, u_2, u_3, u_4, u_5, u_6, u_7\}$ be the universe of 7 patients have data sheets shown in Table I with possible dengue symptoms. If some experts give us the general relation R defined among those patients as follows:

TABLE I. PATIENTS INFORMATION SYSTEM

U	Conditional Attributes (C)			Decision (D)
	Temperature	Flu	Headache	Dengue
u1	Normal	No	No	No
u2	High	No	No	No
u3	Very High	No	No	Yes
u4	High	No	Yes	Yes
u5	Very High	No	Yes	Yes
u6	High	Yes	Yes	Yes
u7	Very High	Yes	Yes	Yes

$$R = \{(u_1, u_1), (u_1, u_7), (u_2, u_2), (u_3, u_3), (u_3, u_6), (u_4, u_4), (u_5, u_5), (u_6, u_6), (u_7, u_7)\}.$$

The topological knowledge base will take the following form:

$$\beta = \{\{u_1, u_7\}, \{u_2\}, \{u_3, u_6\}, \{u_4\}, \{u_5\}, \{u_6\}, \{u_7\}\}$$

For some patients $X = \{u_2, u_3, u_7\}$ the upper and lower approximations based on the topological knowledge base are given by:

$\overline{R}_\beta(X) = \{u_1, u_2, u_3, u_6, u_7\}$, and $\underline{R}_\beta = \{u_2, u_7\}$.

By using the lower and upper approximations, the granules of universe are three disjoint regions as follows:

$$POS_\beta(X) = \underline{R}_\beta(X) = \{u_2, u_7\},$$

$$BND_\beta(X) = \overline{R}_\beta(X) - \underline{R}_\beta(X) = \{u_1, u_3, u_6\},$$

$$NEG_\beta(X) = U - \overline{R}_\beta(X) = \{u_4, u_5\}.$$

According to the topological knowledge base we can easily see that:

$\overline{\tau}_\beta(X) = \{u_1, u_2, u_3, u_7\}$, $\underline{\tau}_\beta(X) = \{u_2, u_3, u_7\}$.

Then we have the following granules of the universe:

1. $POS_\beta(X) = \{u2, u7\}$,

2. $\tau - POS(X) = \{u3\}$,
3. $\tau - BND(X) = \{u1\}$,
4. $\tau - NEG(X) = \{u6\}$,
5. $NEG_\beta(X) = \{u4, u5\}$.

V. New Topological Generalizations of Rough Sets

In this section, we used the topological tool $\delta\beta$-open sets to introduce the concepts of $\delta\beta$-lower and $\delta\beta$-upper approximations. The suggested model helps in decreasing the boundary region of concepts in information systems. Also, we use the topological measure $\alpha_{R_{\delta\beta}}$ is used as a topological accurate measure of data granulation correctness.

For any subset X of a topological space (U,τ). The δ-closure of a subset X is defined by $cl_\delta(X) = \{x \in U : X \cap int(cl(G)) \neq \phi, G \in \tau \text{ and } x \in G\}$. A set X is called δ-closed if $X = cl_\delta(X)$. The complement of a δ-closed set is called δ-open.

Notice that $int_\delta(X) = U \setminus cl_\delta(U \setminus X)$.

A subset X of a topological space (U,τ) is called $\delta\beta$-open if $X \subseteq cl(int(cl_\delta(X)))$.

Let (U,τ) be a topological space and $X \subseteq U$, the following new topological tools of any subset X are defined as follows [1,2,6]:

- Regular open tool if $X = Int(Cl(X))$.
- Semi-open tool if $X \subset Cl(Int(X))$.
- $\alpha -$ open tool if $X \subset Int(Cl(Int(X)))$.
- Pre-open tool if $X \subset Int(Cl(X))$.
- Semi pre open tool ($\beta -$ open) if $X \subset Cl(Int(Cl(X)))$.

The family of all $\delta\beta$-open sets of U is denoted by $\delta\beta O(U)$. The complement of $\delta\beta$-open set is called $\delta\beta$-closed set. The family of $\delta\beta$-closed sets are denoted by $\delta\beta C(U)$.

Let X be a subset of a topological space (U,τ), then we have:

(i) The union of all $\delta\beta$-open sets contained inside X is called the $\delta\beta$-interior of X and is denoted by $\beta int_\delta(X)$.

(ii) The intersection of all $\delta\beta$-closed sets containing X is called the $\delta\beta$-closure of X and is denoted by $\beta cl_\delta(X)$.

Lemma 6.1 For a subset X of a topological space (U,τ) we have:

(i) $\beta int_\delta(X) = X \cap cl(int(cl(X)))$.

(ii) $\beta cl_\delta(X) = X \cup int(cl(int(X)))$.

$\delta\beta$-open sets is stronger than any topological near open sets such as δ-open, regular open, semi-open, $\alpha -$ open, pre-open, β-open.

The following example illustrates the above note.

Example 5.1 Let (U,τ) be a topological space where, $U = \{a,b,c,d,e\}$ and $\tau = \{U, \varphi, \{d\}, \{e\}, \{a,d\}, \{d,e\}, \{a,d,e\}, \{b,c,e\}, \{b,c,d,e\}\}$. We have $\{a,c\} \in \delta\beta O(U)$ but $\{a,c\} \notin \delta O(U)$, $\{b,d,e\} \in \delta\beta O(U)$ but $\{b,d,e\} \notin RO(U)$, $\{a,e\} \in \delta\beta O(U)$ but $\{a,e\} \notin PO(U)$, $\{c\} \in \delta\beta O(U)$ but $\{c\} \notin \beta O(U)$, $\{b\} \in \delta\beta O(U)$ but $\{b\} \notin SO(U)$ and $\{c,d\} \in \delta\beta O(U)$ but $\{c,d\} \notin \alpha O(U)$. Where $\delta O(U)$, $RO(U)$, $SO(U)$, $\alpha O(U)$, $PO(U)$ and $\beta O(U)$ denoted the family of all δ-open, regular open, semi-open, $\alpha -$ open, pre-open and β-open sets of U respectively.

Arbitrary union of $\delta\beta$-open sets is again $\delta\beta$-open set, but the intersection of two $\delta\beta$-open sets may not be $\delta\beta$-open set. Thus the $\delta\beta$-open sets in a space U do not form a topology.

Let U be a finite non-empty universe. The pair $(U, R_{\delta\beta})$ is called a $\delta\beta$-approximation space where $R_{\delta\beta}$ is a general relation used to get a subbase for a topology τ on U which generates the class $\delta\beta O(U)$ of all $\delta\beta$-open sets.

Example 6.2 Let $U = \{a,b,c,d,e\}$ be a universe and a relation R defined by $R = \{(a,a), (a,e), (b,c), (b,d), (c,e), (d,a), (d,e), (e,e)\}$, thus $aR = dR = \{a,e\}$,

$bR = \{c,d\}$ and $cR = eR = \{e\}$. Then the topology associated with this relation is $\tau = \{U, \phi, \{e\}, \{a,e\}, \{c,d\}, \{c,d,e\}, \{a,c,d,e\}\}$ earned $\delta\beta O(U) = P(U) - \{b\}$. So $(U, R_{\delta\beta})$ is a $\delta\beta$ - approximation space.

Let $(U, R_{\delta\beta})$ be a $\delta\beta$ - approximation space. $\delta\beta$ -lower approximation and $\delta\beta$ -upper approximation of any non-empty subset X of U is defined as:

$$\underline{R}_{\delta\beta}(X) = \bigcup\{G \in \delta\beta O(U) : G \subseteq X\},$$

$$\overline{R}_{\delta\beta}(X) = \bigcap\{F \in \delta\beta C(U) : F \supseteq X\}.$$

We see that:
$$\underline{R}(X) \subseteq \underline{R}_{\beta}(X) \subseteq \underline{R}_{\delta\beta}(X) \subseteq X \subseteq \overline{R}_{\delta\beta}(X) \subseteq \overline{R}_{\beta}(X) \subseteq \overline{R}(X).$$

Let $(U, R_{\delta\beta})$ be a $\delta\beta$ - approximation space, $X \subseteq U$. From the relation
$$int(X) \subseteq \beta int(X) \subseteq \delta\beta int(X) \subseteq X \subseteq \delta\beta cl(X) \subseteq \beta cl(X) \subseteq cl(X),$$
The Universe U can be separated into divergent 24 granules with respect to any $X \subseteq U$.

We can distinguish the degree of completeness of granules of U by the topological tool named $\delta\beta$ -accuracy measure defined for any granule $X \subseteq U$ as follows:

$$\alpha_{R_{\delta\beta}}(X) = \frac{|\underline{R}_{\delta\beta}(X)|}{|\overline{R}_{\delta\beta}(X)|} \quad where \quad X \neq \phi.$$

Example 5.2 According to Example 5.1 we can construct the following table (Table II) showing the degree of accuracy measure $\alpha_R(X)$, β -accuracy measure $\alpha_{R_\beta}(X)$ and $\delta\beta$ - accuracy measure $\alpha_{R_{\delta\beta}}(X)$ for some granules of U.

TABLE II. ACCURACY MEASURES OF SOME GRANULES

Some granules	Pawlak's accuracy	β -accuracy	$\delta\beta$ -accuracy
{b, d}	0%	100%	100%
{b, e}	33.3%	66.6%	100%
{a, b, e}	66.6%	100%	100%
{a, c, d}	50%	66.6%	100%
{b, c, d, e}	60%	80%	100%

We see that the degree of accuracy of the granule $\{b,c,d,e\}$ using Pawlak's accuracy measure equal to 60%, using β -accuracy measure equal to 80% and using $\delta\beta$ - accuracy measure equal to 100%. Accordingly $\delta\beta$ - accuracy measure is more precise than Pawlak's accuracy and β -accuracy measures.

VI. CONCLUSIONS AND APPLICATION NOTES

In the near future is the completion of a new paper for the application of the granules concepts of this paper in medicine especially in the field of heart disease in collaboration with specialists in this field. We designed a JAVA application program novelty to generate granules division automatically once you select points covered by the heart scan and the medical relationship among them using topology defined on it. The program works under any operating system but needs to be a great RAM memory and strong processor to end the division of the millions of points to the granules in seconds.

REFERENCES

[1] H.M. Abu-Donia (2012), Multi knowledge based rough approximations and applications Knowledge-Based Systems, Volume 26, , Pages 20-29

[2] D. Andrijevic(1986) , Semi-pre, open sets, Mat. Vesnik. 38, 24-32.

[3] Tutut Herawan, Mustafa Mat Deris (2010), Jemal H. Abawajy , A rough set approach for selecting clustering attribute Knowledge-Based Systems, Volume 23, Issue 3, , Pages 220-231

[4] Jiye Liang (2009) , Junhong Wang, Yuhua Qian , A new measure of uncertainty based on knowledge granulation for rough sets. Information Sciences, 179, 458–470.

[5] E. Lashein (2005) , A.M. Kozae, , A. Abo Khadra, T. Medhat, Rough Set Theory for Topological Spaces, International Journal of Approximate Reasoning 40, 35-43.

[6] T.Y. Lin (1998), Granular Computing on Binary Relations I: data mining and neighborhood systems, II: rough set representations and belief functions, In: Rough Setsin Knowledge Discovery 1, L. Polkowski, A.Skowron (Eds.), Phys.-Verlag, Heidelberg, 107-14.

[7] T.Y. Lin (2002), Y.Y. Yao, L.A. Zadeh, Data Mining, Rough Sets and Granular Computing (Studies in Fuzziness and Soft Computing), Physica-Verlag, Heidelberg.

[8] Guilong Liu (2009), Ying Sai, A comparison of two types of rough sets induced by coverings, International Journal of Approximate Reasoning 50, 521–528.

[9] Yee Leung (2008), Manfred M. Fischer , Wei-Zhi Wu , Ju-Sheng Mi, A rough set approach for the discovery of classification rules in interval-valued information systems, International Journal of Approximate Reasoning 47, 233–246.

[10] Guilong Liu (2010), Rough set theory based on two universal sets and its applications Knowledge-Based Systems, 23(2),110-115

[11] Guilong Liu(2008), Axiomatic systems for rough sets and fuzzy rough sets, , International Journal of Approximate Reasoning 48, 857–867.

[12] A. S. Mashhour (1982), M. E. Abd El-Monsef, S. N. El-Deeb , On pre-continuous and week pre-continuous mappings, Proc. Math. & phys. Soc. Egypt 53, 47-53.

[13] T. Nishino(2005), M. Nagamachi, H. Tanaka, Variable Precision Bayesian Rough Set Model and Its Application to Human Evaluation Data, RSFDGrC 2005, LNAI 3641, Springer Verlag, 294-303.

[14] T. Nishino (2006), M.Sakawa, K. Kato, M. Nagamachi, H.Tanak, Probabilistic Rough Set Model and Its Application to Kansei Engineering, Transactions on Rough Sets V (Inter. J. of Rough Set Society), LNCS 4100, Springer, 190-206.

[15] O. Njasted, On some classes of nearly open sets, Pro. J. Math. 15 (1965) 961-970.

[16] N. Levine (1963), Semi open sets and semi continuity topological spaces, Amer. Math. Monthly 70 ,24-32.

[17] Zhi Pei (2011), Daowu Pei, Li Zheng, Topology vs generalized rough sets, International Journal of Approximate Reasoning 52, 231-239.

[18] Zhi Pei (2011), Daowu Pei, Li Zheng, Covering rough sets based on neighborhoods an approach without using neighborhoods, International Journal of Approximate Reasoning 52 , 461-472.

[19] L. Polkowski and A.Skowron (1998),Towards Adaptive Calculus of Granules, Proceedings of 1998 IEEE Inter. Conf. on Fuzzy Sys., 111-116.

[20] Z. Pawlak, A. Skowron (2007), Rough sets and Boolean reasoning, Information Sciences 177 , 41–73.

[21] Z. Pawlak, A. Skowron (2007), Rough sets: some extensions, Information Sciences 177 , 28–40.

[22] Z. Pawlak, A. Skowron (2007), Rudiments of rough sets, Information Sciences 177, 3–27.

[23] Z. Pawlak (1981), Rough sets, Int. J. Comput. Information Sciences 11, 341–356.

[24] Yuhua Qian, Jiye Liang (2009), Chuangyin Dang , Knowledge structure, knowledge granulation and knowledge distance in a knowledge base, International Journal of Approximate Reasoning 50, 174–188.

[25] Yuhua Qian, Liang Jiye (2010), Yao Yiyu, Dang Chuangyin MGRS: A multi-granulation rough set, Information Sciences 180, 949–970.

[26] Hu Qinghua (2008), Liu Jinfu, Yu Daren, Mixed feature selection based on granulation and approximation, Knowledge-based system, 21, 294–304.

[27] A. S. Salama (2008), Topologies Induced by Relations with Applications, journal of Computer Science 4, 879-889.

[28] A. S. Salama (2008), Two New Topological Rough Operators, J. of Interdisciplinary Math. Vol. 11, No.1, New Delhi Taru Publications-, INDIA 1-10.

[29] A. S. Salama (2010); Topological Solution for missing attribute values in incomplete information tables, Information Sciences 180, 631-639 .

[30] D. Slezak (2004), The Rough Bayesian Model for Distributed Decision Systems, RSCT 2004, LNAI 3066, Springer Verlag, 384-393.

[31] D. Slezak (2005), Rough Sets and Bayes factors, Transactions on Rough Set III, LNCS 3400, 202-229.

[32] D. Slezak (2002) , W.Ziarko, Bayesian Rough Set Model, In: Proc. of the Int. Workshop on Foundation of Data Mining (FDM 2002), December 9, Maebashi, Japan ,131–135.

[33] D. Slezak, W.Ziarko(2003), Variable Precision Bayesian Rough Set Model, RSFDGrC 2003, LNAI 2639, Springer Verlag, 312-315.

[34] Andrzej Skowron (1996), Jaroslaw Stepaniuk , Tolerance Approximation Spaces. Fundam. Inform. 27(2-3): 245-253

[35] Andrzej Skowron (2012), Jaroslaw Stepaniuk, Roman W. Swiniarski, Modeling rough granular computing based on approximation spaces. Information Sciences 184(1): 20-43

[36] D. J. Spiegelhalter (2004), K. R. Abrams, J. P. Myles, " Bayesian Approaches to Clinical Trials and Health-Care Evaluation". John Wiley & Sons Ltd, The Atrium, Southern Gate, Chichester, England.

[37] D.Slezak, W.Ziarko(2003), Attribute Reduction in the Bayesian Version of Variable Precision Rough Set Model, In: Proc. of RSKD, ENTCS, 82, 4-14.

[38] D.Slezak, W.Ziarko (2005),The Investigation of the Bayesian Rough Set Model, International Journal of Approximate Reasoning vol.40, 81-91.

[39] Yanhong She (2012), Xiaoli He , On the structure of the multigranulation rough set model Knowledge-Based Systems, In Press, Uncorrected Proof, Available online 12 June 2012

[40] You- Chen Shyang (2012), Classifying credit ratings for Asian banks using integrating feature selection and the CPDA-based rough sets approach Knowledge-Based Systems, Volume 26, , Pages 259-270

[41] Ronald R. Yager(2009), Comparing approximate reasoning and probabilistic reasoning using the Dempster–Shafer framework, International Journal of Approximate Reasoning 50, 812–821.

[42] E. A. Rady, A. M. Kozae (2004), M. M. E. Abd El-Monsef, Generalized Rough Sets, Chaos, Solitons, & Fractals 21, 49-53.

[43] Y.Y.Yao(1998), Constructive and algebraic methods of theory of rough sets, Information Sciences 109, 21–47.

[44] Y.Y.Yao (1998), Relational interpretations of neighborhood operators and rough set approximation operators, Information Sciences 111, 239–259.

[45] Y.Yang,R.I.John (2008), Generalizations of roughness bounds in rough set operations, International Journal of Approximate Reasoning 48, 868-878.

[46] Y. Yao, Y. Zhao(2008); Attribute reduction in decision-theoretic rough set models. Information Sciences 178, 3356–3373.

[47] Y.Y. Yao(1999), Granular Computing using Neighborhood Systems, in: Advances in Soft Computing: Engineering Design and Manufacturing, R. Roy, T. Furuhashi, and P. K. Chawdhry (Eds.), Springer-Verlag, London, 539-553.

[48] A.M. Zahran(2000), Regularly open sets and a good extension on fuzzy topological spaces, Fuzzy Sets and Systems 116, 353-359.

[49] L.A. Zadeh(1979), "Fuzzy Sets and Information Granularity". In: Advances in Fuzzy Set Theory and Applications, Gupta, N., Ragade, R. and Yager, R. (Eds.), North- Holland, Amsterdam, 3-18.

[50] L. A. Zadeh (1997) , "Towards a Theory of Fuzzy Information Granulation and its Centrality in Human Reasoning and Fuzzy Logic". Fuzzy Sets and Systems, 19, 111-127.

[51] L. A. Zadeh (2006), Generalized theory of uncertainty (GTU)—principal concepts and ideas , Computational Statistics & Data Analysis, 51(1) 15-46

[52] L. A. Zadeh (2002), Toward a perception-based theory of probabilistic reasoning with imprecise probabilities Journal of Statistical Planning and Inference, 105(1), 233-264

A Novel Approach for Discovery Quantitative Fuzzy Multi-Level Association Rules Mining Using Genetic Algorithm

Saad M. Darwish
Department of Information Technology
Institute of Graduate Studies and Research University of Alexandria
Alexandria, Egypt

Abeer A. Amer
Department of Computer and Information Systems
Sadat Academy for Management Sciences (SAMS)
Alexandria, Egypt

Sameh G. Taktak
Department of Computer and Information Systems
Sadat Academy for Management Sciences (SAMS)
Alexandria, Egypt

***Abstract*—Quantitative multilevel association rules mining is a central field to realize motivating associations among data components with multiple levels abstractions. The problem of expanding procedures to handle quantitative data has been attracting the attention of many researchers. The algorithms regularly discretize the attribute fields into sharp intervals, and then implement uncomplicated algorithms established for Boolean attributes. Fuzzy association rules mining approaches are intended to defeat such shortcomings based on the fuzzy set theory. Furthermore, most of the current algorithms in the direction of this topic are based on very tiring search methods to govern the ideal support and confidence thresholds that agonize from risky computational cost in searching association rules. To accelerate quantitative multilevel association rules searching and escape the extreme computation, in this paper, we propose a new genetic-based method with significant innovation to determine threshold values for frequent item sets. In this approach, a sophisticated coding method is settled, and the qualified confidence is employed as the fitness function. With the genetic algorithm, a comprehensive search can be achieved and system automation is applied, because our model does not need the user-specified threshold of minimum support. Experiment results indicate that the recommended algorithm can powerfully generate non-redundant fuzzy multilevel association rules.**

***Keywords*—Quantitative Data Mining; Fuzzy Association Rule Mining; Multilevel Association rule; Optimization Algorithm**

I. INTRODUCTION

In data-mining, discovering association rules in transaction databases is frequently examined. Association rules are widely offered and are beneficial for planning and marketing. For example, they can be managed to implicate supermarket officials of what products the customers have an inclination to purchase together. Taking market basket analysis as an example, the mining problem can be explained as given a database *D* of transactions, each transaction is a set of items; find all rules that relate the carriage of one set of items with that of another set of items [1].

The classical algorithms for mining association rules are formed on binary attributes databases, which have two weaknesses. Firstly, it cannot treat quantitative attributes; secondly, it handles each item with the same weight despite that strange item may have different importance. Also, a binary association rule bears from explicit boundary problems. Besides many real world transactions consist of quantitative attributes. That is why numerous researchers have been serving on the generation of association rules for quantitative data [2] [3].

Beginning approaches for quantitative association rule mining manages distinctive partitioning for transforming other attributes to binary ones which convey from a major problem that results in information damage because of sharp limits. In other words, modern algorithms neglect or exaggerate items beside the boundary. When distributing an attribute in the data into sets comprising individual ranges of values, the users are faced with the sharp boundary problem[2]. Quantitative attributes are discretized by joining the concept of hierarchies. This manner occurs before the event of mining. For example, a concept hierarchy for age may be adopted to reconstruct the initial numeric values of this attribute by ranges [2]. To surmount this problem researchers are laboring for mining association rules for quantitative attributes. They have contributed several algorithms that tackle quantitative algorithm and reveal how they dispense with quantitative data [3] [4].

In general, fuzzy technique overcomes the main drawback of the discretize technique. Fuzzy logic produces linguistic term instead of intervals which is more nearer to the human mind. The disadvantage is that although the loss of information is small but it exists. Furthermore, the needs for fuzzy membership function to be given by an expert, which is not always straightforward and can be biased. Despite that, fuzzy association rule mining approach appeared out of the requirement to mine quantitative data uniformly present in databases efficiently [2]. There are two essential basic criteria for association rules, support(*s*) and confidence(*c*). Since the database is large and users interest about only those frequently purchased items, usually thresholds of support and confidence are predetermined by users to separate those rules that are not so attractive or beneficial. The two thresholds are called minimal support and minimal confidence respectively [5].

Genetic Algorithm (GA) is a heuristic exploration that imitates the process of natural evolving. This heuristic is routinely applied to produce valuable explications to optimization and search obstacles. Genetic Algorithm is based on conceptions of evolution hypothesis as a fundamental policy is that only the strongest beings remain. The genetic algorithms are significant when identifying association rules because they work with the global search to determine the set of items frequency and they are less complex than other algorithms frequently worked in data mining. The genetic algorithms for discovery of association rules have been settled into usage in real problems such as commercial databases and fraud detection [6].

Earlier investigations on data mining directed on locating association rules at a single-concept level. Mining association rules at multiple concept levels may guide to the discovery of more broad and significant knowledge from data. Related item taxonomies are normally predefined in real-world purposes and can be interpreted as hierarchy trees. Terminal nodes on the trees express actual items looking in transactions; internal nodes describe classes or concepts built from lower-level nodes [7]. A simple example is given in Fig. 1. Mining multi-level association rules are motivated by several purposes, such as: (a) the multi-level association rules are more reasonable and are more interpretable for users. (b) The multi-level association rules can supply us solutions for the undesirable and undesired rules. Encouraging applications involve spatial data analysis, emergency event analysis, and network data mining [8] [9].

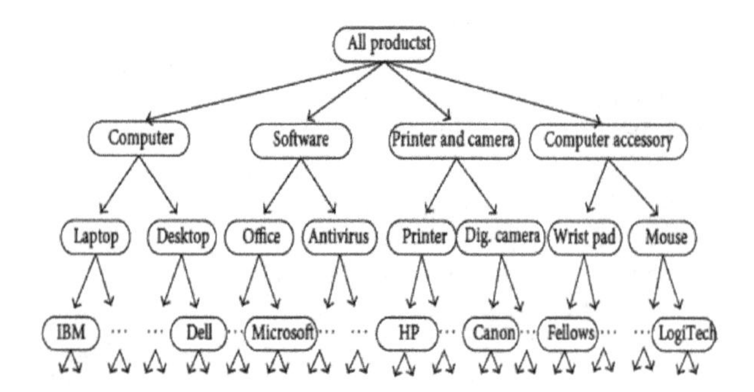

Fig. 1. The predefined taxonomy

A. Motivation & Rationale

If it arrives at quantitative association rule mining, a number of trials have been directed on performance (speed) and effectiveness (number of rules). Less effort has been converged on quality. Modern quantitative multilevel association rule mining algorithms depend on intense looks of the database to obtain regular exemplars beyond different abstraction levels [9]. However, these mining algorithms are often based on the postulate that users can blueprint the minimum support relevant to their databases [10]. Mining quantitative association rules is not a manageable enlargement of mining categorical association rules. Since the search space is unlimited, our aim is to detect a measurable set of exciting solutions (quantitative rules), near to the optimal answers. This illustrates why we have decided to solve this search problem with meta-heuristics routines, mainly genetic algorithms [11].

Regularly, when managing quantitative association rule mining several rules can be identified or inferred and confuse the user. But more importantly, some of these rules could be redundant and produce no new knowledge. Some attempt has been aimed at selling with redundant rules in flat datasets. However, datasets can have a hierarchy/taxonomy or multiple concept levels and thus redundancy in these datasets require to be adjusted. This subject is one of the phases of this research. Currently, the approach being taken is to resolve which rules are redundant and eliminate them, thus diminishing the number of rules a user has to deal with while not decreasing the information content [12]. This Paper offers an adapted version of the Apriori algorithm for mining fuzzy multi-level association rules in large databases for locating familiar item set at distinct levels of abstraction [13].

B. Research Problems

The purpose of this study is to offer to the field of data mining and in precise to multilevel quantitative fuzzy association rule mining. Hence to attain this distinct and well-defined investigation, difficulties are needed. By answering these it is likely to provide in an essential style. For our research, there are many central difficulties that we will converge on and strive to solve. These difficulties are:

- Boolean attributes can be studied as an exceptional instance of categorical attributes and it is an almost applicant to generalize the Boolean data mining algorithms. For qquantitative attributes, despite, the state is not so easy. We either have to somehow convert the quantitative association rules problem into Boolean one or to get different algorithms. Here we shall, in fact, produce an approach to discover quantitative association rules stemmed from a dataset with multiple concept levels.
- The number of rules expands exponentially with the number of items. But this complexity is undertaken with some advanced algorithms which can efficiently cut the search space [1][4]. The work picking this problem principally assists the user when scanning the rule set. Yet, the evolution of further valuable quality measures on the rules through employing genetic algorithm with fitness function (relative confidence of the association rule) to affirm the most intriguing association rules signifies the advanced aims to solve this problem [14].
- Without a priori knowledge, however, ascertaining the right intervals (discretize) for quantitative data mining can be a complex and intricate task. Moreover, these intervals may not be compact and acceptable enough for human experts to quickly gain nontrivial knowledge from those created rules. Fuzzy membership function can help to advise this problem.
- Completely and efficiently identifies no redundant association rules from datasets with a hierarchy.

C. Problem Statement

Market data in real-world usually involve quantitative values, so creating an advanced data-mining algorithm equipped to contract with quantitative data grants a challenge to workers in this study domain [4]. The multilevel association rules mining problem can be defined as follows: there are a collection of items $I = \{i_1, i_2, ..., i_n\}$ and Γ is a classification tree that concisely clarifies the multilevel categorizing relations among items as the field awareness. i_1 is the ancestor of i_2 and i_2 is the descendant of i_1 if there is an edge in Γ from i_1 to i_2. Only leaf nodes are displayed in the database. *D* is a database of transactions where each transaction *T* in *D* is a set of items such that $T \subseteq I$. Each transaction is attached with an identifier T_{ID}. Let *P* indicates the set of positive integers, and I_v symbolizes the set $I \times p$. A couple $< x, v > \in I_v$ means the quantitative attribute *x*, with the related value $v.I_v$ $\{\forall < x,l,u > \in I \times p \times p \mid l \le u\}$, l is the lower limit and u is the upper limit of p. A triple $< x,l,u > \in I_v$ stands for a quantitative x with a value in the interval [*l, u*]. Note that a transaction *T* holds an item $x \in I$ if x is in *T* or x is an ancestor of some items in *T*. In addition, a transaction *T* involves $X \subseteq I$ if *T* bears every item in X.

A multilevel association rule is an inference of the form $X \Rightarrow Y$, where $X \subseteq I, Y \subseteq I$, and $X \cap Y = \phi$. No item in Y is an ancestor of any item in X; that is $Y \cap \text{ancestors(x)} = \phi$. This is because a rule of the style $"x \Rightarrow \text{ancestors(x)} = \phi"$ is slightly true with 100% confidence, which is redundant. Both X and Y can contain items from any level of Γ [9][15][16].

However, there are still some limitations in quantitative association rule mining, such as [3][4][10]: (1) separation of the quantitative attribute, which is adopted in the design, is not accessible for all attribute and every user. (2) users, and even experts, regularly believe difficult to provide those thresholds like the minimum support, the interest level, and the minimum confidence.(3) the search space might be very large when we contact with quantitative attributes. Finally, (4) the rules declared by the algorithm might be too many to manage with.

II. LITERATURE REVIEW

In this section we compare the quantitative association rule mining algorithms taking into account the form of the rules and discuss each technique advantages, disadvantages and what kind of database can be used[17][18]:

1) Discretization:

The elementary intention of this routine is to transform quantities data to Boolean by examining a separation of the numerical attributes into collections of intervals. Then, an algorithm for detecting Boolean association rules can be handled to prepare quantities rules. Two main representations of partitions are included. A fixed partition, where the assortments of intervals are disjoint and another type, where the ends of intervals are overlaid with each other.

The principal benefit of this technique, beyond being the first work done on this track, is that manipulating both categorical and numerical data correspondingly. However, situations (disjoint or overlapped) yield problems; disjoint sets damage from *Min_Sup* and *Min_Conf* thresholds and overlapped sets suffer from the cutting boundary problem. Using intervals rather than the real continuous data will inevitably result in a loss of information. The rules we make will be only an estimation of the best results. Another problem is the enlargement of the attributes dimension; the problem here is the need for more memory and time to treat these data.

2) Adjusted difference analysis:

This algorithm is based on engaging both adjusted difference analysis and discretization to discover rules between two attributes. The two attributes could be any mixture of numerical or categorical. This technique has the capability to identify positive and negative association rules and does not need any user thresholds (support and confidence). Its advantages are that it does not want any user thresholds and it has the talent to obtain a new significant objective measure of the association rules. The disadvantage of this technique is the problems of discretization as in the first technique. Also, this technique is obviously considered to be as generating a special case rule since the generated rules are always between two attributes only.

3) Fuzzy Approach based on integrating fuzzy set and fuzzy logic concepts with Apriori algorith:

It reforms numerical data into fuzzy member between [0,1] with membership function; then operate with the fuzzy member with an adjusted Apriori technique that can manipulate comfortably the extracted rules, which are stated in linguistic terms. These approaches are based on the fuzzy additions to the classical association rules mining by establishing support and confidence of the fuzzy rule. While the mining results are straightforward to interpret by human operators, two shortcomings still insist on implementing such fuzzy approaches to the original problems. One is the computational time for mining from the database, and the other is the precision of deduced rules. More formal description, as well as a survey of the existing methods of quantitative association rule mining, can be found in [20].

In the literature, several researchers have concentrated on fuzzy multilevel association rules mining [3][21-25]. Some of these methods evoked multilevel membership functions by ant colony systems and genetic algorithm without stipulating the actual minimum support. To improve the performance of computing, setting the functions for each item followed by calculating minimum supports is engaged. Other work carried benefits of the OLAP and data mining technology which conducted efficiency and adaptability [9].

Up-to-date, there exist only a few algorithms for quantitative multilevel fuzzy association rule mining (*QMLFRL*). For examples, in [4] the authors advised an *QMLFRL* based on the idea that the minimum support for an item at a higher taxonomic concept is valued as the minimum of the minimum supports of the items pertaining to it and an item minimum support for an itemset is established as the maximum of the minimum supports of the items enclosed in the itemset. Under this limitation, the characteristic of downward-closure is conserved, such that the original Apriori algorithm can be simply prolonged to find fuzzy large item sets.

With the same purpose, the authors in [26] suggested a new method of quantitative association rule extraction that can quantize the attribute by applying a clustering algorithm and learn rules simultaneously. They implemented clustering using all attributes at the same time in advance and deduced the rules from the clusters in the aspect of "association". Based on the numerical experiments, the authors have confirmed that their algorithm outperformed the conventional algorithm based on Cartesian product type quantization in terms of total precision of quantization and rule extraction.

Extra relevant work introduced in [6] to create rules based on the quantitative dataset, utilizing the notion of threshold - frequent item sets that are produced using the genetic algorithm. In this illustrations, crossover & mutation are involved to create numerous unification of the rule and can recognize co-occurrence of item sets. Here three objectives are studied: comprehensibility, interestingness, confidence, so produced rules are established as multi-objective association rules. These objectives serve to decrease search space for fitness function. Finally, optimal rules are formed that is based on distribution approach for the numeric-valued attribute (Right-hand side of a rule reveals the distribution of the values of numeric attributes such as the mean or variance).

The benefit of the preceding systems is that they carry linguistic expressions which make created rules to be much normal for human experts; but they may generate a large number of interesting association rules. Still, for many purposes, it is not periodically simple to ascertain effective association rules (matching the minimum support and confidence) among data items at low (primitive) levels of abstraction due to the sparsity of data in multidimensional space. Other associated problems cover: (1) the shortage of sufficient support for dynamically needed hierarchies; (2) algorithm efficiency cannot meet real application specifications; (3) the association between different concepts levels may be dropped; (4) Their approach enabled users to stipulate various minimum supports to different items. [4][27].

A. Research Contribution

The idea developed in this paper is partly inspired by the existing work on *QMLFRL*, buts it utilizes the genetic algorithm to compute the minimum support and minimum confidence for each level in the Taxonomy regardless of the nature of the data; thus making automatic system. Prior studies have completely investigated single-level association rules mining with GA, such as mining single objective rules and mining multi-objective rules. However, in the big data analysis setting, powerful association rules are regularly in multilevel forms and mining multilevel association rules in big data demands more efficient methods. The GA-based multilevel association rules mining method recommended in this paper is one effort to efficiently discover multilevel association rules in big data.

III. THE PROPOSED MODEL

The advanced mining algorithm combines fuzzy set notions, data mining, and multiple-level taxonomy to determine fuzzy association rules in a given transaction data set deposited as quantitative values. The knowledge obtained is described by fuzzy linguistic terms, and thus simply readable by human beings. This system utilizes a top-down progressively deepening strategy to locate large itemsets [4]. In this paper, we made our primary intention toward automatically detecting minimum support and minimum confidence of each taxonomy' level by constructing a genetic algorithm based heuristic method for practical multilevel association rules mining in big datasets. By using the advantage of the genetic algorithm, which can efficiently ascertain multiple solutions concurrently in a large multidimensional problem without conducting exhaustive searches, our offered method can enhance the mining performance while preserving the wanted accuracy but bypassing the exhausting list of association rule candidates. Definitions linked to multilevel association rules are presented as follows [9][15][16]:

Definition 1: an item set, X, is a set of data items $\{X_{i,}X_j\}$, where $X_{i,}X_j \in I$. The support of an item set X in a set *S*, $\sigma(X/S)$ is the number of transactions (in *S*) which covers X against the total number of transactions in *S*. The confidence of $X \Rightarrow Y$ in *S*, $\varphi(X \Rightarrow Y/S)$, is the fraction of $\sigma(X \cup Y/S)$ in competition with $\sigma(X/S)$, i.e., the possibility that item set *Y* takes place in *S* when item set occurs X in *S*.

Definition 2: An item set X is large in set *S* at level L if the support of X is no less than its matching minimum support threshold σ'_L. The confidence of a rule $X \Rightarrow Y/S$ is high at level L if its confidence is no less than its equivalent minimum confidence threshold φ'_L.

Definition 3: a rule $X \Rightarrow Y/S$ is strong if $X \cup Y/S$ is large at the existing level and the confidence of $X \Rightarrow Y/S$ is high at the current level.

Definition 4: A fuzzy transaction denoted by *T* is given by:

$$\overline{\overline{T}} = \{(x, \mu(x)) \mid \forall x \in I\},\ 0 \le \mu(x) \le 1, \mu : I \rightarrow [0,1], \overline{\overline{T}} \subseteq T$$

where *T* is a general set of transactions, and $\mu(x)$ is degree of membership of *x*.

Definition 5: A soft quantitative transaction set that is symbolized by T'_q. Let (*F, E*) is a soft set over the universe *U* and $X \subseteq E$, *F* means the fuzzy power set of *U*, and *E* is a set of parameters. A set of attributes *X* is said to be supported by a transaction if:

$$T'_q = \{(< x, l, u >, e) \| \forall < x, l, u > \in I \times p \times p \mid l \le u, e \in E\}$$

Mining of association rules essentially focalizes at a single conceptual level. There are applications which lack to locate associations at multiple abstract planes. In a large database of transactions, where each transaction consists of a set of items and a taxonomy (is-a hierarchy) on items, it is expected to find out associations between items at any level of taxonomy. To investigate multilevel association rule mining, anyone wants to afford data at multiple-level association at multiple levels of abstraction and efficient methods for multiple level rule mining. The first specification can be accomplished by producing concept taxonomies from the primitive level concepts to higher levels. The second condition dictates efficient methods for multilevel rule mining [15].

One modification of Apriori to multi-level datasets is the ML_T2L1 procedure [15][24]. The ML_T2L1 algorithm manages a transaction table that has the hierarchy information encoded into it. Each level in the dataset is treated separately. Firstly, level 1 (the highest level in the hierarchy) is examined for large 1-itemsets using Apriori. The list of level 1 large 1-item sets is then employed to refine and clip the transaction dataset of any item that does not have an ancestor in the level 1 large 1-itemset list and eliminate any transaction which has no common items (thus comprises only infrequent items when evaluated using the level 1 large 1-itemset list). From the level 1 large 1-itemset list, level 1 large 2-itemsets are concluded (using the cleaned dataset). Then level 1 large 3- item sets are inferred and so on until there are no more frequent item sets to find at level 1. Since ML_T2L1 specifies that only the items that are descendant from frequent items at level 1 (essentially they must descend from level 1 large 1-itemsets) can be frequent themselves, the level 2 item sets are concluded from the refined transaction table. For level 2, the large 1-itemsets are created, from which the large 2-itemsets are determined and then large 3-itemsets etc. After all the frequent itemsets are found at level 2, the level 3 large 1-itemsets are located (from the same purified dataset) and so on. ML_T2L1 reforms until either all levels are explored using Apriori or no large 1-itemsets are exposed at a level. The principal steps of the proposed system are as follows [8-10][14][24][29-31]:

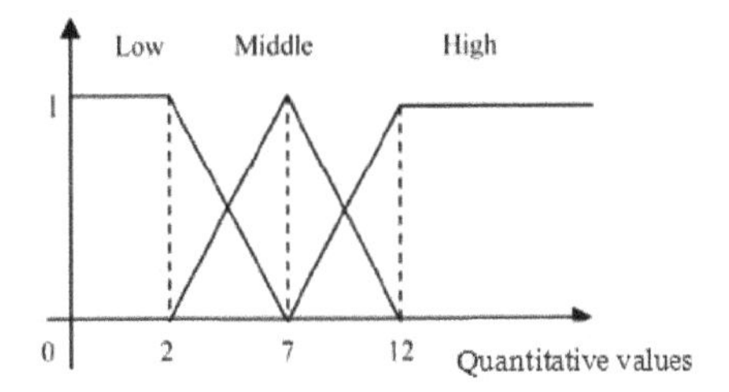

Fig. 2. The membership functions of items in Γ

Input: A group of N quantitative transaction data D, a pre-determined catalog Γ with the original items $\{i_1, i_2, ..., i_n\}$, a set of membership functions for each item in deferent levels. In our case, all the membership functions have the same style as shown in Fig. 2; but the x-axis is determined for each element in Γ based on the higher quantitative value associated with it. Finally, the parameter set minimum support α_k and minimum confidence λ_k that are acquired by genetic algorithm.

Output: A collection of fuzzy multiple-level association rules below the restrictions of optimal minimum support and confidence.

Step 1: Translate the predefined taxonomy using an arrangement of numbers and the symbol "*" by the formula, $C = \rho * 10 + i$, where i is the position number of the node at current level l, C signifies the code for the ith node at current level and ρ is the code of parent of the ith node at the present level.

Step 2: Interpret the item terms in the transaction data agreeing to the encoding scheme. Then set $k = 1$, $r = 1$ where k, $1 \le k \le x$ is the recent level number, x is the number of level in a given taxonomy and r denotes the number of items kept in the current frequent item sets.

Step 3: Cluster the items with the same first k digits in each transaction D_i, and add the quantities of the items in the similar sets in D_i. Symbolize the total of the j-th group I_j^k for D_i as v_{ij}^k.

Step 4: We explored several membership function for various data items for that each data item has its own features and its own membership function, then transform the value v_{ij}^k of each transaction D_i for each encoded group I_j^k into a fuzzy set f_{ijl}^k (Eq.1) by plotting v_{ij}^k on the specified membership function, where I_j^k is the j-th item on level k, v_{ij}^k is the quantitative value of I_j^k in D_i, h_j^k is the number of fuzzy areas for I_j^k, R_{jl}^k $(1 \le l \le h_j^k)$ is the l-th fuzzy region of I_j^k, f_{ijl}^k is v_{ij}^k fuzzy membership value in R_{jl}^k

$$\left(\frac{f_{ij1}^k}{R_{j1}^k} + \frac{f_{ij2}^k}{R_{j2}^k} + ... + \frac{f_{ijh}^k}{R_{jh}^k}\right). \quad (1)$$

Step 5: Assemble the fuzzy regions (linguistic terms) with membership values larger than zero to create the candidate set C_1^k; Calculate the scalar cardinality S_{jl}^k of each fuzzy region R_{jl}^k in the transaction data as $S_{jl}^k = \sum_{i=1}^{n} f_{ijl}^k$.

Step 6: Investigate if the value S_{jl}^k of each region R_{jl}^k in C_1^k is larger than or equals to the threshold α_k which is the optimal minimum support for level k obtained from implementing the genetic algorithm to the set of transactions included in this level according to Γ (see algorithm 1). If R_{jl}^k matches the threshold, place it into the large 1-itemset L_1^k for level k. That is:

$$L_1^k = \{R_{jl}^k | S_{jl}^k \ge \alpha_k, R_{jl}^k \in C_1^k\}. \quad (2)$$

Step 7: if L_1^k is null, let $k = k + 1$ and go to step 3; else, create the applicant set C_2^K from $L_1^1, L_1^2, ..., L_1^K$ to catch "level-crossing" large itemsets. The created applicant set C_2^K has to fulfill the following conditions: (1) Each 2-itemset in C_2^K must comprise at least one item in L_1^k. (2)The two regions in a 2-itemset may not have the same item name. (3) The two item names in a 2-itemset may not be with the hierarchy relation in the taxonomy. (4) Both of the support values of the two large

1–itemsets including a candidate 2-itemset must be larger than or equal to the minimum support $\alpha_{k=2}$.

Step 8: If L_1^k is null, then increase k by one, r =1 and go to step3 else set r=r+1.

(a) If r = 2 create the candidate set C_2^K, where C_2^K is the set of candidate itemset with 2 items on level k from $L_1^1, L_1^2, L_1^3,..., L_1^k$ to learn "level-crossing" of frequent itemset. Each 2-itemset in C_2^k must contain at least one item in the L_1^k and the next item should not be its ancestor in the taxonomy. All possible 2-itemsets are composed in C_2^k.

(b) If $r > 2$, produce the candidate set C_r^k, where C_r^k is the set of candidate itemset with r-items on level k from L_{r-1}^k in a way similar to that in the preceding steps.

Step 9: For each acquired candidate r-itemset S with items $(S_1, S_2,....S_r)$ in C_r^k:

a) Calculate the fuzzy value of S in each transaction datum D_i by the minimum operator as $f_{is} = \min(f_{is1}, f_{is2},..., f_{isr})$

b) Estimate the scalar cardinality of S in all the transaction data as $count_s = \sum_{i=1}^{n} f_{is}$.

c) If $count_s$ is larger than or equals to the pre-defined minimum support α_k place S into L_r^k.

Step 10: If L_r^k equal null then increase K by one and go to the next step; if not increase r by one and go to step 8.

Step 11: If $k > x$ then go to the next step, else set r =1 and go to step 3.

Step 12: create the fuzzy association rules for all frequent r- itemset including $S = (S_1, S_2, ..., S_r)$, r >2 as follows:

- Catch all the rules $A \rightarrow B$ where $A \subset S$, $B \subset S$ and $A \cap B = \phi$, $A \cup B = S$.
- Calculate the confidence value of all association rules by $\frac{\sum_{i=1}^{n} min(f_{iS})}{\sum_{i=1}^{n} min(f_{iA})}$.

Step 13: Choice the rules that have confidence values not less than predefined confidence threshold λ_k, where λ_k is the predefined minimum confidence value for level k found by applying genetic algorithm.

Step 14: eliminate redundant rules from multi-level datasets. Herein, Rule R_1 is redundant to rule R_2 if (1) the itemset X_1 is made up of items where at least one item in X_1 is descendant from the items in X_2 and (2) the item set X_2 is entirely made up of items where at least one item in X_2 is an ancestor of the items in X_1 and (3) the other non-ancestor items in X_2 are all present in item set X_1. The additional state (4) the confidence of $R_1(C_1)$ is less than or equal to the confidence of $R_2(C_2)$.

A. Parameters extraction using genetic algorithm

A genetic algorithm is a class of investigating algorithm that is employed to automatically set the optimal minimum support and minimum confidence for each taxonomy's level. It explores a solution space for an optimal answer to a problem [28]. The algorithm generates a "population" of feasible solutions to the problem and makes them "evolve" over many generations to locate valid and better solution. The algorithm begins with a collection of solutions (represented by chromosomes) called a population. Solutions from one population are selected and managed to establish a new population. The framework of the basic genetic algorithm is as follows (see Fig. 3).

```
Procedure genetic algorithm
begin (1)
    t = 0;
    inicialize P(t);
    evaluate P(t);
    While (Not termination-condition) do
    begin (2)
        t = t + 1;
        select P(t) from P(t − 1);
        recombine P(t);
        evaluate P(t);
    end (2)
end (1)
```

Fig. 3. Structure of the genetic algorithm [29]

1) [Start] create arbitrary samples of n chromosomes (appropriate results for the problem)

2) [Fitness] assess the fitness (qualification) function f(x) of every chromosome x in the population

3) [New population] generate a new resident by iterating the subsequent steps until the new population is complete.

Selection: pick two parent chromosomes from a population according to their fitness (the better fitness, the higher possibility to be chosen)

Crossover: with a crossover probability, crossover the parents to produce a new generation (children). If no crossover was conducted, offspring is an accurate reflection of parents.

Mutation: with a mutation probability, the GA mutates a new generation at each location (site on the chromosome).

Accepting: store distinct generation in a new population.

4) [Replace] manage recently produced population for a more route of the algorithm

5) [Test] if the end condition is satisfied, stops, and returns the best solution in current population

6) [Loop] go to step 2.

Through repetition t, the GA keeps a population $p(t)$ of results $r_1^t,....,r_N^t$, where r_i^t characterizes rule set that is arbitrarily created for each level. Each solution r_i^t is gauged by the function $E(\bullet)$ and $E(r_i^t)$ is a degree of fitness of the solution. The fitness value determines the relevant power of an individual to remain and create offspring in the next production. In the next iteration (t+1) a new resident is designed on the foundation of the operations (2) and (3) [29].

B. Data Encoding

Given a randomly generated association rules for each level, the system uses the Michigan approach for encoding, in which a chromosome is a collection of all used rules; here the population consists of many rule collections. Coding in the Michigan method is binary coding, in which "1" means that a knowledge base rule will be in a knowledge base, whereas "0" means it will not be used. The key benefit of this technique is that the entire rule base is coded; therefore, it is not necessary to do the quantitative analysis of indispensable rules to see if the method functions properly, because, unlike the Pitts method, all possible rules take part in the working time of the genetic algorithm. The considerable size of the chromosome is a disadvantage. The dimension of the chromosome is dependent on the volume of the rule base and it increases exponentially depending on the number of itemsets [10][14][30].

C. Generic Operators

The frequently employed genetic operators are reproduction, crossover, and mutation. To achieve genetic operators, one must pick individuals in the population to be worked on. The collection plan is mainly based on the fitness level of the individuals exhibited in the population. For election; the system manages roulette wheel sampling fashion. In this procedure, the parents for crossover and mutations are chosen based on their fitness, i.e. if a candidate has more fitness function value more will be its opportunity to get elected. The implementation of roulette wheel sampling is performed by first normalizing the values of all applicants so that, their chances sprawl between 0 and 1, and then by applying random number function, a random number is estimated, and then matching to this value and the fitness normalized value, the candidate is elected [14].

As an individual is picked, reproduction operators only imitate it from the current population into the new population (i.e., the new generation) without transposition. The crossover operator begins with two selected individuals and then the crossover point (an integer between 1 and L-1, where L is the length of strings) is picked arbitrarily. The third genetic operator, mutation, offers random variations in the arrangements in the population, and it may irregularly have useful results: departing from a local optimum. In our GA, mutation is just to oppose every bit of the strings, i.e., changes a 1 to 0 and vice versa, with probability p_m [30].

The algorithm stops fulfilling when the decay situation is reached – i.e. when the best and worst producing chromosome in the population disagrees by less than 0.1%. It also ends execution when the total number of generations defined by the user has arrived. Besides, the algorithm bypasses forming the initial population completely randomly because it may appear in rules that will include no training data instance whereby having very low fitness. Furthermore, a population with rules that are insured to comprise at least one training instance can lead to over–fitting the data. It was shown that employing non–random initialization can reach to an elevation in the quality of the solution and can drastically decrease the runtime [24]. We, therefore, devised an initialization method which involves picking a training instance to serve as a "seed" for rule generation based on the alteration of itemsets within each level [31].

In general, genetic operator assists in controlling the heterogeneity of the population and also in blocking early concurrence to local optima [14]. Our intention is to explore fascinating association rules. Consequently, the fitness function is vital for ascertaining the interestingness of chromosome, and it does influence the convergence of the genetic algorithm. In this case, the proposed system examines two different fitness functions. The first one considers the identical confidence of the corresponding association rule as illustrated in Eq. 3, whereas the second fitness function joins the support (*sup*) and *confidence* (*conf*) attributes, which are required to define an association rule (see Eq. 4) [9][10][14]. Parameters α and β are the significant factors to equilibrium the weight of the *support* and *confidence* in the fitness function, $\alpha+\beta=1$. To mine confirmed association rules from the big database with our GA approach, the threshold of the fitness function has to be predefined; in our case, $\alpha=\beta=0.5$.

$$f_1 : rconf(X \rightarrow Y) = \frac{\sup(X \cup Y) - \sup(X) \times \sup(Y)}{\sup(X)(1-\sup(Y))} \quad (3)$$

$$f_2(x \rightarrow y) = \alpha \times \sup(x \rightarrow y) + \beta \times conf(x \rightarrow y) \quad (4)$$

By adopting the recommended system, rather of producing an untold number of interesting rules in conventional mining models, only the most interesting rules are declared according to the interestingness measure determined by the fitness function. The main motivation for using GA in the learning of high-level prediction rules is that they conduct a global search and cope better with attribute cooperation than the greedy rule selection algorithms [14].

In brief, the proposed evolutionary method for quantitative association rule mining is particularly prompted by (1) partition of quantitative attribute is not accessible for every attribute and every user, (2) users, and even experts, usually feel tedious to define the minimum-support, (3) the search space might be very large when we face quantitative attributes, and (4) the rules passed might be too many to deal with [10]. However, mining association rules are not adequate of benefits; it has some defects too, first of all, the algorithmic complexity. The number of rules increases exponentially with the number of items. But this complexity is undertaken with some advanced algorithms which can efficiently clip the search space. Secondly, the obstacle of attaining rules from rules, i.e. selecting interesting rules from the set of rules.

The suggested work undertakes the second problem that essentially assist the user when scanning the rule set, and valuable quality measures on the rules are adopted based on genetic algorithm. Usually, when handling association rule mining many rules can be found or inferred and confuse the user. But more importantly, some of these rules could be redundant and yield no new knowledge. Some attempts have been pointed at dealing with redundant rules in flat datasets, however, datasets can have a hierarchy/taxonomy or compound concept levels and thus redundancy in these datasets require to be concentrated on. This issue is one of the features of this study.

IV. Experiments and Results

In this section, we perform some experiments that have been conducted out to examine the performance of the proposed approach and confirm enhancements over the traditional method without optimization. The experiment was conducted with MATLAB software. All the experiments are handled on a laptop computer with the following specifications: Processor Intel(R) Core (TM) i5-2520MCPU@2.50GHz 2.50GHz, memory 4.00GB, and System type 64-bit operating system, x64-based processor.

A. Dataset

The dataset was hired as in [8] and can be viewed as a benchmark because it is used for comparison. This is a market basket dataset that consists of the items and amounts of items marketed in every purchasing container. This dataset consists of 1000 sales receipts of a food material repository based on the predefined taxonomy from 7 items (10000 transactions). The predefined taxonomy in the first level holds 7 nodes that describe the items worked in the test, the second level comprises 14 nodes that describe the taste or different types of a particular stock and in the third level, it also consists of 48 nodes that express the manufacturing companies and factories. The database transactions carry the name of the product and the quantity of such purchased merchandise. One item may not be employed twice in one transaction.

B. Methodology

A comparison was made between the conventional approach for mining multi-level fuzzy quantitative association rules [8] and the approach proposed in this paper that uses GA optimization technique. The objective was to find new detailed knowledge by (1) an enhancement in pronouncing multi-level optimal support and confidence that is employed to obtain interesting rules. (2) eliminating redundant rules that were encountered in the traditional approach. Both algorithms utilize a top-down progressively deepening approach to infer large itemsets and also consolidate fuzzy boundaries instead of explicit boundary intervals.

The mined rules from the proposed system are closer to the reality, and it gives the ability to mine association rules at different levels based on the optimal re-calculated mining parameters (*min_sup*, *min_conf*); unlike the traditional method that depends on the experts to determine these parameters manually. Employing GA to find these parameters makes the proposed system is context-independent and more general. In the experiments, thresholds for *min-sup* and *min-conf* were set at 0.28 and 1.7 respectively for the traditional algorithm for each taxonomy level.

In the first experiment, we examine if correct association rules can be specified with a fixed number of initial generations and in a bounded time period. Using our dataset, the initial population size ranges from 30 to 100. The results are displayed in Table 1. With a limited population, most strong association rules could be inferred in our dataset. We can decide that if the population is too small, the realization of the GA-based algorithm will be similar to the random algorithm. But if the population is too large, despite we can get full association rules immediately, but the computational complexity rises fast. However, as we can see, there is a good stability that, with a restricted population and a limited time period, most valid association rules have been mined. Therefore, we choose 50 as the default population for the dataset, which works well in our algorithm.

TABLE I. Relation Between the Number of Initial GA Population and the Number of Multilevel Association Rules (Using F1 with Michigan Encoding, Generation No. =10)

No. of Initial population	30	50	70	100
No. of association rule (Redundant)	3706	607	607	607
No. of association rule (non-redundant)	1733	349	349	349

One reason of the stability of the number of extracted rules with initial population contains 50 chromosomes is that the proposed system utilizes Michigan approach for rule coding in which a chromosome is a collection of all used rules. So, the lowest number of initial populations will contain specific rules permutations. In Michigan approach, each chromosome contains a comprehensive representation of the rules.

In the next collection of experiments, we confirmed how worthy the extracted association rules are from either the GA-based algorithm or the traditional algorithm without GA. To measure its value with the 10000 transactions, we use the formula of the fitness function f_1 using the following configuration mutation rate = 0.1, crossover rate = 0.9, generation No.=10,and initial population=50. The results in dataset are shown in Table 2.

TABLE II. Comparative Study

Methods	No. of non-redundant rules	Calculated *min-sup*	Calculated *min-conf*	Time (Sec)
Proposed Method with GA	349	L1=0.95 L2=0.67 L3=0.27	L1=1.12 L2=1.45 L3=1.92	500
Traditional method without GA [8]	2281	0.28 (L=1 to 3)	1.7 (L=1 to 3)	400

TABLE III. Comparative Study between the Two Fitness Functions

Fitness Function	levels	Computed Fitness	No. of association rules (non-redundant)
f_1 (Eq. 3)	level 1	1.90	349
	level 2	1.97	
	level 3	2.06	
f_2 (Eq.4) $\alpha=\beta=0.5$	level 1	1.07	2249
	level 2	1.26	
	level 3	1.47	

TABLE IV. PROPOSED SYSTEM EVALUATION UNDER DIFFERENT PARAMETERS OF GA USING F_1

Parameters ratio	No. of association rules (redundant)	No. of association rules (non-redundant)
Mutation= 0.9 crossover= 0.1	751	419
Mutation=0.8 crossover=0.2	751	419
Mutation=0.7 crossover=0.3	607	349

Compatible with the outcomes above, with the advance of the time boundary, GA-based approach can catch high relevant association rules with a little more time than traditional method (25% increases in time). But in terms of quality, the proposed system extracts more interesting rules; only about 17 % of the total number of rules extracted from the comparative system. In general, a large number of extracted rules inside market basket analysis will hamper the decision-maker. The proposed system offers the most interesting rules subject to the fitness function, which is accountable for acting the assessment that imitates how optimal the solution is: the higher the number, the better the solution.

The third set of experiments was executed to compare how relevant the mined association rules are from either the fitness function that considers the relative confidence of the corresponding association rule (Eq. 3) or the fitness function that considers both the support and confidence attributes (Eq. 4). The experiment is conducted under the previous configuration for GA. The results in Table 3 reveal that the use of f_1 generates a further mined association rules rate improvement of 83% reduction in the number of extracted rules. From this experiment, we realized that the fitness function represents a critical issue in the success of genetic algorithm; this is clearly shown in the case of f_2 . Using f_2 did not bring any advantages to the GA; thus, we get the same number of extracted rules that have been obtained from the traditional method (about 2249).

The performance improvement of GA using f_1 comes from the correct extraction of interesting rules; because of calculating the support of union for items inside each rule in addition to the support of each item separately. Unlike the second function that uses the support and confidence of each rule, which represents the standard case used by many of the existing mining algorithms (e.g. Apriori).

Having compared our system to different multi-level quantitative mining algorithm, we will next explore the influence that GA factor settings have on our system, which incorporates both mutation rate and crossover rate. To retain the number of factor setting blending small, we will only fluctuate the setting for one parameter at a time while holding the setting for another parameter to its default value. In Table 4, we vary mutation rate from 0.7 to 0.9 and look at the number of association rules (non-redundant) achieved by our system on the used dataset. From the table, we can see that decreasing mutation rate will diminish the number of extracted rule (17% lower in the number of rules). This decreasing is noticeable. This decrease is due to the fact that mutation is managed to preserve genetic heterogeneity from one generation of a population to the next. In GA, mutation operators are frequently employed to give exploration and crossover operators are extensively employed to supervise the population to focalize on one the good solutions encounter so far (exploitation). Consequently, while crossover attempts to concentrate to a special point in the landscape, the mutation does its best to evade convergence and investigate more areas.

V. CONCLUSION

Really, mining quantitative association rules is an optimization obstacle rather than being an uncomplicated discretization one. In this paper, we have introduced a new genetic-based algorithm to mine multilevel association rules in big quantitative date sets that deals with quantitative attributes by accurate fuzzification the values -partitioning the values of the attribute. The proposed system uses fuzzy set concepts, multi-Level taxonomy, different pre-calculated minimum supports for each level and different membership function for each item to discover fuzzy association rules in a given transaction data set.

In our algorithm, the minimum supports and minimum confidences for each level for fuzzy quantitative association rule are defined by the genetic algorithm optimization. In this case, the employed GA combines a population initialization technique that guarantees the production of high- quality individuals; individually planned breeding operators that confirm the removal of inadequate genotypes; an adaptive mutation probability to ensure genetic heterogeneity of the population; and uniqueness testing based on both support and confidence that is employed to hold only high quality and interesting rules.

The proposed system gives the user with rules according to two interestingness metrics, which can quickly be extended if need by changing the fitness function. The results report that: In terms of mining of association rules, the proposed method keep higher precision compared with the traditional methods and the extracted rules are more close to reality. This is because of adopting various membership functions for every individual item, optimized minimum supports, and minimum confidences, and finally, the non-redundant algorithm to enhance the quality and application of the rules. Future work includes employing GA to tune the fuzzy membership function for each item.

REFERENCES

[1] J. Han, M. Kamber, and J. Pei, “Data mining concepts and techniques,” Elsevier Inc., USA, 2012.

[2] T. Hong, Y. Tung, S. Wang, Y. L. Wu, and M. T. Wu, "A multi-level ant colony mining algorithm for membership functions, "Information Sciences, vol. 182, no. 1, 2012, pp. 3-14.

[3] A. Gosain, and M. Bhugra, "A comprehensive survey of association rules on quantitative data in data mining, " the IEEE Conference on Information and Communication Technologies, India, 2013, pp.1003-1008.

[4] Y. C. Lee, T. P. Hong, and T. C. Wang, "Mining fuzzy multiple-level association rules with multiple minimum supports, " Expert Systems with Applications, vol. 34, no. 1, 2008, pp.459-468.

[5] K. Poornamala, and R. Lawrance "A general survey on frequent pattern mining using genetic algorithm, " ICTACT Journal on Soft Computing, vol. 3, issue 1, 2012, pp.440-444.

[6] D. Kanani, and S. Mishra, "An optimize association rule mining using genetic algorithm," International Journal of Computer Applications, vol. 119, issue 14, 2015, pp.11-15.

[7] E. Mahmoudi, E. Sabetnia, M. Torshiz, M. Jalal, and G. Tabrizi, "multi-level fuzzy association rules mining via determining minimum supports and membership functions, " the Second International Conference on Intelligent Systems Modeling, and Simulation, Iran, 2013pp.55-61.

[8] A. Kousari, S. Mirabedini, and E. Ghasemkhani, "Improvement of mining fuzzy multiple-level association rules from quantitative data," Journal of Software Engineering and Applications, vol. 5, no. 3, 2009,pp. 190-199.

[9] X. Yang, M. Zeng, Q. Liu, and X. Wang, "A genetic algorithm based multilevel association rules mining for big datasets" Mathematical Problems in Engineering, Vol. 2014, 2014, pp.1-10.

[10] X. Yan, C. Zhang, and S. Zhang, "genetic algorithm-based strategy for identifying association rules without specifying actual minimum support, " Expert Systems with Applications vol. 36, no. 2, 2009, pp. 3066-3076.

[11] S. Aouissi, A. Vrain, and C. Nortet, "Quantminer: a genetic algorithm for mining quantitative association rules", the 20th International Conference on Artificial Intelligence , 2007, pp. 1035–1040.

[12] G. Shaw, Y. Xu, and S. Geva, "Utilizing non-redundant association rules from multi-level datasets" the IEEE International Conference on Web Intelligence and Intelligent Agent Technology, vol. 03, Australia, 2008, pp.681-684.

[13] P. Gautam, and K. R. Pardasani, "Algorithm for efficient multilevel association rule mining", International Journal on Computer Science and Engineering, vol. 2, no.5, 2010, pp.1700-1704

[14] S. Manish, A. Agrawal, and A. Lad, "Optimization of association rule mining using improved genetic algorithms," the International Conference on Systems, Man and Cybernetics, vol. 4, USA, 2004, pp.3725-3729.

[15] V. Ramana, M. Rathnamma, and A. Reddy, "Methods for mining cross level association rule in taxonomy data structures" International Journal of Computer Applications, vol. 7, no.3, 2010, pp.28-35.

[16] S. Saraf, N. Adlakha, and S. Sharma, " Soft set approach for mining quantitative fuzzy association patterns in databases, " International Journal of Advanced Research in Computer Science and Software Engineering, vol. 3, issue 11, 2013, pp.359-369.

[17] M. Delgado, N. Manín, M. J. Martín-Bautista, D. Sánchez, and M. -A. Vila" Mining fuzzy association rules: an overview, "Studies in Fuzziness and Soft Computing, Springer Berlin Heidelberg , vol. 164, 2005, pp.351-373.

[18] G. Attila, "A fuzzy approach for mining quantitative association rules, " ACTA CYBERN, vol. 15, no.2, 2001, pp.305-320.

[19] W. Toshihiko, and H. Takahashi, "A study on quantitative association rules mining algorithm based on clustering algorithm, " International Journal of Biomedical Soft Computing and Human Sciences, vol. 16, no.2, 2010, pp.59-67.

[20] R. Sridevi, and E. Ramaraj, "A general survey on multidimensional and quantitative association rule mining algorithms, " International Journal of Engineering Research and Applications, vol. 3, issue 4,' 2013, pp.1442-1448.

[21] P. Gautam, N. Khare, and K. R. Pardasani," A model for mining multilevel fuzzy association rule in database, " Journal of Computing, vol. 2, issue 1, 2010, pp.58-64.

[22] O. Oladipupo, C. Ayo, and C.Uwadia, " a fuzzy association rule mining expert-driven Approach to Knowledge Acquisition," African Journal of Computing and ICT, vol. 5, no. 5, 2012,pp.53-60.

[23] S. Prakash, M. Vijayakumar, and R.Parvathi, " A Novel Method of Mining Association Rule with Multilevel Concept Hierarchy", International Journal of Computer Applications, vol. 5, no. 5, 2011, pp.26-29.

[24] G. Shaw, Y. Xu, and S. Geva, "Eliminating redundant association rules in multi-level datasets, " the 4th International Conference on Data Mining, USA, 2008pp.1-8.

[25] E. Ayetiran, and A. Adeyemo, "A data mining-based response model for target selection in direct marketing," International Journal Information Technology and Computer Science, vol. 4, no. 1, 2012, pp. 9-18.

[26] W. Toshihiko, and H. Takahashi, "A study on quantitative association rules mining algorithm based on clustering algorithm," Journal of the Biomedical Fuzzy Systems Association, vol.16, no.2, 2010, pp.59-67.

[27] Y. Wan, Y. Liang, and L. Ding, "mining multilevel association rules from primitive frequent itemsets," Journal of Macau University of Science and Technology, vol.3, issue 1, 2009,pp.10-19.

[28] R. Haldulakar, J. Agrawal, "Optimization of Association Rule Mining through Genetic Algorithm," International Journal on Computer Science and Engineering, vol. 3, no. 3, 2011, pp.1252-1259.

[29] M. Kayaa, R. Alhajj, "Genetic algorithm based framework for mining fuzzy association rules, " Fuzzy Sets and Systems, vol. 152, 2005, pp. 587–601.

[30] S. Tiwari, M. K. Rao" Optimization in association rule mining using distance weight vector and genetic algorithm , "International Journal of Advanced Technology & Engineering Research, vol. 4, Issue 1, 2014, pp.79-84.

[31] P. Wakabi-Waiswa, and V. Baryamureeba, "Mining high quality association rules using genetic algorithms", In Proceedings of the twenty second Midwest Artificial Intelligence and Cognitive Science Conference, USA, 2009, pp. 73-78.

A Model for Facial Emotion Inference Based on Planar Dynamic Emotional Surfaces

Ruivo, J. P. P.
Escola Politécnica
Universidade de S˜ao Paulo
S˜ao Paulo, Brazil

Negreiros, T.
Escola Politécnica
Universidade de S˜ao Paulo
S˜ao Paulo, Brazil

Barretto, M. R. P.
Escola Politécnica
Universidade de S˜ao Paulo
S˜ao Paulo, Brazil

Tinen, B.
Escola Politécnica
Universidade de S˜ao Paulo
S˜ao Paulo, Brazil

***Abstract*—Emotions have direct influence on the human life and are of great importance in relationships and in the way interactions between individuals develop. Because of this, they are also important for the development of human-machine interfaces that aim to maintain a natural and friendly interaction with its users. In the development of social robots, which this work aims for, a suitable interpretation of the emotional state of the person interacting with the social robot is indispensable. The focus of this paper is the development of a mathematical model for recognizing emotional facial expressions in a sequence of frames. Firstly, a face tracker algorithm is used to find and keep track of faces in images; then the found faces are fed into the model developed in this work, which consists of an instantaneous emotional expression classifier, a Kalman filter and a dynamic classifier that gives the final output of the model.**

Keywords*—*emotion recognition, facial emotion, Kalman filter, machine learning

I. Introduction

Emotions influence the human behavior and the way individuals interact and relate to other members of society. They permeate one's daily life and determine how people react to the various situations they encounter in their routines.

Studies indicate that people with impairments to express or recognize feelings end up having great difficulty keeping even casual relationships [1]. Emotions also help the body prepare for specific external events. For example, the fear people may experience when they see a large object coming fastly towards them stimulates blood circulation in their legs, allowing them to act promptly and respond trying to avoid the object.

Computer interfaces that can understand the emotional state of its users can communicate more naturally compared to interfaces without this capability. Affective computing comes to deal with the integration of the concept of emotion in the computational area [2].

Emotions are characterized by signs in voice, speech and body movements, which are recognized regardless of culture, possibly being a legacy of human evolution and not a result of personal experiences of the individual [3]. Particularly in the face, the most obvious signs are presented in the regions of the mouth, eyes and eyebrows. Ekman and Friesen showed evidence for the hypothesis of universality of emotional facial expressions in intercultural studies with illiterate populations of Papua New Guinea and investigated the influence of the cultural phenomena [3].

Works from Ekman [4] [5] propose the existence of six major universal emotions: joy, sadness, surprise, fear, anger and disgust. An emotional display can either be classified as belonging to one category, such as joy, or more than one category, forming composite emotions, such as the mixture of fear and angry, or joy and surprise.

This study aims to identify five basic emotional states: Happiness, Sadness, Anger and Fear, plus the Neutral state, which could be understood as the absence of emotions. The model proposed in this work does not try to describe short-lived or rapidly changing emotions (micro expressions, in the works of Ekman), but focuses on trying to detect lasting emotional states people may be subject to. The dynamic model for emotion recognition presented in this work is a novel model based on the work of [6].

The rest of this paper is organized as follows. In Section II are reviewed previous works regarding automatic emotion inference. The adopted methodology, including face detection, feature extraction, instantaneous emotion recognition and dynamic emotion recognition is then presented at Section III. Section IV describes the results obtained. Finally, the conclusions and future work directions are presented on Section V.

II. Bibliographic Review

There are three main approaches to emotions classification: discrete model, dimensional model and the approach based on evaluation mechanisms [7].

The discrete model arranges emotions in categories, like the basic emotions of Ekman. Categorization of emotions is an intuitive and practical way to identify them, even if a large number of classes is necessary in order to classify all of the known affective states. Many of the works developed in the area utilizes this approach [8] [7] [9] [10].

The dimensional model seeks to describe the emotions by means of some criteria or dimensions. Two key dimensions are valence and arousal [10] [11]. Valence transmits how the person feels under the influence of a certain emotion, and can assume continuous values ranging from extreme sadness, for negative valence, to extreme happiness, for positive valence. Arousal is associated to the possibility of an individual to take or to perform an action under influence of an emotion, and can assume continuous values ranging from an extremely passive attitude, for negative arousal, to an extremely active attitude, for positive arousal. Some authors [12] suggest other dimensions for the model, such as dominance. Dominance is related to the control somenone has over a situation while under the influence of an emotion, and can assume continuous values ranging from total lack of control to total control of the situation. The dimensional model avoids the need for an extensive list of categories. Emotions are identified depending on its position on the model's axes. However, because of the limited number of dimensions this approach deals with, the projection of an emotion to the model's axes could cause loss of information [7].

The evaluation approach classifies emotional displays based on a set of assessments of the event that caused such display. For a given emotion, it is evaluated how relevant it is the event that elicited the emotion, what are its implications, the individual's ability to deal with these implications, and what is the significance of that event for the society the individual inhabits [13]. This approach is less simple and intuitive when compared to the others, as it requires a detailed analysis of the situations that elicited the emotions.

Pantic [14] suggests automatic recognition of facial emotion expressions to be done in three main steps: face detection, extraction of relevant features of the face and emotion classification

Face detection is a crucial step in the recognition of expressed emotions, and comprises of locating faces in still images or image sequences. In several works, such images are obtained under conditions that helps face detection algorithms, like the capture of the face in frontal orientation, without occlusions, and under uniform lighting conditions. However, in real situations, these conditions rarely can be reproduced, which makes the problem more challenging. Consequently, an ideal method of facial detection should deal with problems such as the different scales and orientations the human face may take, besides having to consider possible partial occlusions of the face and changes in the lighting conditions.

Extraction of face relevant characteristics has the purpose of generating a feature vector to be used for the emotion identification. It seeks to describe the face through certain categorical or numerical information that should contribute to the recognition of the emotional state of the analyzed person. These characteristics may be based on features of the human face such as eyebrows, nose and mouth, or may be based on mathematical models. These models, in turn, may follow an analytical approach, in which the face is represented by a set of points or patterns of interest that contain specific regions of the face; or they may follow a holistic approach, in which the face is seen as a unit, with its particular shape and texture. Hybrid approaches also exist, in which features of the two above-mentioned approaches are combined. Different scales and orientations of the face, as well as partial occlusion and noise, hamper the execution of this step.

The extracted features vector should then be used to estimate the expressed emotion via a classification algorithm. In this step, any of the approaches presented for emotion classification may be used; however, much of the work done in the area uses the discrete approach [15]. The classification of the facial emotion expressions is done by machine learning algorithms trained with the feature vectors extracted from the samples of one or more training databases. Examples of these algorithms are Support Vector Machines (SVMs), Decision Trees and Neural Networks (NNs).

The present work introduces a fourth step to the process proposed by [14] and includes the usage of a continuous emotional classifier model, following the line of work of [9] and [16]. This step was introduced so that the model would be able to detect long-lasting emotional states rather than instantaneous emotional displays; also, it should help with the minimization of the influences of natural noises, like laugh and speech, that deform the face and difficult the determination of someone's facial emotion expression.

Figure 1 presents a flow diagram of the steps aforementioned.

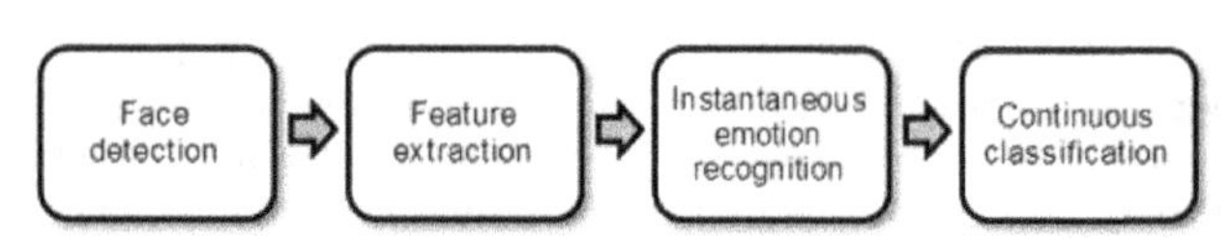

Fig. 1: Flow diagram for the proposed model.

One way to describe one's facial emotion expression is to use the Facial Actions Coding System (FACS) [17]. This system defines 44 Action Units (AU), each one representing the facial movements caused by muscle activity in a specific region of the face. Studies show that a particular subset of 15 of these AUs have greater relevance in the communication between humans [18].

FACS can be understood as an abstraction layer of the underlying facial muscle activity. Through the identification of the level of activity of the relevant AUs, one can infer the related muscles' activities and the corresponding facial expression. FACS defines, for example, involuntary and sincere expression of happiness as the activation of AUs numbers 6 and 12, that is, the lifting of the cheeks and the lateral and vertical extent of the lips, respectively. A forced (faked) expression shows only activation of AU 12 instead. This differentiation is possible because AU 12, which is the contraction of the zygomatic major muscle, is voluntary, while AU 6, contraction of the orbicularis oculi muscle, occurs involuntarily.

Furthermore, FACS brings into consideration the duration and intensity of AUs. Spontaneous muscle activations are in the range 250ms to 5 seconds, depending on the AU [19]. Rules for determining the intensity of each AU are also determined on FACS, for example, as the degree of elevation of the corner of the lips to the AU 12 or the wrinkle density over the nose for AU 44.

As noted in [9] [20], two different categories of properties could be extracted from faces: geometric properties and appearance properties.

Methods based on geometric properties look for characteristic regions of the face, such as eye contour, representing the shape and geometry of the features to be studied. For the extraction of data in video, one approach is the optical flow, as in [21], with tracking of characteristic points. Another approach are three-dimensional methods [22], which were developed along with the development of three-dimensional videos. In the solution presented in [23], the Active Shape

Model [24] and a Kalman filter were used to locate specific areas such as mouth and eyes in each frame of a video.

Appearance-based methods, however, search for changes in texture, such as wrinkles on the face. These methods can be used to describe the whole face or specific regions of interest [25] [18].

Following Figure 1, the next step, emotion classification, can be based on neural networks (NN), support vector machines (SVM) or hidden Markov chains (HMM) [9] [22] [22] [26], among other algorithms [18] [23].

It should also be noted that the humans' emotions detection system is not perfect, and emotions are not always interpreted correctly [14]. Donato [21] shows that people who had no training were able to correctly identify emotions in about 80% of a set of photos, but trained people, such as those passing throught FACS training, have a hit rate of about 90%. For Russell [27], however, a number of studies show that the rate of recognition by individuals varies according to the experimental conditions, ranging from about 55% to about 95%; also, negative emotions, such as anger and sadness, have a significantly lower accuracy recognition rate than positive ones.

The instantaneous emotion recognition model presented in this work is based on the work of Loconsole et al. [28]. In the referred work, an emotion classifier (namely, a random forest) based on geometric facial features is trained and used to differentiate images of faces expressing five emotional states: Joy, Sorrow, Surprise, Fear, Disgust and Anger. The authors analyze the accuracy their model achieved with and without calibration with neutral faces and considering different quantities of learned facial expressions. Also, they compare the accuracy of their model with that of other authors' models, and conclude their model achieved higher accuracy for the experiments made.

III. Methodology

This section briefly introduces the methods and techniques used to implement each of the steps shown in the diagram of Figure 1.

A. Face Detection

In this step, the Chehra Face Tracker is used [29]. This tracker detects and keeps track of faces in input images. It can be classified as a discriminative tracker, as it uses facial landmarks and discriminative functions to describe the current state of the face of a person, rather than a generative tracker, which would seek parameters that would maximize the probability of the deformable model to reconstruct a given face [29].

The Chehra Face Tracker uses an incremental parallel cascade of linear regressions to train the model, which has a better performance on face tracking in videos when compared to both the parallel cascade of linear regressions and the sequential cascade of linear regressions, showing better adaptation over time and robustness to environment changes on the face [29].

The tracker is capable of handling new training samples without having to retrain the model from scratch. It can also automatically tailor the model to the subject being tracked and to the imaging conditions, hence becoming person-specific over time [29].

B. Feature Extraction

Once the face tracker is able to fit the face model on one of the found faces in the image, one can proceed to extract features of interest from it.

The process of choosing what features to extract is not trivial, as the chosen feature set should be one that describes the studied concepts (in this case, the five facial emotion expressions: Happiness, Sadness, Anger and Fear, plus the Neutral state), so the trained classifier may have a better chance of learning how to properly differentiate amongst samples of these concepts. Loconsole [28] presents a feature set which is intended to differentiating among facial displays of Ekman's six basic emotions. This set comprises of two kinds of features: linear features and eccentricity features. While the linear features are determined by calculating the normalized linear distances between two given landmarks outputted by the face tracking model, the eccentricity features are given by the eccentricity measures of ellipses fitted over groups of three facial landmarks.

In the present work, Loconsole feature set is adopted with some new features added to it. The added features were chosen based on facial cues Ekman found to be of relevance in the process of facial emotion recognition [4]. The complete set of features adopted is described in Table I (refer to Figure 2 for the landmark's labels referenced in the table).

Table I: Extracted feature set

Name	Measure	By
F1	$\overline{UEBl_{m7y}UEl_{m3y}}/DEN$	[28]
F2	$\overline{U_{m1y}SN_y}/DEN$	[28]
F3	$\overline{D_{m2y}SN_y}/DEN$	[28]
F4	$\overline{EBlr_{Mx}EBrl_{Mx}}/DEN$	Us
F5	$\overline{A_{My}D_{m2y}}/DEN$	Us
F6	$\overline{B_{My}D_{m2y}}/DEN$	Us
F7	$\overline{A_{My}U_{m1y}}/DEN$	Us
F8	$\overline{B_{My}U_{m1y}}/DEN$	Us
F9	$\overline{EBlr_{My}Elr_{My}}/DEN$	Us
F10	$\overline{EBrl_{My}Erç_{My}}/DEN$	Us
F11	$\angle(A_m, D_{m2}, B_m)$	Us
F12	$\angle(A_M, U_{m1}, B_M)$	Us
F14	$\angle(EBll_M, EBl_{aux}, EBlr_M)$	Us
F13	$\angle(EBrr_M, EBr_{aux}, EBrl_M)$	Us
F15	$\angle(EBllm_m, EBlr_M, EBl_{aux})$	Us
F16	$\angle(EBrr_M, EBrl_M, EBr_{aux})$	Us
F17	$Ecc(A_M, B_M, D_{m2})$	[28]
F18	$Ecc(A_M, B_M, D_{m2})$	[28]
F19	$Ecc(Ell_M, Elr_M, UEl_{m3})$	[28]
F20	$Ecc(Ell_M, Err_M, DEl_{m4})$	[28]
F21	$Ecc(Erl_M, Err_M, UEr_{mr})$	[28]
F22	$Ecc(Erl_M, Err_M, UEr_{m6})$	[28]
F23	$Ecc(EBll_M, EBlr_M, UEBl_{m7})$	[28]
F24	$Ecc(EBrl_M, EBrr_M, UEBr_{m8})$	[28]

In Table I, $\overline{(P_1P_2)}$ represents the linear distance between points P_1 and P_2, and the indices x and y are used to represent the horizontal and vertical points' coordinates, respectively. The notation $\angle(P_1, P_2, P_3)$ represents the internal angle between points P_1, P_2 and P_3, in radians. Finally, $Ecc(P_1, P_2, P_3)$ represents the eccentricity of an ellipse fitted over the points P_1, P_2 and P_3. The measure of eccentricity of an ellipse is given by the formula below (refer to Figure 3).

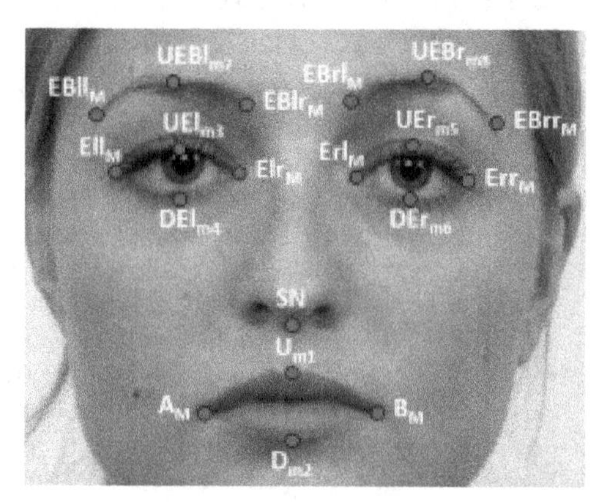

Fig. 2: The facial landmarks considered for the feature extraction process (taken from [28]).

$$Ecc(P_1, P_2, P_3) = \sqrt{\frac{\left(\frac{P_{1x} - P_{3x}}{2}\right)^2 + \left(\frac{P_{1y} - P_{2y}}{1}\right)^2}{\left(\frac{P_{1x} - P_{3x}}{2}\right)^2}} \tag{1}$$

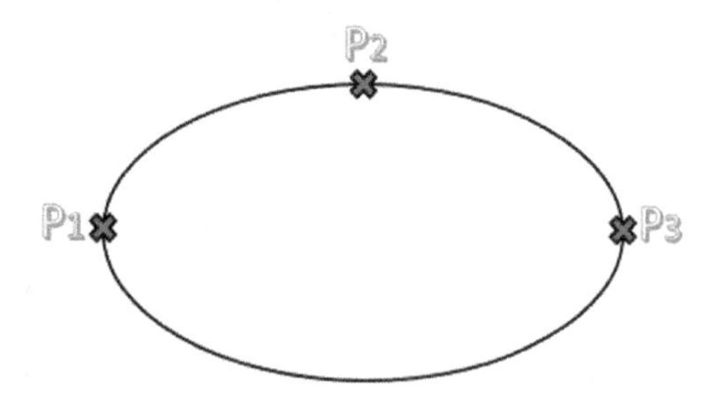

Fig. 3: An ellipse and the necessary points to the calculation of its eccentricity.

Feature F4 is a measure of the horizontal distance between the inner points of the eyebrows. This distance should be smaller in angry faces (which usually present the inner points of the eyebrows closer together) and bigger in surprised faces (which usually present the inner points of the eyebrows farther apart), for example.

Features F5 and F6 are measures of the vertical distances between the leftmost and the rightmost points of the mouth and the bottommost point of the mouth, respectively. These features should be helpful in differentiating facial expressions that present open mouths (like an angry expression, with exposed teeth) and closed mouths (like in a neutral expression). Also, they should be helpful in detecting if the analysed face is currently speaking or not.

Features F7 and F8 are similar to F5 and F6, but measure the vertical distances between the leftmost and the rightmost points of the mouth and the topmost point of the mouth. They have the same purpose features F5 and F6 have.

Features F9 and F10 are measures of the vertical distance between the inner points of the eyebrows and the inner points of the eyes. These features should help to differentiate facial expressions that present the inner corners of the eyebrows lifted (like in a surprised expression) from facial expressions that present the inner corners of the eyebrows lowered (like in an angered expression).

Feature F11 is the measure of the inner angle formed by the leftmost and rightmost points of the mouth with the bottommost point of the mouth. Feature F12 is the measure of the inner angle formed by the leftmost and the rightmost points of the mouth with the topmost point of the mouth. Together, they should be helpful in describing if the mouth is closed of opened, similarly to the features F5 to F8.

Features F13 and F14 are the measures of the inner angles formed by the corner of the eyebrows with the central point of each eyebrow. They should be helpful in describing if the eyebrows are arched (like in a surprised facial expression) or flat (like in an angered expression).

Features F15 and F16 are the measures of the inner angles formed by the outter corner and center points of the eyebrows with the inner corners of the eyebrows. They have the same purpose of the features F13 and F14.

Some of the points used to calculate the features aren't directly output by the face tracker algorithm adopted in this work, and must be calculated before the features can be computed. These points are: UEl_{m3}, UEr_{m5}, EBl_{aux} and EBr_{aux}. The Equations 2, 3, 4 and 5 describe how each of these points are obtained. EBl_{aux} and EBr_{aux} are not facial landmarks, but auxiliar points used in conjunction with the landmarks to calculate some of the chosen features.

$$UEl_{m3} = \left(\frac{Ell_{Mx} + Elr_{Mx}}{2}, \frac{Ell_{My} + Elr_{My}}{2}\right) \tag{2}$$

$$UEr_{m5} = \left(\frac{Erl_{Mx} + Err_{Mx}}{2}, \frac{Erl_{My} + Err_{My}}{2}\right) \tag{3}$$

$$\begin{aligned} UEr_{m5} = (&Elr_{Mx} - Erl_{Mx} + EBlr_{Mx}, \\ &Elr_{My} - Erl_{My} + EBlr_{My}) \end{aligned} \tag{4}$$

$$\begin{aligned} UEr_{m5} = (&Erl_{Mx} - Elr_{Mx} + EBrl_{Mx}, \\ &Erl_{My} - Elr_{My} + EBrl_{My}) \end{aligned} \tag{5}$$

C. Databases

Once the feature set is chosen, the next step is to choose one or more databases to extract these features from. These databases should contain samples of all of the concepts the machine learning algorithm is expected to learn.

In the present work, both Cohn-Kanade Plus[30] and MMI Facial Expression [31] Databases are used to train the instantaneous facial emotion classifier model.

The Cohn-Kanade Plus (or CK+) Database comprises of 486 sets of pictures from 97 posers. Each set contains a sequence of pictures depicting a person acting the onset of a particular target emotion and each sequence is labeled as a sample of that particular represented target emotion. All of the sets start with a neutral expression and evolve into a particular target emotion expression.

The CK+ Database contains, but is not limited to, sequences of all of the studied basic emotions, that is: Happiness, Sadness, Anger and Fear; but doesn't contain sets labeled as Neutral. For the purpose of this work, for each selected set, the first picture of the sequence is taken as a Neutral sample and the last picture of the sequence is taken as a sample of the sequence's target emotion. To avoid one emotion being predominant over the others in the training set, which could degrade the quality of the training process, the limit of samples for each target emotion is set to be the number of samples available for the scarcer target emotion. After the features are extracted from the chosen sets, 129 samples are generated by this process.

The MMI Facial Expression Database comprises of over 2900 videos and images of 75 posers. Only part of these videos are labeled as samples of basic emotion, so just a subset of the database is effectively utilized in this work. The selected videos show humans acting a full emotional cycle of a particular target emotion, that is, all of the three phases of the emotional display are represented: onset, apex and offset. All of the selected videos start with a Neutral face expression, which progresses to a target emotion expression and then regresses back to the Neutral display.

Similarly to the CK+ Database, the MMI Facial Expression Database contains, but is not limited to, videos of all of the studied basic emotions, but doesn't contain samples of Neutral displays. Since the videos aren't labeled at a frame-level and considering there is no preliminary indication of which of the frames represent the emotion's apex, one must first manually annotate the frames' target emotions before they can extract the features from them.

That said, all of the 74 videos chosen from this database were annotated in the following manner: the authors would watch the videos and pinpoint four instants of interest. The first instant (referred to as t_1 from here forth) represents the start of the emotional onset in the video; the second instant (t_2) represents the emotional onset's ending and the beginning of the apex; the third instant (t_3) represents the apex's ending and the beginning of the emotional offset; finally, the fourth instant (t_4) represents the emotional offset's ending.

With these instants annotated, a frame-level categorization of the videos is created: the frames before t_1 and after t_4 (inclusive) are classified as Neutral samples; the frames between t_2 and t_3 (inclusive) are classified as that video's target emotion samples; finally, the frames between t_1 and t_2 and t_3 and t_4 are classified partially as Neutral samples and partially as that video's target emotion samples.

However, not all of the generated samples were used to train the classifier. The first and the last frames of each video were chosen to compose the Neutral set of the database; also, windows of size $n = 10frames$ were built around the center of the apex region (that is, around the middle frame between t_2 and t_3) in each video, and all of the frames within these windows were taken as samples of that video's target emotion. The value of n was chosen empirically, and aimed to stablish a balance between the quantity of Neutral samples and the quantity of the other four emotions' samples. Also, care was taken so the created windows would never exceed their boundaries, that is, a window would never start at an instant before t_2 nor would it end after t_3.

After the features are extracted from the chosen pictures, 809 samples are generated by the described process.

It's worth saying both of the adopted databases contain videos and images of faces in profile and in other non-frontal orientations. However, different head orientations may cause the selected features to vary considerably for samples of the same target emotion. This could hamper the classifier's learning process and, for that reason, only videos and images containing emotional displays in frontal-oriented faces are used to train the classifier.

Finally, one should take note that all the sample images contained in these databases were acted, and not naturally elicited.

D. Instantaneous Facial Emotion Recognizer

The instantaneous facial emotion recognizer is a machine learning algorithm trained over the training set extracted through the previously described procedure.

The Random Forest learning algorithm is adopted in this work, as it was shown to have good accuracy on the work of Loconsole [28] when compared to other algorithms. The learner's accuracy and other statistics of interest are presented further in Section IV

The information fed into the dynamic classifier, however, is not simply the category output by the instantaneous classifier for a given sample, but rather, a measure of confidence that the classifier has for that sample to belong to each of the considered classes. The confidence measure used was the normalization of the number of votes each class received by the weak learners. Suppose, as an example, that a particular sample is classified by a random forest containing 100 random trees, and that 70 trees vote for the sample to belong to the Happiness class and the rest of the trees vote for it to belong to the Neutral class; in that case, the confidence measure for the sample to belong to the Happiness class would be 70%, the confidence measure for the sample to belong to the Neutral class would be 30% and the confidence measure for the sample to belong to the other classes would be 0. So, given a sample S_1, the output of the instantaneous classifier that is fed into the dynamic model is a vector of the form $V_1 = (Pr_{1n}, Pr_{1h}, Pr_{1s}, Pr_{1a}, Pr_{1f})$, where $Pr_{1n}, Pr_{1h}, Pr_{1s}, Pr_{1a}$ and Pr_{1f} are the confidence levels for S_1 to belong to the Neutral, Happiness, Sadness, Anger and Fear categories, respectively.

E. Kalman Filter

After the instaneous facial emotion classifier is properly trained, its outputs can be fed into the dynamic classifier, which will output the model's final prediction for the samples. However, aiming to eliminate high frequency noises, these outputs are firstly processed by a Kalman filter before they are inserted into the dynamic model. This section describes this filter and highlights the advantages of its usage.

As a natural consequence of the use of video frames to analyze the facial features of a person, different sources of noise can affect the classification algorithm.

It is assumed that the emotions are represented by the data initially fed in the training phase, which are gathered under controlled conditions; thus, effects such as face deformation resulting from speech, light source variations and unexpected face motions should be minimized. Furthermore, the objective of the model is to enhance the presentation of the slow and continuous emotions in spite of the instantaneous ones, sp a low pass filter should be used.

Kalman filtering is the solution proposed to this model, being a filter that has a good performance on linear systems with zero mean Gaussian noise on both the model and in the process of data acquisition. The empirical evidence presented in [9] supports this choice.

Being x_s the state variable of a linear system and y, the output of the filter for a single emotion, the filtered signal related to one of the emotions being analyzed, 6 and 7 describe the Kalman filter.

$$\dot{x_s} = x_s \tag{6}$$

$$\dot{F_a} = y = \frac{Kx_s}{\tau} \tag{7}$$

In the above equations, K is the filter's gain and τ is the time constant. There are two steps for the filtering, the first being the prediction step and the second the update step. The update is only run when new information from the sensors – in this case the output of the instantaneous emotion analyzer – is available. If the delay between data acquisitions is higher than the delay between filter steps calculations there will be some steps in which only the prediction steps will be run.

The prediction step is described by 8 and 9, where $x_{s,t}$ is the current state $x_{s,t-1}$ is the previous state, w is the noise covariance, p the covariance of the state variable on the t state. Note that the update step always assumes that the state variable has not changed, only the covariance of the system.

$$x_{s,t} = x_{s,t-1} \tag{8}$$

$$p = p + w/\tau^2 \tag{9}$$

The update step is described by equations 10, 11 and 12, where m is the residual covariance, v is the covariance of the observation noise and r_t and y_t are the filter input and output at instant t. This input corresponds to the output of the instantaneous emotion classifier. The state variable now has its value updated and, consequently, the output of the Kalman filter has its value proportionally changed.

$$m = \frac{\frac{pK}{\tau}}{p\left(\frac{K}{\tau}\right)^2 + v} \tag{10}$$

$$x_{s,t} = x_{s,t} + m(r_t - y_t) \tag{11}$$

$$p = (1 - \frac{mk}{\tau})p \tag{12}$$

Note that these equations describe the filtering process for a single class (that is, the filtering of outputs of a particular emotion). The full model is represented by applying these equations for each emotion separately.

However, neither w nor v are known, and have to be estimated by an optimization algorithm, which is described in Section III-G.

F. Dynamic Model

After the instantaneous output is filtered, it is ready to be fed into the dynamic model.

The dynamic model proposed here does not aim to describe rapid emotional variations a person may be subject to, but rather, it tries to describe more lasting emotional states. Suppose, for illustration purposes, that a man is talking to a dear friend of him that he has not seen for a while. One may expect the overall conversation to elicit a pleasant emotion. However, during this conversarion, he happens to see a person throwing trash in the street; it infuriates him for a while, but he rapidly get back to talking to his friend and forgets the sight that angered him. If pictures of his face were fed into the proposed model during this entire event, one should expect the model to detect the overall pleasant emotion of the conversation (that is, if it was pleasant enough so that his facial expression indicated so); however, his temporary enragement should not modify the output of the model.

The dynamic model is based on the work of [9], and utilizes the concept of Dynamic Emotional Surfaces (DESs).

As name indicates, DESs are surfaces that aim to describe the dynamics of transitions between different emotional states. In this work, a planar surface is adopted, and it is partitioned in four quadrants, one for each of the considered basic emotions: Happiness, Sadness, Anger and Surprise. Centered in the intersection of the four areas, there is the Neutral area, which represents the absence of emotions. Figure 4 illustrates the model's DES.

Located on the $+45°$ and $-45°$ diagonals of this plane, there are four Emotional Attractors (EAs), one for each of

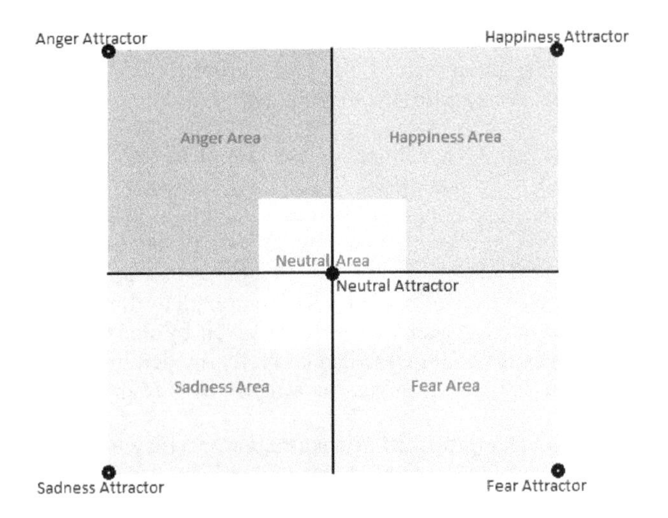

Fig. 4: A representation of the planar DES used in this work.

the considered non-Neutral emotions, and each located in its corresponding quadrant. Refer to Figure 4. The Happiness, Fear, Sadness and Anger attractors are located on the points $PA_{Happiness}$, PA_{Fear}, $PA_{Sadness}$ and PA_{Anger}, respectively, and the Neutral attractor is located on the point $PA_{Neutral}$.

Let to slide upon the plane, there are Emotional Particles (EPs), one for each analyzed subject. The location of a particle in a given instant indicates the model's output emotion for that instant, according to the equation below, where $P = (P_x, P_y)$ is an EP's position and $f(P)$ is the model's output in the considered instant.

$$f(P) = \begin{cases} \text{Happiness}, & \text{if } P_x > K_{nr} \text{ and } P_y > K_{nr} \\ \text{Sadness}, & \text{if } P_x < -K_{nr} \text{ and } P_y < -K_{nr} \\ \text{Anger}, & \text{if } P_x < -K_{nr} \text{ and } P_y > K_{nr} \\ \text{Fear}, & \text{if } P_x > K_{nr} \text{ and } P_y < -K_{nr} \end{cases} \quad (13)$$

In the equation 13, K_{nr} is a constant that determines the width and height of the Neutral area.

The EAs are responsible for pulling EPs towards them. The stronger the confidence level the Kalman filter outputs for a given emotion, the stronger the pull velocity for that emotion's attractor will be. If at the instant $\bar{t}$ Kalman filter outputs a confidence level of $Pr_E(\bar{t})$ for emotion E, then E's attractor velocity, $VA_E(\bar{t})$, is given by Equation 14.

$$VA_E(\bar{t}) = K_{avm} Pr_E(\bar{t}) \quad (14)$$

The parameter K_{avm} is the attractors velocity modulator parameter, which value, like the Kalman filter parameters w and v, is also found via an optimization algorithm.

The dynamics for EPs are described by equations 15 and 16, where $P(t)$ and $V(t)$ are particles' position and velocity, $Pr_E(t)$ is the confidence measure for emotion E and $VA_E(t)$ is the attractor's E pull velocity, all at instant t.

$$P(t) = P(t-1) + V(t) \quad (15)$$

$$V(t) = \begin{cases} VA_{Neutral}, & \text{if } max(Pr_E(t)) = Pr_{Neutral}(t) \\ \sum_{E=\bar{e}} VA_E, & \text{if not} \end{cases} \quad (16)$$

where $\bar{e}$ is the subset $\{Happiness, Sadness, Anger, Fear\}$. Also, the position of the particle is never let to exceed the rectangle delimited by the four non-Neutral EAs.

The noise smoothing introduced by the Kalman filter and the intrinsic inertia presented by the proposed model make so that natural facial noises - like the mouth movements caused by laughter or speech - should have its influence on the predictions diminished, when in comparison to the instantaneous classifier.

G. Parameters Optimization

As the previous sections explained, some of the model parameters can not be known *a priori*, and are better defined via an optimization process. These parameters are: Kalman filter's noise covariance (w), Kalman filter's covariance of the observation noise(v) and DES's attractors velocity modulator (K_{avm}).

The optimization process here adopted is based on the simulated annealing algorithm, and can be described by the pseudo-code presented below.

In the above pseudo-code, T_0, T_{room} and T_{curr} are the initial, room and current temperatures of the optimizer, in that order; p_curr, p_{la} and p_{sol} are the current iteration's parameters, the last accepted solution's parameters and the final solution parameters, respectivelty; e_0, e_{curr}, e_{la} and e_{sol} are the initial energy, the current iteration's energy, the last accepted solution's energy and the final solution's energy, in that order; dr is the temperature decay rate and Pr_{acc} is the probability that a solution will be accepted by the algorithm.

Note that an iteration's energy, e_{curr} is obtained by the function $calculateEnergy(p_{curr}, dataset)$, which considers both the current value of the parameters being optimized and a dataset chosen for the optimization. The MMI Facial Emotion Database's previously selected 74 videos were used to extract the energy measure; however, this time they were considered in their full-length. The adopted energy measure is the number of frames the model misclassified in the iteration.

A proposed solution is always accepted if it causes the system's energy to decrease in comparison to the last accepted solution's energy. However, even if a solution causes the energy to increase, it has a chance of being accepted that is proportional to the iteration's current temperature and inversely proportional to the energy increase it causes. This measure helps the optimizer to avoid getting stuck in local minima.

If a solution is accepted, its parameters and energy are stored to serve as comparison data for the next iteration. However, a solution is only stored as a final solution if its energy is smaller than the last accepted final solution.

// Initializations:
$T_0 = 200°C$;
$T_{room} = 20°C$;
$T_{curr} = T_0$;
$p_{curr} = randomizeParameters()$;
$p_{la} = p_{curr}$;
$p_{sol} = p_c$;
$e_0 = +\infty$;
$e_{curr} = e_0$;
$e_{la} = e_0$;
$e_{sol} = e_0$;
$dr = 0.99995$;
// Iterations:
while $(T_{curr} > T_{room})$ **do**
  $e_c = calculateEnergy(p_{curr}, dataset)$;
  if $(e_{curr} < e_{la})$ **then**
    $Pr_{acc} = 1$;
  else
    $Pr_{acc} = e^{(e_{la} - e_{curr})/T_{curr}}$;
  end
  if $(Rnd(0,1) > Pr_{acc})$ **then**
    $p_{la} = p_{curr}$;
    $e_{la} = e_{curr}$;
    if $(e_{la} > e_{sol})$ **then**
      $p_{sol} = p_{la}$;
      $e_{sol} = e_{la}$;
    else
      $p_{curr} = p_{la}$;
    end
  end
  $p_{curr} = moveAround(T_{curr})$;
  $T_{curr} = T_{curr} \times dr$;
end

At the end of every iteration, the parameters are varied through the function $moveAround(T_{curr})$, which takes into consideration the iteration's temperature - the higher the temperature, the more the parameters are allowed to variate -, and the optimizer temperature is made to decay by a constant rate dr.

IV. Tests and Results

This section presents the results of the tests realized on the model. These tests are presented separately for the instantaneous facial emotion classifier, for the parameters optimization algorithm and for the dynamic facial emotion classifier.

A. Tests on the Instantaneous Facial Emotion Classifier

Tests were made to measure the quality of the instantaneous classifier. Since a poorly trained classifier may compromise the overall performance of the model, the quality of its outputs should be analyzed with caution.

The random forest learning algorithm discards the need for procedures like cross-validation, bootstrap or separate test sets for estimating the classifier's accuracy. During the training of each of the weak learners (that is, of each tree of the forest), an out-of-bag set (or "oob set", containing roughly 1/3 of the complete training set) is created for that learner. The oob set is used to validate the accuracy of that particular tree. After all the trees have finished training, the following procedure is used to calculate an estimation of the accuracy of the learner for samples stranger to the training set:

1) Each sample contained in the complete training set is considered separately;
2) All trees that contain a particular sample in their oob sets are used to classify that sample, and a vote counting is used to decide to what class it belongs to. The procedure is repeated for all samples of the complete training set;
3) The random forest's accuracy is given by the number of samples of the set classified correctly divided by the number of samples classified incorrectly by that process....

Through the described procedure, an accuracy estimate of approximately 90% was obtained.

The analysis of the learner's confusion matrix allows one to observe how its predictions are distributed amongst the different classes. Table II presents the confusion matrix for the trained random forest.

Table II: The confusion matrix for the trained random forest

	Neutral	Anger	Fear	Happiness	Sadness
Neu.	201	5	4	8	20
Ang.	0	168	1	0	5
Fea.	0	0	168	0	1
Hap.	0	0	2	189	0
Sad.	0	1	0	0	113

One can notice that the Neutral class is the one with more misclassified samples, even if considering relative numbers. Also, more than half of the misclassified Neutral samples are categorized as Sadness samples, which suggests that the boundaries between these two classes is the less obvious for the classifier, at least on the considered dataset.

B. Tests on the Parameters Optimization Algorithm

Tests were made with values of K_{nr} (which determines the height and width of the neutral area of the plane) varying from 1 to 5, with unitary increments. Also, for all the tests, the attractors were positioned on the points $PA_{Happiness} = (10, 10)$, $PA_{Fear} = (10, -10)$, $PA_{Sadness} = (-10, -10)$, $PA_{Anger} = (-10, 10)$ and $PA_{Neutral} = (0, 0)$. Figure 5 shows the model's accuracy history for the best optimization achieved - that is, for the optimization that reached the lowest energy on the used dataset.

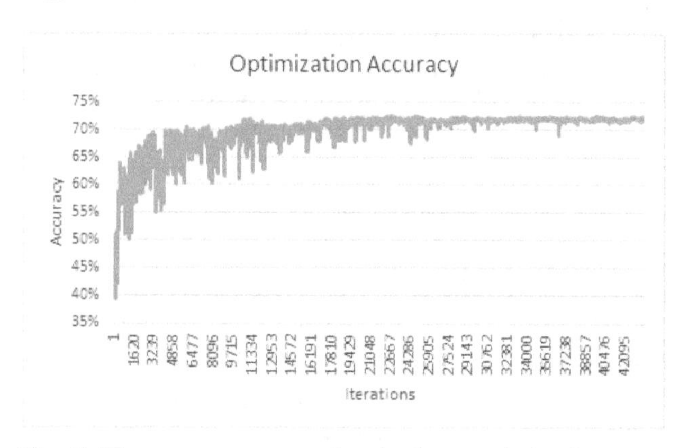

Fig. 5: The accuracy curve for the best optimization case.

It is possible to see an overall increase of the optimization accuracy as the iterations progress; also, the accuracy curve seems to converge to a value of about 72% by the end of the process. The accuracy achieved with the instantaneous model for the same dataset is of 64%. This fact suggests that the use of the dynamic model was beneficial even for a dataset with videos that do not contain too many facial noises caused by factors like laughter or speech.
The best accuracy was reached for a value 1 of K_{nr}.

C. Tests on the Dynamic Facial Emotion Classifier

To test the developed dynamic classifier, a test was run on the video "S43_an_2" of the eNTERFACE'05 Database [32], the same analyzed in the work of [9].
This video depicts the face of an angered person as she irritatedly proclaims a certain sentence. The presence of facial noises in the video is relevant for the experiment, as it allows for the analysis of how well the dynamic model is able to deal with such noises. Also, this is the first experiment that utilizes a video entirely stranger to the datasets used for training the instantaneous classifier and for the optimization process.
Because the video "S43_an_2" is simply classified as an Anger video, and since there is no information about whether any other emotional displays are considered to be present in it, all of its frames were considered as Anger samples and fed into the model.
Figures 6 and 7 present the dynamic model output and the instantaneous classifier output for each frame of the video, respectively. The accuracy achieved with the dynamic model was of 89%, while the accuracy achieved with the instantaneous classifier was of 64%. This result suggests that the dynamic model successfuly dealt with a considerable portion of the facial noises presented in the video. Note that not only the dynamic model achieved a higher accuracy on the video, but its outputs seem to be more reliable. With exception of the last frame, all frames in the video were classified as Anger or Neutral frames by the dynamic model, and there are less variations between different emotional states; the classifications attributed by the instantaneous model, however, flicker more rapidly and between a larger number of emotional states. One could argue that the result achieved by the dynamic model is more useful than the one achieved by the instantaneous classifier if it was to be used to control an automated system like a social robot - maybe the social robot would not be able to react as well to a flickering input as it would react to a more stable one.

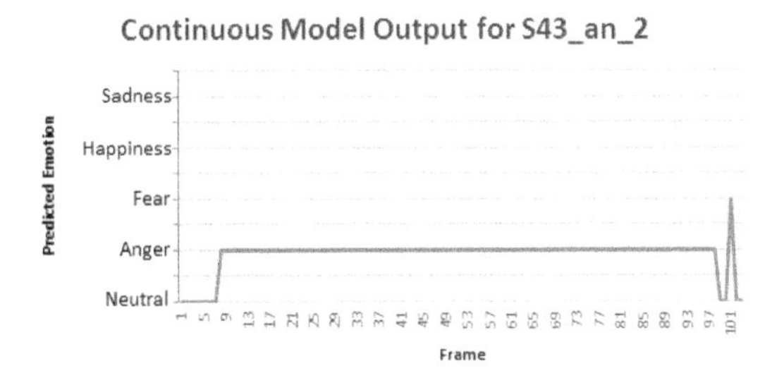

Fig. 6: The output of the continuous model for the video "$S43\_an\_2$".

Finally, Figure 8 presents the trajectory on which the EP traveled throughout the video. Note that the particle rapidly

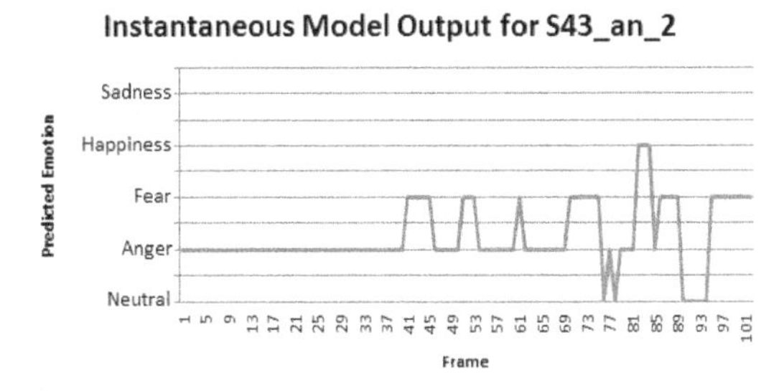

Fig. 7: The output of the instantaneous classifier for the video "$S43\_an\_2$".

progresses to the Anger area, where it remains until the latter parts of the video, regressing back to the neutral area and then to the Fear area by the end of the video. The transition to the Fear area is probably due to the considerably large number of Fear predictions outputted by the instantaneous classifier in the latter parts of the video.

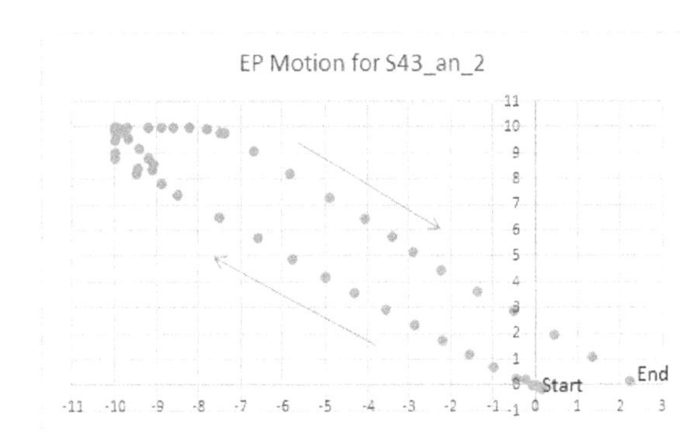

Fig. 8: Motion of the EP throughout the video "$S43\_an\_2$".

V. Conclusions

In the present work, an innovative dynamic emotion recognition model was presented. This model comprises of the conjugation of a machine learning algorithm, a Kalman filter and an original dynamic model that aims to describe durable emotional states and to minimize facial noises like deformations caused by laughter and speech. A simulated annealing algorithm was utilized to optimize the model's parameters.
The model has shown good performance when compared to the instantaneous emotion classifier trained in the present work: while the former achieved an accuracy rate of 72% over the chosen dataset, the latter presented an accuracy rate of just 64%, on the same dataset.
When tests on a sample stranger to the datasets utilized to train the instantaneous classifier and to optimize the model's parameters, the dynamic model once again outmatched the instantaneous model: not only it achieved a higher accuracy rate (89% against 64%), but it also provided a much more stable output.
As target objectives for future works, the following tasks are proposed:

1) Execute more tests on the dynamic model, in order to better analyze its accuracy and the way it describes the progression of emotional expressions in faces;

2) Utilize larger datasets to train the instantaneous model and to optimize the dynamic model;
3) Utilize datasets that contain faces deformed by natural facial noises, like laugther or speech, for the training and optimization of the model;
4) Study possible changes the proposed planar DES may need to better describe the way emotions manifest themselves in human faces;
5) Increase the number of considered emotions and study how the DES should be changed to accommodate this change.

REFERENCES

[1] P. Ekman, "Basic emotions," in *Handbook of Cognition and Emotion*, T. Dalgleish and M. J. Power, Eds. Chichester, UK: John Wiley Sons, Ltd, 2005, ch. 3, p. 45–60.

[2] R. W. Picard, "Affective computing," MIT Media Lab, Perceptual Computing, Cambridge, MA, Tech. Rep. 295, 1995.

[3] P. Ekman, E. R. Sorenson, and W. V. Friesen, "Pan-cultural elements in facial displays of emotion," *Science*, vol. 164, no. 3875, pp. 86–88, 1969.

[4] P. Ekman, "A linguagem das emoçöes," *São Paulo: Lua de Papel*, 2011.

[5] P. Ekman and W. V. Friesen, *Unmasking the face: A guide to recognizing emotions from facial clues*. Ishk, 2003.

[6] R. A. M. Gonçalves, "Um modelo matemático para inferência computacional de estado emocional a partir de detectores de expressões faciais," M. Eng. thesis, Universidade de São Paulo, 2012.

[7] Z. Zeng, M. Pantic, G. Roisman, and T. S. Huang, "A survey of affect recognition methods: Audio, visual, and spontaneous expressions," *Pattern Analysis and Machine Intelligence, IEEE Transactions on*, vol. 31, no. 1, pp. 39–58, 2009.

[8] M. Zarkowski, "Identification-driven emotion recognition system for a social robot," in *Methods and Models in Automation and Robotics (MMAR), 2013 18th International Conference on*. IEEE, 2013, pp. 138–143.

[9] R. A. M. Gonçalves, D. R. Cueva, M. R. Pereira-Barretto, and F. G. Cozman, "A model for inference of emotional state based on facial expressions," *Journal of the Brazilian Computer Society*, vol. 19, no. 1, pp. 3–13, 2013.

[10] A. Rabie, B. Wrede, T. Vogt, and M. Hanheide, "Evaluation and discussion of multi-modal emotion recognition," in *Computer and Electrical Engineering, 2009. ICCEE'09. Second International Conference on*, vol. 1. IEEE, 2009, pp. 598–602.

[11] M. K. Greenwald, E. W. Cook, and P. J. Lang, "Affective judgment and psychophysiological response: Dimensional covariation in the evaluation of pictorial stimuli," *Journal of psychophysiology*, 1989.

[12] J. A. Russell and A. Mehrabian, "Evidence for a three-factor theory of emotions," *Journal of research in Personality*, vol. 11, no. 3, pp. 273–294, 1977.

[13] D. Sander, D. Grandjean, and K. R. Scherer, "A systems approach to appraisal mechanisms in emotion," *Neural networks*, vol. 18, no. 4, pp. 317–352, 2005.

[14] M. Pantic and L. J. M. Rothkrantz, "Automatic analysis of facial expressions: The state of the art," *Pattern Analysis and Machine Intelligence, IEEE Transactions on*, vol. 22, no. 12, pp. 1424–1445, 2000.

[15] P. Ekman and W. V. Friesen, "Facial action coding system," 1977.

[16] B. Tinen, "Sistema de identificação de emoçães por expressões faciais com operação ao vivo," Eng. thesis, Universidade de São Paulo, 2014.

[17] P. Ekman and W. Friesen, "Facial action coding system: A technique for the measurement of facial movements," *Consulting Psychologist*, vol. 2, 1978.

[18] M. Valstar and M. Pantic, "Fully automatic facial action unit detection and temporal analysis," in *Computer Vision and Pattern Recognition Workshop, 2006. CVPRW'06. Conference on*. IEEE, 2006, pp. 149–149.

[19] B. Fasel and J. Luettin, "Automatic facial expression analysis: a survey," *Pattern recognition*, vol. 36, no. 1, pp. 259–275, 2003.

[20] A. K. Jain and S. Z. Li, *Handbook of Face Recognition*. Secaucus, NJ, USA: Springer-Verlag New York, Inc., 2005.

[21] G. Donato, M. S. B. J. C. Hager, P. Ekman, and T. J. Sejnowski, "Classifying facial actions," *Pattern Analysis and Machine Intelligence, IEEE Transactions on*, vol. 21, no. 10, pp. 974–989, 1999.

[22] S.-S. Liu, Y.-T. Tian, and D. Li, "New research advances of facial expression recognition," in *Machine Learning and Cybernetics, 2009 International Conference on*, vol. 2. IEEE, 2009, pp. 1150–1155.

[23] J. Hamm, C. G. K. R. C. Gur, and R. Verma, "Automated facial action coding system for dynamic analysis of facial expressions in neuropsychiatric disorders," *Journal of neuroscience methods*, vol. 200, no. 2, pp. 237–256, 2011.

[24] T. F. Cootes, J. C. Taylor, D. H. Cooper, and J. Graham, "Active shape models-their training and application," *Computer vision and image understanding*, vol. 61, no. 1, pp. 38–59, 1995.

[25] S. Du, Y. Tao, and A. M. Martinez, "Compound facial expressions of emotion," *Proceedings of the National Academy of Sciences*, vol. 111, no. 15, pp. E1454–E1462, 2014.

[26] M.-H. Yang, D. J. Kriegman, and N. Ahuja, "Detecting faces in images: A survey," *Pattern Analysis and Machine Intelligence, IEEE Transactions on*, vol. 24, no. 1, pp. 34–58, 2002.

[27] J. A. Russell, "Is there universal recognition of emotion from facial expressions? a review of the cross-cultural studies." *Psychological bulletin*, vol. 115, no. 1, p. 102, 1994.

[28] C. Loconsole, C. R. Miranda, G. Augusto, A. Frisoli, and V. C. Orvalho, "Real-time emotion recognition-novel method for geometrical facial features extraction." in *VISAPP (1)*, 2014, pp. 378–385.

[29] A. Asthana, S. Zafeiriou, S. Cheng, and M. .Pantic, "Incremental face alignment in the wild," *Science*, 2014.

[30] P. Lucey, J. F. Cohn, T. Kanade, J. Saragih, Z. Ambadar, and I. Matthews, "The extended cohn-kanade dataset (ck+): A complete dataset for action unit and emotion-specified expression," in *Computer Vision and Pattern Recognition Workshops (CVPRW), 2010 IEEE Computer Society Conference on*. IEEE, 2010, pp. 94–101.

[31] M. Pantic, M. Valstar, R. Rademaker, and L. Maat, "Web-based database for facial expression analysis," *Multimedia and Expo, 2005. ICME 2005. IEEE International Conference on*, 2005.

[32] O. Martin, I. Kotsia, B. Macq, and I. Pitas, "The enterface'05 audio-visual emotion database," in *Data Engineering Workshops, 2006. Proceedings. 22nd International Conference on*. IEEE, 2006, pp. 8–8.

Permissions

All chapters in this book were first published in IJARAI, by The Science and Information Organization; hereby published with permission under the Creative Commons Attribution License or equivalent. Every chapter published in this book has been scrutinized by our experts. Their significance has been extensively debated. The topics covered herein carry significant findings which will fuel the growth of the discipline. They may even be implemented as practical applications or may be referred to as a beginning point for another development.

The contributors of this book come from diverse backgrounds, making this book a truly international effort. This book will bring forth new frontiers with its revolutionizing research information and detailed analysis of the nascent developments around the world.

We would like to thank all the contributing authors for lending their expertise to make the book truly unique. They have played a crucial role in the development of this book. Without their invaluable contributions this book wouldn't have been possible. They have made vital efforts to compile up to date information on the varied aspects of this subject to make this book a valuable addition to the collection of many professionals and students.

This book was conceptualized with the vision of imparting up-to-date information and advanced data in this field. To ensure the same, a matchless editorial board was set up. Every individual on the board wen through rigorous rounds of assessment to prove their worth. After which they invested a large part of their time researching and compiling the most relevant data for our readers.

The editorial board has been involved in producing this book since its inception. They have spent rigorous hours researching and exploring the diverse topics which have resulted in the successful publishing of this book. They have passed on their knowledge of decades through this book. To expedite this challenging task, the publisher supported the team at every step. A small team of assistant editors was also appointed to further simplify the editing procedure and attain best results for the readers.

Apart from the editorial board, the designing team has also invested a significant amount of their time in understanding the subject and creating the most relevant covers. They scrutinized every image to scout for the most suitable representation of the subject and create an appropriate cover for the book.

The publishing team has been an ardent support to the editorial, designing and production team. Their endless efforts to recruit the best for this project, has resulted in the accomplishment of this book. They are a veteran in the field of academics and their pool of knowledge is as vast as their experience in printing. Their expertise and guidance has proved useful at every step. Their uncompromising quality standards have made this book an exceptional effort. Their encouragement from time to time has been an inspiration for everyone.

The publisher and the editorial board hope that this book will prove to be a valuable piece of knowledge for researchers, students, practitioners and scholars across the globe.

List of Contributors

Kohei Arai, Takuya Fukamachi and Hiroshi Okumura
Graduate School of Science and Engineering Saga University Saga City, Japan

Shuji Kawakami
JAXA, Japan Tsukuba City, Japan

Hind Rustum Mohammed
CS dept. Faculty of Computer Science and Mathematics University of Kufa Najaf, Iraq

Husein Hadi Alnoamani
MS dept. Faculty of Computer Science and Mathematics University of Kufa Najaf, Iraq

Ali Abdul Zahraa Jalil
Student CS dept. Faculty of Computer Science and Mathematics University of Kufa, Najaf, Iraq

Peter Simon Sapaty
Institute of Mathematical Machines and Systems National Academy of Sciences Kiev, Ukraine

Ali Altalbe
Deanship of e-Learning and Distance Education King Abdulaziz University Jeddah, Saudi Arabia

Sri Mulyana, Sri Hartati, Retantyo Wardoyo and Edi Winarko
Department of Computer Sciences and Electronics Gadjah Mada University Yogyakarta, Indonesia

Masanori Sakashita
Department of Information Science Saga University Saga City, Japan

Osamu Shigetomi
Saga Prefectural Agricultural Research Institute Saga Prefecture Saga City, Japan

Yuko Miura
Saga Prefectural Agricultural Research Institute Saga Prefecture Saga City, Japan

Li Jing
College of Mobile Telecommunications, Chongqing University of Posts and Telecommunications Chongqing, China

Zhou Wenwen
Chongqing University of Posts and Telecommunications Chongqing, China

Zahra Nazari and Dongshik Kang
Department of Information Engineering University of the Ryukyus Okinawa, Japan

Fatimah Altuhaifa
College of Computer Engineer & Science Prince Mohammed bin Fahd University Alkhobar, Saudi Arabia

Keun-Chang Kwak
Dept. of Control and Instrumentation Engineering Chosun University, 375 Seosuk-Dong Gwangju, Korea

Ahmed T. Sadiq Al-Obaidi
Department of Computer Sciences University of Technology Baghdad, Iraq

Ahmed Badre Al-Deen Majeed
Quality Assurance Department University of Baghdad Baghdad, Iraq

Ying Qian and Zhi Li
The lab of Graphics and Multimedia Chongqing University of Posts and Telecommunications Chongqing, China

Nayden Valkov Nenkov
Faculty of Mathematics and Informatics University of Shumen "Episkop Konstantin Preslavsky" Shumen, Bulgaria

Mariana Mateeva Petrova
Faculty of Mathematics and Informatics St. Cyril and St. Methodius University of Veliko Turnovo, Bulgaria

Umar Manzoor
Faculty of Computing and Information Technology

Mohammed A. Balubaid
Industrial Engineering Department, Engineering Faculty, King Abdulaziz University, Jeddah, Saudi Arabia

Bassam Zafar
Faculty of Computing and Information Technology King Abdulaziz University, Jeddah, Saudi Arabia

Hafsa Umar and M. Shoaib Khan
National University of Computer and Emerging Sciences, Islamabad, Pakistan

Yuta Koike, Kei Sawai and Tsuyoshi Suzuki
Department of Information and Communication Engineering, Tokyo Denki University Tokyo, Japan

Anat Reiner-Benaim and Anna Grabarnick
Department of Statistics University of Haifa Haifa, Israel

Edi Shmueli
Intel Corporation Haifa, Israel

H. Mahersia and M. Zaroug
Department of Computer Science, College of science and arts of Baljurashi, Albaha University, Albaha, Kingdom of Saudi Arabia

L. Gabralla
Faculty of Computer Science & Information Technology University of Science &Technology, Khartoum, Sudan

Wen-Kung Tseng
Graduate Institute of Vehicle Engineering National Changhua University of Education Changhua City, Taiwan

Jakub Ševčík, Ondrej Kainz, Peter Feciľak and František Jakab
Dept. of Computers and Informatics FEI TU of Kosice Kosice, Slovak Republic

Ahmed Fouad Ali
Faculty of Computers and Information Dept. of Computer Science, Suez Canal University Ismailia, Egypt

Oana Diana Eva
Faculty of Electronics, Telecommunications and Information Technology
"Gheorghe Asachi" Technical University of Iasi Iasi, Romania

Anca Mihaela Lazar
Faculty of Medical Bioengineering "Grigore T. Popa" University of Medicine and Pharmacy Iasi, Romania

Rastislav Eštók, Miroslav Michalko, Ondrej Kainz and František Jakab
Dept. of Computers and Informatics FEI TU of Kosice Kosice, Slovak Republic

Dr. Mamta Madan
Professor Vivekananda Institute of Professional Studies GGSIP University,India

Meenu Chopra
Assistant Professor Vivekananda Institute of Professional Studies GGSIP University, India

Dania Aljeaid, Caroline Langensiepen and Xiaoqi Ma
School of Science and Technology Nottingham Trent University Nottingham, United Kingdom

Sami Nassimi, Noora Mohamed, Walaa AbuMoghli and Mohamed WaleedFakhr
Electrical and Electronic Engineering University of Bahrain Isa Town, Bahrain

Huda Jalil Dikhil, Mohammad Shkoukani and Suhail Sami Owais
Dept. of Comoputer Sciecne Applied Science Private University Amman, Jordan

Md. Shahin Reza, Md. Rashedul Hasan and Oreste Bursi
Department of Civil, Environment and Mechanical Engineering University of Trento, Italy

Feroz Farazi
Department of Information Engineering and Computer Science University of Trento, Italy

Ernesto D'Avanzo
Department of Political, Social and Communication Sciences University of Salerno,Italy

Omae Malack Oteri, Langat Philip Kibet and Ndung'u Edward N.
Department of Telecommunication and Information Engineering Jomo Kenyatta University of Agriculture & Technology Nairobi, Kenya

Beaudelaire Saha Tchinda and Daniel Tchiotsop
Laboratoire d'Automatique et d'Informatique Appliquée (LAIA), IUT-FV de Bandjoun, Université de Dschang-Cameroun, B.P. 134 Bandjoun

René Tchinda
Laboratoire d'Ingénierie des Systèmes Industriels et de l'Environnement (LISIE), IUT-FV de Bandjoun, Université de Dschang- Cameroun, B.P. 134 Bandjoun

Didier Wolf
CRAN UMR CNRS 7039, ENSEM Université de Lorraine, Nancy, France

Michel Noubom
Département des sciences Biomédicales, Faculté des sciences, Université de Dschang

A. S. Salama
Department of Mathematics, Faculty of Science, Tanta University, Tanta, Egypt
Department of Mathematics, Faculty of Science and Humanities Shaqra University Al-Dawadmi, KSA

Saad M. Darwish
Department of Information Technology Institute of Graduate Studies and Research University of Alexandria Alexandria, Egypt

Abeer A. Amer and Sameh G. Taktak
Department of Computer and Information Systems Sadat Academy for Management Sciences (SAMS) Alexandria, Egypt

Ruivo, J. P. P., Negreiros, T., Barretto, M. R. P., Tinen and B.
Escola Politécnica Universidade de São Paulo São Paulo, Brazil

Index

Printed in the USA
CPSIA information can be obtained
at www.ICGtesting.com
JSHW052024301024
72690JS00004B/154